Series on Advances in Statistical Mechanics – Vol. 17

Chaos

From Simple Models to Complex Systems

Series on Advances in Statistical Mechanics – Vol. 17

Chaos

From Simple Models to Complex Systems

Massimo Cencini • Fabio Cecconi
INFM - Consiglio Nazionale delle Ricerche, Italy

Angelo Vulpiani
University of Rome "Sapienza", Italy

World Scientific

NEW JERSEY • LONDON • SINGAPORE • BEIJING • SHANGHAI • HONG KONG • TAIPEI • CHENNAI

Published by

World Scientific Publishing Co. Pte. Ltd.

5 Toh Tuck Link, Singapore 596224

USA office: 27 Warren Street, Suite 401-402, Hackensack, NJ 07601

UK office: 57 Shelton Street, Covent Garden, London WC2H 9HE

British Library Cataloguing-in-Publication Data
A catalogue record for this book is available from the British Library.

Series on Advances in Statistical Mechanics — Vol. 17
CHAOS
From Simple Models to Complex Systems

ISBN-13 978-981-4277-65-5
ISBN-10 981-4277-65-7

Preface

The discovery of chaos and the first contributions to the field date back to the late 19th century with Poincaré's pioneering studies. Even though several important results were already obtained in the first half of the 20th century, it was not until the '60s that the modern theory of chaos and dynamical systems started to be formalized, thanks to the works of E. Lorenz, M. Hénon and B. Chirikov. In the following 20–25 years, chaotic dynamics gathered growing attention, which led to important developments, particularly in the field of dynamical systems with few degrees of freedom. During the mid '80s and the beginning of the '90s, the scientific community started considering systems with a larger number of degrees of freedom, trying to extend the accumulated body of knowledge to increasingly complex systems. Nowadays, it is fair to say that low dimensional chaotic systems constitute a rather mature field of interest for the wide community of physicists, mathematicians and engineers. However, notwithstanding the progresses, the tools and concepts developed in the low dimensional context often become inadequate to explain more complex systems, as dimensionality dramatically increases the complexity of the emerging phenomena. To date, various books have been written on the topic. Texts for undergraduate or graduate courses often restrict the subject to systems with few degrees of freedom, while discussions on high dimensional systems are usually found in advanced books written for experts. This book is the result of an effort to introduce dynamical systems accounting for applications and systems with different levels of complexity. The first part (Chapters 1 to 7) is based on our experience in undergraduate and graduate courses on dynamical systems and provides a general introduction to the basic concepts and methods of dynamical systems. The second part (Chapters 8 to 14) encompasses more advanced topics, such as information theory approaches and a selection of applications, from celestial and fluid mechanics to spatiotemporal chaos. The main body of the text is then supplemented by 32 additional call-out boxes, where we either recall some basic notions, provide specific examples or discuss some technical aspects. The topics selected in the second part mainly reflect our research interests in the last few years. Obviously, the selection process forced us to omit or just briefly mention a few interesting topics, such as random dynamical systems, control, transient chaos, non-attracting chaotic sets, cellular automata and chaos in quantum physics.

The intended audience of this book is the wide and heterogeneous group of science students and working scientists dealing with simulations, modeling and data analysis of complex systems. In particular, the first part provides a self-consistent undergraduate/graduate physics or engineering course in dynamical systems. Chapters from 2 to 9 are also supplemented with exercises (whose solutions can be found at: http://denali.phys.uniroma1.it/~chaosbookCCV09) and suggestions for numerical experiments. A selection of the advanced topics may be used to either focus on some specific aspects or to develop PhD courses. As the coverage is rather broad, the book can also serve as a reference for researchers.

We are particularly indebted to Massimo Falcioni, who, in many respects, contributed to this book with numerous discussions, comments and suggestions. We are very grateful to Alessandro Morbidelli for the careful and critical reading of the part of the book devoted to celestial mechanics. We wish to thank Alessandra Lanotte, Stefano Lepri, Simone Pigolotti, Lamberto Rondoni, Alessandro Torcini and Davide Vergni for providing us with useful remarks and criticisms, and for suggesting relevant references. We also thank Marco Cencini, who gave us language support in some parts of the book.

We are grateful to A. Baldassarri, J. Bec, G. Benettin, E. Bodenschatz, G. Boffetta, E. Calzavarini, H. Hernandez-Garcia, H. Kantz, C. Lopez, E. Olbrich and A. Torcini for providing us with some of the figures. We would also like to thank several collaborators and colleagues who, during the past years, have helped us in developing our ideas on the matter presented in this book, in particular M. Abel, R. Artuso, E. Aurell, J. Bec, R. Benzi, L. Biferale, G. Boffetta, M. Casartelli, P. Castiglione, A. Celani, A. Crisanti, D. del-Castillo-Negrete, M. Falcioni, G. Falkovich, U. Frisch, F. Ginelli, P. Grassberger, S. Isola, M. H. Jensen, K. Kaneko, H. Kantz, G. Lacorata, A. Lanotte, R. Livi, C. Lopez, U. Marini Bettolo Marconi, G. Mantica, A. Mazzino, P. Muratore-Ginanneschi, E. Olbrich, L. Palatella, G. Parisi, R. Pasmanter, M. Pettini, S. Pigolotti, A. Pikovsky, O. Piro, A. Politi, I. Procaccia, A. Provenzale, A. Puglisi, L. Rondoni, S. Ruffo, A. Torcini, F. Toschi, M. Vergassola, D. Vergni and G. Zaslavsky. We wish to thank the students of the course of Physics of Dynamical Systems at the Department of Physics of the University of Rome La Sapienza, who, during last year, used a draft of the first part of this book and provided us with useful comments and highlighted several misprints; in particular, we thank M. Figliuzzi, S. Iannaccone, L. Rovigatti and F. Tani. Finally, it was a pleasure to thank the staff of World Scientific and, in particular, the scientific editor Prof. Davide Cassi for his assistance and encouragement, and the production specialist Rajesh Babu, who helped us with some aspects of LaTeX.

We dedicate this book to Giovanni Paladin, who had a long collaboration with A.V. and assisted M.C. and F.C. at the beginning of the their career.

M. Cencini, F. Cecconi and A. Vulpiani
Rome, Spring 2009

Introduction

All truly wise thoughts have been thought already thousands of times; but to make them truly ours, we must think them over again honestly, till they take root in our personal experience.

Johann Wolfgang von Goethe (1749–1832)

Historical note

The first attempt to describe the physical reality in a quantitative way, presumably, dates back to the Pythagoreans, with their effort to explain the tangible world by means of integer numbers. The establishment of mathematics as the proper language to decipher natural phenomena lagged behind until the 17th century, when Galileo inaugurated modern physics with his major work (1638): *Discorsi e dimostrazioni matematiche intorno a due nuove scienze* (Discourses and mathematical demonstrations concerning two new sciences). Half a century later, in 1687, Newton published the *Philosophiae Naturalis Principia Mathematica* (The Mathematical Principles of Natural Philosophy) which laid the foundations of classical mechanics. The publication of the *Principia* represents the *summa* of the scientific revolution, in which Science, as we know it today, was born.

From a conceptual point of view, the main legacy of Galileo and Newton is the idea that Nature obeys unchanging laws which can be formulated in mathematical language, therefrom physical events can be predicted with certainty. These ideas were later translated in the philosophical proposition of *determinism*, as expressed in a rather vivid way by Laplace (1814) in his book *Essai philosophique sur les probabilités* (Philosophical Essay on Probability):

> *We must consider the present state of Universe as the effect of its past state and the cause of its future state. An intelligence that would know all forces of nature and the respective situation of all its elements, if furthermore it was large enough to be able to analyze all these data,*

> *would embrace in the same expression the motions of the largest bodies of Universe as well as those of the slightest atom: nothing would be uncertain for this intelligence, all future and all past would be as known as present.*

The above statement was widely recognized as the landmark of scientific thinking: a good scientific theory must describe a natural phenomenon by using mathematical methods; once the temporal evolution equations of the phenomenon are known and the initial conditions are determined, the state of the system can be known at each future time by solving such equations. Nowadays, the quoted text is often cited and criticized in some popular science books as too naive. In contrast with how often asserted, it should be emphasized that Laplace was not as naive about the true relevance of the determinism. Actually, he was aware of the practical difficulties of a strictly deterministic approach to many everyday life phenomena which exhibit unpredictable behaviors as, for instance, the weather. How do we reconcile Laplace's deterministic assumption with the "irregularity" and "unpredictability" of many observed phenomena? Laplace himself gave an answer to this question, in the same book, identifying the origin of the irregularity in our imperfect knowledge of the system:

> *The curve described by a simple molecule of air or vapor is regulated in a manner just as certain as the planetary orbits; the only difference between them is that which comes from our ignorance. Probability is relative, in part to this ignorance, in part to our knowledge.*

A fairer interpretation of Laplace's image of "mathematical intelligence" probably lies in his desire to underline the importance of prediction in science, as it transparently appears from a famous anecdote quoted by Cohen and Stewart (1994). When Napoleon received Laplace's masterpiece *Méchanique Céleste* told him *M. Laplace, they tell me you have written this large book on the system of the universe, and have never even mentioned its Creator.* And Laplace answered *I did not need to make such assumption.* So that Napoleon replied: *Ah! That is a beautiful assumption, it explains many things*, and Laplace: *This hypothesis, Sire, does explain everything, but does not permit to predict anything. As a scholar, I must provide you with works permitting predictions.*

The main reason for the almost unanimous consensus of 19th century scientists about determinism has to be, perhaps, searched in the great successes of Celestial Mechanics in making accurate predictions of planetary motions. In particular, we should mention the spectacular discovery of Neptune after its existence was predicted — theoretically deduced — by Le Verrier and Adams using Newtonian mechanics. Nevertheless, still within the 19th century, other phenomena not as regular as planet motions were active subject of research, from which statistical physics originated. For example, in 1873, Maxwell gave a conference with the significant title: *Does the progress of Physical Science tend to give any advantage to the opinion of Necessity (or Determinism) over that of the Contingency of Events and the Freedom of the Will?*

The great Scottish scientist realized that, in some cases, system details are so fine that lie beyond any possibility of control. Since *the same antecedents never again concur, and nothing ever happens twice*, he criticized as empirically empty the well recognized law *from the same antecedents the same consequences follow.* Actually, he went even further by recognizing the possible failure of the weaker version *from like antecedents like consequences follow*, as instability mechanisms can be present.

Ironically, the first[1] clear example of what we know today as Chaos — a paradigm for deterministic irregular and unpredictable phenomena — was found in Celestial Mechanics, the science of regular and predictable phenomena *par excellence.* This is the case of the longstanding *three-body problem* — i.e. the motion of three gravitationally interacting bodies such as, e.g. Moon-Earth-Sun [Gutzwiller (1998)] — which was already in the nightmares of Newton, Euler, Lagrange and many others. Given the law of gravity, the initial positions and velocities of the three bodies, the subsequent positions and velocities are determined by the equations of mechanics. In spite of the deterministic nature of the system, Poincaré (1892, 1893, 1899) found that the evolution can be chaotic, meaning that small perturbations in the initial state, such as a slight change in one body's initial position, might lead to dramatic differences in the later states of the system.

The deep implication of these results is that determinism and predictability are distinct problems. However, Poincaré's discoveries did not receive the due attention for a quite long time. Probably, there are two main reasons for such a delay. First, in the early 20th century, scientists and philosophers lost interest in classical mechanics[2] because they were primarily attracted by two new revolutionary theories: relativity and quantum mechanics. Second, an important role in the recognition of the importance and ubiquity of Chaos has been played by the development of the computer, which came much after Poincaré's contribution. In fact, only thanks to the advent of computer and scientific visualization was possible to (numerically) compute and see the staggering complexity of chaotic behaviors emerging from nonlinear deterministic systems.

A widespread view claims that the line of scientific research opened by Poincaré remained neglected until 1963, when meteorologist Lorenz rediscovered deterministic chaos while studying the evolution of a simple model of the atmosphere. Consequently, often, it is claimed that the new paradigm of deterministic chaos begun in

[1]In 1898 chaos was noticed also by Hadamard who found that a negative curvature system displaying sensitive dependence on the initial conditions.

[2]It is interesting to mention the case of the young Fermi who, in 1923, obtained interesting results in classical mechanics from which he argued (erroneously) that Hamiltonian systems, in general, are ergodic. This conclusion has been generally accepted (at least by the physics community) Following Fermi's 1923 work, even in the absence of a rigorous demonstration, the ergodicity problem seemed, at least to physicists, essentially solved. It seems that Fermi was not very worried of the lacking of rigor of his "proof", likely the main reason was his (and more generally of the large part of the physics community) interest in the development of quantum physics.

the sixties. This is not true, as mathematicians never forgot the legacy of Poincaré, although it was not so well known by physicists. Although this is not the proper place for precise historical[3] considerations, it is important to give, at least, an idea of the variegated history of dynamical systems and its interconnections with other fields before the (re)discovery of chaos, and its modern developments. The schematic list below, containing the most relevant contributions, serves to this aim:

[early 20th century] Stability theory and qualitative analysis of differential equations, which started with Poincaré and Lyapunov and continues with Birkhoff and the soviet school.

[starting from the '20s] Control theory with the work of Andronov, van der Pol and Wiener.

[mid '20s and '40s-'50s] Investigation of nonlinear models for population dynamics and ecological systems by Volterra and Lotka and, later, the study of the logistic map by von Neumann and Ulam.

['30s] Birkhoff and von Neumann studies of ergodic theory. The seminal work of Krylov on mixing and the foundations of statistical mechanics.[4]

[1948–1960] Information theory born already mature with Shannon's work and was introduced in dynamical systems theory, during the fifties, by Kolmogorov and Sinai.

[1955] Fermi-Pasta-Ulam (FPU) numerical experiment on nonlinear Hamiltonian systems showed that ergodicity is a non-generic property.

[1954–1963] The KAM theorem for the regular behavior of almost integrable Hamiltonian systems, which was proposed by Kolmogorov and subsequently completed by Arnold and Moser.

This, non exhaustive, list demonstrates how claiming chaos as a new paradigmatic theory born in the sixties is not supported by facts.[5]

It is worth concluding this brief historical introduction by mentioning some of the most important steps which lead to "modern" (say after 1960) development of dynamical systems in physics.

The pioneering contributions of Lorenz, Hénon and Heiles, and Chirikov, showing that even simple low dimensional deterministic systems can exhibit irregular and unpredictable behaviors, brought chaos to the attention of the physics community. The first clear evidence of the physical relevance of chaos to important phenomena, such as turbulence, came with the works of Ruelle, Takens and Newhouse on the onset of chaos. Afterwords, brilliant experiments on the onset of chaos in Rayleigh-Bénard convection (Libchaber, Swinney, Gollub and Giglio) confirmed

[3]For throughout introduction to dynamical systems history see the nice work of Aubin and Dalmedico (2002).

[4]His thesis *Mixing processes in phase space* appeared posthumously in 1950, when it was translated in English [Krylov (1979)] the book came as a big surprise in the West.

[5]For a detailed discussion about the use and abuse of chaos see *Science of Chaos or Chaos in Science?* by Bricmont (1995).

the theoretical predictions, boosting the interest of physicists in nonlinear dynamical systems. Another crucial moment for the development of dynamical systems theory was the disclosure of the connections among chaos, critical phenomena and scaling subsequent to the works of Feigenbaum[6] on the universality of the period doubling mechanism for the transition to chaos. The thermodynamic formalism, originally proposed by Ruelle and then "translated" in more physical terms with the introduction of multifractals and periodic orbits expansion, disclosed the deep connection between chaos and statistical mechanics. Fundamental in providing the suitable (practical) tools for the investigation of chaotic dynamical systems were: the introduction of efficient numerical methods for the computation of Lyapunov exponents (Benettin, Galgani, Giorgilli and Strelcyn), the fractal dimension (Grassberger and Procaccia), and the embedding technique, pioneered by Takens, which constitutes a bridge between theory and experiments.

The physics of chaotic dynamical systems benefited of many contributions from mathematicians which were very active after 1960 among whom we should remember Bowen, Ruelle, Sinai and Smale.

Overview of the book

The book is divided into two parts.

Part I: **Introduction to Dynamical Systems and Chaos** (Chapters 1–7) aims to provide basic results, concepts and tools on dynamical systems, encompassing stability theory, classical examples of chaos, ergodic theory, fractals and multifractals, characteristic Lyapunov exponents and the transition to chaos.

Part II: **Advanced Topics and Applications: From Information Theory to Turbulence** (Chapters 8–14) introduces the reader to the applications of dynamical systems in celestial and fluid mechanics, population biology and chemistry. It also introduces more sophisticated tools of analysis in terms of information theory concepts and their generalization, together with a review of high dimensional systems from chaotic extended systems to turbulence.

Chapters are organized in main text and call-out *boxes*, which serve as appendices with various scopes. Some boxes are meant to make the book self-consistent by recalling some basic notions, e.g. Box B.1 and B.6 are devoted to Hamiltonian dynamics and Markov Chains, respectively. Some others present examples of technical or pedagogical interest, e.g. Box B.14 deals with the resonance overlap criterion while Box B.23 shows an example of use of discrete mapping to describe Halley comet dynamics. Most of boxes focuses on technical aspects or deepening of some aspects which are only briefly considered in the main text. Furthermore, Chapters from 2 to 9 end with a few exercises and suggestions for numerical experiences meant helping to master the presented concepts and tools.

[6]Actually also other authors obtained independently the same results, see Derrida *et al.* (1979).

Chapters are organized as follows.

The first three Chapters are meant to be a gentle introduction to chaos, and set the language and notation used in the rest of the book. In particular, Chapter 1 aims to introduce newcomers to the main aspects of chaotic dynamics with the aid of a specific example, namely the nonlinear pendulum, in terms of which the distinction between determinism and predictability is clarified. The definition of dissipative and conservative (Hamiltonian) dynamical systems, the basic language and notation, together with a brief account of linear and nonlinear stability analysis are presented in Chapter 2. Three classical examples of chaotic behavior — the logistic map, the Lorenz system and the Hénon-Heiles model — are reviewed in Chapter 3

With Chapter 4 it starts the formal treatment of chaotic dynamical systems. In particular, the basic notions of ergodic theory and mixing are introduced, and concepts such as invariant and natural measure discussed. Moreover, the analogies between chaotic systems and Markov Chains are emphasized. Chapter 5 defines and explains how to compute the basic tools and indicators for the characterization of chaotic systems such the multifractal description of strange attractors, the stretching and folding mechanism, the characteristic Lyapunov exponents and the finite time Lyapunov exponents.

The first part of the book ends with Chapter 6 and 7 which discuss, emphasizing the universal aspects, the problem of the transition from order to chaos in dissipative and Hamiltonian systems, respectively.

The second part of the book starts with Chapter 8 which introduces the Kolmogorov-Sinai entropy and deals with information theory and, in particular, its connection with algorithmic complexity, the problem of compression and the characterization of "randomness" in chaotic systems. Chapter 9 extends the information theory approach introducing the ε-entropy which generalizes Shannon and Kolmogorov-Sinai entropies to a coarse-grained description level. With similar purposes, it is also discussed the Finite Size Lyapunov Exponents, an extension to the usual Lyapunov exponents accounting for finite perturbations.

Chapter 10 reviews the practical and theoretical issues inherent to computer simulations and experimental data analysis of chaotic systems. In particular, it accounts for the effects of round-off errors and the problem of discretization in digital computations. As for the data analysis, the main methods and their limitations are discussed. Further, it is discussed the longstanding issue of distinguishing chaos from noise and model building from time series.

Chapter 11 is devoted to some important applications of low dimensional Hamiltonian and dissipative chaotic systems encompassing celestial mechanics, transport in fluids, population dynamics, chemistry and the problem of synchronization.

High dimensional systems with their complex spatiotemporal behaviors and connection to statistical mechanics are discussed in Chapters 12 and 13. In the former, after briefly reviewing the systems of interest, we focus on three main aspects: the generalizations of the Lyapunov exponents needed to account for the spatiotemporal evolution of perturbations; the description of some phenomena in terms of non-equilibrium statistical mechanics; the description of high dimensional systems at a coarse-grained level and its connection to the problem of model building. The latter Chapter focuses on fluid mechanics with emphasis on turbulence. In particular, we discuss the statistical mechanics description of perfect fluids, the phenomenology of two- and three-dimensional turbulence, the general problem of the reduction of partial differential equations to systems with a finite number of degrees of freedom and various aspects of the predictability problem in turbulent flows.

At last, in Chapter 14 starting from the seminal paper by Fermi, Pasta and Ulam (FPU) we discuss a specific research issue, namely the relationship between statistical mechanics and the chaotic properties of the underlying dynamics. This Chapter will give us the opportunity to reconsider some subtle issues which stand at the foundation of statistical mechanics. Especially, the discussion on FPU numerical experiments has a great pedagogical value in showing how, in a typical research program, only with a clever combination of theory, computer simulations, probabilistic arguments and conjectures is possible a real progress.

The book ends with an epilogue containing some general considerations on the role of models, computer simulations and the impact of chaos in the scientific research activity in the last decades.

Hints on how to use/read this book

Some possible paths to the use of this book are:

A) For a basic course aiming to introduce chaos and dynamical system: the first five Chapters and parts of Chapter 6 and 7, depending if the emphasis of the course is on dissipative or Hamiltonian systems, part of Chapter 8 for the Kolmogorov-Sinai entropy;

B) For an advanced general course: the first part, Chapters 8 and 10.

C) For advanced topical courses: the first part and a selection of the second part, for instance

- C.1) Chapters 8 and 9 for an information theory, or computer science, oriented course;
- C.2) Chapters 8-10 for researchers and/or graduate students, interested in the treatment of experimental data and modeling;
- C.3) Section 11.3 for a tour on chaos in chemistry and biology;
- C.4) Chapters 12, 13 and 14 if the main interest is in high dimensional systems;
- C.5) Section 11.2 and Chapter 13 for a tour on chaos and fluid mechanics;
- C.6) Sections 12.4 and 13.2 plus Chapter 14 for a tour on chaos and statistical mechanics.

We encourage all who wish to comment on the book to contact us through the book homepage URL: http://denali.phys.uniroma1.it/~chaosbookCCV09/ where errata and solutions to the exercises will be maintained.

Contents

Advanced Topics and Applications: From Information Theory to Turbulence

PART 1

Introduction to Dynamical Systems and Chaos

Chapter 1

First Encounter with Chaos

> If you do not expect the unexpected you will not find it, for it is not to be reached by search or trail.
>
> *Heraclitus (ca. 535–475 BC)*

This Chapter is meant to provide a simple and heuristic illustration of some basic features of chaos. To this aim, we exemplify the distinction between determinism and predictability, which stands at the essence of deterministic chaos, with the help of a specific example — the nonlinear pendulum.

1.1 Prologue

In the search for accurate ways of measuring time, the famous Dutch scientist Christian Huygens in 1656, exploiting the regularity of pendulum oscillations, made the first pendulum clock. Being able to measure time accumulating an error of something less than a minute per day (an accuracy never achieved before), such a clock represented a great technological advancement. Even though nowadays pendulum clocks are not used anymore, everybody would subscribe the expression predictable (or regular) as a pendulum clock. Generally, the adjectives predictable and regular would be referred to the evolution of any mechanical system ruled by Newton's laws, which are deterministic. This is not only because the pendulum oscillations look very regular but also because, in the common sense, we tend to confuse or associate the two terms deterministic and predictable. In this Chapter, we will see that even the pendulum may give rise to surprising behaviors, which impose to reconsider the meaning of predictability and determinism.

1.2 The nonlinear pendulum

Let's start with the simple case of a planar pendulum consisting of a mass m attached to a pivot point O by means of a mass-less and inextensible wire of length L, as illustrated in Fig. 1.1a. From any elementary course of mechanics, we know that two forces act on the mass: gravity $\boldsymbol{F}_g = m\boldsymbol{g}$ (where $\boldsymbol{g}$ is the gravity acceleration of modulus g and directed in the negative vertical direction) and the tension $\boldsymbol{T}$ parallel to the wire and directed toward the pivot point O. For the sake of simplicity, we momentarily neglect friction exerted by air molecules on the moving bead. By exploiting Newton's law $\boldsymbol{F} = m\boldsymbol{a}$, we can straightforwardly write the equations of pendulum evolution. The only variables we need to describe the pendulum state are the angle θ between the wire and the vertical, and the angular velocity $\mathrm{d}\theta/\mathrm{d}t$. We are then left with a second order differential equation for θ:

$$\frac{\mathrm{d}^2\theta}{\mathrm{d}t^2} + \frac{g}{L}\sin\theta = 0\,. \tag{1.1}$$

It is rather easy to imagine the pendulum undergoing small amplitude oscillations as a devise for measuring time. In such a case the approximation $\sin\theta \approx \theta$ recovers the usual (linear) equation of an harmonic oscillator:

$$\frac{\mathrm{d}^2\theta}{\mathrm{d}t^2} + \omega_0^2\theta = 0\,, \tag{1.2}$$

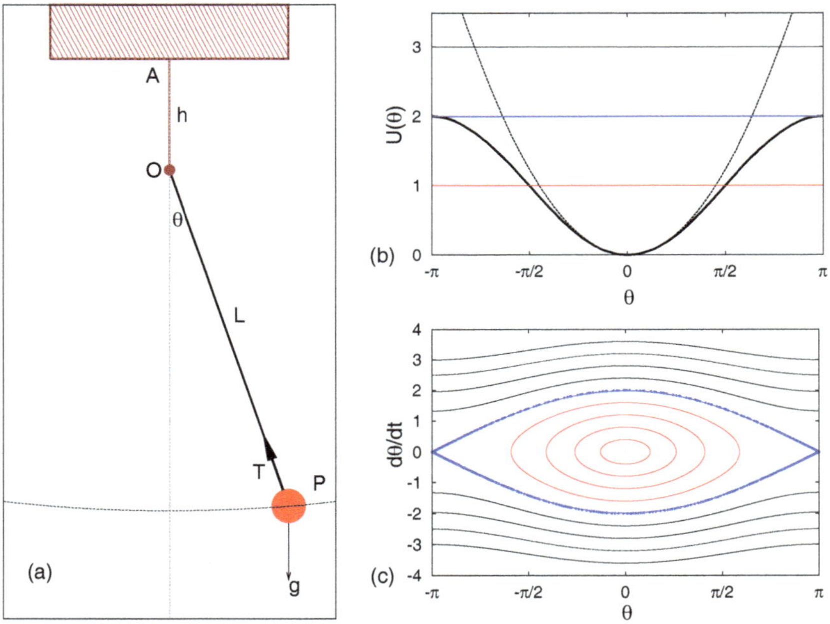

Fig. 1.1 Nonlinear pendulum. (a) Sketch of the pendulum. (b) The potential $U(\theta) = mgL(1 - \cos(\theta))$ (thick black curve), and its approximation $U(\theta) \approx mgL\theta^2/2$ (dashed curve) valid for small oscillations. The three horizontal lines identify the energy levels corresponding to qualitatively different trajectories: oscillations (red), the separatrix (blue) and rotations (black). (c) Trajectories corresponding to various initial conditions. Colors denote different classes of trajectories as in (b).

where $\omega_0 = \sqrt{g/L}$ is the fundamental frequency. The above equation has periodic solutions with period $2\pi/\omega_0$, hence, properly choosing the pendulum length L, we can fix the unit to measure time. However, for larger oscillations, the full nonlinearity of the sin function should be considered, it is then natural to wonder about the effects of such nonlinearity.

The differences between Eq. (1.1) and (1.2) can be easily understood introducing pendulum energy, the sum of kinetic K and potential U energy:

$$H = K + U = \frac{1}{2}mL^2\left(\frac{\mathrm{d}\theta}{\mathrm{d}t}\right)^2 + mgL(1 - \cos\theta)\,, \tag{1.3}$$

that is conserved, as no dissipation mechanism is acting. Figure 1.1b depicts the pendulum potential energy $U(\theta)$ and its harmonic approximation $U(\theta) \approx mgL\theta^2/2$. It is easy to realize that the new features are associated with the presence of a threshold energy (in blue) below which the mass can only oscillate around the rest position, and above which it has energy high enough to rotate around the pivot point (of course, in Fig. 1.1a one should remove the upper wall to observe it). Within the linear approximation, rotation is not permitted, as the potential energy barrier for observing rotation is infinite.

The possible trajectories are exemplified in Fig. 1.1c, where the blue orbit separates (hence the name *separatrix*) two classes of motions: oscillations (closed orbits) in red and rotations (open orbits) in black. The separatrix physically corresponds to the pendulum starting with zero velocity from the unstable equilibrium positions $(\theta, \mathrm{d}\theta/\mathrm{d}t) = (\pi, 0)$ and performing a complete turn so to come back to it with zero velocity, in an infinite time. Periodic solutions follows from energy conservation $H(\theta, \mathrm{d}\theta/\mathrm{d}t) = E$ and Eq. (1.3) leading to the relation $\mathrm{d}\theta/\mathrm{d}t = f(E, \cos\theta)$ between angular velocity $\mathrm{d}\theta/dt$ and θ. As $\cos\theta$ is cyclic, it follows the periodicity of $\theta(t)$.

Then, apart from enriching a bit the possible behaviors, the presence of nonlinearities does not change much what we learned from the simple harmonic pendulum.

1.3 The damped nonlinear pendulum

Now we add the effect of air drag on the pendulum. According to Stokes' law, this amounts to include a new force proportional to the mass velocity, and always acting against its motion. Equation (1.1) with friction becomes

$$\frac{\mathrm{d}^2\theta}{\mathrm{d}t^2} + \gamma\frac{\mathrm{d}\theta}{\mathrm{d}t} + \frac{g}{L}\sin\theta = 0\,, \tag{1.4}$$

γ being the viscous drag coefficient, usually depending on the bead size, air viscosity etc. Common experience suggests that, waiting a sufficiently long time, the pendulum ends in the rest state with the mass lying just down the vertical from the pivot point, independently of its initial speed. In mathematical language this means that, the friction term dissipates energy making the rest state $(\theta, \mathrm{d}\theta/\mathrm{d}t) = (0, 0)$ an attracting point for Eq. (1.4) (as exemplified in Fig. 1.2).

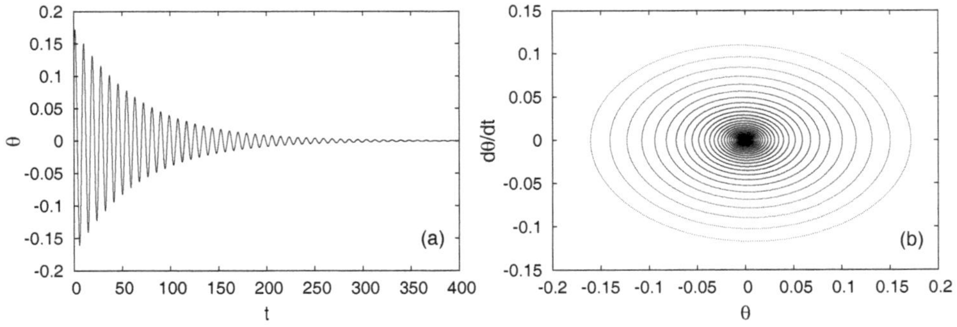

Fig. 1.2 Damped nonlinear pendulum: (a) angle versus time for $\gamma=0.03$; (b) evolution in phase space, i.e. $\mathrm{d}\theta/\mathrm{d}t$ vs θ.

Summarizing, nonlinearity is not sufficient to make pendulum motion nontrivial. Further, the addition of dissipation alone makes trivial the system evolution.

1.4 The vertically driven and damped nonlinear pendulum

It is now interesting to see what happens if an external driving is added to the nonlinear pendulum with friction to maintain its state of motion. For example, with reference to Fig. 1.1a, imagine to have a mechanism able to modify the length h of the segment $\overrightarrow{\mathrm{AO}}$, and hence to drive the pendulum by bobbing its pivot point O. In particular, suppose that h varies periodically in time as $h(t) = h_0 \cos(\omega t)$, where h_0 is the maximal extension of $\overrightarrow{\mathrm{AO}}$ and ω the frequency of bobbing.

Let's now understand how Eq. (1.4) modifies to account for the presence of such an external driving. Clearly, we know how to write Newton's equation in the reference frame attached to the pivot point O. As it moves, such a reference frame is non-inertial and any first course of mechanics should have taught us that *fictitious* forces appear. In the case under consideration, we have that $\boldsymbol{r}_\mathrm{A} = \boldsymbol{r}_\mathrm{O} + \overrightarrow{\mathrm{AO}} = \boldsymbol{r}_\mathrm{O} + h(t)\hat{\boldsymbol{y}}$, where $\boldsymbol{r}_\mathrm{O} = \overrightarrow{\mathrm{OP}}$ is the mass vector position in the non-inertial (pivot point) reference frame, $\boldsymbol{r}_\mathrm{A} = \overrightarrow{\mathrm{AP}}$ that in the inertial (laboratory) one, and $\hat{\boldsymbol{y}}$ is the unit vector identifying the vertical direction. As a consequence, in the non-inertial reference frame, the acceleration is given by $\boldsymbol{a}_\mathrm{O} = \mathrm{d}^2\boldsymbol{r}_\mathrm{O}/\mathrm{d}t^2 = \boldsymbol{a}_\mathrm{A} - \mathrm{d}^2 h/\mathrm{d}t^2 \hat{\boldsymbol{y}}$. Recalling that, in the inertial reference frame, the *true* forces are gravity $m\boldsymbol{g} = -mg\hat{\boldsymbol{y}}$ and tension, the net effect of bobbing the pivot point, in the non-inertial reference frame, is to modify gravity force as $mg\hat{\boldsymbol{y}} \to m(g + \mathrm{d}^2h/\mathrm{d}t^2)\hat{\boldsymbol{y}}$.[1] We can thus write the equation for θ as

$$\frac{\mathrm{d}^2\theta}{\mathrm{d}t^2} + \gamma' \frac{\mathrm{d}\theta}{\mathrm{d}t} + (\alpha - \beta\cos t)\sin\theta = 0 \tag{1.5}$$

[1]Notice that if the pivot moves of uniform motion, i.e. $\mathrm{d}^2h/\mathrm{d}t^2 = 0$, the usual pendulum equation are recovered because the fictitious force is not present anymore and the reference frame is inertial.

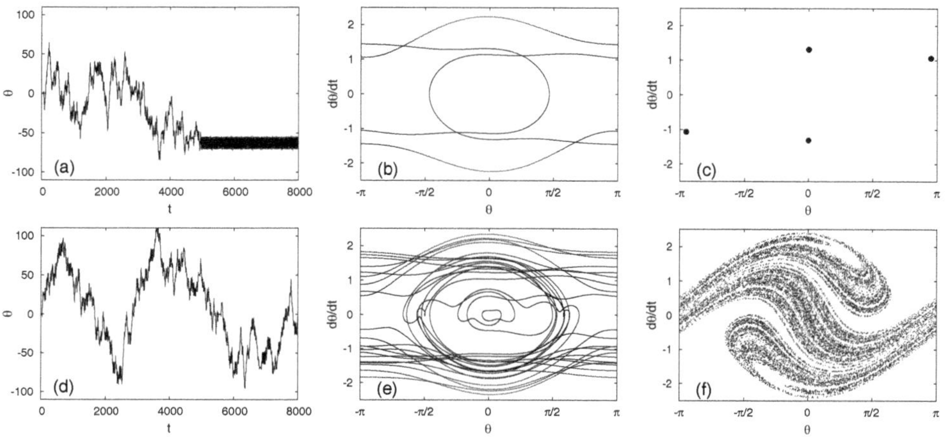

Fig. 1.3 Driven-damped nonlinear pendulum: (a) θ vs t for $\alpha = 0.5$, $\beta = 0.63$ and $\gamma' = 0.03$ with initial condition $(\theta, \mathrm{d}\theta/\mathrm{d}t) = (0, 0.1)$; (b) the same trajectory shown in phase space using the cyclic representation of angle in $[-\pi; \pi]$; (c) stroboscopic map showing that the trajectory has period 4. (d-f) Same as (a-c) for $\alpha = 0.5$, $\beta = 0.70$ and $\gamma' = 0.03$. In (e) only a portion of the trajectory is shown due to its tendency to fill the domain.

where, for the sake of notation simplicity, we rescaled time with the frequency of the external driving $t\omega \to t$, obtaining the new parameters $\gamma' = \gamma/\omega$, $\alpha = g/(L\omega^2)$ and $\beta = h_0/L$. In such normalized units, the period of the vertical driving is $T_0 = 2\pi$. Equation (1.5) is rather interesting[2] because of the explicit presence of time which enlarges the "effective" dimensionality of the system to $2 + 1$, namely angle and angular velocity plus time.

Equation (1.5) may be analyzed by, for instance, fixing γ' and α and varying β, which parametrizes the external driving intensity. In particular, with $\alpha = 0.5$ and $\gamma' = 0.03$, qualitatively new solutions can be observed depending on β. Clearly, if $\beta = 0$, we have again the damped pendulum (Fig. 1.2). The behavior complicates a bit increasing β. In particular, Bartuccelli *et al.* (2001) showed that for values of $0 < \beta < 0.55$ all orbits, after some time, collapse onto the same periodic orbit characterized by the period $T_0 = 2\pi$, corresponding to that of the forcing. This is somehow similar to the case of the nonlinear dissipative pendulum, but it differs as the asymptotic state is not the rest state but a periodic one.

Let's now see what happens for $\beta > 0.55$. In Fig. 1.3a we show the evolution of angle θ (here represented without folding it in $[0 : 2\pi]$) for $\beta = 0.63$. After a rather long transient, where the pendulum rotates in an erratic/random way (portion of the graph for $t \lesssim 4500$), the motion sets onto a periodic orbit. As shown in Fig. 1.3b, such a periodic orbit draws a pattern in the $(\theta, \mathrm{d}\theta/dt)$-plane more complicated than those found for the simple pendulum (Fig. 1.1c). To understand

[2]We mention that by approximating $\sin\theta \approx \theta$, Eq. (1.5) becomes the Mathieu equation, a prototype example of ordinary differential equation exhibiting *parametric resonance* [Arnold (1978)], which will not be touched in this book.

the period of the depicted trajectory, one can use the following strategy. Imagine to look at the trajectory in a dark room, and to switch on the light only at times $t_0, t_1, \ldots$ chosen in such a way that $t_n = nT_0 + t^*$ (with an arbitrary reference t^*, which is not important). As in a disco stroboscopic lights (whose basic functioning principle is the same) give us static images of dancers, we do not see anymore the temporal evolution of the trajectory as a continuum but only the sequence of pendulum positions at times $t_1, t_2, \ldots, t_n \ldots$. In Fig. 1.3c, we represent the states of the pendulum as points in the $(\theta, \mathrm{d}\theta/\mathrm{d}t)$-plane, when such a stroboscopic view is used. We can recognize only four points, meaning that the period is $4T_0$, amounting to four times the forcing period.

In the same way we can analyze the trajectories for larger and smaller β's. Doing so, one discovers that for $\beta > 0.55$ the orbits are all periodic but with increasing period $2T_0$, $4T_0$ (as for the examined case), $8T_0, \ldots, 2^n T_0$. This *period-doubling* sequence stops at a critical value $\beta_d = 0.64018$ above which no regularities can be observed. For $\beta > \beta_d$, any portion of the time evolution $\theta(t)$ (see, e.g., Fig. 1.3d) displays an aperiodic irregular behavior similar to the transient one of the previous case. Correspondingly, the $(\theta, \mathrm{d}\theta/\mathrm{d}t)$-plane representation of it (Fig. 1.3e) becomes very complicated and inter-winded. Most importantly, no evidences of periodicity can be found, as the stroboscopic map depicted in Fig. 1.3f demonstrates.

We have thus to accept that even an "innocent" (deterministic) pendulum may give rise to an irregular and aperiodic motion. The fact that Huygens could use the pendulum for building a clock now appears even more striking. Notice that if the driving would have been added to an harmonic damped oscillator, the resulting dynamical behavior would have been much simpler than the one here observed (giving rise to the well known resonance phenomenon). Therefore, nonlinearity is necessary to have the complicated features of Fig. 1.3d–f.

1.5 What about the predictability of pendulum evolution?

Figure 1.3d may give the impression that the pendulum rotates and oscillates in a random and unpredictable way, questioning about the possibility to predict the motions originating from a deterministic system, like the pendulum. However, we can think that it is only our inability to describe the trajectory in terms of known functions to cause such a difficulty to *predict*. Following this point of view, the unpredictability would be only apparent and not substantial.

In order to make concrete the above line of reasoning, we can reformulate the problem of predicting the trajectory of Figure 1.3d in the following way. Suppose that two students, say Sally and Adrian, are both studying Eq. (1.5). If Sally produced on her computer Fig. 1.3d, then Adrian, knowing the initial condition, should be able to reproduce the same figure. Thanks to the theorem of existence and uniqueness, holding for Eq. (1.5), Adrian is of course able to reproduce Sally's result. However, let's suppose, for the moment, that they do not know such a theorem and let's ask Sally and Adrian to play the game.

They start considering the periodic trajectory of Fig. 1.3b which, looking predictable, will constitute the benchmark case. Sally, discarding the initial behavior, tells to Adrian as a starting point of the trajectory the values of the angle and angular velocity at $t_0 = 6000$, where the transient dynamics died out, i.e. $\theta(t_0) = -68.342110$ and $d\theta/dt = 1.111171$. By mistake, she sends an email to Adrian typing -68.342100 and 1.111181, committing an error of $\mathcal{O}(10^{-5})$ in both the angle and angular velocity. Adrian takes the values and, using his code, generates a new trajectory starting from this initial condition. Afterwords, they compare the results and find that, despite the small error, the two trajectories are indistinguishable.

Later, they realize that two slightly different initial conditions were used. As the prediction was anyway possible, they learned an important lesson: at practical level a prediction is so if it works even with an imperfect knowledge of the initial condition. Indeed, while working with a real system, the knowledge of the initial state will always be limited by unavoidable measurements errors. In this respect the pendulum behavior of Fig. 1.3b is a good example of predictable system.

Next they repeat the prediction experiment for the trajectory reported in Fig. 1.3d. Sally decides to follow exactly the same procedure as above. Therefore, she opts, also in this case, for choosing the initial state of the pendulum after a certain time lapse, in particular at time $t_0 = 6000$ where $\theta(t_0) = -74.686836$ and $d\theta/dt = -0.234944$. Encouraged by the test case, bravely but confidently, she intentionally transmits to Adrian a wrong initial state: $\theta(t_0) = -74.686826$ and $d\theta/dt = -0.234934$: differing again of $\mathcal{O}(10^{-5})$ in both angle and velocity. Adrian computes the new trajectory, and goes to Sally for the comparison, which looks as in Fig. 1.4. The trajectories now almost coincide at the beginning but then become completely different (eventually coming close and far again and again).

Surprised Sally tries again by giving an initial condition with a smaller error to Adrian: nothing changes but the time at which the two trajectories depart from each other. At last, Sally decides to check whether Adrian has a bug in his code and gives him the true initial condition, hoping that the trajectory will be different. But Adrian is as good as Sally in programming and their trajectories now coincide.[3]

Sally and Adrian made no error, they were just too confident about the possibility to predict a deterministic evolution. They did not know about *chaos*, which can momentarily defined as: a property of motion characterized by an aperiodic evolution, often appearing so irregular to resemble a random phenomenon, with a strong dependence on initial conditions.

We conclude by noticing that also the simple nonlinear pendulum (1.1) may display sensitivity to initial conditions, but only for very special ones. For instance,

[3]We will learn later that even giving the same initial condition does not guarantee that the results coincide. If, for example, the time step for the integration is different, the computer or the compiler are different, or other conditions that we will see are not fulfilled.

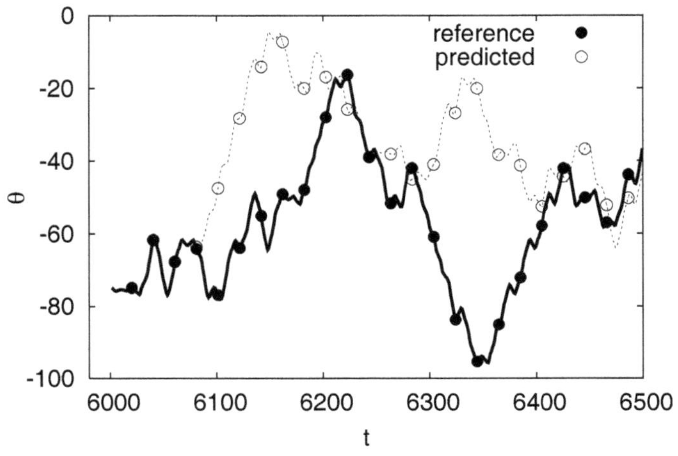

Fig. 1.4 θ versus t for Sally's reference trajectory and Adrian's "predicted" one, see text.

if the pendulum of Fig. 1.1 is prepared in two different initial conditions such that it is slightly displaced on the left/right from the vertical but at the opposite of the rest position, in other words $\theta(0) = \pi \pm \epsilon$ with ϵ a small as wanted but positive value. The bead will go on the left (+) or on the right (−). This is because the point $(\pi, 0)$ is an unstable equilibrium point.[4] Thus chaos can be regarded as a situation in which all the possible states of a system are, in a still vague sense, "unstable".

1.6 Epilogue

The nonlinear pendulum example practically exemplifies the abstract meaning of determinism and predictability discussed in the Introduction. On the one side, quoting Laplace, if we were the *intelligence that knows all forces acting on the pendulum* (the equations of motion) and *the respective situation of all its elements* (perfect knowledge of the initial conditions) then *nothing would be uncertain*: at least with the computer, we can perfectly predict the pendulum evolution. On the other hand, again quoting Laplace, *the problem may come from our ignorance* (on the initial conditions). More precisely, in the simple pendulum a small error on the initial conditions remains small, so that the prediction is not (too severely) spoiled by our ignorance. On the contrary, the imperfect knowledge on the present state of the nonlinear driven pendulum amplifies to a point that the future state cannot be predicted beyond a finite time horizon. This sensitive dependence on the initial state constitutes, at least for the moment, our working definition of chaos. The quantitative meaning of this definition together with the other aspects of chaos will become clearer in the next Chapters of the first part of this book.

[4]We will learn in the next Chapter that this is an unstable hyperbolic fixed point.

Chapter 2

The Language of Dynamical Systems

The book of Nature is written in the mathematical language.
Galileo Galilei (1564–1642)

The pendulum of Chapter 1 is a simple instance of *dynamical system*. We define dynamical system any mathematical model or rule which determines the future evolution of the variables describing the state of the system from their initial values. We can thus generically call dynamical system any evolution law. In this definition we exclude the presence of randomness, namely we restrict to *deterministic dynamical systems*. In many natural, economical, social or other kind of phenomena, it makes sense to consider models including an intrinsic or external source of randomness. In those cases one speaks of *random dynamical systems* [Arnold (1998)]. Most of the book will focus on deterministic laws. This Chapter introduces the basic language of dynamical systems, building part of the dictionary necessary for their study. Refraining from using a too formalized notation, we shall anyway maintain the due precision. This Chapter also introduces linear and nonlinear stability theories, which constitute useful tools in approaching dynamical systems.

2.1 Ordinary Differential Equations (ODE)

Back to the nonlinear pendulum of Fig. 1.1a, it is clear that, once its interaction with air molecules is disregarded, the *state* of the pendulum is determined by the values of the angle θ and the angular velocity $\mathrm{d}\theta/\mathrm{d}t$. Similarly, at any given time t, the state of a generic system is determined by the values of all variables which specify its state of motion, i.e., $\boldsymbol{x}(t) = (x_1(t), x_2(t), x_3(t), \ldots, x_d(t))$, d being the system dimension. In principle, $d = \infty$ is allowed and corresponds to partial differential equations (PDE) but, for the moment, we focus on finite dimensional dynamical systems and, in the first part of this book, low dimensional ones. The set of all possible states of the system, i.e. the allowed values of the variables x_i $(i = 1, \ldots, d)$, defines the *phase space* of the system. The pendulum of Eq. (1.1) corresponds to $d = 2$ with $x_1 = \theta$ and $x_2 = \mathrm{d}\theta/dt$ and the phase space is a cylinder as θ and $\theta + 2\pi k$ (for any integer k) identify the same angle. The trajectories depicted in Fig. 1.1c represent the *phase-space portrait* of the pendulum.

The state variable $\boldsymbol{x}(t)$ is a point in phase space evolving according to a system of ordinary differential equations (ODEs)

$$\frac{\mathrm{d}\boldsymbol{x}}{\mathrm{d}t} = \boldsymbol{f}(\boldsymbol{x}(t)), \tag{2.1}$$

which is a compact notation for

$$\begin{aligned}\frac{\mathrm{d}x_1}{\mathrm{d}t} &= f_1(x_1(t), x_2(t), \cdots, x_d(t)),\\ &\vdots\\ \frac{\mathrm{d}x_d}{\mathrm{d}t} &= f_d(x_1(t), x_2(t), \cdots, x_d(t)).\end{aligned}$$

More precisely, Eq. (2.1) defines an *autonomous* ODE as the functions f_i's do not depend on time. The driven pendulum Eq. (1.5) explicitly depends on time and is an example of *non-autonomous* system, whose general form is

$$\frac{\mathrm{d}\boldsymbol{x}}{\mathrm{d}t} = \boldsymbol{f}(\boldsymbol{x}(t), t). \tag{2.2}$$

The d-dimensional non-autonomous system (2.2) can be written as a $(d+1)$-dimensional autonomous one by defining $x_{d+1} = t$ and $f_{d+1}(\boldsymbol{x}) = 1$.

Here, we restrict our range of interests to the (very large) subclass of (smooth) differentiable functions, i.e. we assume that

$$\frac{\partial f_j(\boldsymbol{x})}{\partial x_i} \equiv \partial_i f_j(\boldsymbol{x}) \equiv \mathbb{L}_{ji}$$

exists for any $i, j = 1, \ldots, d$ and any point $\boldsymbol{x}$ in phase space; $\mathbb{L}$ is the so-called stability matrix (see Sec. 2.4). We thus speak of *smooth dynamical systems*,[1] for which the *theorem of existence and uniqueness* holds. Such a theorem, ensuring the existence and uniqueness[2] of the solution $\boldsymbol{x}(t)$ of Eq. (2.1) once the initial condition $\boldsymbol{x}(0)$ is given, can be seen as a mathematical reformulation of Laplace sentence quoted in the Introduction. As seen in Chapter 1, however, this does not imply

[1]Having restricted the subject of interest may lead to the wrong impression that *non-smooth dynamical systems* either do not exist in nature or are not interesting. This is not true. Consider the following example

$$\frac{\mathrm{d}x}{\mathrm{d}t} = \frac{3}{2}x^{1/3},$$

which is non-differentiable in $x = 0$, $h = 1/3$ is called Hölder exponent. Choosing $x(0) = 0$ one can verify that both $x(t) = 0$ and $x(t) = t^{3/2}$ are valid solutions. Although bizarre or unfamiliar, this is not impossible in nature. For instance, the above equation models the evolution of the distance between two particles transported by a fully developed turbulent flow (see Sec. 11.2.1 and Box B.26).

[2]For smooth functions, often called Lipschitz continuous used for the non-differentiable ones, the theorem of existence holds (in general) up to a finite time. Sometimes it can be extended up to infinite time, although this is not always possible [Birkhoff (1966)]. For instance, the equation $\mathrm{d}x/\mathrm{d}t = -x^2$ with initial condition $x(0)$ has the unique solution $x(t) = x(0)/(1 - x(0)t)$ which diverges in a finite time $t^* = 1/x(0)$.

that the trajectory $\boldsymbol{x}(t)$ can be predicted, at a practical level, which is the one we — finite human beings — have to cope with.

If the functions f_i's can be written as $f_i(\boldsymbol{x}) = \sum_{j=1,d} A_{ij} x_j$ (with A_{ij} constant or time-dependent functions) we speak of a linear system, whose solutions may be analyzed with standard mathematical tools (see, e.g. Arnold, 1978). Although finding the solutions of such linear equations may be nontrivial, they cannot originate chaotic behaviors as observed in the nonlinear driven pendulum.

Up to now, apart from the pendulum, we have not discussed other examples of dynamical systems which can be described by ODEs as Eq. (2.1). Actually there are many of them. The state variables x_i may indicate the concentration of chemical reagents and the functions f_i the reactive rates, or the prices of some good while f_i's describe the inter-dependence among the prices of different but related goods. Electric circuits are described by the currents and voltages of different components which, typically, nonlinearly depend on each other. Therefore, dynamical systems theory encompasses the study of systems from chemistry, socio-economical sciences, engineering, and Newtonian mechanics described by $\boldsymbol{F} = m\boldsymbol{a}$, i.e. by the ODEs

$$\begin{aligned} \frac{\mathrm{d}\boldsymbol{q}}{\mathrm{d}t} &= \boldsymbol{p} \\ \frac{\mathrm{d}\boldsymbol{p}}{\mathrm{d}t} &= \boldsymbol{F}\,, \end{aligned} \tag{2.3}$$

where $\boldsymbol{q}$ and $\boldsymbol{p}$ denote the coordinates and momenta, respectively. If $\boldsymbol{q}, \boldsymbol{p} \in \mathbb{R}^N$ the phase space, usually denoted by Γ, has dimension $d = 2 \times N$. Equation (2.3) can be rewritten in the form (2.1) identifying $x_i = q_i\,;\ x_{i+N} = p_i$ and $f_i = p_i\,;\ f_{i+N} = F_i$, for $i = 1, \ldots, N$. Interesting ODEs may also originate from approximation of more complex systems such as, e.g., the Lorenz (1963) model:

$$\begin{aligned} \frac{\mathrm{d}x_1}{\mathrm{d}t} &= -\sigma x_1 + \sigma x_2 \\ \frac{\mathrm{d}x_2}{\mathrm{d}t} &= -x_2 - x_1 x_3 + r\, x_1 \\ \frac{\mathrm{d}x_3}{\mathrm{d}t} &= -b x_3 + x_1 x_2\,, \end{aligned}$$

where σ, r, b are control parameters, and x_i's are variables related to the state of fluid in an idealized Rayleigh-Bénard cell (see Sec. 3.2).

2.1.1 *Conservative and dissipative dynamical systems*

We can identify two general classes of dynamical systems. To introduce them, let's imagine to have N pendulums as that in Fig. 1.1a and to choose a slightly different initial state for any of them. Now put all representative points in phase space Γ forming an *ensemble*, i.e. a spot of points, occupying a Γ-volume, whose distribution is described by a probability density function (pdf) $\rho(\boldsymbol{x}, t\!=\!0)$ normalized in such a way that $\int_\Gamma \mathrm{d}\boldsymbol{x}\, \rho(\boldsymbol{x}, 0) = 1$. How does such a pdf evolve in time? The number of pendulums cannot change so that $\mathrm{d}N/\mathrm{d}t = 0$. The latter result can be expressed via the continuity equation

$$\frac{\partial \rho}{\partial t} + \sum_{i=1}^{d} \frac{\partial (f_i \rho)}{\partial x_i} = 0\,, \tag{2.4}$$

where $\rho \boldsymbol{f}$ is the flux of representative points in a volume $\mathrm{d}\boldsymbol{x}$ around $\boldsymbol{x}$. Equation (2.4) can be rewritten as

$$\partial_t \rho + \sum_{i=1}^{d} f_i \partial_i \rho + \rho \sum_{i=1}^{d} \partial_i f_i = \partial_t \rho + \boldsymbol{f} \cdot \boldsymbol{\nabla} \rho + \rho \boldsymbol{\nabla} \cdot \boldsymbol{f} = 0\,, \tag{2.5}$$

where $\partial_t = \partial/\partial t$ and $\boldsymbol{\nabla} = (\partial_1, \ldots, \partial_d)$. We can now distinguish two classes of systems depending on the vanishing or not of the divergence $\boldsymbol{\nabla} \cdot \boldsymbol{f}$:
If $\boldsymbol{\nabla} \cdot \boldsymbol{f} = 0$, Eq. (2.5) describes the evolution of an ensemble of points advected by an incompressible velocity field $\boldsymbol{f}$, meaning that phase-space volumes are *conserved.* The velocity field $\boldsymbol{f}$ deforms the spot of points maintaining constant its volume. We thus speak of *conservative dynamical systems.*
If $\boldsymbol{\nabla} \cdot \boldsymbol{f} < 0$, phase-space volumes contract and we speak of *dissipative dynamical systems.*[3]

The pendulum (1.5) without friction ($\gamma = 0$) is an example of conservative[4] system. In general, in the absence of dissipative forces, any Newtonian system is conservative. This can be seen recalling that a Newtonian system is described by a Hamiltonian $H(\boldsymbol{q}, \boldsymbol{p}, t)$. In terms of H the equations of motion (2.3) read (see Box B.1 and Gallavotti (1983); Goldstein *et al.* (2002))

$$\begin{aligned} \frac{\mathrm{d}q_i}{\mathrm{d}t} &= \frac{\partial H}{\partial p_i} \\ \frac{\mathrm{d}p_i}{\mathrm{d}t} &= -\frac{\partial H}{\partial q_i}\,. \end{aligned} \tag{2.6}$$

Identifying $x_i = q_i\,;\ x_{i+N} = p_i$ for $i = 1, \ldots, N$ and $f_i = \partial H/\partial p_i\,;\ f_{i+N} = -\partial H/\partial q_i$, immediately follows $\boldsymbol{\nabla} \cdot \boldsymbol{f} = 0$ and Eq. (2.5) is nothing but the Liouville theorem. In Box B.1, we briefly recall some notions of Hamiltonian systems which will be useful in the following.

In the presence of friction ($\gamma \neq 0$ in Eq. (1.5)), we have that $\boldsymbol{\nabla} \cdot \boldsymbol{f} = -\gamma$: phase-space volumes are contracted at any point with a constant rate $-\gamma$. If the driving is absent ($\beta = 0$ in Eq. (1.5)) the whole phase space contracts to a single point as in Fig. 1.2.

The set of points asymptotically reached by the trajectories of dissipative systems lives in a space of dimension $D < d$, i.e. smaller that the original phase-space

[3] Of course, there can be points where $\boldsymbol{\nabla} \cdot \boldsymbol{f} > 0$, but the interesting cases are when on average along the trajectories $\boldsymbol{\nabla} \cdot \boldsymbol{f}$ is negative. Cases where the average is positive are not very interesting because it implies an unbounded motion in phase space.

[4] Note that if $\beta = 0$ energy (1.3) is also conserved, but conservative here refers to the preservation of phase-space volumes.

dimension d. This is a generic feature and such a set is called *attractor*. In the damped pendulum the attractor consists of a single point. Conservative systems do not possess an attractor, and evolve occupying the available phase-space. As we will see, due to this difference, chaos appears and manifests in a very different way for these two classes of systems.

Box B.1: Hamiltonian dynamics

This Box reviews some basic notions on Hamiltonian dynamics. The demanding reader may find an exhaustive treatment in dedicated monographs (see, e.g. Gallavotti (1983); Goldstein *et al.* (2002); Lichtenberg and Lieberman (1992)).

As it is clear from the main text, many fundamental models of physics are Hamiltonian dynamical systems. It is thus not surprising to find applications of Hamiltonian dynamics in such diverse contexts as celestial mechanics, plasma physics and fluid dynamics.

The state of a Hamiltonian system with N degrees of freedom is described by the values of $d = 2 \times N$ state variables: the generalized coordinates $\boldsymbol{q} = (q_1, \dots, q_N)$ and the generalized momenta $\boldsymbol{p} = (p_1, \dots, p_N)$, $\boldsymbol{q}$ and $\boldsymbol{p}$ are called canonical variables. The evolution of the canonical variables is determined by the Hamiltonian $H(\boldsymbol{q}, \boldsymbol{p}, t)$ through Hamilton equations

$$\begin{aligned} \frac{\mathrm{d}q_i}{\mathrm{d}t} &= \frac{\partial H}{\partial p_i} \\ \frac{\mathrm{d}p_i}{\mathrm{d}t} &= -\frac{\partial H}{\partial q_i} \, . \end{aligned} \tag{B.1.1}$$

It is useful to use the more compact *symplectic notation*, which is helpful to highlight important symmetries and properties of Hamiltonian dynamics. Let' s first introduce $\boldsymbol{x} = (\boldsymbol{q}, \boldsymbol{p})$ such that $x_i = q_i$ and $x_{N+i} = p_i$ and consider the matrix

$$\mathbb{J} = \begin{pmatrix} \mathbb{O}_N & \mathbb{I}_N \\ -\mathbb{I}_N & \mathbb{O}_N \end{pmatrix} , \tag{B.1.2}$$

where $\mathbb{O}_N$ and $\mathbb{I}_N$ are the null and identity $(N \times N)$-matrices, respectively. Equation (B.1.1) can thus be rewritten as

$$\frac{\mathrm{d}\boldsymbol{x}}{\mathrm{d}t} = \mathbb{J} \boldsymbol{\nabla}_{\boldsymbol{x}} H \, , \tag{B.1.3}$$

$\boldsymbol{\nabla}_{\boldsymbol{x}}$ being the column vector with components $(\partial_{x_1}, \dots, \partial_{x_{2N}})$.

A: Symplectic structure and Canonical Transformations

We now seek for a change of variables $\boldsymbol{x} = (\boldsymbol{q}, \boldsymbol{p}) \to \boldsymbol{X} = (\boldsymbol{Q}, \boldsymbol{P})$, i.e.

$$\boldsymbol{X} = \boldsymbol{X}(\boldsymbol{x}) \, , \tag{B.1.4}$$

which preserves the Hamiltonian structure, in other words, such that the new Hamiltonian $H' = H(\boldsymbol{x}(\boldsymbol{X}))$ rules the evolution of $\boldsymbol{X}$, namely

$$\frac{\mathrm{d}\boldsymbol{X}}{\mathrm{d}t} = \mathbb{J}\boldsymbol{\nabla}_{\boldsymbol{X}} H' \,. \tag{B.1.5}$$

Transformations satisfying such a requirement are called *canonical transformations*. In order to be canonical the transformation Eq. (B.1.4) should fulfill a specific condition, which can be obtained as follows. We can compute the time derivative of (B.1.4), exploiting the chain rule of differentiation and (B.1.3), so that:

$$\frac{\mathrm{d}\boldsymbol{X}}{\mathrm{d}t} = \mathbb{M}\mathbb{J}\mathbb{M}^T \boldsymbol{\nabla}_{\boldsymbol{X}} H \tag{B.1.6}$$

where $\mathbb{M}_{ij} = \partial X_i/\partial x_j$ is the Jacobian matrix of the transformation and $\mathbb{M}^T$ its transpose. From (B.1.5) and (B.1.6) it follows that the Hamiltonian structure is preserved, and hence the transformation is canonical, if and only if the matrix $\mathbb{M}$ is a *symplectic matrix*,[5] defined by the condition

$$\mathbb{M}\mathbb{J}\mathbb{M}^T = \mathbb{J} \,. \tag{B.1.7}$$

The above derivation is restricted to the case of time-independent canonical transformations but, with the proper modifications, can be generalized. Canonical transformations are usually introduced by the generating functions approach instead of the symplectic structure. It is not difficult to show that the two approaches are indeed equivalent [Goldstein *et al.* (2002)]. Here, for brevity, we presented only the latter.

The modulus of the determinant of any symplectic matrix is equal to unity, $|\det(\mathbb{M})| = 1$, as it follows from definition (B.1.7):

$$\det(\mathbb{M}\mathbb{J}\mathbb{M}^T) = \det(\mathbb{M})^2 \det(\mathbb{J}) = \det(\mathbb{J}) \Longrightarrow |\det(\mathbb{M})| = 1 \,.$$

Actually it can be proved that $\det(\mathbb{M}) = +1$ always [Mackey and Mackey (2003)]. An immediate consequence of this property is that canonical transformations preserve[6] phase-space volumes as $\int \mathrm{d}\boldsymbol{x} = \int \mathrm{d}\boldsymbol{X}\,|\det(\mathbb{M})|$.

It is now interesting to consider a special kind of canonical transformation. Let $\boldsymbol{x}(t) = (\boldsymbol{q}(t), \boldsymbol{p}(t))$ be the canonical variables at a given time t, then consider the map $\mathcal{M}^\tau$ obtained evolving them according to Hamiltonian dynamics (B.1.1) till time $t+\tau$ so that

$$\boldsymbol{x}(t+\tau) = \mathcal{M}^\tau(\boldsymbol{x}(t))$$

with $\boldsymbol{x}(t+\tau) = (\boldsymbol{q}(t+\tau), \boldsymbol{p}(t+\tau))$.

The change of variable $\boldsymbol{x} \to \boldsymbol{X} = \boldsymbol{x}(t+\tau)$ can be proved (the proof is here omitted for brevity see, e.g., Goldstein *et al.* (2002)) to be a canonical transformation, in other words the Hamiltonian flow preserves its structure. As a consequence, the Jacobian matrix $\mathbb{M}_{ij} = \partial X_i/\partial x_j = \partial \mathcal{M}^\tau_i(\boldsymbol{x}(t))/\partial x_j(t)$ is symplectic and $\mathcal{M}^\tau$ is called a *symplectic map* [Meiss (1992)]. This implies Liouville theorem according which Hamiltonian flows behave as incompressible velocity fields.

[5]It is not difficult to see that symplectic matrices form a group: the identity belong to it and easily one can prove that the inverse exists and is symplectic too. Moreover, the product of two symplectic matrices is a symplectic matrix.

[6]Actually they preserve much more as for example the Poincaré invariants $\mathcal{I} = \oint_{C(t)} \mathrm{d}\boldsymbol{q}\cdot\boldsymbol{p}$, where $C(t)$ is a closed curve in phase space, which moves according to the Hamiltonian dynamics [Goldstein *et al.* (2002); Lichtenberg and Lieberman (1992)].

This example should convince the reader that there is no basic difference between Hamiltonian flows and symplectic mappings. Moreover, the Poincaré map (Sec. 2.1.2) of a Hamiltonian system is symplectic. Finally, we observe that the numerical integration of a Hamiltonian flow amounts to build up a map (time is always discretized), therefore it is very important to use algorithms preserving the symplectic structure — *symplectic integrators* — (see also Sec. 2.2.1 and Lichtenberg and Lieberman (1992)).

It is worth remarking that the Hamiltonian/Symplectic structure is very "fragile" as it is destroyed by arbitrary transformations or perturbations of Hamilton equations.

B: Integrable systems and Action-Angle variables

In the previous section, we introduced the canonical transformations and stressed their deep relationship with the symplectic structure of Hamiltonian flows. It is now natural to wonder about the practical usefulness of canonical transformations. The answer is very easy: under certain circumstances finding an appropriate canonical transformation means to have solved the problem. For instance, this is the case of time-independent Hamiltonians $H(\boldsymbol{q},\boldsymbol{p})$, if one is able to find a canonical transformation $(\boldsymbol{q},\boldsymbol{p}) \to (\boldsymbol{Q},\boldsymbol{P})$ such that the Hamiltonian expressed in the new variables only depends on the new momenta, i.e. $H(\boldsymbol{P})$. Indeed, from Hamilton equations (B.1.1) the momenta are conserved remaining equal to their initial value, $P_i(t) = P_i(0)$ any i, so that the coordinates evolve as $Q_i(t) = Q_i(0) + \partial H/\partial P_i|_{\boldsymbol{P}(0)}\, t$. When this is possible the Hamiltonian is said to be *integrable* [Gallavotti (1983)]. Necessary and sufficient condition for integrability of a N-degree of freedom Hamiltonian is the existence of N independent integral of motions, i.e. N functions F_i $(i = 1, \ldots, N)$ preserved by the dynamics $F_i(\boldsymbol{q}(t),\boldsymbol{p}(t)) = f_i = \text{const}$; usually $F_1 = H$ denotes the Hamiltonian itself. More precisely, in order to be integrable the N integrals of motion should be *in involution*, i.e. to commute one another $\{F_i, F_j\} = 0$ for any $i, j = 1, \ldots, N$. The symbol $\{f, g\}$ stays for the Poisson brackets which are defined by

$$\{f,g\} = \sum_{i=1}^{N} \left(\frac{\partial f}{\partial q_i}\frac{\partial g}{\partial p_i} - \frac{\partial f}{\partial p_i}\frac{\partial g}{\partial q_i} \right), \quad \text{or} \quad \{f,g\} = (\nabla_{\boldsymbol{x}} f)^T \mathbb{J} \nabla_{\boldsymbol{x}} g\,, \tag{B.1.8}$$

where the second expression is in symplectic notation, the superscript ^{T} denotes the transpose of a column vector, i.e. a row vector.

Integrable Hamiltonians give rise to periodic or *quasiperiodic* motions, as will be clarified by the following discussions.

It is now useful to introduce a peculiar type of canonical coordinates called *action and angle variables*, which play a special role in theoretical developments and in devising perturbation strategies for non-integrable Hamiltonians.

We consider an explicit example: a one degree of freedom Hamiltonian system independent of time, $H(q,p)$. Such a system is integrable and has periodic trajectories in the form of closed orbits (oscillations) or rotations, as illustrated by the nonlinear pendulum considered in Chapter 1. Since energy is conserved, the motion can be solved by quadratures (see Sec. 2.3). However, here we follow a slightly different approach. For periodic trajectories, we can introduce the action variable as

$$\mathrm{I} = \frac{1}{2\pi}\oint \mathrm{d}q\, p\,, \tag{B.1.9}$$

where the integral is performed over a complete period of oscillation/rotation of the orbit

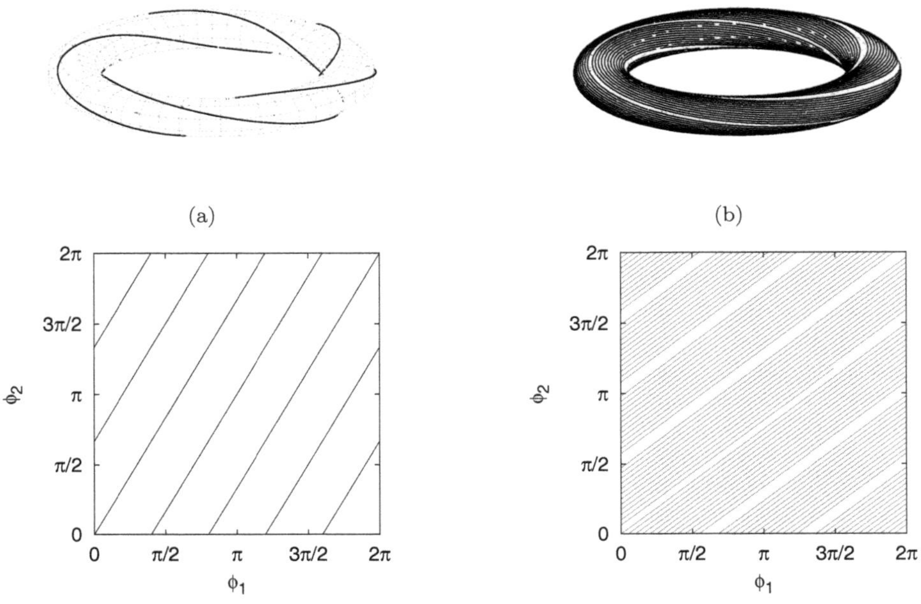

Fig. B1.1 Trajectories on a two-dimensional torus. (Top) Three-dimensional view of the torus generated by (B.1.10) in the case of (a) periodic (with $\phi_{1,2}(0) = 0$, $\omega_1 = 3$ and $\omega_2 = 5$) and (b) quasiperiodic (with $\phi_{1,2}(0) = 0$, $\omega_1 = 3$ and $\omega_2 = \sqrt{5}$) orbit. (Bottom) Two-dimensional view of the top panels with the torus wrapped in the periodic square $[0\!:\!2\pi] \times [0\!:\!2\pi]$.

(the ratio for the name action should be found in its similarity with the classical action used in Hamilton principle [Goldstein *et al.* (2002)]). Energy conservation, $H(q,p) = E$, implies $p = p(q,E)$ and, as a consequence, the action I in Eq. (B.1.9) is a function of E only, we can thus write $H = H(\mathrm{I})$. The variable conjugate to I is called angle ϕ and one can show that the transformation $(q,p) \to (\phi, \mathrm{I})$ is canonical. The term angle is obvious once Hamilton equations (B.1.1) are used to determine the evolution of I and ϕ:

$$\begin{aligned} \frac{\mathrm{dI}}{\mathrm{d}t} &= 0 && \longrightarrow \quad \mathrm{I}(t) = \mathrm{I}(0) \\ \frac{\mathrm{d}\phi}{\mathrm{d}t} &= \frac{\mathrm{d}H}{\mathrm{dI}} = \omega(\mathrm{I}) && \longrightarrow \quad \phi(t) = \phi(0) + \omega(\mathrm{I}(0))\,t\,. \end{aligned}$$

The canonical transformation $(q,p) \to (\phi, \mathrm{I})$ also shows that ω is exactly the angular velocity of the periodic motion[7] i.e. if the period of the motion is $\mathcal{T}$ then $\omega = 2\pi/\mathcal{T}$.

The above method can be generalized to N-degree of freedom Hamiltonians, namely we can write the Hamiltonian in the form $H = H(\mathbf{I}) = H(\mathrm{I}_1, \ldots, \mathrm{I}_N)$. In such a case the

[7]This is rather transparent for the specific case of an Harmonic oscillator $H = p^2/(2m) + m\omega_0^2 q^2/2$. For a given energy $E = H(q,p)$ the orbits are ellipses of axis $\sqrt{2mE}$ and $\sqrt{2E/(m\omega_0^2)}$. The integral (B.1.9) is equal to the area spanned by the orbit divided by 2π, hence the formula for the area of an ellipse yields $\mathrm{I} = E/\omega_0$ from which it is easy to see that $H = H(\mathrm{I}) = \omega_0 \mathrm{I}$, and clearly $\omega_0 = \mathrm{d}H/\mathrm{dI}$ is nothing but the angular velocity.

trajectory in phase space is determined by the N values of the actions $\mathrm{I}_i(t) = \mathrm{I}_i(0)$ and the angles evolve according to $\phi_i(t) = \phi_i(0) + \omega_i t$, with $\omega_i = \partial H/\partial \mathrm{I}_i$, in vector notation $\boldsymbol{\phi}(t) = \boldsymbol{\phi}(0) + \boldsymbol{\omega} t$. The $2N$-dimensional phase space is thus reduced to a $N-$dimensional torus. This can be seen easily in the case $N = 2$. Suppose to have found a canonical transformation to action-angle variables so that:

$$\begin{aligned} \phi_1(t) &= \phi_1(0) + \omega_1 t \\ \phi_2(t) &= \phi_2(0) + \omega_2 t\,, \end{aligned} \tag{B.1.10}$$

then ϕ_1 and ϕ_2 evolve onto a two-dimensional torus (Fig. B1.1) where the motion can be either periodic (Fig. B1.1a) whenever ω_1/ω_2 is rational, or quasiperiodic (Fig. B1.1b) when ω_1/ω_2 is irrational. From the bi-dimensional view, periodic and quasiperiodic orbits are sometimes easier to visualize. Note that in the second case the torus is, in the course of time, completely covered by the trajectory as in Fig. B1.1b. The same phenomenology occurs for generic N. In Chapter 7, we will see that quasiperiodic motions, characterized by irrational ratios among the ω_i's, play a crucial role in determining how chaos appears in (non-integrable) Hamiltonian systems.

2.1.2 *Poincaré Map*

Visualization of the trajectories for $d > 3$ is impossible, but one can resort to the so-called *Poincaré section (or map)* technique, whose construction can be done as follows. For simplicity of representation, consider a three dimensional autonomous system $\mathrm{d}\boldsymbol{x}/\mathrm{d}t = \boldsymbol{f}(\boldsymbol{x})$, and focus on one of its trajectories. Now define a plane (in general a (d–1)-surface) and consider all the points $\boldsymbol{P}_n$ in which the trajectory crosses the plane from the same side, as illustrated in Fig. 2.1. The Poincaré map of the flow $\boldsymbol{f}$ is thus defined as the map $\boldsymbol{G}$ associating two successive crossing points, i.e.

$$\boldsymbol{P}_{n+1} = \boldsymbol{G}(\boldsymbol{P}_n)\,, \tag{2.7}$$

which can be simply obtained by integrating the original ODE from the time of the n-intersection to that of the $(n+1)$-intersection, and so it is always well defined. Actually also its inverse $\boldsymbol{P}_{n-1} = \boldsymbol{G}^{-1}(\boldsymbol{P}_n)$ is well defined by simply integrating backward the ODE, therefore the map (2.7) is *invertible.*

The stroboscopic map employed in Chapter 1 to visualize the pendulum dynamics can be seen as a Poincaré map, where time t is folded in $[0\!:\!2\pi]$, which is possible because time enters the dynamics through a cyclic function.

Poincaré maps allow a d-dimensional phase space to be reduced to a $(d-1)$-dimensional representation which, as for the pendulum example, permits to identify the periodicity (if any) of a trajectory also when its complete phase-space behavior is very complicated. Such maps are also valuable for more refined analysis than the mere visualization, because preserve the stability properties of points and curves.

We conclude remarking that building an appropriate Poincaré map for a generic system is not an easy task, as choosing a good plane or (d–1)-surface of intersection requires experience.

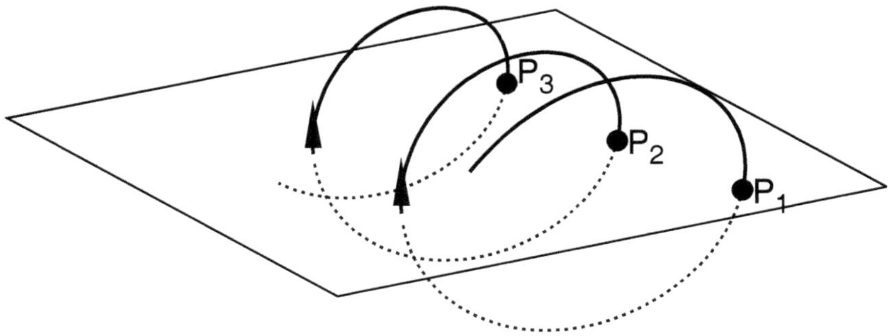

Fig. 2.1 Poincaré section for a generic trajectory, sketch of its construction for the first three intersection points $\boldsymbol{P}_1$, $\boldsymbol{P}_2$ and $\boldsymbol{P}_3$.

2.2 Discrete time dynamical systems: maps

The Poincaré map can be seen as a *discrete time dynamical systems*. There are situations in which the evolution law of a system is intrinsically discrete as, for example, the generations of biological species. It is thus interesting to consider also such discrete time dynamical systems or maps. It is worth remarking from the outset that there is no specific difference between continuous and discrete time dynamical systems, as the Poincaré map construction suggests. In principle, also systems in which the state variable $\boldsymbol{x}$ assumes discrete values[8] may be considered, as e.g. *Cellular Automata* [Wolfram (1986)]. When the number of possible states is finite and the evolution rule is deterministic, only periodic motions are possible, though complex behaviors may manifest in a different way [Wolfram (1986); Badii and Politi (1997); Boffetta *et al.* (2002)].

Discrete time dynamical systems can be written as the map:

$$\boldsymbol{x}(n+1) = \boldsymbol{f}(\boldsymbol{x}(n)), \tag{2.8}$$

which is a shorthand notation for

$$\begin{aligned} x_1(n+1) &= f_1(x_1(n), x_2(n), \cdots, x_d(n)), \\ &\vdots \\ x_d(n+1) &= f_d(x_1(n), x_2(n), \cdots, x_d(n)), \end{aligned} \tag{2.9}$$

the index n is a positive integer, denoting the iteration, generation or step number.

[8]At this point, the reader may argue that computer integration of ODEs entails a discretization of the states due to the finite floating point representation of real numbers. This is indeed true and we refer the reader to Chapter 10, where this point will be discussed in details.

In analogy with ODEs, for smooth functions f_i's, a theorem of existence and uniqueness exists and we can distinguish conservative or *volume preserving* maps from dissipative or *volume contracting* ones. Continuous time dynamical systems with $\boldsymbol{\nabla} \cdot \boldsymbol{f} = 0$ are conservative, we now seek for the equivalent condition for maps. Consider an infinitesimal volume $\mathrm{d}^d\boldsymbol{x}$ around a point $\boldsymbol{x}(n)$, i.e. an hypercube identified by $\boldsymbol{x}(n)$ and $\boldsymbol{x}(n)+\mathrm{d}x\,\hat{\mathbf{e}}_j$, $\hat{\mathbf{e}}_j$ being the unit vector in the direction j. After one iteration of the map (2.8) the vertices of the hypercube evolve to $x_i(n+1) = f_i(\boldsymbol{x}(n))$ and $x_i(n+1)+\sum_j \partial_j f_i|_{\boldsymbol{x}(n)}\,\mathrm{d}x\,\hat{\mathbf{e}}_j = x_i(n+1)+\sum_j \mathbb{L}_{ij}(\boldsymbol{x}(n))\,\mathrm{d}x\,\hat{\mathbf{e}}_j$ so that the volumes at iteration $n+1$ and n are related by:

$$\mathrm{Vol}(n+1) = |\det(\mathbb{L})|\mathrm{Vol}(n)\,.$$

If $|\det(\mathbb{L})| = 1$, the map preserves volumes and is conservative, while, if $|\det(\mathbb{L})| < 1$, volumes are contracted and it is dissipative.

2.2.1 *Two dimensional maps*

We now briefly discuss some examples of maps. For simplicity, we consider two-dimensional maps, which can be seen as transformations of the plane into itself: each point of the plane $\boldsymbol{x}(n) = (x_1(n), x_2(n))$ is mapped to another point $\boldsymbol{x}(n+1) = (x_1(n+1), x_2(n+1))$ by a transformation T

$$T: \quad \begin{cases} x_1(n+1) = f_1(x_1(n), x_2(n)) \\ x_2(n+1) = f_2(x_1(n), x_2(n))\,. \end{cases}$$

Examples of such transformations (in the linear realm) are translations, rotations, dilatations or a combination of them.

2.2.1.1 *The Hénon Map*

An interesting example of two dimensional mapping is due to Hénon (1976) – *the Hénon map.* Though such a mapping is a pure mathematical example, it contains all the essential properties of chaotic systems. Inspired by some Poincaré sections of the Lorenz model, Hénon proposed a mapping of the plane by composing three transformations as illustrated in Fig. 2.2a-d, namely:
T_1 a nonlinear transformation which folds in the x_2-direction (Fig. 2.2a→b)

$$T_1 : \quad x_1^{(1)} = x_1 \quad x_2^{(1)} = x_2 + 1 - a x_1^2$$

where a is a tunable parameter;
T_2 a linear transformation which contracts in the x_1-direction (Fig. 2.2b →c)

$$T_2 : \quad x_1^{(2)} = b x_1^{(1)} \quad x_2^{(2)} = x_2^{(1)}\,,$$

b being another free parameter with $|b| < 1$;
T_3 operates a rotation of $\pi/2$ (Fig. 2.2c → d)

$$T_3 : \quad x_1^{(3)} = x_2^{(2)} \quad x_2^{(3)} = x_1^{(2)}\,.$$

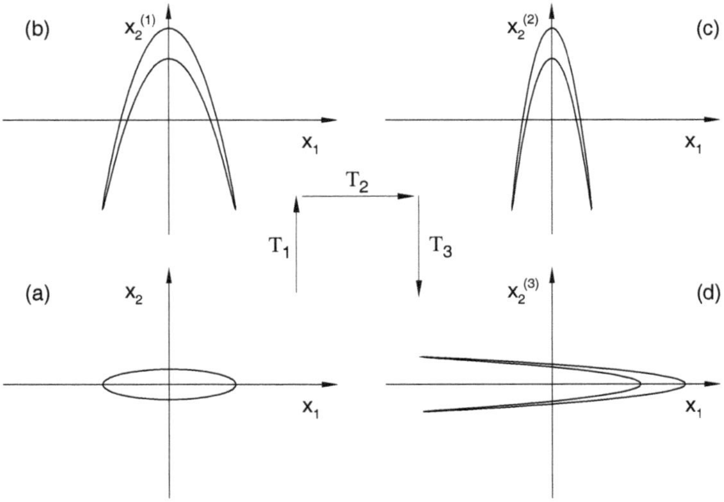

Fig. 2.2 Sketch of the action of the three transformations T_1, T_2 and T_3 composing the Hénon map (2.10). The ellipse in (a) is folded preserving the area by T_1 (b), contracted by T_2 (c) and, finally, rotated by T_3 (d). See text for explanations.

The composition of the above transformations $T = T_3T_2T_1$ yields the Hénon map[9]

$$\begin{aligned} x_1(n+1) &= x_2(n) + 1 - ax_1^2(n) \\ x_2(n+1) &= bx_1(n)\,, \end{aligned} \tag{2.10}$$

whose action contracts areas as $|\det(\mathbb{L})| = |b| < 1$. The map is clearly invertible as

$$\begin{aligned} x_1(n) &= b^{-1}x_2(n+1) \\ x_2(n) &= x_1(n+1) - 1 + ab^{-1}x_2^2(n+1)\,, \end{aligned}$$

and hence it is a one-to-one mapping of the plane into itself.

Hénon studied the map (2.10) for several parameter choices finding a richness of behaviors. In particular, chaotic motion was found to take place on a set in phase space named after his work Hénon *strange attractor* (see Chap. 5 for a more detailed discussion).

Nowadays, Hénon map and the similar in structure Lozi (1978) map

$$\begin{aligned} x_1(n+1) &= x_2(n) + 1 - a|x_1(n)| \\ x_2(n+1) &= bx_1(n)\,. \end{aligned}$$

are widely studied examples of dissipative two-dimensional maps. The latter possesses nice mathematical properties which allow many rigorous results to be derived [Badii and Politi (1997)].

[9] As noticed by Hénon himself, the map (2.10) incidentally is also the simplest bi-dimensional quadratic map having a constant Jacobian i.e. $|\det(\mathbb{L})| = |b|$.

At the core of Hénon mapping there is the simultaneous presence of *stretching* and *folding* mechanisms which are the two basic ingredients of chaos, as will become clear in Sec. 5.2.2.

2.2.1.2 *Two-dimensional symplectic maps*

For their importance, here, we limit the discussion to a specific class of conservative maps, namely to *symplectic maps* [Meiss (1992)]. These are $d = 2N$ dimensional maps $\boldsymbol{x}(n+1) = \boldsymbol{f}(\boldsymbol{x}(n))$ such that the stability matrix $\mathbb{L}_{ij} = \partial f_i / \partial x_j$ is symplectic, that is $\mathbb{L}\mathbb{J}\mathbb{L}^T = \mathbb{J}$, where $\mathbb{J}$ is

$$\mathbb{J} = \begin{pmatrix} \mathbb{O}_N & \mathbb{I}_N \\ -\mathbb{I}_N & \mathbb{O}_N \end{pmatrix}$$

$\mathbb{O}_N$ and $\mathbb{I}_N$ being the null and identity $(N \times N)$-matrices, respectively. As discussed in Box B.1, such maps are intimately related to Hamiltonian systems.

Let's consider, as an example with $N = 1$, the following transformation [Arnold and Avez (1968)]:

$$x_1(n+1) = x_1(n) + \ x_2(n) \quad \text{mod } 1, \tag{2.11}$$

$$x_2(n+1) = x_1(n) + 2x_2(n) \quad \text{mod } 1, \tag{2.12}$$

where mod indicates the modulus operation. Three observations are in order. First, this map acts not in the plane but on the torus $[0:1] \times [0:1]$. Second, even though it looks like a linear transformation, it is not! The reason for both is in the modulus operation. Third, a direct computation shows that $\det(\mathbb{L}) = 1$ which for $N = 1$ (i.e. $d = 2$) is a necessary and sufficient condition for a map to be symplectic. On the contrary, for $N \geq 2$, the condition $\det(\mathbb{L}) = 1$ is necessary but not sufficient for the matrix to be symplectic [Mackey and Mackey (2003)].

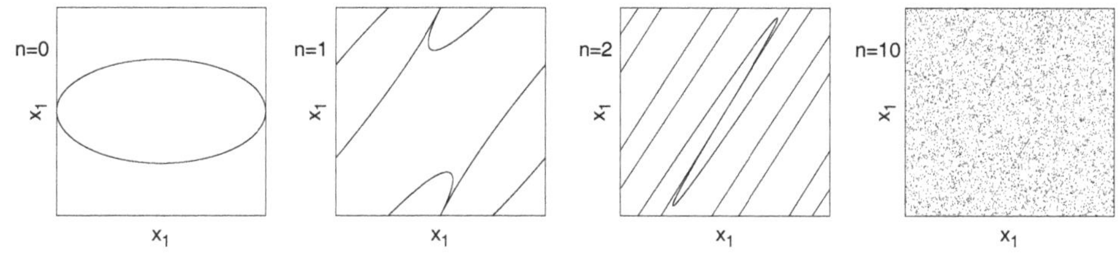

Fig. 2.3 Action of the cat map (2.11)–(2.12) on an elliptic area after $n = 1, 2$ and $n = 10$ iterations. Note how the pattern becomes more and more "random" as n increases.

The multiplication by 2 in Eq. (2.12) causes stretching while the modulus implements folding.[10] Successive iterations of the map acting on points, initially lying on a smooth curve, are shown in Fig. 2.3. More and more foliated and inter-winded

[10] Again stretching and folding are the basic mechanisms.

structures are generate till, for $n > 10$, a seemingly random pattern of points uniformly distributed on the torus is obtained.

This is the so-called *Arnold cat map* or simply cat map.[11] The cat map, as clear from the figure, has the property of "randomizing" any initially regular spot of points. Moreover, points which are at the beginning very close to each other quickly separate, providing another example of *sensitive dependence on initial conditions.*

We conclude this introduction to discrete time dynamical systems by presenting another example of symplectic map which has many applications, namely the *Standard map* or Chirikov-Taylor map, from the names of whom mostly contributed to its understanding. It is instructive to introduce the standard map in the most general way, so to see, once again, the link between Hamiltonian systems and symplectic maps (Box B.1).

We start considering a simple one degree of freedom Hamiltonian system with $H(p,q) = p^2/2m + U(q)$. From Eq. (2.6) we have:

$$\begin{aligned} \frac{\mathrm{d}q}{\mathrm{d}t} &= \frac{p}{m} \\ \frac{\mathrm{d}p}{\mathrm{d}t} &= -\frac{\partial U}{\partial q} \,. \end{aligned} \tag{2.13}$$

Now suppose to integrate the above equation on a computer by means of the simplest (lowest order) algorithm, where time is discretized $t = n\Delta t$, Δt being the time step. Accurate numerical integrations would require Δt to be very small, however such a constraint can be relaxed as we are interested in the discrete dynamics by itself.

With the notation $q(n) = q(t)$, $q(n+1) = q(t+\Delta t)$, and the corresponding for p, the most obvious way to integrate Eq. (2.13) is:

$$q(n+1) = q(n) + \Delta t\, \frac{p(n)}{m} \tag{2.14}$$

$$p(n+1) = p(n) - \Delta t\, \left.\frac{\partial U}{\partial q}\right|_{q(n)} \,. \tag{2.15}$$

However, "obvious" does not necessarily mean "correct": a trivial computation shows that the above mapping does not preserve the areas, indeed $|\det(\mathbb{L})| = |1 + \frac{1}{m}(\Delta t)^2 \partial^2 U/\partial q^2|$ and since Δt may be finite $(\Delta t)^2$ is not small. Moreover, even if in the limit $\Delta t \to 0$ areas are conserved the map is not symplectic. The situation changes if we substitute $p(n)$ with $p(n+1)$ in Eq. (2.14)

$$q(n+1) = q(n) + \Delta t\, \frac{p(n+1)}{m} \tag{2.16}$$

$$p(n+1) = p(n) - \Delta t\, \left.\frac{\partial U}{\partial q}\right|_{q(n)} \,, \tag{2.17}$$

[11] Where is the cat? According to someone the name comes from Arnold, who first introduced it and used a curve with the shape of a cat instead of the ellipse, here chosen for comparison with Fig. 2.2. More reliable sources ascribe the name cat to C-property Automorphism on the Torus, which summarizes the properties of a class of map among which the Arnold cat map is the simplest instance.

which is now symplectic. For very small Δt, Eqs. (2.16)-(2.17) define the lowest order symplectic-integration scheme [Allen and Tildesly (1993)].

The map defined by Eqs. (2.17) and (2.16) can be obtained by straightforwardly integrating a peculiar type of time-dependent Hamiltonians [Tabor (1989)]. For instance, consider a particle which periodically experiences an impulsive force in a time interval $\nu\mathcal{T}$ (with $0 < \nu < 1$), and moves freely for an interval $(1-\nu)\mathcal{T}$, as given by the Hamiltonian

$$H(p,q,t) = \begin{cases} \dfrac{U(q)}{\nu} & n\mathcal{T} < t < (n+\nu)\mathcal{T} \\ \dfrac{p^2}{(1-\nu)m} & (n+\nu)\mathcal{T} < t < (n+1)\mathcal{T} \,. \end{cases}$$

The integration of Hamilton equations (2.6) in $n\mathcal{T} < t < (n+1)\mathcal{T}$ exactly retrieves (2.16) and (2.17) with $\Delta t = \mathcal{T}$. A particular choice of the potential, namely $U(q) = K\cos(q)$, leads to the *standard map*:

$$\begin{aligned} q(n+1) &= q(n) + p(n+1) \\ p(n+1) &= p(n) + K\sin(q(n))\,, \end{aligned} \tag{2.18}$$

where we put $\mathcal{T} = 1 = m$. By defining q modulus 2π, the map is usually confined to the cylinder $q, p \in [0:2\pi] \times \mathbb{R}$.

The standard map can also be derived by integrating the Hamiltonian of the kicked rotator [Ott (1993)], which is a sort of pendulum without gravity and forced with periodic Dirac-δ shaped impulses. Moreover, it finds applications in modeling transport in accelerator and plasma physics. We will reconsider this map in Chapter 7 as prototype of how chaos appears in Hamiltonian systems.

2.3 The role of dimension

The presence of nonlinearity is not enough for a dynamical systems to observe chaos, in particular such a possibility crucially depends on the system dimension d.

Recalling the pendulum example, we observed that the autonomous case ($d = 2$) did not show chaos, while the non-autonomous one ($d = 2 + 1$) did it. Generalizing this observation, we can expect that $d = 3$ is the critical dimension for continuous time dynamical systems to generate chaotic behaviors. This is mathematically supported by a general result known as the *Poincaré-Bendixon theorem* [Poincaré (1881); Bendixon (1901)]. This theorem states that, in $d = 2$, the fate of any orbit of an autonomous systems is either periodicity or asymptotically convergence to a point $\boldsymbol{x}^*$. We shall see in the next section that the latter is an asymptotically stable fixed point for the system dynamics. For the sake of brevity we do not demonstrate such a theorem, it is anyway instructive to show that it is trivially true for autonomous Hamiltonian dynamical systems. One degree of freedom, i.e. $d = 2$, Hamiltonian systems are always *integrable* and chaos is ruled out. As energy is a constant of motion, $H(p,q) = p^2/(2m) + U(q) = E$, we can write $p = \pm\sqrt{2m[E - U(q)]}$ which, together Eq. (2.6), allows the problem to be solved by *quadratures*

$$t = \int_{q_0}^{q} \mathrm{d}q' \sqrt{\frac{m}{2[E - U(q')]}} \,. \tag{2.19}$$

Thus, even if the integral (2.19) may often require numerical evaluation, the problem is solved. The above result can be obtained also noticing that by means of a proper canonical transformation, a one degree of freedom Hamiltonian systems can always be expressed in terms of the action variable only (see Box B.1).

What about discrete time systems? An invertible d-dimensional discrete time dynamical system can be seen as a Poincaré map of a $(d+1)$-dimensional ODE, therefore it is natural to expect that $d = 2$ is the critical dimension for observing chaos in maps. However, non-invertible maps, such as the logistic map

$$x(t+1) = rx(t)(1 - x(t))\,,$$

may display chaos also for $d = 1$ (see Sec. 3.1).

2.4 Stability theory

In the previous sections we have seen several examples of dynamical systems, the question now is how to understand the behavior of the trajectories in phase space. This task is easy for one-degree of freedom Hamiltonian systems by using simple qualitative analysis, it is indeed intuitive to understand the phase-space portrait once the potential (or only its qualitative form) is assigned. For example, the pendulum phase-space portrait in Fig. 1.1c could be drawn by anybody who has seen the potential in Fig. 1.1b even without knowing the system it represents. The case of higher dimensional systems and, in particular, dissipative ones is less obvious.

We certainly know how to solve simple linear ODEs [Arnold (1978)] so the hope is to qualitatively extract information on the (local) behavior of a nonlinear system by linearizing it. This procedure is particularly meaningful close to the fixed points of the dynamics, i.e. those points $\boldsymbol{x}^*$ such that $\boldsymbol{f}(\boldsymbol{x}^*) = 0$ for ODEs or $\boldsymbol{f}(\boldsymbol{x}^*) = \boldsymbol{x}^*$ for maps. Of course, a trajectory with initial conditions $\boldsymbol{x}(0) = \boldsymbol{x}^*$ is such that $\boldsymbol{x}(t) = \boldsymbol{x}^*$ for any t (t may also be discrete as for maps) but what is the behavior of trajectories starting in the neighborhood of $\boldsymbol{x}^*$?

The answer to this question requires to study the *stability* of a fixed point. In general a fixed point $\boldsymbol{x}^*$ is said to be *stable* if any trajectory $\boldsymbol{x}(t)$, originating from its neighborhood, remains close to $\boldsymbol{x}^*$ for all times. Stronger forms of stability can be defined, namely: $\boldsymbol{x}^*$ is *asymptotically locally (or Lyapunov) stable* if for any $\boldsymbol{x}(0)$ in a neighborhood of $\boldsymbol{x}^*$ $\lim_{t\to\infty} \boldsymbol{x}(t) = \boldsymbol{x}^*$, and *asymptotically globally stable* if for any $\boldsymbol{x}(0)$, $\lim_{t\to\infty} \boldsymbol{x}(t) = \boldsymbol{x}^*$, as for the pendulum with friction. The knowledge of the stability properties of a fixed point provides information on the local structure of the system phase portrait.

2.4.1 *Classification of fixed points and linear stability analysis*

Linear stability analysis is particularly easy in $d = 1$. Consider the ODE $\mathrm{d}x/\mathrm{d}t = f(x)$, and be x^* a fixed point $f(x^*) = 0$. The stability of x^* is completely determined by the sign of the derivative $\lambda = \mathrm{d}f/\mathrm{d}x|_{x^*}$. Following a trajectory $x(t)$ initially displaced by δx_0 from x^*, $x(0) = x^* + \delta x_0$, the displacement $\delta x(t) = x(t) - x^*$ evolves in time as:

$$\frac{\mathrm{d}\delta x}{\mathrm{d}t} = \lambda\, \delta x\,,$$

so that, before nonlinear effects come into play, we can write

$$\delta x(t) = \delta x(0)\, \mathrm{e}^{\lambda t}\,. \tag{2.20}$$

It is then clear that, if $\lambda < 0$, the fixed point is *stable* while it is *unstable* for $\lambda > 0$. The best way to visualize the local flow around x^* is by imagining that f is a velocity field, as sketched in Fig. 2.4. Note that one dimensional velocity fields can always be expressed as derivatives of a scalar function $V(x)$ — the potential — therefore it is immediate to identify points with $\lambda < 0$ as the minima of such potential and those with $\lambda > 0$ as the maxima, making the distinction between stable and unstable very intuitive.

The linear stability analysis of a generic d-dimensional system is not easy as the local structure of the phase-space flow becomes more and more complex as the dimension increases. We focus on $d = 2$, which is rather simple to visualize and yet instructive. Consider the fixed points $f_1(x_1^*, x_2^*) = f_2(x_1^*, x_2^*) = 0$ of the two-dimensional continuous time dynamical system

$$\frac{\mathrm{d}x_1}{\mathrm{d}t} = f_1(x_1, x_2)\,, \quad \frac{\mathrm{d}x_2}{\mathrm{d}t} = f_2(x_1, x_2)\,.$$

Linearization requires to compute the stability matrix

$$\mathbb{L}_{ij}(\boldsymbol{x}^*) = \left.\frac{\partial f_i}{\partial x_j}\right|_{\boldsymbol{x}^*} \quad \text{for} \quad i, j = 1, 2\,.$$

A generic displacement $\delta\boldsymbol{x} = (\delta x_1, \delta x_2)$ from $\boldsymbol{x}^* = (x_1^*, x_2^*)$ will evolve, in the linear approximation, according to the dynamics:

$$\frac{\mathrm{d}\delta x_i}{\mathrm{d}t} = \sum_{j=1}^{2} \mathbb{L}_{ij}(\boldsymbol{x}^*)\delta x_j\,. \tag{2.21}$$

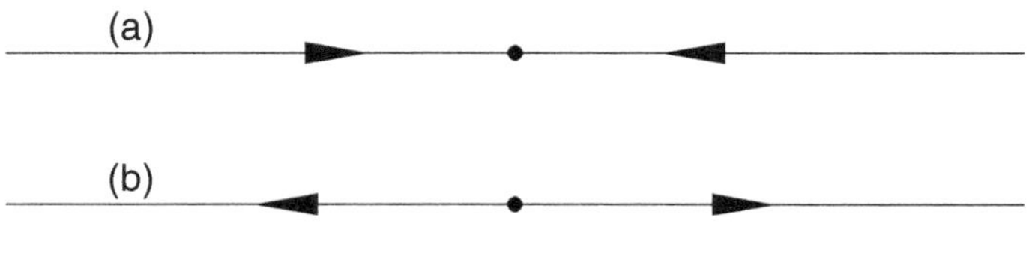

Fig. 2.4 Local phase-space flow in $d = 1$ around a stable (a) and an unstable (b) fixed point.

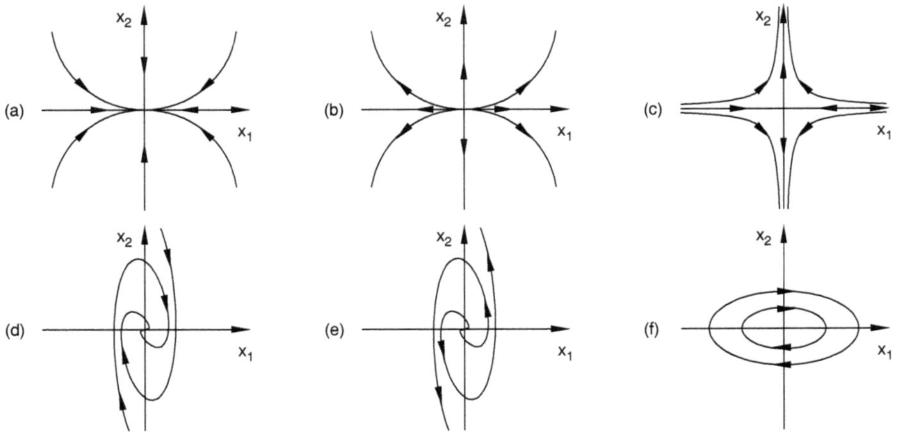

Fig. 2.5 Sketch of the local phase-space flow around the fixed points in $d = 2$, see Table 2.1 for the corresponding eigenvalues properties and classification.

Table 2.1 Classification of fixed points (second column) in $d = 2$ for non-degenerate eigenvalues. For the case of ODEs see the second column and Fig. 2.5 for the corresponding illustration. The case of maps correspond to the third column.

Case	Eigenvalues (ODE)	Type of fixed point	Eigenvalues (maps)
(a)	$\lambda_1 < \lambda_2 < 0$	stable node	$\rho_1 < \rho_2 < 1$ & $\theta_1 = \theta_2 = k\pi$
(b)	$\lambda_1 > \lambda_2 > 0$	unstable node	$1 < \rho_1 < \rho_2$ & $\theta_1 = \theta_2 = k\pi$
(c)	$\lambda_1 < 0 < \lambda_2$	hyperbolic fixed point	$\rho_1 < 1 < \rho_2$ & $\theta_1 = \theta_2 = k\pi$
(d)	$\lambda_{1,2} = \mu \pm i\omega$ & $\mu < 0$	stable spiral point	$\theta_1 = -\theta_2 \neq \pm k\pi/2$ & $\rho_1 = \rho_2 < 1$
(e)	$\lambda_{1,2} = \mu \pm i\omega$ & $\mu > 0$	unstable spiral point	$\theta_1 = -\theta_2 \neq \pm k\pi/2$ & $\rho_1 = \rho_2 > 1$
(f)	$\lambda_{1,2} = \pm i\omega$	elliptic fixed point	$\theta_1 = -\theta_2 = \pm(2k+1)\pi/2$ & $\rho_{1,2} = 1$

As customary in linear ODE (see, e.g. Arnold (1978)), for finding the solution of Eq. (2.21) we first need to compute the eigenvalues λ_1 and λ_2 of the two-dimensional stability matrix $\mathbb{L}$, which amounts to solve the secular equation:

$$\det[\mathbb{L} - \lambda\mathbb{I}] = 0\,.$$

For the sake of simplicity, we disregard here the degenerate case $\lambda_1 = \lambda_2$ (see Hirsch *et al.* (2003); Tabor (1989) for an extended discussion). By denoting with $\mathbf{e}_1$ and $\mathbf{e}_2$ the associated eigenvalues ($\mathbb{L}\mathbf{e}_i = \lambda_i \mathbf{e}_i$), the most general solution of Eq. (2.21) is

$$\delta \boldsymbol{x}(t) = c_1 \mathbf{e}_1 e^{\lambda_1 t} + c_2 \mathbf{e}_2 e^{\lambda_2 t}\,, \tag{2.22}$$

where each constant c_i is determined by the initial conditions. Equation (2.22) generalizes the $d = 1$ result (2.20) to the two-dimensional case.

We have now several cases according to the values of λ_1 and λ_2, see Table 2.1 and Fig. 2.5. If both the eigenvalues are real and negative/positive we have a *stable/unstable node*. If they are real and have different sign, the point is said to be *hyperbolic* or a *saddle*. The other possibility is that they are complex conjugate then: if the real part is negative/positive we call the corresponding point a *stable/unstable spiral*;[12] if the real part vanishes we have an *elliptic* point or *center*. The classification originates from the typical shape of the local flow around the points as illustrated in Fig. 2.5. The eigenvectors associated to eigenvalues with real and positive/negative eigenvalues identify the unstable/stable directions.

The above presented procedure is rather general and can be applied also to higher dimensions. The reader interested to local analysis of three-dimensional flows may refer to Chong *et al.* (1990).

Within the linearized dynamics, a fixed point is asymptotically stable if all the eigenvalues have negative real parts $\Re\{\lambda_i\} < 0$ (for each $i = 1, \ldots, d$) and unstable if there is at least an eigenvalue with positive real part $\Re\{\lambda_i\} > 0$ (for some i), the fixed point becomes a *repeller* when all eigenvalues are positive. If the real part of all eigenvalues is zero the point is a center or *marginal*. Moreover, if d is even and all eigenvalues are imaginary it is said to be an elliptic point.

So far, we considered ODEs, it is then natural to seek for the extension of stability analysis to maps, $\boldsymbol{x}(n+1) = \boldsymbol{f}(\boldsymbol{x}(n))$. In the discrete time case, the fixed points are found by solving $\boldsymbol{x}^* = \boldsymbol{f}(\boldsymbol{x}^*)$ and Eq. (2.21), for $d = 2$, reads

$$\delta x_i(n+1) = \sum_{j=1}^{2} \mathbb{L}_{ij}(\boldsymbol{x}^*)\delta x_j(n)\,,$$

while Eq. (2.22) takes the form (we exclude the case of degenerate eigenvalues):

$$\delta \boldsymbol{x}(n) = c_1\,\lambda_1^n\,\mathbf{e}_1 + c_2\,\lambda_2^n \mathbf{e}_2\,. \tag{2.23}$$

The above equation shows that, for discrete time systems, the stability properties depend on whether λ_1 and λ_2 are in modulus smaller or larger than unity. Using the notation $\lambda_i = \rho_i e^{i\theta_i}$, if all eigenvalues are inside the unit circle ($\rho_i \leq 1$ for each i) the fixed point is stable. As soon as, at least, one of them crosses the circle ($\rho_j > 1$ for some j) it becomes unstable. See the last column of Table 2.1. For general d-dimensional maps, the classification asymptotically stable/unstable remains the same but the boundary of stability/instability is now determined by $\rho_i = 1$.

In the context of discrete dynamical systems, symplectic maps are characterized by some special feature because the linear stability matrix $\mathbb{L}$ is a symplectic matrix, see Box B.2.

Box B.2: A remark on the linear stability of symplectic maps

The linear stability matrix $\mathbb{L}_{ij} = \partial f_i/\partial x_j$ associated to a symplectic map verifies Eq. (B.1.7) and thus is a symplectic matrix. Such a relation constraints the structure of the map and, in particular, of the matrix $\mathbb{L}$. It is easy to prove that if λ is an eigenvalue of $\mathbb{L}$ then $1/\lambda$ is an eigenvalue too. This is obvious for $d = 2$, as we know that

[12] A spiral point is sometimes also called a focus.

$\det(\mathbb{L}) = \lambda_1\lambda_2 = 1$. We now prove this property in general [Lichtenberg and Lieberman (1992)]. First, let's recall that $\mathbb{A}$ is a symplectic matrix if $\mathbb{A}\mathbb{J}\mathbb{A}^T = \mathbb{J}$, which implies that

$$\mathbb{A}\mathbb{J} = \mathbb{J}(\mathbb{A}^T)^{-1} \tag{B.2.1}$$

with $\mathbb{J}$ as in (B.1.2). Second, we have to call back a theorem of linear algebra stating that if λ is an eigenvalue of a matrix $\mathbb{A}$, it is also an eigenvalue of its transpose $\mathbb{A}^T$

$$\mathbb{A}^T\mathbf{e} = \lambda\mathbf{e}$$

$\mathbf{e}$ being the eigenvector associated to λ. Applying $(\mathbb{A}^T)^{-1}$ to both sides of the above expression we find

$$(\mathbb{A}^T)^{-1}\mathbf{e} = \frac{1}{\lambda}\mathbf{e}\,. \tag{B.2.2}$$

Finally, multiplying Eq. (B.2.2) by $\mathbb{J}$ and using Eq. (B.2.1), we end with

$$\mathbb{A}\,(\mathbb{J}\,\mathbf{e}) = \frac{1}{\lambda}\,(\mathbb{J}\,\mathbf{e})\,,$$

meaning that $\mathbb{J}\,\mathbf{e}$ is an eigenvector of $\mathbb{A}$ with eigenvalue $1/\lambda$. As a consequence, a $(d\!=\!2N)$-dimensional symplectic map has $2N$ eigenvalues such that

$$\lambda_{i+N} = \frac{1}{\lambda_i} \qquad i = 1, \ldots, N\,.$$

As we will see in Chapter 5 this symmetry has an important consequence for the Lyapunov exponents of chaotic Hamiltonian systems.

2.4.2 *Nonlinear stability*

Linear stability, though very useful, is just a part of the history. Nonlinear terms, disregarded by linear analysis, can indeed induce nontrivial effects and lead to the failure of linear predictions. As an example consider the following ODEs:

$$\begin{aligned}\frac{dx_1}{dt} &= \quad x_2 + \alpha x_1(x_1^2 + x_2^2)\\ \frac{dx_2}{dt} &= -x_1 + \alpha x_2(x_1^2 + x_2^2)\,,\end{aligned} \tag{2.24}$$

clearly $\boldsymbol{x}^* = (0,0)$ is a fixed point with eigenvalues $\lambda_{1,2} = \pm i$ independent of α, which means an elliptic point. Thus trajectories starting in its neighborhood are expected to be closed periodic orbits in the form of ellipses around $\boldsymbol{x}^*$. However, Eq. (2.24) can be solved explicitly by multiplying the first equation by x_1 and the second by x_2 so to obtain

$$\frac{1}{2}\frac{dr^2}{dt} = \alpha r^4\,,$$

with $r = \sqrt{x_1^2 + x_2^2}$, which is solved by

$$r(t) = \frac{r(0)}{\sqrt{1 - 2\alpha r^2(0)\, t}}\,.$$

It is then clear that: if $\alpha < 0$, whatever $r(0)$ is, $r(t)$ asymptotically approaches the fixed point $r^* = 0$ which is therefore stable; while if $\alpha > 0$, for any $r(0) \neq 0$, $r(t)$ grows in time, meaning that the point is unstable. Actually the latter solution diverges at the critical time $1/(2\alpha r^2(0))$.

Usually, nonlinear terms are non trivial when the fixed point is *marginal*, e.g. a center with pure imaginary eigenvalues, while when the fixed point is an attractor, repeller or a saddle the flow topology around it remains locally unchanged. Anyway nonlinear terms may also give rise to other kinds of motion, not permitted in linear systems, as *limit cycles*.

2.4.2.1 *Limit cycles*

Consider the ODEs:

$$\begin{aligned} \frac{\mathrm{d}x_1}{\mathrm{d}t} &= x_1 - \omega x_2 - x_1(x_1^2 + x_2^2) \\ \frac{\mathrm{d}x_2}{\mathrm{d}t} &= \omega x_1 + x_2 - x_2(x_1^2 + x_2^2)\,, \end{aligned} \tag{2.25}$$

with fixed point $\boldsymbol{x}^* = (0, 0)$ of eigenvalues $\lambda_{1,2} = 1 \pm i\omega$, corresponding to an unstable spiral. For any $\boldsymbol{x}(0)$ in a neighborhood of 0, the distance from the origin of the resulting trajectory $\boldsymbol{x}(t)$ grows in time so that the nonlinear terms soon becomes dominant. These terms have the form of a nonlinear friction $-x_{1,2}(x_1^2+x_2^2)$ pushing back the trajectory toward the origin. Thus the competition between the linear pulling away from the origin and the nonlinear pushing toward it should balance in a trajectory which stays at a finite distance from the origin, circulating around it. This is the idea of a limit cycle.

The simplest way to understand the dynamics (2.25) is to rewrite it in polar coordinates $(x_1, x_2) = (r\cos\theta, r\sin\theta)$:

$$\begin{aligned} \frac{\mathrm{d}r}{\mathrm{d}t} &= r(1 - r^2) \\ \frac{\mathrm{d}\theta}{\mathrm{d}t} &= \omega\,. \end{aligned}$$

The equations for r and θ are decoupled, and the dynamical behavior can be inferred analyzing the radial equation solely, the angular one being trivial. Clearly, $r^* = 0$ corresponding to $(x_1^*, x_2^*) = (0, 0)$ is an unstable fixed point and $r^* = 1$ to an attracting one. The latter corresponds to the stable limit cycle defined by the circular orbit $(x_1(t), x_2(t)) = (\cos(\omega t), \sin(\omega t))$ (see Fig. 2.6a). The limit cycle can also be unstable (Fig. 2.6b) or half-stable (Fig. 2.6c) according to the specific radial dynamics.

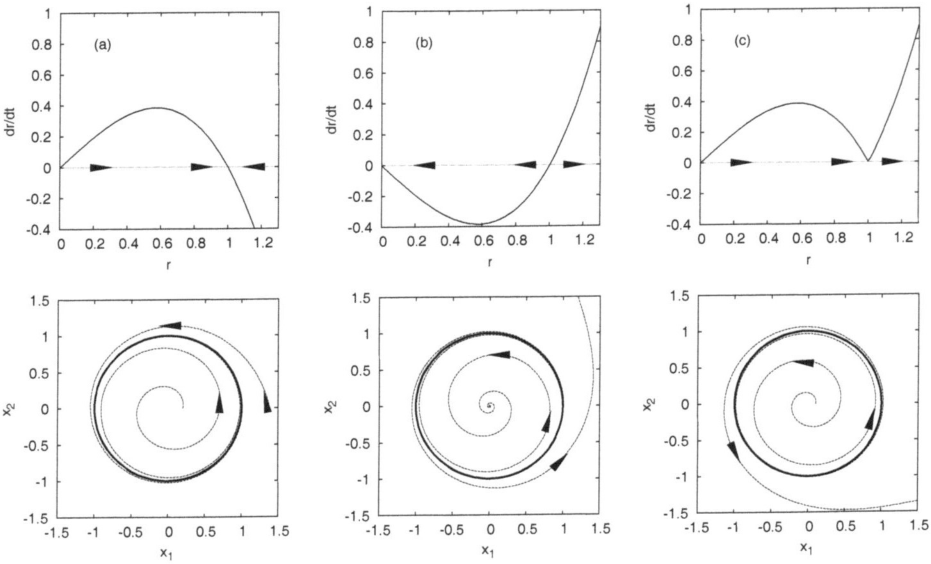

Fig. 2.6 Typical limit cycles. (Top) Radial dynamics (Bottom) corresponding limit cycle. (a) $\mathrm{d}r/\mathrm{d}t = r(1-r^2)$ attracting or stable limit cycle; (b) $\mathrm{d}r/\mathrm{d}t = -r(1-r^2)$ repelling or unstable limit cycle; (c) $\mathrm{d}r/\mathrm{d}t = r|1-r^2|$ saddle or half-stable limit cycle. For the angular dynamics we set $\omega = 4$.

This method, with the necessary modifications (see Box B.12), can be used to show that also the Van der Pol oscillator [van der Pol (1927)]

$$\begin{aligned} \frac{\mathrm{d}x_1}{\mathrm{d}t} &= x_2 \\ \frac{\mathrm{d}x_2}{\mathrm{d}t} &= -\omega^2 x_1 + \mu(1-x_1^2)x_2 \end{aligned} \tag{2.26}$$

possesses limit cycles around the fixed point in $\boldsymbol{x}^* = (0,0)$.

In autonomous ODE, limit cycles can appear only in $d \geq 2$, we saw another example of them in the driven damped pendulum (Fig. 1.3a–c). In general it is very difficult to determine if an arbitrary nonlinear system admits limit cycles and, even if its existence can be proved, it is usually very hard to determine its analytical expression and stability properties.

However, the demonstration that a given system do not possess limit cycles is sometimes very easy. This is, for instance, the case of systems which can be expressed as gradients of a single-valued scalar function — the potential — $V(\boldsymbol{x})$,

$$\frac{\mathrm{d}\boldsymbol{x}}{\mathrm{d}t} = -\boldsymbol{\nabla} V(\boldsymbol{x})\,.$$

An easy way to understand that no limit cycles or, more in general, closed orbits can occur in *gradient systems* is to proceed by reduction to *absurdum*. Suppose that a closed trajectory of period $\mathcal{T}$ exists, then in one cycle the potential variation should be zero $\Delta V = 0$, V being monodrome. However, an explicit computation gives:

$$\int_t^{t+T} dt \, \frac{dV}{dt} = \int_t^{t+T} dt \, \frac{d\boldsymbol{x}}{dt} \cdot \boldsymbol{\nabla} V = - \int_t^{t+T} dt \left| \frac{d\boldsymbol{x}}{dt} \right|^2 < 0 \,, \qquad (2.27)$$

which contradicts $\Delta V = 0$. As a consequence no closed orbits can exist.

Closed orbits, but not limit cycles, can exist for energy conserving Hamiltonian systems, those orbits are typical around elliptic points like for the simple pendulum at low energies (Fig. 1.1c). The fact that they are not limit cycles is a trivial consequence of energy conservation.

2.4.2.2 *Lyapunov Theorem*

It is worth concluding this Chapter by mentioning the Lyapunov stability criterion, which provides the sufficient condition for the asymptotic stability of a fixed point, beyond linear theory. We enunciate the theorem without proof (for details see Hirsch *et al.* (2003)). Consider an autonomous ODE having $\boldsymbol{x}^*$ as a fixed point:

> *If, in a neighborhood of $\boldsymbol{x}^*$, there exists a positive defined function $\Phi(\boldsymbol{x})$ (i.e., $\Phi(\boldsymbol{x}) > 0$ for $\boldsymbol{x} \neq \boldsymbol{x}^*$ and $\Phi(\boldsymbol{x}^*) = 0$) such that $d\Phi/dt = d\boldsymbol{x}/dt \cdot \boldsymbol{\nabla}\Phi = \boldsymbol{f} \cdot \boldsymbol{\nabla}\Phi \leq 0$ for any $\boldsymbol{x} \neq \boldsymbol{x}^*$ then $\boldsymbol{x}^*$ is stable. Furthermore, if $d\Phi/dt$ is strictly negative the fixed point is asymptotically stable.*

Unlike linear theory where a precise protocol exists (to determine the matrix $\mathbb{L}$, its eigenvalues and so on), in nonlinear theory there are no general methods to determine the Lyapunov function Φ. The presence of integrals of motion can help to find Φ, as it happens in Hamiltonian systems. In such a case, fixed points are solutions of $p_i = 0$ and $\partial U/\partial q_i = 0$, and the Lyapunov function is noting but the energy (minus its value on the fixed point), and one has the well known Laplace theorem: if the energy potential has a minimum the fixed point is stable. By using as a Lyapunov function Φ the potential energy, the damped pendulum (1.4) represents another simple example in which the theorem is satisfied in the strong form, implying that the rest state globally attracts all trajectories.

We end this brief excursion on the stability problem noticing that systems admitting a Lyapunov function cannot evolve into closed orbits, as trivially obtained by using Eq. (2.27).

2.5 Exercises

Exercise 2.1: Consider the following systems and specify whether: A) chaos can or cannot be present; B) the system is conservative or dissipative.

$$(1)\quad \begin{aligned} x(t+1) &= x(t) + y(t) && \mod 1 \\ y(t+1) &= 2x(t) + 3y(t) && \mod 1\,; \end{aligned}$$

$$(2)\quad x(t+1) = \begin{cases} x(t) + 1/2 & x(t) \in [0:1/2] \\ x(t) - 1/2 & x(t) \in [1/2:1] \end{cases}$$

(3) $\frac{dx}{dt} = y\,, \qquad \frac{dy}{dt} = -\alpha y + f(x - \omega t),$ where f is a periodic function, and $\alpha > 0$.

Exercise 2.2: Find and draw the Poincaré section for the forced oscillator

$$\frac{dx}{dt} = y\,, \qquad \frac{dy}{dt} = -\omega x + F\cos(\Omega t)\,,$$

with $\omega^2 = 8$, $\Omega = 2$ and $F = 10$.

Exercise 2.3: Consider the following periodically forced system,

$$\frac{dx}{dt} = y\,, \qquad \frac{dy}{dt} = -\omega x - 2\mu y + F\cos(\Omega t)\,.$$

Convert it in a three-dimensional autonomous system and compute the divergence of the vector field, discussing the conservative and dissipative condition.

Exercise 2.4: Show that in a system satisfying Liouville theorem, $dx_n/dt = f_n(\boldsymbol{x})$, with $\sum_{n=1}^{N} \partial f_n(\boldsymbol{x})/\partial x_n = 0$, asymptotic stability is impossible.

Exercise 2.5: Discuss the qualitative behavior of the following ODEs

(1) $\frac{dx}{dt} = x(3 - x - y)\,, \qquad \frac{dy}{dt} = y(x-1)$

(2) $\frac{dx}{dt} = x^2 - xy - x\,, \qquad \frac{dy}{dt} = y^2 + xy - 2y$

Hint: *Start from fixed points and their stability analysis.*

Exercise 2.6:

A rigid hoop of radius R hangs from the ceiling and a small ring can move without friction along the hoop. The hoop rotates with frequency ω about a vertical axis passing through its center as in the figure on the right. Show that if $\omega < \omega_0 = \sqrt{g/R}$ the bottom of the hoop is a stable fixed point, while if $\omega > \omega_0$ the stable fixed points are determined by the condition $\cos\theta^* = g/(R\omega^2)$.

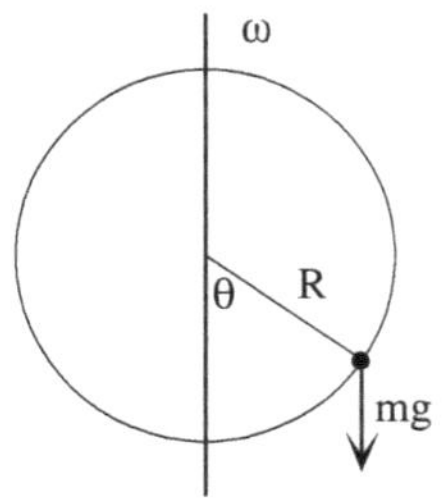

Exercise 2.7: Show that the two-dimensional map:

$$x(t+1) = x(t) + f(y(t))\,, \qquad y(t+1) = y(t) + g(x(t+1))$$

is symplectic for any choice of the functions $g(u)$ and $f(u)$.
Hint: *Consider the evolution of an infinitesimal displacement* $(\delta x(t), \delta y(t))$.

Exercise 2.8: Show that the one-dimensional non-invertible map

$$x(t+1) = \begin{cases} 2x(t) & x(t) \in [0:1/2]; \\ c & x(t) \in [1/2:1] \end{cases}$$

with $c < 1/2$, admits superstable periodic orbits, i.e. after a finite time the trajectory becomes periodic.
Hint: *Consider two classes of initial conditions* $x(0) \in [1/2:1]$ *and* $x(0) \in [0:1/2]$.

Exercise 2.9: Discuss the qualitative behavior of the system

$$\mathrm{d}x/\mathrm{d}t = xg(y)\,, \qquad \mathrm{d}y/\mathrm{d}t = -yf(x)$$

under the conditions that $f(x)$ and $g(x)$ are differentiable decreasing functions such that $f(0) > 0, g(0) > 0$, moreover there is a point (x^*, y^*), with $x^*, y^* > 0$, such that $g(x^*) = f(y^*) = 0$. Compare the dynamical behavior of the system with that of the Lotka-Volterra model (Sec. 11.3.1).

Exercise 2.10: Consider the autonomous system

$$\frac{\mathrm{d}x}{\mathrm{d}t} = yz\,, \qquad \frac{\mathrm{d}y}{\mathrm{d}t} = -2xz\,, \qquad \frac{\mathrm{d}z}{\mathrm{d}t} = xy$$

(1) show that $x^2 + y^2 + z^2 = \text{const}$;

(2) discuss the stability of the fixed points, inferring that the qualitative behavior on the sphere defined by $x^2 + y^2 + z^2 = 1$;

(3) Discuss the generalization of the above system:

$$\frac{\mathrm{d}x}{\mathrm{d}t} = ayz\,, \qquad \frac{\mathrm{d}y}{\mathrm{d}t} = bxz\,, \qquad \frac{\mathrm{d}z}{\mathrm{d}t} = cxy$$

where a, b, c are non-zero constants with the constraint $a + b + c = 0$.

Hint: *Use conservation laws of the system to study the phase portrait.*

Chapter 3

Examples of Chaotic Behaviors

Classical models tell us more than we at first can know.
Karl Popper (1902–1994)

In this Chapter, we consider three systems which played a crucial role in the development of dynamical systems theory: the logistic map introduced in the context of mathematical ecology; the model derived by Lorenz (1963) as a simplification of thermal convection; the Hénon and Heiles (1964) Hamiltonian system introduced to model the motion of a star in a galaxy.

3.1 The logistic map

Dynamical systems constitute a mathematical framework common to many disciplines, among which ecology and population dynamics. As early as 1798, the Reverend Malthus wrote *An Essay on the Principle of Population* which was a very influential book for later development of population dynamics, economics and evolution theory.[1] In this book, it was introduced a growth model which, in modern mathematical language, amounts to assume that the differential equation $\mathrm{d}x/\mathrm{d}t = rx$ describes the evolution of the number of individuals x of a population in the course of time, r being the reproductive power of individuals. The Malthusian growth model, however, is far too simplistic as it predicts, for $r > 0$, an unbounded exponential growth $x(t) = x(0)\exp(rt)$, which is unrealistic for finite-resources environments. In 1838 the mathematician Verhulst, inspired by Malthus' essay, proposed to use the *Logistic equation* to model the self-limiting growth of a biological population: $\mathrm{d}x/\mathrm{d}t = rx(1 - x/K)$ where K is the carrying capacity — the maximum number of individuals that the environment can support. With $x/K \to x$, the above equation can be rewritten as

$$\frac{\mathrm{d}x}{\mathrm{d}t} = f_r(x) = rx(1-x)\,, \tag{3.1}$$

[1] It is cited as a source of inspiration by Darwin himself.

where $r(1-x)$ is the normalized reproductive power, accounting for the decrease of reproduction when too many individuals are present in the same limited environment. The logistic equation thus represents a more realistic model. By employing the tools of linear analysis described in Sec. 2.4, one can readily verify that Eq. (3.1) possesses two fixed points: $x^* = 0$ unstable as $r > 0$ and $x^* = 1$ which is stable. Therefore, asymptotically the population stabilizes to a number of individuals equal to the carrying capacity.

The reader may now wonder: Where is chaos? As seen in Sec. 2.3, a one-dimensional ordinary differential equation, although nonlinear, cannot sustain chaos. However, a differential equation to describe population dynamics is not the best model as populations grow or decrease from one generation to the next one. In other terms, a discrete time model, connecting the n-th generation to the next $n+1$-th, would be more appropriate than a continuous time one. This does not make a big difference in the Malthusian model as $x(n+1) = rx(n)$ still gives rise to an exponential growth ($r > 1$) or extinction ($0 < r < 1$) because $x(n) = r^n x(0) = \exp(n \ln r) x(0)$. However, the situation changes for the discretized logistic equation or logistic map:

$$x(n+1) = f_r(x(n)) = rx(n)(1-x(n)), \tag{3.2}$$

which, as seen in Sec. 2.3, being a one-dimensional but non-invertible map may generate chaotic orbits. Unlike its continuous version, the logistic map is well defined only for $x \in [0:1]$, limiting the allowed values of r to the range $[0:4]$.

The logistic map is able to produce erratic behaviors resembling random noise for some values of r. For example, already in 1947 Ulam and von Neumann proposed its use as a random number generator with $r = 4$, even though a mathematical understanding of its behavior came later with the works of Ricker (1954) and Stein and Ulam (1964). These works together with other results are reviewed in a seminal paper by May (1976).

Let's start the analysis of the logistic map (3.2) in the linear stability analysis framework. Before that, it is convenient to introduce a graphical method allowing us to easily understand the behavior of trajectories generated by any one-dimensional map. Figure 3.1 illustrates the iteration of the logistic map for $r = 0.9$ via the following graphical method

(1) draw the function $f_r(x)$ and the line bisecting the square $[0:1] \times [0:1]$;
(2) draw a vertical line from $(x(0), 0)$ up to intercepting the graph of $f_r(x)$ in $(x(0), f_r(x(0)) = x(1))$;
(3) from this point draw a horizontal line up to intercepting the bisecting line;
(4) repeat the procedure from (2) with the new point.

The graphical method (1)–(4) enables to easily understand the qualitative features of the evolution $x(0), \ldots x(n), \ldots$. For instance, for $r = 0.9$, the bisecting line intersects the graph of $f_r(x)$ only in $x^* = 0$, which is the stable fixed point as $\lambda(0) = |\mathrm{d}f_r/\mathrm{d}x|_0| < 1$, which is the slope of the tangent to the curve in 0 (Fig. 3.1).

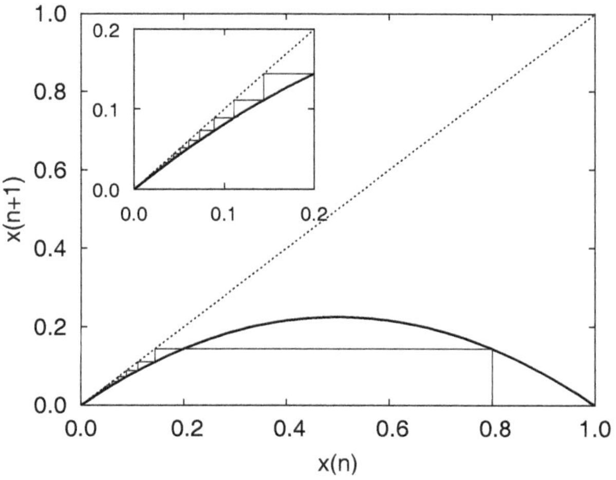

Fig. 3.1 Graphical solution of the logistic map (3.2) for $r = 0.9$, for a description of the method see text. The inset shows a magnification of the iteration close to the fixed point $x^* = 0$.

Starting from, e.g., $x(0) = 0.8$ one can see that few iterations of the map lead the trajectory $x(n)$ to converge to $x^* = 0$, corresponding to population extinction. For $r > 1$, the bisecting line intercepts the graph of $f_r(x)$ in two (fixed) points (Fig. 3.2)

$$x^* = f_r(x^*) \quad \Longrightarrow \quad x_1^* = 0\,,\; x_2^* = 1 - \frac{1}{r}\,.$$

We can study their stability either graphically or evaluating the map derivative

$$\lambda(x^*) = |f_r'(x^*)| = |r(1 - 2x^*)|\,, \tag{3.3}$$

where, to ease the notation, we defined $f_r'(x^*) = \mathrm{d}f_r(x)/\mathrm{d}x|_{x^*}$. For $1 < r < 3$, the fixed point $x_1^* = 0$ is unstable while $x_2^* = 1 - 1/r$ is (asymptotically) stable. This means that all orbits, whatever the initial value $x(0) \in\,]0:1[$, will end at x_2^*, i.e. population dynamics is attracted to a stable and finite number of individuals. This is shown in Fig. 3.2a, where we plot two trajectories $x(t)$ starting from different initial values. What does happen to the population for $r > r_1 = 3$? For such values of r, the fixed point becomes unstable, $\lambda(x_2^*) > 1$. In Fig. 3.2b, we show the iterations of the logistic map for $r = 3.2$. As one can see, all trajectories end in a period-2 orbit, which is the discrete time version of a limit cycle (Sec. 2.4.2). Thanks to the simplicity of the logistic map, we can easily extend linear stability analysis to periodic orbits. It is enough to consider the second iterate of the map

$$f_r^{(2)}(x) = f_r(f_r(x)) = r^2 x(1-x)(1 - rx + rx^2)\,, \tag{3.4}$$

which connects the population of the grandmothers with that of the granddaughters, i.e. $x(n+2) = f_r^{(2)}(x(n))$. Clearly, a period-2 orbit corresponds to a fixed point of such a map. The quartic polynomial (3.4) possesses four roots

$$x^* = f_r^{(2)}(x^*) \quad \Longrightarrow \quad \begin{cases} x_1^* = 0\,, & x_2^* = 1 - \frac{1}{r} \\ x_{3,4}^* = \frac{(r+1) \pm \sqrt{(r+1)(r-3)}}{2r} & : \end{cases} \tag{3.5}$$

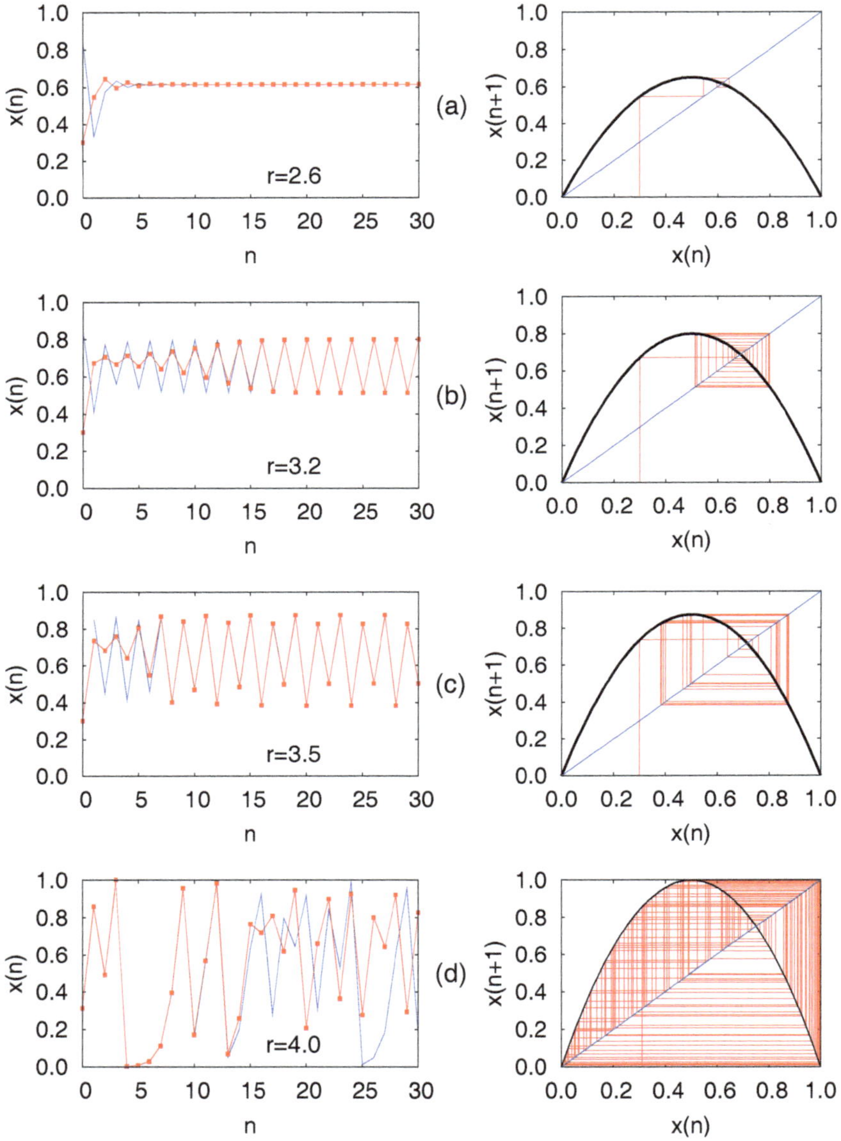

Fig. 3.2 Left: (a) evolution of two trajectories (red and blue) initially at distance $|x'(0) - x(0)| \approx 0.5$ which converge to the fixed point for $r = 2.6$; (b) same of (a) but for an attracting period-2 orbit at $r = 3.2$; (c) same of (a) but for an attracting period-4 orbit at $r = 3.5$; (d) evolution of two trajectories (red and blue), initially very close $|x'(0) - x(0)| = 4 \times 10^{-6}$, in the chaotic regime for $r = 4$. Right: graphical solution of the logistic map as explained in the text.

two coincide with the original ones ($x^*_{1,2}$), as an obvious consequence of the fact that $f_r(x^*_{1,2}) = x^*_{1,2}$, and two ($x^*_{3,4}$) are new. The change of stability of the fixed points

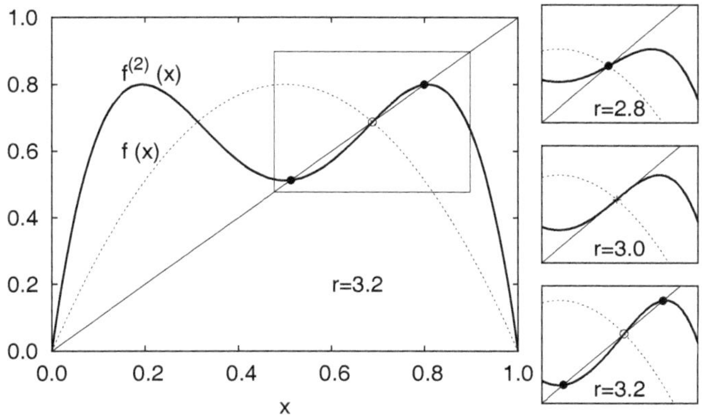

Fig. 3.3 Second iterate $f_r^{(2)}(x)$ (solid curve) of the Logistic map (dotted curve). Note the three intercepts with the bisecting line, i.e. the three fixed points x_2^* (unstable open circle) and $x_{3,4}^*$ (stable in filled circles). The three panels on the right depict the evolution the intercepts from $r < r_1 = 3$ to $r > r_1$ as in label.

is shown on the right of Fig. 3.3. For $r < 3$, the stable fixed point is $x_2^* = 1 - 1/r$. At $r = 3$, as clear from Eq. (3.5), x_3^* and x_4^* start to be real and, in particular, $x_3^* = x_4^* = x_2^*$. We can now compute the stability eigenvalues through the formula

$$\lambda^{(2)}(x^*) = \left| \left. \frac{\mathrm{d}f_r^{(2)}}{\mathrm{d}x} \right|_{x^*} \right| = |f_r'(f_r(x^*)) \cdot f_r'(x^*)| = \lambda(f_r(x^*))\lambda(x^*) \,, \tag{3.6}$$

where the last two equalities stem from the chain rule[2] of differentiation. One thus finds that: for $r = 3$, $\lambda^{(2)}(x_2^*) = (\lambda(x_2^*))^2 = 1$ i.e. the point is marginal, the slope of the graph of $f_r^{(2)}$ is 1; for $r > 3$, it is unstable (the slope exceeds 1) so that x_3^* and x_4^* become the new stable fixed points.

For $r_1 < r < r_2 = 3.448\ldots$, the period-2 orbit is stable as $\lambda^{(2)}(x_3^*) = \lambda^{(2)}(x_4^*) < 1$. From Fig. 3.2c we understand that, for $r > r_2$, period-4 orbits become the stable and attracting solutions. By repeating the above procedure to the 4th-iterate $f^{(4)}(x)$, it is possible to see that the mechanism for the appearance of period-4 orbits from period-2 ones is the same as the one illustrated in Fig. 3.3. Step by step several critical values r_k with $r_k < r_{k+1}$ can be found: if $r_k < r < r_{k+1}$, after an initial transient, $x(n)$ evolves on a period-2^k orbit [May (1976)].

The change of stability, at varying a parameter, of a dynamical system is a phenomenon known as *bifurcation*. There are several types of bifurcations which

[2]Formula (3.6) can be straightforwardly generalized for computing the stability of a generic period-T orbit $x^*(1), x^*(2), \ldots, x^*(T)$, with $f^{(T)}(x^*(i)) = x^*(i)$ for any $i = 1, \ldots, T$. Through the chain rule of differentiation the derivative of the map $f^{(T)}(x)$ at any of the points of the orbit is given by

$$\left. \frac{\mathrm{d}f^{(T)}}{\mathrm{d}x} \right|_{x^*(1)} = f'(x^*(1))\, f'(x^*(2)) \cdots f'(x^*(T)) \,.$$

constitute the basic mechanisms through which more and more complex solutions and finally chaos appear in dissipative dynamical systems (see Chapter 6). The specific mechanism for the appearance of the period-2^k orbits is called *period doubling* bifurcation. Remarkably, as we will see in Sec. 6.2, the sequence r_k has a limit: $\lim_{k\to\infty} r_k = 3.569945\ldots = r_\infty < 4$.

For $r > r_\infty$, the trajectories display a qualitative change of behavior as exemplified in Fig. 3.2d for $r = 4$, which is called the Ulam point. The graphical method applied to the case $r = 4$ suggests that, unlike the previous cases, no stable periodic orbits exist,[3] and the trajectory looks random, giving support to the proposal of Ulam and von Neumann (1947) to use the logistic map to generate random sequences of numbers on a computer. Even more interesting is to consider two initially close trajectories and compare their evolution with that of trajectories at $r < r_\infty$. On the one hand, for $r < r_\infty$ (see the left panel of Fig. 3.2a–c) two trajectories $x(n)$ and $x'(n)$ starting from distant values (e.g. $\delta x(0) = |x(0) - x'(0)| \approx 0.5$, any value would produce the same effect) quickly converge toward the same period-2^k orbit.[4] On the other hand, for $r = 4$ (left panel of Fig. 3.2d), even if $\delta x(0)$ is infinitesimally small, the two trajectories quickly become "macroscopically" distinguishable, resembling what we observed for the driven-damped pendulum (Fig. 1.4). This is again chaos at work: emergence of very irregular, seemingly random trajectories with sensitive dependence on the initial conditions.[5]

Fortunately, in the specific case of the logistic map at the Ulam point $r = 4$, we can easily understand the origin of the sensitive dependence on initial conditions. The idea is to establish a change of variable transforming the logistic in a simpler map, as follows. Define $x = \sin^2(\pi\theta/2) = [1 - \cos(\pi\,\theta)]/2$ and substitute it in Eq. (3.2) with $r = 4$, so to obtain $\sin^2(\pi\theta(n+1)/2) = \sin^2(\pi\theta(n))$ yielding to

$$\pi\theta(n+1))/2 = \pm\pi\theta(n) + k\pi, \tag{3.7}$$

where k is any integer. Taking $\theta \in [0:1]$, it is straightforward to recognize that Eq. (3.7) defines the map

$$\theta(n+1) = \begin{cases} 2\theta(n) & 0 \le \theta < \frac{1}{2} \\ 2 - 2\theta(n) & \frac{1}{2} \le \theta \le 1 \end{cases} \tag{3.8}$$

or, equivalently, $\theta(n+1) = g(\theta(n)) = 1 - 2|\theta(t) - 1/2|$ which is the so-called *tent map* (Fig. 3.4a).

Intuition suggests that the properties of the logistic map with $r = 4$ should be the same as those of the tent map (3.8), this can be made more precise introducing the concept of *Topological Conjugacy* (see Box B.3). Therefore, we now focus on the behavior of a generic trajectory under the action of the tent map (3.8), for which

[3] There is however an infinite number of unstable periodic orbits, as one can easily understand plotting the n-iterates of the map and look for the intercepts with the bisectrix.

[4] Note that the periodic orbit may be shifted of some iterations.

[5] One can check that making $\delta x(0)$ as small as desired simply shifts the iteration at which the two orbits become macroscopically distinguishable.

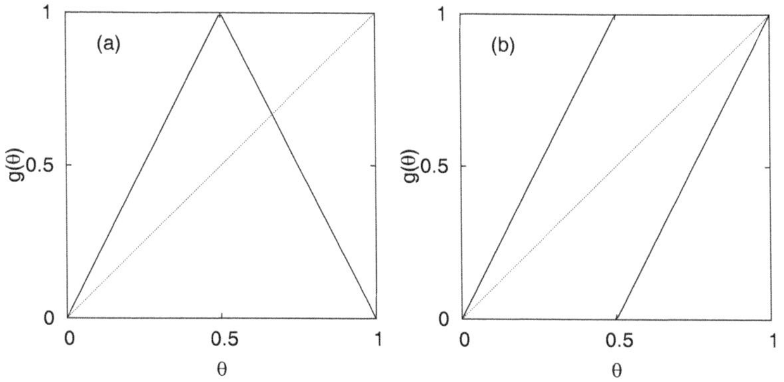

Fig. 3.4 (a) Tent map (3.8). (b) Bernoulli shift map (3.9).

chaos appears in a rather transparent way, so to infer the properties of the logistic map for $r = 4$.

To understand why chaos, meant as sensitive dependence on initial conditions, characterizes the tent map, it is useful to warm up with an even simpler instance, that is the *Bernoulli Shift map*[6] (Fig. 3.4b)

$$\theta(n+1) = 2\,\theta(n) \quad \mathrm{mod}\, 1\,, \mathrm{i.e.} \quad \theta(n+1) = \begin{cases} 2\theta(n) & 0 \le \theta(n) < \frac{1}{2} \\ 2\theta(n) - 1 & \frac{1}{2} \le \theta(n) < 1\,, \end{cases} \tag{3.9}$$

which is composed by a branch of the tent map, for $\theta < 1/2$, and by its reflection with respect to the line $g(\theta) = 1/2$, for $1/2 < \theta < 1$. The effect of the iteration of the Bernoulli map is trivially understood by expressing a generic initial condition in binary representation

$$\theta(0) = \sum_{i=1}^{\infty} \frac{a_i}{2^i} \equiv [a_1, a_2, \ldots]$$

where $a_i = 0, 1$. The action of map (3.9) is simply to remove the most significant digit, i.e. the binary shift operation

$$\theta(0) = [a_1, a_2, a_3, \ldots] \to \theta(1) = [a_2, a_3, a_4, \ldots] \to \theta(2) = [a_3, a_4, a_5, \ldots]$$

so that, given $\theta(0)$, $\theta(n)$ is nothing but $\theta(0)$ with the first $(n-1)$ binary digits removed.[7] This means that any small difference in the less significant digits will be

[6]The Bernoulli map and the tent map are also topologically conjugated but through a complicated non differentiable function (see, e.g., Beck and Schlögl, 1997).

[7]The reader may object that when $\theta(0)$ is a rational number, the resulting trajectory $\theta(n)$ should be rather trivial and non-chaotic. This is indeed the case. For example, if $\theta(0) = 1/4$ i.e. in binary representation $\theta(0) = [0, 1, 0, 0, 0, \ldots]$ under the action of (3.9) will end in $\theta(n > 1) = 0$, or $\theta(0) = 1/3$ corresponding to $\theta(0) = [0, 1, 0, 1, 0, 1, 0, \ldots]$ will give rise to a period-2 orbit, which expressed in decimal is $\theta(2k) = 1/3$ and $\theta(2k+1) = 2/3$ for any integer k. Due to the fact that rationals are infinitely many, one may wrongly interpret the above behavior as an evidence

amplified by the shift operation by a factor 2 at each iteration. Therefore, considering two trajectories, $\theta(n)$ and $\theta'(n)$ initially almost equal but for an infinitesimal amount $\delta\theta(0) = |\theta(0) - \theta'(0)| \ll 1$, their distance or the error we commit by using one to predict the other will grow as

$$\delta\theta(n) = 2^n \delta\theta(0) = \delta\theta(0)\, e^{n \ln 2}\,, \tag{3.10}$$

i.e. exponentially fast with a rate $\lambda = \ln 2$ which is the *Lyapunov exponent* — the suitable indicator for quantifying chaos, as we will see in Chapter 5.

Let us now go back to the tent map (3.8). For $\theta(n) < 1/2$ it acts as the shift map, while for $\theta(n) > 1/2$ the shift is composed with another unary operation that is *negation*, $\neg$ in symbols, which is defined by $\neg 0 = 1$ and $\neg 1 = 0$. For example, consider the initial condition $\theta(0) = 0.875 = [1, 1, 1, 0, 0, 0, \ldots]$ then $\theta(1) = 0.25 = [0, 0, 1, 1, 1, \ldots] = [\neg 1, \neg 1, \neg 0, \neg 0 \ldots]$. In general, one has $\theta(0) = [a_1, a_2, \ldots] \to \theta(1) = [a_2, a_3, \ldots]$ if $\theta(0) < 1/2$ (i.e. $a_1 = 0$) while $\to \theta(1) = [\neg a_2, \neg a_3, \ldots]$ if $\theta(0) > 1/2$ (i.e. $a_1 = 1$). Since $\neg^0$ is the identity ($\neg^0 a = a$), we can write

$$\theta(1) = [\neg^{a_1} a_2, \neg^{a_1} a_3, \ldots]$$

and therefore

$$\theta(n) = [\neg^{(a_1+a_2+\ldots+a_n)} a_{n+1}, \neg^{(a_1+a_2+\ldots+a_n)} a_{n+2}, \ldots]\,.$$

It is then clear that Eq. (3.10) also holds for the tent map and hence, thanks to the topological conjugacy (Box B.3), the same holds true for the logistic map.

The tent and shift maps are *piecewise linear maps* (see next Chapter), i.e. with constant derivative within sub-intervals of $[0\!:\!1]$. It is rather easy to recognize (using the graphical construction or linear analysis) that for chaos to be present at least one of the slopes of the various pieces composing the map should be in absolute value larger than 1.

Before concluding this section it is important first to stress that the relation between the logistic and the tent map holds only for $r = 4$ and second to warn the reader that the behavior of the logistic map, in the range $r_\infty < r < 4$, is a bit more complicated than one can expect. This is clear by looking at the so-called *bifurcation diagram (or tree)* of the logistic map shown in Fig. 3.5. The figure is obtained by plotting, for several r values, the M successive iterations of the map (here $M = 200$) after a transient of N iterates (here $N = 10^6$) is discarded. Clearly, such a bifurcation diagram allows periodic orbits (up to period M, of course) to be identified. In the diagram, the higher density of points corresponds to values of r for which either periodic trajectories of period $> M$ or chaotic ones are present. As

of the triviality of the map. However, we know that, although infinitely many, rationals have zero Lebesgue measure, while irrationals, corresponding to the *irregular* orbits, have measure 1 in the unit interval $[0 : 1]$. Therefore, for almost all initial conditions the resulting trajectory will be irregular and chaotic in the sense of Eq. (3.10). We end this footnote remarking that rationals correspond to infinitely many (unstable) periodic orbits embedded in the dynamics of the Bernoulli shift map. We will come back to this observation in Chapter 8 in the context of *algorithmic complexity*.

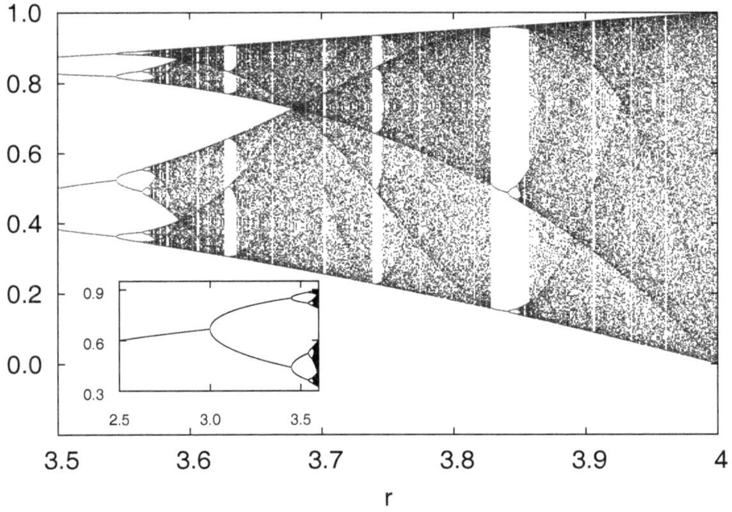

Fig. 3.5 Logistic map bifurcation tree for $3.5 < r < 4$. The inset shows the period-doubling region, $2.5 < r < 3.6$. The plot is obtained as explained in the text.

readily seen in the figure, for $r > r_\infty$, there are several windows of regular (periodic) behavior separated by chaotic regions. A closer look, for instance, makes possible to identify also regions with stable orbits of period-3 for $r \approx 3.828\ldots$, which then bifurcate to period-6, 12 etc. orbits. For understanding the origin of such behavior one has to study the graphs of $f_r^{(3)}(x)$, $f_r^{(6)}(x)$ etc.

We will come back to the logistic map and, in particular, to the period doubling bifurcation in Sec. 6.2.

Box B.3: Topological conjugacy

In this Box we briefly discuss an important technical issue. Just for the sake of notation simplicity, consider the one-dimensional map

$$x(0) \to x(t) = S^t x(0) \quad \text{where} \quad x(t+1) = g(x(t)) \tag{B.3.1}$$

and the (invertible) change of variable

$$x \to y = h(x)$$

where $\mathrm{d}h/\mathrm{d}x$ does not change sign. Of course, we can write the time evolution of $y(t)$ as

$$y(0) \to y(t) = \tilde{S}^t y(0) \quad \text{where} \quad y(t+1) = f(y(t)) \;, \tag{B.3.2}$$

the function $f(\bullet)$ can then be expressed in terms of $g(\bullet)$ and $h(\bullet)$:

$$f(\bullet) = h(g(h^{(-1)}(\bullet)))\,,$$

where $h^{(-1)}(\bullet)$ is the inverse of h. In such a case one says that the dynamical systems (B.3.1) and (B.3.2) are topologically conjugate, i.e. there exists a homeomorphism between x and y. If two dynamical systems are topologically conjugate they are nothing but two equivalent versions of the same system and there is a one-to-one correspondence between their properties [Eckmann and Ruelle (1985); Jost (2005)].[8]

3.2 The Lorenz model

One of the first and most studied example of chaotic system was introduced by meteorologist Lorenz in 1963. As detailed in Box B.4, Lorenz obtained such a set of equations investigating Rayleigh-Bénard convection, a classic problem of fluid mechanics theoretically and experimentally pioneered by Bénard (1900) and continued with Lord Rayleigh (1916). The description of the problem is as follows. Consider a fluid, initially at rest, constrained by two infinite horizontal plates maintained at constant temperature and at a fixed distance from each other. Gravity acts on the system perpendicularly to the plates. If the upper plate is maintained hotter than the lower one, the fluid remains at rest and in a state of *conduction*, i.e. a linear temperature gradient establishes between the two plates. If the temperatures are inverted, gravity induced buoyancy forces tend to rise toward the top the hotter and thus lighter fluid that is at the bottom.[9] This tendency is contrasted by viscous and dissipative forces of the fluid so that the conduction state may persist. However, as the temperature difference exceeds a certain amount, the conduction state is replaced by a steady *convection* state: the fluid motion consists of steady counter-rotating vortices (rolls) which transport upwards the hot/light fluid in contact with the bottom plate and downwards the cold/heavy fluid in contact with the upper one (see Box B.4). The steady convection state remains stable up to another critical temperature difference above which it becomes unsteady, very irregular and hardly predictable.

At the beginning of the '60s, Lorenz became interested to this problem. He was mainly motivated by the well reposed hope that the basic mechanisms of the irregular behaviors observed in atmospheric physics could be captured by "conceptual" models and thus avoiding the technical difficulties of a too detailed description of the phenomenon. By means of a truncated Fourier expansion, he reduced the

[8] In Chapter 5 we shall introduce the Lyapunov exponents and the information dimension while in Chapter 8 the Kolmogorov-Sinai entropy. These are mathematically well defined indicators which quantify the chaotic behavior of a system. All such numbers do not change under topological conjugation.

[9] We stress that this is not an academic problem but it corresponds to typical phenomena taking place in the atmosphere.

partial differential equations describing the Rayleigh-Bénard convection to a set of three ordinary differential equations, $\mathrm{d}\boldsymbol{R}/\mathrm{d}t = \boldsymbol{F}(\boldsymbol{R})$ with $\boldsymbol{R} = (X, Y, Z)$, which read (see the Box B.4 for details):

$$
\begin{aligned}
\frac{\mathrm{d}X}{\mathrm{d}t} &= -\sigma X + \sigma Y \\
\frac{\mathrm{d}Y}{\mathrm{d}t} &= -XZ + rX - Y \\
\frac{\mathrm{d}Z}{\mathrm{d}t} &= XY - bZ\,.
\end{aligned}
\tag{3.11}
$$

The three variables physically are linked to the intensity of the convection (X), the temperature difference between ascending and descending currents (Y) and the deviation of the temperature from the linear profile (Z). Same signs of X and Y denotes that warm fluid is rising and the cold one descending. The constants σ, r, b are dimensionless, positive defined parameters linked to the physical problem: σ is the Prandtl number measuring the ratio between fluid viscosity and thermal diffusivity; r can be regarded as the normalized imposed temperature difference (more precisely it is the ratio between the value of the Rayleigh number and its critical value), and is the main control parameter; finally, b is a geometrical factor. Although the behavior of Eq. (3.11) is quantitatively different from the original problem (i.e. atmospheric convection), Lorenz's right expectation was that the qualitative features should roughly be the same.

As done for the logistic map, we can warm up by performing the linear stability analysis. The first step consists in computing the stability matrix of Eq. (3.11)

$$
\mathbb{L} = \begin{pmatrix} -\sigma & \sigma & 0 \\ (r-Z) & -1 & -X \\ Y & X & -b \end{pmatrix}.
$$

As commonly found in nonlinear systems, the matrix elements depend on the variables, and thus linear analysis is informative only if we focus on fixed points. Before computing the fixed points, we observe that

$$
\boldsymbol{\nabla}\cdot\boldsymbol{F} = \frac{\partial}{\partial X}\frac{\mathrm{d}X}{\mathrm{d}t} + \frac{\partial}{\partial Y}\frac{\mathrm{d}Y}{\mathrm{d}t} + \frac{\partial}{\partial Z}\frac{\mathrm{d}Z}{\mathrm{d}t} = \mathrm{Tr}\,(\mathbb{L}) = -(\sigma + b + 1) < 0 \tag{3.12}
$$

meaning that phase-space volumes are uniformly contracted by the dynamics: an ensemble of trajectories initially occupying a certain volume converges exponentially fast, with constant rate $-(\sigma + b + 1)$, to a subset of the phase space having zero volume. The Lorenz system is thus dissipative. Furthermore, it is possible to show that the trajectories do not explore the whole space but, at times long enough, stay in a bounded region of the phase space.[10]

[10]To show this property, following Lorenz (1963), we introduce the change of variables $X_1 = X$, $X_2 = Y$ and $X_3 = Z - r - \sigma$, with which Eq. (3.11) can be put in the form $\mathrm{d}X_i/\mathrm{d}t =$

Elementary algebra shows that the fixed points of Eq. (3.11), i.e. the roots of $\boldsymbol{F}(\boldsymbol{R}^*) = \boldsymbol{0}$, are:

$$\boldsymbol{R}^*_o = (0,0,0) \qquad \boldsymbol{R}^*_\pm = (\pm\sqrt{b(r-1)}, \pm\sqrt{b(r-1)}, r-1)$$

the first represents the conduction state, while $\boldsymbol{R}^*_\pm$, which are real for $r \geq 1$, two possible states of steady convection with the $\pm$ signs corresponding to clockwise/anticlockwise rotation of the convective rolls. The secular equation $\det(\mathbb{L}(\boldsymbol{R}^*) - \lambda\mathbb{I}) = 0$ yields the eigenvalues $\lambda_i(\boldsymbol{R}^*)$ $(i = 1,2,3)$. Skipping the algebra, we summarize the result of this analysis:

- For $0 < r < 1$, $\boldsymbol{R}^*_0 = (0,0,0)$ is the only real fixed point and, moreover, it is stable being all the eigenvalues negative — stable conduction state;
- For $r > 1$, one of the eigenvalues associated with $\boldsymbol{R}^*_0$ becomes positive while $\boldsymbol{R}^*_\pm$ have one real negative and two complex conjugate eigenvalues — conduction is unstable and replaced by convection. For $r < r_c$, the real part of such complex conjugate eigenvalues is negative — steady convection is stable — and, for $r > r_c$, positive — steady convection is unstable — with
$$r_c = \frac{\sigma(\sigma + b + 3)}{(\sigma - b - 1)}\,.$$
Because of their physical meaning, r, σ and b are positive numbers, and thus the above condition is relevant only if $\sigma > (b+1)$, otherwise the steady convective state is always stable.

What does happen if $\sigma > (b+1)$ and $r > r_c$? Linear stability theory cannot answer this question and the best we can do is to resort to numerical analysis of the equations — as Lorenz did in 1963. Following him, we fix $b = 8/3$ and $\sigma = 10$ and $r = 28$, well above the critical value $r_c = 24.74\ldots$. For illustrative purposes, we perform two numerical experiments by considering two trajectories of Eq. (3.11) starting from far away or very close initial conditions.

The result of the first numerical experiment is shown in Fig. 3.6. After a short transient, the first trajectory, originating from P_1, converges toward a set in phase space characterized by alternating circulations of seemingly random duration around the two unstable steady convection states $\boldsymbol{R}^*_\pm = (\pm 6\sqrt{2}, \pm 6\sqrt{2}, 27)$. Physically speaking, this means that the convection irregularly switches from clockwise to anticlockwise circulation. The second trajectory, starting from the distant point P_2, always remains distinct from the first one but qualitatively behaves in the same way visiting, in the course of time, the same subset in phase space. Such a

$\sum_{jk} a_{ijk} X_j X_k + \sum_j b_{ij} X_j + c_i$ with a_{ijk}, b_{ij} and c_j constants. Furthermore, we notice that $\sum_{ijk} a_{ijk} X_i X_j X_k = 0$ and $\sum_{ij} b_{ij} X_i X_j > 0$. If we define the "energy" function $Q = (1/2)\sum X_i^2$ and denote with e_i the roots of the linear equation $\sum_j (b_{ij} + b_{ji}) e_j = c_i$, then from the equations of motion we have

$$\frac{\mathrm{d}Q}{\mathrm{d}t} = \sum_{ij} b_{ij} e_i e_j - \sum_{ij} b_{ij} (X_i - e_i)(X_j - e_j)\,.$$

From the above equation it is easy to see that $\mathrm{d}Q/\mathrm{d}t < 0$ outside a sufficiently large domain, so that trajectories are asymptotically confined in a bounded region.

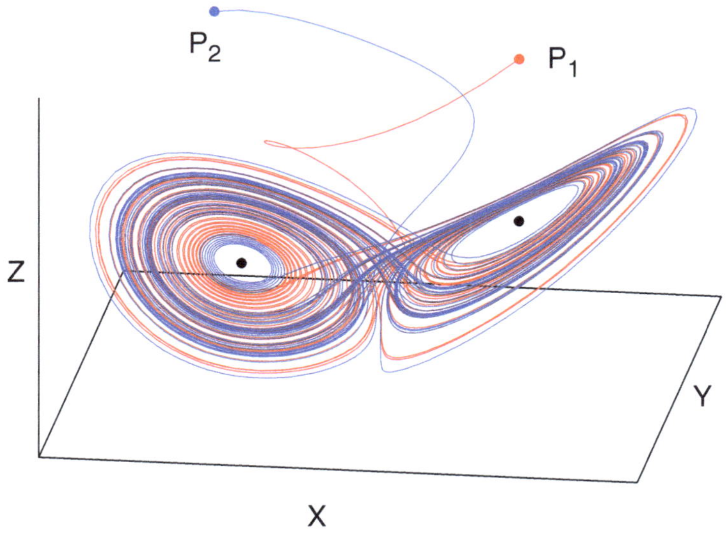

Fig. 3.6 Lorenz model: evolution of two trajectories starting from distant points P_1 and P_2, which after a transient converge, remaining distinct, toward the same subset of the phase space — the Lorenz attractor. The two black dots around which the two orbits circulate are the fixed points $\boldsymbol{R}^* = (\pm 6\sqrt{2}, \pm 6\sqrt{2}, 27)$ of the dynamics for $r = 28$, $b = 8/3$ and $\sigma = 10$.

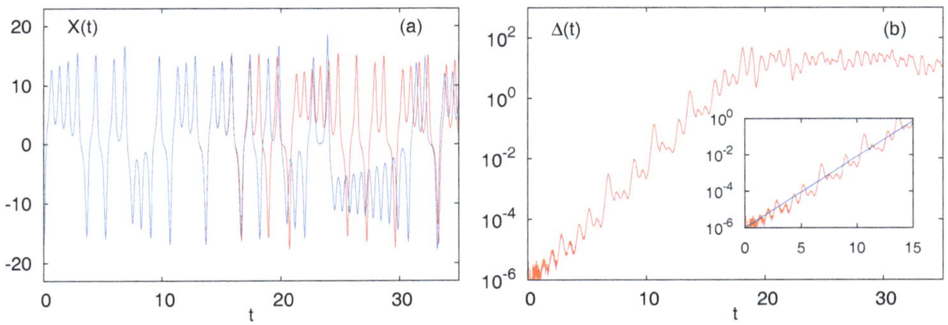

Fig. 3.7 Lorenz model: (a) evolution of reference $X(t)$ (red) and perturbed $X'(t)$ (blue) trajectories, initially at distance $\Delta(0) = 10^{-6}$. (b) Evolution of the separation between the two trajectories. Inset: zoom in the range $0 < t < 15$ in semi-log scale. See text for explanation.

subset, attracting all trajectories, is the *strange attractor* of the Lorenz equations.[11] The attractor is indeed very weird as compared to the ones we encountered up to now: fixed points or limit cycles. Moreover, it is characterized by complicated

[11] Note that it is nontrivial from mathematical point of view to establish whether a set is strange attractor. For example, Smale's 14th problem, which is about proving that the Lorenz attractor is indeed a strange attractor, was solved only very recently [Tucker (2002)].

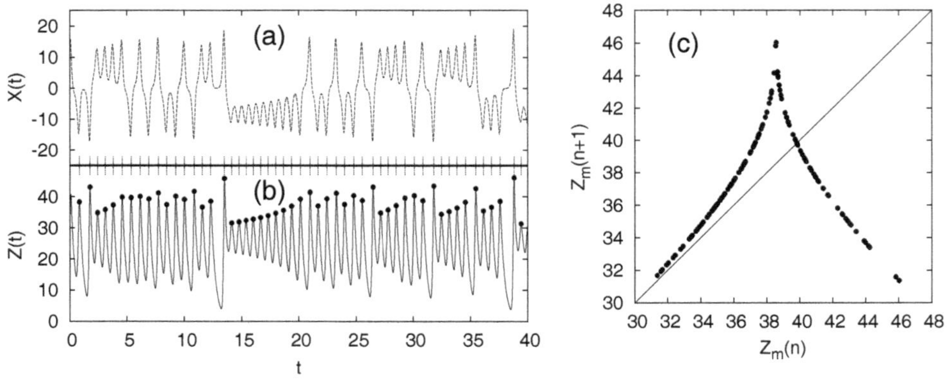

Fig. 3.8 Lorenz model: (a) time evolution of $X(t)$, (b) $Z(t)$ for the same trajectory, black dots indicate local maxima. Vertical tics between (a) and (b) indicate the time locations of the maxima Z_m. (c) Lorenz return map, see text for explanations.

geometrical properties whose quantitative treatment requires concepts and tools of *fractal geometry*,[12] which will be introduced in Chapter 5.

Having seen the fate of two distant trajectories, it is now interesting to contrast it with that of two initially infinitesimally close trajectories. This is the second numerical experiments which is depicted in Fig. 3.7a,b and was performed as follows. A reference trajectory was obtained from a generic initial condition, by waiting enough time for it to settle onto the attractor of Fig. 3.6. Denote with $t = 0$ the time at the end of such a transient, and with $\boldsymbol{R}(0) = (X(0), Y(0), Z(0))$ the initial condition of the reference trajectory. Then, we consider a new trajectory starting at $\boldsymbol{R}'(0)$ very close to the reference one, such that $\Delta(0)=|\boldsymbol{R}(0)-\boldsymbol{R}'(0)|=10^{-6}$. Both trajectories are then evolved and Figure 3.7a shows $X(t)$ and $X'(t)$ as a function of time. As one can see, for $t < 15$, the trajectories are almost indistinguishable but at larger times, in spite of a qualitatively similar behavior, become "macroscopically" distinguishable. Moreover, looking at the separation $\Delta(t)=|\boldsymbol{R}(t)-\boldsymbol{R}'(t)|$ (Fig. 3.7b) an exponential growth can be observed at the initial stage (see inset), after which the separation becomes of the same order of the signal $X(t)$ itself, as the motions take place in a bounded region their distance cannot grow indefinitely. Thus also for the Lorenz system the erratic evolution of trajectories is associated with *sensitive dependence on initial conditions.*

Lorenz made another remarkable observation demonstrating that the chaotic behavior of Eq. (3.11) can be understood by deriving a chaotic one-dimensional map, return map, from the system evolution. By comparing the time course of $X(t)$ (or $Y(t)$) with that of $Z(t)$, he noticed that sign changes of $X(t)$ (or $Y(t)$) — i.e. the random switching from clockwise to anticlockwise circulation — occur concomitantly with $Z(t)$ reaching local maxima Z_m, which overcome a certain threshold value.

[12]See also Sec. 3.4 and, in particular, Fig. 3.12.

This can be readily seen in Fig. 3.8a,b where vertical bars have been put at the times where Z reaches local maxima to facilitate the eye. He then had the intuition that the nontrivial dynamics of the system was encoded by that of the local maxima Z_m. The latter can be visualized by plotting $Z_m(n+1)$ versus $Z_m(n)$ each time t_n Z reaches a local maxima, i.e. $Z_m(n) = Z(t_n)$. The resulting one dimensional map, shown in Fig. 3.8c, is rather interesting. First, the points are not randomly scattered but organized on a smooth one-dimensional curve. Second, such a curve, similarly to the logistic map, is not invertible and so chaos is possible. Finally, the slope of the tangent to the map is larger than 1 everywhere, meaning that there cannot be stable fixed points either for the map itself or for its kth-iterates. From what we learn in the previous section it is clear that such map will be chaotic.

We conclude mentioning that if r is further increased above $r = 28$, similarly to the logistic map for $r > r_\infty$, several investigators have found regimes with alternating periodic and chaotic behaviors.[13] Moreover, the sequence of events (bifurcation) leading to chaos depends on the parameter range, for example, around $r = 166$, an interesting transition to chaos occurs (see Chapter 6).

Box B.4: Derivation of the Lorenz model

Consider a fluid under the action of a constant gravitational acceleration $\boldsymbol{g}$ directed along the $z-$axis, and contained between two horizontal, along the $x-$axis, plates maintained at constant temperatures T_U and T_B at the top and bottom, respectively. For simplicity, assume that the plates are infinite in the horizontal direction and that their distance is H. The fluid density is a function of the temperature $\rho = \rho(T)$. Therefore, if $T_U = T_B$, ρ is roughly constant in the whole volume while, if $T_U \neq T_B$, it is a function of the position. If $T_U > T_B$ the fluid is stratified with cold/heavy fluid at the bottom and hot/light one at the top. From the equations of motion [Monin and Yaglom (1975)] one derives that the fluid remains at rest establishing a stable thermal gradient, i.e. the temperature depends on the altitude z

$$T(z) = T_B + z\frac{T_U - T_B}{H}\,, \tag{B.4.1}$$

this is the *conduction state*. If $T_U < T_B$, the density profile is unstable due to buoyancy: the lighter fluid at the bottom is pushed toward the top while the cold/heavier one goes toward the opposite direction. This is contrasted by viscous forces. If $\Delta T = T_B - T_U$ exceeds a critical value the conduction state becomes unstable and replaced by a *convective state*, in which the fluid is organized in counter-rotating rolls (vortices) rising the warmer and lighter fluid and bringing down the colder and heavier fluid as sketched in Fig. B4.1.

This is the Rayleigh-Bénard convection which is controlled by the Rayleigh number:

$$Ra = \frac{\rho_0 g \alpha H^3 |T_U - T_B|}{\kappa\nu}\,, \tag{B.4.2}$$

[13] In this respect, the behavior of the Lorenz model depart from actual Rayleigh-Bénard problem. Much more Fourier modes need to be included in the description to approximate the behavior of the PDE ruling the problem.

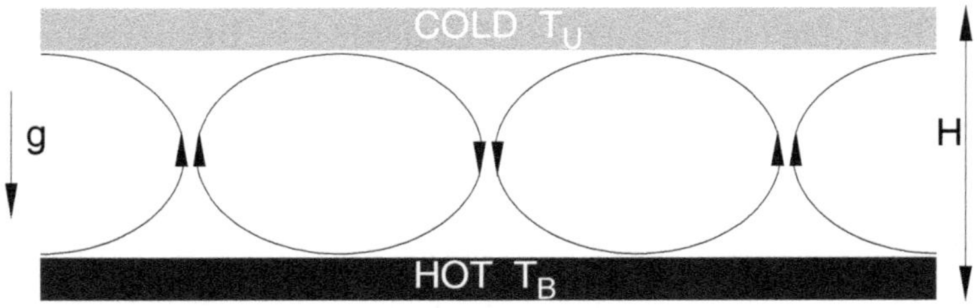

Fig. B4.1 Two-dimensional sketch of the steady Raleigh-Bénard convection state.

where κ is the coefficient of thermal diffusivity and ν the fluid viscosity. The average density is denoted by ρ_0 and α is the thermal dilatation coefficient, relating the density at temperatures T and T_0 by $\rho(T) = \rho(T_0)[1 - \alpha(T - T_0)]$, which is the linear approximation valid for not too high temperature differences.

Experiments and analytical computations show that if $Ra \leq Ra_c$ conduction solution (B.4.1) is stable. For $Ra > Ra_c$ the steady convection state (Fig. B4.1) becomes stable. However, if Ra exceeds Ra_c by a sufficiently large amount the steady convection state becomes also unstable and the fluid is characterized by a rather irregular and apparently unpredictable convective motion. Being crucial for many phenomena taking place in the atmosphere, in stars or Earth magmatic mantle, since Lord Rayleigh, many efforts were done to understand the origin of such convective irregular motions.

If the temperature difference $|T_B - T_U|$ is not too large the PDEs for the temperature and the velocity can be written within the Boussinesq approximation giving rise to the following equations [Monin and Yaglom (1975)]

$$\partial_t \boldsymbol{u} + \boldsymbol{u} \cdot \boldsymbol{\nabla} \boldsymbol{u} = -\frac{\boldsymbol{\nabla} p}{\rho_0} + \nu \Delta \boldsymbol{u} + g\alpha\Theta \tag{B.4.3}$$

$$\partial_t \Theta + \boldsymbol{u} \cdot \boldsymbol{\nabla} \Theta = \kappa \Delta \Theta + \frac{T_U - T_B}{H} u_z \,, \tag{B.4.4}$$

supplemented by the incompressibility condition $\boldsymbol{\nabla} \cdot \boldsymbol{u} = 0$, which is still making sense if the density variations are small; $\Delta = \boldsymbol{\nabla} \cdot \boldsymbol{\nabla}$ denotes the Laplacian. The first is the Navier-Stokes equation where p is the pressure and the last term is the buoyancy force. The second is the advection diffusion equation for the deviation Θ of the temperature from the conduction state (B.4.1), i.e. denoting the position with $\boldsymbol{r} = (x, y, z)$, $\Theta(\boldsymbol{r}, t) = T(\boldsymbol{r}, t) - T_B + (T_B - T_U)z/H$. The Rayleigh number (B.4.2) measures the ratio between the nonlinear and Boussinesq terms, which tend to destabilize the thermal gradient, and the viscous/dissipative ones, which would like to maintain it. Such equations are far too complicated to allow an easy identification of the mechanism at the basis of the irregular behaviors observed in experiments.

A first simplification is to consider the two-dimensional problem, i.e. on the (x, z)-plane as in Fig. B4.1. In such a conditions the fluid motion is described by the so-called stream-function $\psi(\boldsymbol{r}, t) = \psi(x, z, t)$ (now we call $\boldsymbol{r} = (x, z)$) defined by

$$u_x = \frac{\partial \psi}{\partial z} \quad \text{and} \quad u_z = -\frac{\partial \psi}{\partial x} \,.$$

The above equations ensure fluid incompressibility. Equations (B.4.3)–(B.4.4) can thus be rewritten in two-dimensions in terms of ψ. Already Lord Rayleigh found solutions of the form:

$$\psi = \psi_0 \sin\left(\frac{\pi a x}{H}\right) \sin\left(\frac{\pi z}{H}\right)$$
$$\Theta = \Theta_0 \cos\left(\frac{\pi a x}{H}\right) \sin\left(\frac{\pi z}{H}\right),$$

where ψ_0 and θ_0 are constants and a is the horizontal wave length of the rolls. In particular, with a linear stability analysis, he found that if Ra exceeds a critical value

$$Ra_c = \frac{\pi^4(1+a^2)^3}{a^2}$$

such solutions become unstable making the problem hardly tractable from an analytical viewpoint. One possible approach is to expand ψ and θ in the Fourier basis with the simplification of putting the time dependence only in the coefficients, i.e.

$$\begin{aligned}\psi(x,z,t) &= \sum_{m,n=1}^{\infty} \psi_{mn}(t) \sin\left(\frac{m\pi a x}{H}\right) \sin\left(\frac{n\pi z}{H}\right) \\ \Theta(x,z,t) &= \sum_{m,n=1}^{\infty} \Theta_{mn}(t) \cos\left(\frac{m\pi a x}{H}\right) \sin\left(\frac{n\pi z}{H}\right).\end{aligned} \tag{B.4.5}$$

However, substituting such an expansion in the original PDEs leads to an infinite number of ODEs, so that Saltzman (1962), following a suggestion of Lorenz, started to study a simplified version of this problem by truncating the series (B.4.5). One year later, Lorenz (1963) considered the simplest possible truncation which retains only three coefficients namely the amplitude of the convective motion $\psi_{11}(t) = X(t)$, the temperature difference between ascending and descending fluid currents $\theta_{11}(t) = Y(t)$ and the deviation from the linear temperature profile $\theta_{02}(t) = Z(t)$. The choice of the truncation was not arbitrary but suggested by the symmetries of the equations. He thus finally ended up with a set of three ODEs — the Lorenz equations:

$$\frac{\mathrm{d}X}{\mathrm{d}t} = -\sigma X + \sigma Y\,, \quad \frac{\mathrm{d}Y}{\mathrm{d}t} = -XZ + rX - Y\,, \quad \frac{\mathrm{d}Z}{\mathrm{d}t} = XY - bZ\,, \tag{B.4.6}$$

where σ, r, b are dimensionless parameters related to the physical ones as follows: $\sigma = \nu/\kappa$ is the Prandtl number, $r = Ra/Ra_c$ the normalized Rayleigh number and $b = 4(1+a^2)^{-1}$ is a geometrical factor linked to the rolls wave length. Unit time in (B.4.6) means $\pi^2 H^{-2}(1+a^2)\kappa$ in physical time units.

The Fourier expansion followed by truncation used by Saltzman and Lorenz is known as Galerkin approximation [Lumley and Berkooz (1996)], which is a very powerful tool often used in the numerical treatment of PDEs (see also Chap. 13).

3.3 The Hénon-Heiles system

Hamiltonian systems, as a consequence of their conservative dynamics and symplectic structure, are quite different from dissipative ones, in particular, for what concerns the way chaos shows up. It is thus here interesting to examine an example of Hamiltonian system displaying chaos. We consider a two-degree of freedom autonomous system, meaning that the phase space has dimension $d = 4$. Motions, however, take place on a three-dimensional hypersurface due to the constraint of energy conservation. This example will also give us the opportunity to become acquainted with the Poincaré section technique (Sec. 2.1.2).

We consider the Hamiltonian system introduced by Hénon and Heiles (1964) in celestial mechanics context. They were interested in understanding if an axis-symmetric potential, which models in good approximation a star in a galaxy, possesses a third integral of motion, besides energy and angular momentum. In particular, at that time, the main question was if such an integral of motion was isolating, i.e. able to constrain the orbit into specific subspaces of phase space. In other terms, they wanted to unravel which part of the available phase space would be filled by the trajectory of the star in the long time asymptotics.

After a series of simplifications Hénon and Heiles ended up with the following two-degree of freedom Hamiltonian:

$$H(Q, q, P, p) = \frac{1}{2}P^2 + \frac{1}{2}p^2 + U(Q, q) \tag{3.13}$$

$$U(Q, q) = \frac{1}{2}\left(Q^2 + q^2 + 2Q^2 q - \frac{2}{3}q^3\right) \tag{3.14}$$

where (Q, P) and (q, p) are the canonical variables. The evolution of Q, q, P, q can be obtained via the Hamilton equations (2.6). Of course, the four-dimensional dynamics can be visualized only through an appropriate Poincaré section.

Actually, the star moves on the three-dimensional constant-energy hypersurface embedded in the four-dimensional phase space, so that we only need three coordinates, say Q, q, p, to locate it, while the fourth, P, can be obtained solving $H(Q, q, P, p) = E$. As $P^2 \geq 0$ we have that the portion of the three-dimensional hypersurface actually explored by the star is given by:

$$\frac{1}{2}p^2 + U(Q, q) \leq E\,. \tag{3.15}$$

Going back to the original question, if no other isolating integral of motion exists the region of non-zero volume (3.15) will be filled by a single trajectory of the star. We can now choose a plane and represent the motion by looking at the intersection of the trajectories with it, identifying the Poincaré map. For instance, we can consider the map obtained by taking all successive intersections of a trajectory with the plane $Q = 0$ in the upward direction, i.e. with $P > 0$. In this way the original four-dimensional phase space reduces to the two-dimensional (q, p)-plane defined by

$$Q = 0 \quad \text{and} \quad P > 0\,.$$

Before analyzing the above defined Poincaré section, we observe that the Hamiltonian (3.13) can be written as the sum of an integrable Hamiltonian plus a perturbation $H = H_0 + \epsilon H_1$ with

$$H_0 = \frac{1}{2}(P^2 + p^2) + \frac{1}{2}(Q^2 + q^2) \qquad \text{and} \qquad \epsilon H_1 = Q^2 q - \frac{1}{3}q^3\,,$$

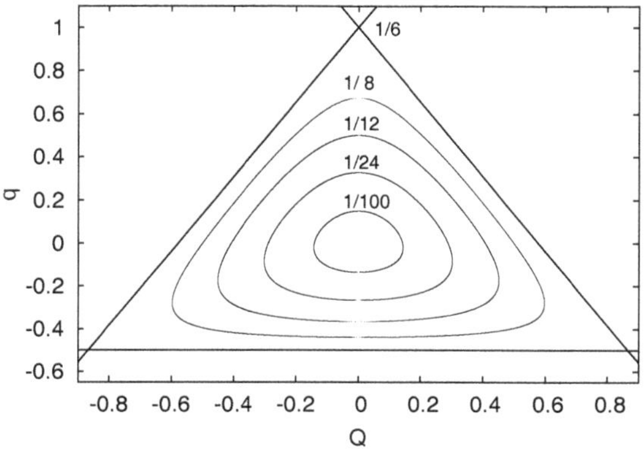

Fig. 3.9 Isolines of the Hénon-Heiles potential $U(q, Q)$ close to the origin.

where H_0 is the Hamiltonian of two uncoupled harmonic oscillators, H_1 represents a nonlinear perturbation to it, and ϵ quantifies the strength of the perturbation. From Eq. (3.13) one would argue that $\epsilon = 1$, and thus that ϵ is not a tunable parameter. However, the actual deviation from the integrable limit depends on the energy level considered: if $E \ll 1$ the nonlinear deviations from the harmonic oscillators limit are very small, while they become stronger and stronger as E increases. In this sense the control parameter is the energy itself, i.e. E plays the role of ϵ.

A closer examination of Eq. (3.14) shows that, for $E \leq 1/6$, the potential $U(Q, q)$ is trapping, i.e. trajectories cannot escape. In Fig. 3.9 we depict the isolines of $U(Q, q)$ for various values of the energy $E \leq 1/6$. For small energy they resemble those of the harmonic oscillator, while, as energy increases, the nonlinear terms in H_1 deform the isolines up to become an equilateral triangle for $E = 1/6$.[14]

We now study the Poincaré map at varying the strength of the deviation from the integrable limit, i.e. at increasing the energy E. From Eq. (3.15), we have that the motion takes place in the region of the (q, p)-plane defined by

$$p^2/2 + U(0, q) \leq E \,, \tag{3.16}$$

which is bounded as the potential is trapping. In order to build the phase portrait of the system, once the energy E is fixed, one has to evolve several trajectories and plot them exploiting the Poincaré section. The initial conditions for the orbits can be chosen selecting $q(0)$ and $p(0)$ and then fixing $Q(0) = 0$ and $P(0) = \pm\sqrt{[2E - p^2(0) - 2U(0, q(0))]}$. If a second isolating invariant exists, the Poincaré map would consist of a succession of points organized in regular curves, while its absence would lead to the filling of the bounded area defined by (3.16).

Figure 3.10 illustrates the Poincaré sections for $E = 1/12, 1/8$ and $1/6$, which correspond to small, medium and large nonlinear deviations from the integrable case. The scenario is as follows.

[14] As easily understood by noticing that $U(Q, q) = 1/6$ on the lines $q = -1/2$ and $q = \pm\sqrt{3}Q + 1$.

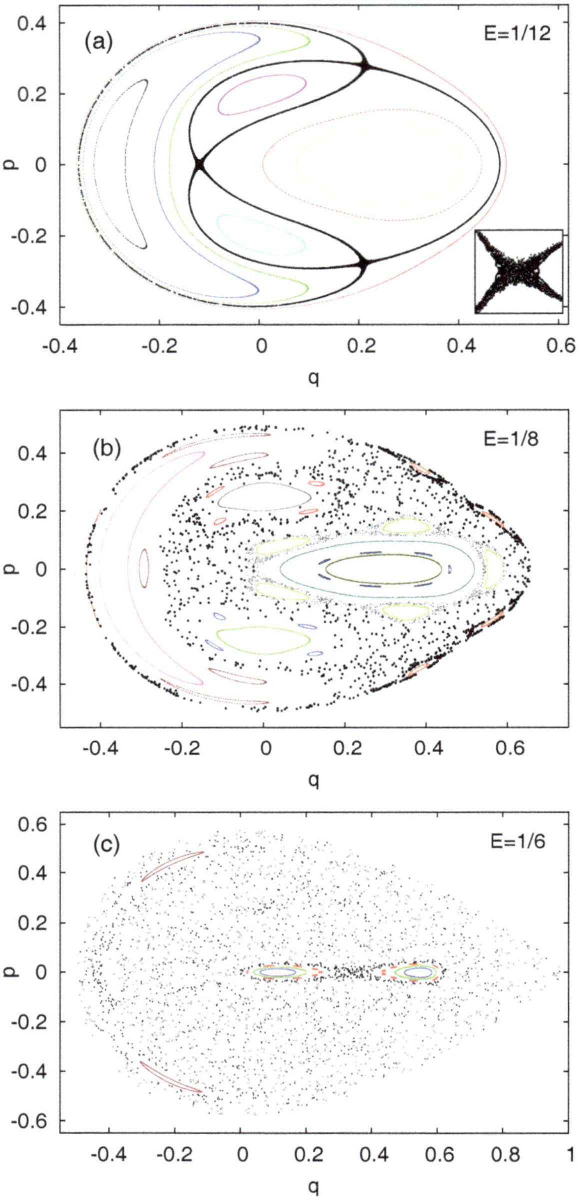

Fig. 3.10 Poincaré section, defined by $Q = 0$ and $P > 0$, of the Hénon-Heiles system: (a) at $E = 1/12$, (b) $E = 1/8$, (c) $E = 1/6$. Plot are obtained by using several trajectories, in different colors. The inset in (a) shows a zoom of the area around $q \approx -0.1$ and $p \approx 0$.

For $E = 1/12$ (Fig. 3.10a), the points belonging to the same trajectory lie exactly on a curve meaning that motions are regular (quasiperiodic or periodic orbits, the latter is when the Poincaré section consists of a finite number of points). We depicted a few trajectories starting from different initial conditions, as one can see the region of the (q,p)-plane where the motions take place is characterized by closed orbits of different nature separated by a self-intersecting trajectory — the separatrix, in black on the figure. We already encountered a separatrix in studying the nonlinear pendulum in Chapter 1 (see Fig. 1.1), in general separatrices either connect different fixed points (*heteroclinic* orbits) as here[15] or form a closed loop containing a single fixed point (*homoclinic* orbit) as in the pendulum. As we will see in Chap. 7, such curves are key for the appearance of chaos in Hamiltonian systems. This can be already appreciated from Fig. 3.10a: apart from the separatrix all trajectories are well defined curves which form a one-parameter family of curves filling the area (3.16); only the separatrix has a slightly different behavior. The blow-up in the inset reveals that, very close to the points of self-intersection, the Poincaré map does not form a smooth curve but fills, in a somehow irregular manner, a small area. Finally, notice that the points at the center of the small four loops correspond to stable periodic orbits of the system. In conclusion, for such energy values, most of trajectories are regular. Therefore, even if another (global) integral of motion besides energy is absent, for a large portion of the phase space, it is like if it exists.

Then we increase energy up to $E = 1/8$ (Fig. 3.10b). Closed orbits still exist near the locations of the lower energy loops (Fig. 3.10a), but they do no more fill the entire area, and a new kind of trajectories appears. For example, the black dots depicted in Fig. 3.10b belong to a single trajectory: they do not define a regular curve and "randomly" jump on the (q,p)-plane filling the space between the closed regular curves. Moreover, even the regular orbits are more complicated than before as, e.g., the five small loops surrounding the central closed orbits on the right, as the color suggests, are formed by the same trajectory. The same holds for the small four loops surrounding the symmetric loops toward the bottom and the top. Such orbits are called chains of islands, and adding more trajectories one would see that there are many of them having different sizes. They are isolated (hence the name islands) and surrounded by a sea of random trajectories (see, e.g., the gray spots around the five dark green islands on the right). The picture is thus rather different and more complex than before: the available phase space is partitioned in regions with regular orbits separated by finite portions, densely filled by trajectories with no evident regularity.

Further increasing the energy $E = 1/6$ (Fig. 3.10c), there is another drastic change. Most of the available phase space can be filled by a single trajectory (in Fig. 3.10c we show two of them with black and gray dots). The "random" character of such point distribution is even more striking if one plots the points one after the other as they appear, then one will see that they jump from on part to another of the domain without regularity. However, still two of the four sets of regular

[15] In the Poincaré map, the three intersection points correspond to three unstable periodic orbits.

trajectories observed at lower energies survive also here (see the bottom/up red loops, or the blue loops on the right surrounded by small chain of islands in green and orange). Notice also that the black trajectory from time to time visits an eight-like shaped region close to the two loops on the center-right of the plot, alternating such visits with random explorations of the available phase space. For this value of energy, the Poincaré section reveals that the motions are organized in a sea of seemingly random trajectories surrounding small islands of regular behavior (much smaller islands than those depicted in the figure are present and a finer analysis is necessary to make them apparent).

Trained by the logistic map and the Lorenz equations, it will not come as a surprise to discover that trajectories starting infinitesimally close to the random ones display sensitive dependence on the initial conditions — exponentially fast growth of their distance — while trajectories infinitesimally close to the regular ones remain close to each other.

It is thus clear that chaos is present also in the Hamiltonian system studied by Hénon and Heiles, but its appearance at varying the control parameter — the energy — is rather different from the (dissipative) cases examined before. We conclude by anticipating that the features emerging from Fig. 3.10 are not specific of the Hénon-Heiles Hamiltonian but are generic for Hamiltonian systems or symplectic maps (which are essentially equivalent as discussed in Box B.1 and Sec. 2.2.1.2).

3.4 What did we learn and what will we learn?

The three examined classical examples of dynamical systems gave us a taste of chaotic behaviors and how they manifest in nonlinear systems. In closing this Chapter, it is worth extracting the general aspects of the problem we are interested in, on the light of what we have learned from the above discussed systems. These aspects will then be further discussed and made quantitative in the next Chapters.

Necessity of a statistical description. We have seen that deterministic laws can generate erratic motions resembling random processes. This is from several points of view the more important lesson we can extract from the analyzed models. Indeed it forces us to reconsider and overcome the counterposition between deterministic and probabilistic worlds. As it will become clear in the following, the irregular behaviors of chaotic dynamical systems call for a probabilistic description even if the number of degrees of freedom involved is small. A way to elucidate this point is by realizing that, even if any trajectory of a deterministic chaotic system is fully determined by the initial condition, chaos is always accompanied by a certain degree of *memory loss* of the initial state. For instance, this is exemplified in Fig. 3.11, where we show the correlation function,

$$C(\tau) = \langle x(t+\tau)x(t)\rangle - \langle x(t)\rangle^2 \,, \tag{3.17}$$

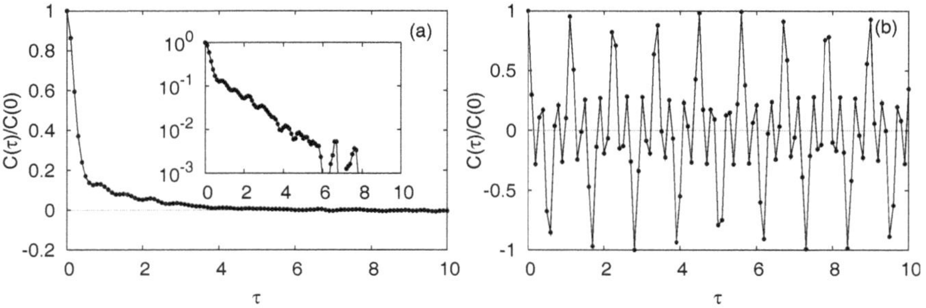

Fig. 3.11 (a) Normalized correlation function $C(\tau)/C(0)$ vs τ computed following the X variable of the Lorenz model (3.11) with $b = 8/3$, $\sigma = 10$ and $r = 28$. As shown in the inset, it decays exponentially at least for long enough times. (b) As in (a) for $b = 8/3$, $\sigma = 10$ and $r = 166$. For such a value of r the model is not chaotic and the correlation function does not decay. See Sec. 6.3 for a discussion about the Lorenz model for r slightly larger than 166.

computed along a generic trajectory of the Lorenz model for $r = 28$ (Fig. 3.11a) and for another value in which it is not chaotic (Fig. 3.11b). This function (see Box B.5 for a discussion on the precise meaning of Eq. (3.17)) measures the degree of "similarity" between the state at time $t + \tau$ with that at previous time t. For chaotic systems it quickly decreases toward 0, meaning completely different states (see inset of Fig. 3.11a). Therefore, in the presence of chaos, past is rapidly forgotten as typically happens in random phenomena. Thus, we must abandon the idea to describe a single trajectory in phase space and must consider the statistical properties of the set of all possible (or better the *typical*[16]) trajectories. With a *motto* we can say that we need to build a statistical mechanics description of chaos — this will be the subject of the next Chapter.

Predictability and sensitive dependence on initial conditions. All the previous examples share a common feature: a high degree of unpredictability is associated with erratic trajectories. This not only because they look random but mostly because infinitesimally small uncertainties on the initial state of the system grow very quickly — actually exponentially fast. In real world, this error amplification translates into our inability to predict the system behavior from the unavoidable imperfect knowledge of its initial state. The logistic map for $r = 4$ helped us a lot in having an intuition of the possible origin of such sensitivity on the initial conditions, but we need to define an operative and quantitative strategy for its characterization in generic systems. Stability theory introduced in the previous Chapter is insufficient in that respect, and will be generalized in Chapter 5, defining the Lyapunov exponents, which are the suitable indicators.

Fractal geometry. The set of points towards which the dynamics of chaotic dissipative systems is attracted can be rather complex, as in the Lorenz example (Fig. 3.6). The term *strange attractor* has indeed been coined to specify the

[16]The precise meaning of the term typical will become clear in the next Chapter.

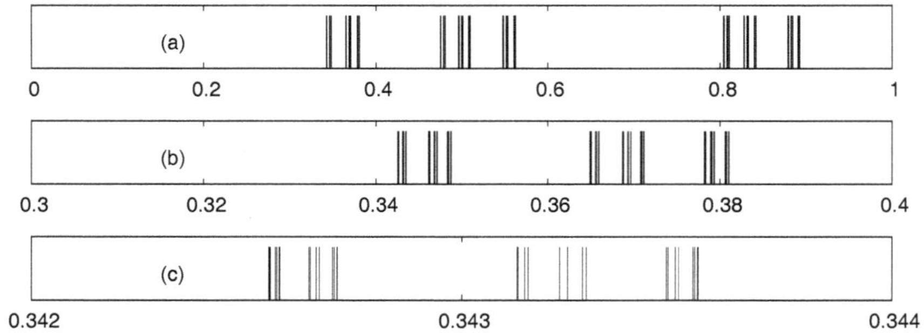

Fig. 3.12 (a) Feigenbaum strange attractor, obtained by plotting a vertical bar at each point $x \in [0:1]$ visited by the logistic map $x(n+1) = rx(n)(1-x(n))$ for $r = r_\infty = 3.569945\ldots$, which is the limiting value of the period doubling transition. (b) Zoom of region $[0.3:0.4]$. (c) Zoom of the region $[0.342:0.344]$. Note the self-similar structure. This set is non-chaotic as small displacements are not exponentially amplified. Further magnifications do not spoil the richness of structure of the attractor.

peculiarities of such a set. Sets as that of Fig. 3.6 are common to many nonlinear systems, and we need to understand how their geometrical properties can be characterized. However, it should be said from the outset that the existence of strange attracting sets is not at all a distinguishing feature of chaos. For instance, they are absent in chaotic Hamiltonian systems and can be present in non-chaotic dissipative systems. As an example of the latter we mention the logistic map for $r = r_\infty$, value at which the map possesses a "periodic" orbit of infinite period (basically meaning aperiodic) obtained as the limit of period-2^k orbits for $k \to \infty$. The set of points of such orbit is called Feigenbaum attractor, and is an example of strange non-chaotic attractor [Feudel *et al.* (2006)]. As clear from Fig. 3.12, Feigenbaum attractor is characterized by peculiar geometrical properties: even if the points of the orbits are infinite they occupy a zero measure set of the unit interval and display remarkable self-similar features revealed by magnifying the figure. As we will see in Chapter 5, fractal geometry constitutes the proper tool to characterize these strange chaotic Lorenz or non-chaotic Feigenbaum attractors.

Transition to chaos. Another important issue concerns the specific ways in which chaos sets in the evolution of nonlinear systems. In the logistic map and the Lorenz model (actually this is a generic feature of dissipative systems), chaos ends a series of bifurcations, in which fixed points and/or periodic orbits change their stability properties. On the contrary, in the Hénon-Heiles system, and generically in non-integrable conservative systems, at changing the nonlinearity control parameter there is not an abrupt transition to chaos as in dissipative systems: portion of the phase space characterized by chaotic motion grow in volume at the expenses of regular regions. Is any system becoming chaotic in a different way? What are the typical routes to chaos? Chapters 6 and 7 will be devoted to the transition to chaos in dissipative and Hamiltonian systems, respectively.

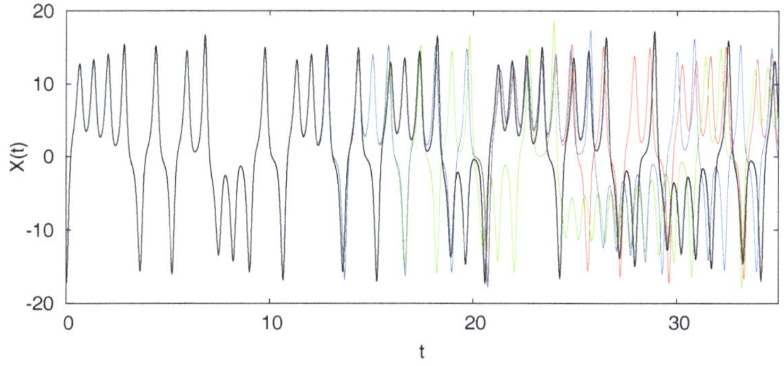

Fig. 3.13 $X(t)$ versus time for the Lorenz model at $r = 28$, $\sigma = 10$ and $b = 8/3$: in red the reference trajectory, in green that obtained by displacing of an infinitesimal amount the initial condition, in blue by a tiny change in the integration step with the same initial condition as in the reference trajectory, in black evolution of same initial condition of the red one but with r perturbed by a tiny amount.

Sensitivity to small changes in the evolution laws and numerical computation of chaotic trajectories. In discussing the logistic map, we have seen that, for $r \in [r_\infty : 4]$, small changes in r causes dramatic changes in the dynamics, as exemplified by the bifurcation diagram (Fig. 3.5). A small variation in the control parameter corresponds to a small change in the evolution law. It is then natural to wonder about the meaning of the evolution law, or technically speaking about the *structural stability* of nonlinear systems. In Fig. 3.13 we show four different trajectories of the Lorenz equations obtained introducing with respect to a reference trajectory an infinitesimal error on the initial condition, or on the integration step, or on the value of model parameters. The effects of the introduced error, regardless of where it is located, is very similar: all trajectories look the same for a while becoming macroscopically distinguishable after a time, which depends on the initial deviations from the reference trajectory or system. This example teaches us that the sensitivity is not only on the initial conditions but also on the evolution laws and on the algorithmic implementation of the models. These are issues which rise several questions about the possibility to employ such systems as model of natural phenomena and the relevance of chaos on experiments performed either in a laboratory or in *silico*, i.e. with a computer. Furthermore, how can we decide if a system is chaotic on the basis of experimental data? We shall discuss most of these issues in Chapter 10, in the second part of the book.

Box B.5: Correlation functions

A simple, but important and efficient way, to characterize a signal $x(t)$ is via its correlation (or auto-correlation) function $C(\tau)$. Assuming the system statistically stationary, we define the correlation function as

$$C(\tau) = \langle (x(t+\tau) - \langle x \rangle)(x(t) - \langle x \rangle) \rangle = \lim_{T\to\infty} \frac{1}{T} \int_0^T \mathrm{d}t\, x(t+\tau)x(t) - \langle x \rangle^2 ,$$

where

$$\langle x \rangle = \lim_{T\to\infty} \frac{1}{T} \int_0^T \mathrm{d}t\, x(t) .$$

In the case of discrete time systems a sum replaces the integral.

After Sec. 4.3, where the concept of ergodicity will be introduced, we will see that the brackets $\langle [\cdots] \rangle$ may indicate also averages over a suitable probability distribution.

The behavior of $C(\tau)$ gives a first indication of the character of the system. For periodic or quasiperiodic motion $C(\tau)$ cannot relax to zero: there exist arbitrarily long values of τ such that $C(\tau)$ is close to $C(0)$ as exemplified in (Fig. 3.11b). On the contrary, in systems whose behavior is "irregular", as in stochastic processes or in the presence of deterministic chaos, $C(\tau)$ approaches zero for large τ. When $0 < \int_0^\infty \mathrm{d}\tau\, C(\tau) = A < \infty$ one can define a characteristic time $\tau_c = A/C(0)$ characterizing the typical time scale over which the system "loses memory" of the past.[17] It is interesting, and important from an experimental point of view, to recall that, thanks to the Wiener-Khinchin theorem, the Fourier transform of the correlation function is the power spectral density, see Sec. 6.5.1.

3.5 Closing remark

We would like to close this Chapter by stressing that all the examples so far examined, which may look academical or, merely, intriguing mathematical toys, were originally considered for their relevance to real phenomena and, ultimately, for describing some aspects of Nature. For example, Lorenz starts the celebrated work on his model system with the following sentence

> *Certain hydrodynamical systems exhibit steady-state flow patterns, while other oscillate in a regular periodic fashion. Still others vary in an irregular, seemingly haphazard manner, and, even when observed for long periods of time do not appear to repeat their previous history.*

This quotation should warn the reader that, although we will often employ abstract mathematical models, the driving motivation for the study of chaos in physical sciences finds its roots in the necessity to explain naturally occurring phenomena.

3.6 Exercises

Exercise 3.1: Study the stability of the map $f(x) = 1 - ax^2$ at varying a with $x \in [-1:1]$, and numerically compute its bifurcation tree using the method described for the logistic map.

[17]The simplest instance is an exponential decay $C(\tau) = C(0)\mathrm{e}^{-\tau/\tau_c}$.

Hint: *Are you sure that you really need to make computations?*

Exercise 3.2: Consider the logistic map for $r^* = 1+\sqrt{8}$. Study the bifurcation diagram for $r > r^*$, which kind of bifurcation do you observe? What does happen at the trajectories of the logistic map for $r \lesssim r^*$ (e.g. $r = r^* - \epsilon$, with $\epsilon = 10^{-3}$, 10^{-4}, 10^{-5})? (If you find it curious look at the second question of Ex.3.4 and then to Ex.6.4).

Exercise 3.3: Numerically study the bifurcation diagram of the sin map $x(t+1) = r\sin(\pi x(t))$ for $r \in [0.6:1]$, is it similar to the one of the logistic map?

Exercise 3.4: Study the behavior of the trajectories (attractor shape, time series of $x(t)$ or $z(t)$) of the Lorenz system with $\sigma = 10$, $b = 8/3$ and let r vary in the regions:

(1) $r \in [145:166]$;

(2) $r \in [166:166.5]$ (after compare with the behavior of the logistic map seen in Ex.3.2);

(3) $r \approx 212$;

Exercise 3.5: Draw the attractor of the Rössler system

$$\frac{\mathrm{d}x}{\mathrm{d}t} = -y - z\,, \qquad \frac{\mathrm{d}y}{\mathrm{d}t} = x + ay \qquad \frac{\mathrm{d}z}{\mathrm{d}t} = b + z(x-c)$$

for $a=0.15$, $b=0.4$ and $c=8.5$. Check that also for this strange attractor there is sensitivity to initial conditions.

Exercise 3.6: Consider the two-dimensional map

$$x(t+1) = 1 - a|x(t)|^m + y(t)\,, \qquad y(t+1) = bx(t)$$

for $m = 1$ and $m = 2$ it reproduces the Hénon and Lozi map, respectively. Determine numerically the attractor generated with $(a = 1.4, b = 0.3)$ in the two cases. In particular, consider an ensemble initial conditions $(x^{(k)}(0), y^{(k)}(0))$, ($k = 1, \ldots, N$ with $N = 10^4$ or $N = 10^5$) uniformly distributed on a circle of radius $r = 10^{-2}$ centered in the point $(x_c, y_c) = (0,0)$. Plot the iterates of this ensemble of points at times $t = 1, 2, 3, \ldots$ and observe the relaxation onto the Hénon (Fig. 5.1) and Lozi attractors.

Exercise 3.7: Consider the following two-dimensional map

$$x(t+1) = y(t)\,, \qquad y(t+1) = -bx(t) + dy(t) - y^3(t)\,.$$

Display the different attractors in a plot $y(t)$ vs d, obtained by setting $b = 0.2$ and varying $d \in [2.0:2.8]$. Discuss the bifurcation diagram. In particular, examine the attractor at $d = 2.71$.

Exercise 3.8: Write a computer code to reproduce the Poincaré sections of the Hénon-Heiles system shown in Fig. 3.10.

Exercise 3.9: Consider the two-dimensional map [Hénon and Heiles (1964)]

$$x(t+1) = x(t) + a(y(t) - y^3(t))\,, \qquad y(t+1) = y(t) - a(x(t+1) - x^3(t+1))$$

show that it is symplectic and numerically study the behavior of the map for $a = 1.6$ choosing a set of initial conditions in $(x, y) \in [-1:1] \times [-1:1]$. Does the phase-portrait look similar to the Poincaré section of the Hénon-Heiles system?

Exercise 3.10: Consider the forced van der Pol oscillator

$$\frac{\mathrm{d}x}{\mathrm{d}t} = y\,, \qquad \frac{\mathrm{d}y}{\mathrm{d}t} = -x + \mu(1-x)y + A\cos(\omega_1 t)\cos(\omega_2 t)$$

Set $\mu = 5.0$, $F = 5.0$, $\omega_1 = \sqrt{2} + 1.05$. Determine numerically the asymptotic evolution of the system for $\omega_2 = 0.002$ and $\omega_2 = 0.0006$. Discuss the features of the two attractors by using a Poincaré section.
Hint: *Integrate numerically the system via a Runge-Kutta algorithm*

Exercise 3.11: Given the dynamical laws $x(t) = x_0 + x_1\cos(\omega_1 t) + x_2\cos(\omega_2 t)$, compute its auto-correlation function:

$$C(\tau) = \langle x(t)x(t+\tau)\rangle = \lim_{T\to\infty}\frac{1}{T}\int_0^T \mathrm{d}t\, x(t)x(t+\tau).$$

Hint: *Apply the definition and solve the integration over time.*

Exercise 3.12: Numerically compute numerically the correlation function $C(t) = \langle x(t)x(0)\rangle - \langle x(t)\rangle^2$ for:

(1) Hénon map (see Ex.3.6) with $a = 1.4, b = 0.3$;

(2) Lozi map (see Ex.3.6) with $a = 1.4, b = 0.3$;

(3) Standard map (see Eq. (2.18)) with $K = 8$, for a trajectory starting from the chaotic sea.

Chapter 4

Probabilistic Approach to Chaos

The true logic of the world is in the calculus of probabilities.
James Clerk Maxwell (1831-1879)

From an historical perspective, the first instance of necessity to use probability in deterministic systems was statistical mechanics. There, the probabilistic approach is imposed by the desire of extracting a few collective variables for the thermodynamic description of macroscopic bodies, composed by a huge number of (microscopic) degrees of freedom. Brownian motion epitomizes such a procedure: reducing the huge number ($\mathcal{O}(10^{23})$) of fluid molecules plus a colloidal particle to only the few degrees of freedom necessary for the description of the latter plus noise [Einstein (1956); Langevin (1908)].

In chaotic deterministic systems, the probabilistic description is not linked to the number of degrees of freedom (which can be just one as for the logistic map) but stems from the intrinsic erraticism of chaotic trajectories and the exponential amplification of small uncertainties, reducing the control on the system behavior.[1]

This Chapter will show that, in spite of the different specific rationales for the probabilistic treatment, deterministic and intrinsically random systems share many technical and conceptual aspects.

4.1 An informal probabilistic approach

In approaching the probabilistic description of chaotic systems, we can address two distinct questions that we illustrate by employing the logistic map (Sec. 3.1):

$$x(t+1) = f_r(x(t)) = r\,x(t)(1 - x(t))\,. \tag{4.1}$$

In particular, the two basic questions we can rise are:

[1]We do not enter here in the epistemological problem of the distinction between ontic (i.e. intrinsic to the nature of the system under investigation) and epistemic (i.e. depending on the lack of knowledge) interpretation of the probability in different physical cases [Primas (2002)].

(1) What is the probability to find the trajectory $x(t)$ in an infinitesimal segment $[x : x + \mathrm{d}x]$ of the unit interval? This amounts to study the probability density function (pdf) defined by

$$\rho(x; x(0)) = \lim_{T\to\infty} \frac{1}{T} \sum_{t=1}^{T} \delta(x - x(t)) , \tag{4.2}$$

which, in principle, may depend on the initial condition $x(0)$. On a computer, such a pdf can be obtained by partitioning the unit interval in N bins of size $\Delta x = 1/N$ and by measuring the number of times n_k such that $x(t)$ visit the k-th bin. Hence, the histogram is obtained from the frequencies:

$$\nu_k = \lim_{t\to\infty} \frac{n_k}{t} , \tag{4.3}$$

as shown, e.g., in Fig. 4.1a. The dependence on the initial condition $x(0)$ will be investigated in the following.
(2) Consider an ensemble of trajectories with initial conditions distributed according to an arbitrary probability $\rho_0(x)\mathrm{d}x$ to find $x(0)$ in $[x : x + \mathrm{d}x]$. Then the problem is to understand the time evolution[2] of the pdf $\rho_t(x)$ under the effect of the dynamics (4.1), i.e. to study

$$\rho_0(x) , \rho_1(x) , \rho_2(x) , \ldots , \rho_t(x) , \ldots , \tag{4.4}$$

an illustration of such an evolution is shown in Fig. 4.1b. Does $\rho_t(x)$ have a limit for $t \to \infty$ and, if so, how fast the limiting distribution $\rho_\infty(x)$ is approached? How does $\rho_\infty(x)$ depend on the initial density $\rho_0(x)$? and also is $\rho_\infty(x)$ related in some way to the density (4.2)?

Some of the features shown in Fig. 4.1 are rather generic and deserve a few comments. Figure 4.1b shows that, at least for the chosen $\rho_0(x)$, the limiting pdf $\rho_\infty(x)$ exists. It is obvious that, to be a limiting distribution of the sequence (4.4), $\rho_\infty(x)$ should be *invariant* under the action of the dynamics (4.1): $\rho_\infty(x) = \rho^{inv}(x)$. Figure 4.1b is also interesting as it shows that the *invariant density* is approached very quickly: $\rho_t(x)$ does not evolve much soon after the 3th or 4th iterate. Finally and remarkably, a direct comparison with Fig. 4.1a should convince the reader that $\rho^{inv}(x)$ is the same as the pdf obtained following the evolution of a single trajectory.

Actually the density obtained from (4.2) is invariant by construction, so that its coincidence with the limiting pdf of Fig. 4.1b sounds less surprising. However, in principle, the problem of the dependence on the initial condition is still present for both approach (1) and (2), making the above observation less trivial than it appears. We can understand this point with the following example. As seen in Sec. 3.1, also in the most chaotic case $r = 4$, the logistic map possesses infinitely many regular solutions in the form of unstable periodic orbits. Now suppose to

[2]This is a natural question for a system with sensitive dependence on the initial conditions: e.g., one is interested on the fate of a spot of points starting very close. In a more general context, we can consider any kind of initial distribution but $\rho_0(x) = \delta(x - x(0))$, as it would be equivalent to evolve a unique trajectory, i.e. $\rho_t(x) = \delta(x - x(t))$ for any t.

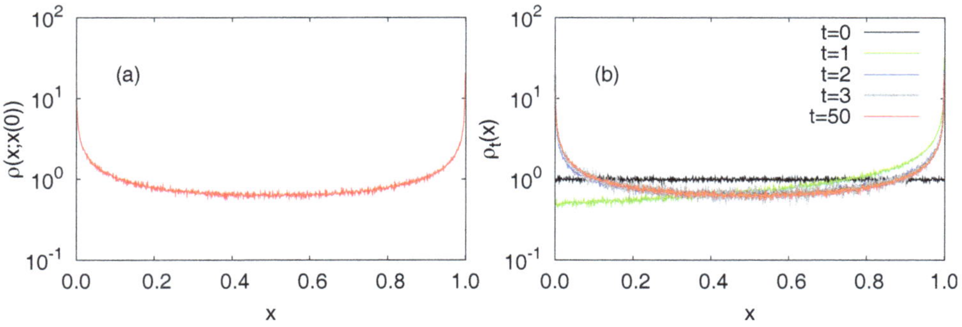

Fig. 4.1 (a) Histogram (4.3) for the logistic map at $r = 4$, obtained with 1000 bins of size $\Delta x = 10^{-3}$ and following for 10^7 iterations a trajectory starting from a generic $x(0)$ in $[0:1]$. (b) Time evolution of $\rho_t(x)$, $t=1,2,3$ and $t=50$ are represented. The histograms have been obtained by using 10^3 bins and $N=10^6$ trajectories with initial conditions uniformly distributed. Notice that for $t \geq 2-3$ $\rho_t(x)$ does not evolve much: ρ_3 and ρ_{50} are almost indistinguishable. A direct comparison with (a) shows that $\rho_\infty(x)$ coincides with $\rho(x;x(0))$.

study the problem (1) by choosing as initial condition a point $x(0) = x_0$ belonging to a period-n unstable orbit. This can be done by selecting as initial condition any solution of the equation $f_r^{(n)}(x) = x$ which is not solution of $f_r^{(k)}(x) = x$ for any $k < n$. It is easily seen that Eq. (4.2) assumes the form

$$\rho(x;x(0)) = \frac{\delta(x-x_0)+\delta(x-x_1)+\ldots+\delta(x-x_{n-1})}{n}, \tag{4.5}$$

where x_i, for $i = 0,\ldots,n-1$, defines the period-n orbit under consideration. Such a density is also invariant, as it is preserved by the dynamics.

The procedure leading to (4.5) can be repeated for any unstable periodic orbit of the logistic map. Moreover, any properly normalized linear combination of such invariant densities is still an invariant density. Therefore, there are many (infinite) invariant densities for the logistic map at $r = 4$. But the one shown in Fig. 4.1a is a special one: it did not require any fine tuning of the initial condition, and actually choosing any initial condition (but for those belonging to unstable periodic orbits) leads to the same density. Somehow, that depicted in Fig. 4.1a is the *natural density* selected by the dynamics and, as we will discuss in sequel, it cannot be obtained by any linear combination of other invariant densities. In the following we formalize the above observations which have general validity in chaotic systems.

We end this informal discussion showing the histogram (4.3) obtained from a generic initial condition of the logistic map at $r = 3.8$ (Fig. 4.2b), another value corresponding to chaotic behavior, and at $r = r_\infty$ (Fig. 4.2a), value at which an infinite period attracting orbit is realized (Fig. 3.12). These histograms appear very ragged due to the presence of singularities. In such circumstances, a density $\rho(x)$ cannot be defined and we can only speak about the measure $\mu(x)$ which, if sufficiently regular (differentiable almost everywhere), is related to ρ by $d\mu(x) =$

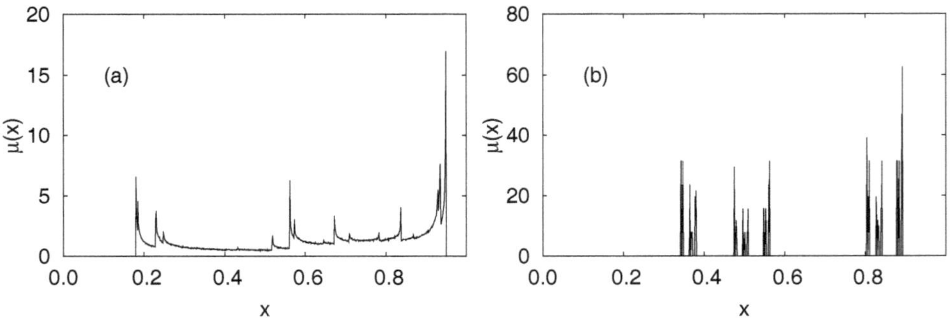

Fig. 4.2 (a) Histogram (4.3) for the logistic map at $r = 3.8$ with 1000 bins, obtained from a generic initial condition. Increasing the number of bins and the amount of data would increase the number of spikes and their heights. (b) Same as (a) for $r = r_\infty = 3.569945\ldots$.

$\rho(x)\mathrm{d}x$. At the Feigenbaum point r_∞, the support of the measure is a fractal set.[3] Measures singular with respect to the Lebesgue measure are indeed rather common in dissipative dynamical systems. Therefore, in the following, when appropriate, we will use the term *invariant measure* μ^{inv} instead of invariant density. Rigorously speaking, given a map $x(n+1) = f(x(n))$ the invariant measure μ^{inv} is defined by

$$\mu^{inv}(f^{-1}(B)) = \mu^{inv}(B) \quad \text{for any measurable set} \quad B\,, \tag{4.6}$$

meaning that the measure of the set B and that of its preimage[4] $f^{-1}(B) \equiv \{x \in f^{-1}(B) \text{ if } y = f(x) \in B\}$ should coincide.

4.2 Time evolution of the probability density

We can now reconsider more formally some of the observations made in the previous section. Let's start with a simple example, namely the Bernoulli map (3.9):

$$x(t+1) = g(x(t)) = \begin{cases} 2x(t) & 0 \le x(t) < 1/2 \\ 2x(t) - 1 & 1/2 \le x(t) \le 1\,, \end{cases}$$

which amplifies small errors by a factor 2 at each iteration (see Eq. (3.10)). How does an initial probability density $\rho_0(x)$ evolve in time?

First, we notice that given an initial density $\rho_0(x)$ for any set A of the unit interval, $A \subset [0:1]$, the probability $\mathrm{Prob}[x(0) \in A]$ is equal to the measure of the set, i.e. $\mathrm{Prob}[x(0) \in A] = \mu_0(A) = \int_A \mathrm{d}x\, \rho_0(x)$. Now, in order to answer the above question we can seek what is the probability to find the first iterate of the map $x(1)$ in a subset of the unit interval, i.e. $\mathrm{Prob}[x(1) \in B]$. As suggested by the simple construction of Fig. 4.3, we have

$$\mathrm{Prob}[x(1) \in B] = \mathrm{Prob}[x(0) \in B_1] + \mathrm{Prob}[x(0) \in B_2] \tag{4.7}$$

[3] See the discussion of Fig. 3.12 and Chapter 5.

[4] The use of the inverse map finds its rationale in the fact that the map may be non-invertible, see e.g. Fig. 4.3 and the related discussion.

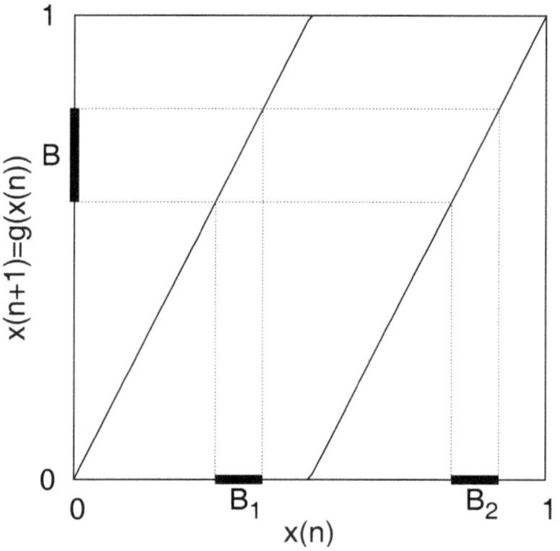

Fig. 4.3 Graphical method for finding the preimages B_1 and B_2 of the set B for the Bernoulli map. Notice that if x is the midpoint of the interval B, then $x/2$ and $x/2 + 1/2$ will be the midpoints of the intervals B_1 and B_2, respectively.

where B_1 and B_2 are the two preimages of B, i.e. if $x \in B_1$ or $x \in B_2$ then $g(x) \in B$. Taking $B \equiv [x : x + \Delta x]$ and performing the limit $\Delta x \to 0$, the above equation implies that the density evolves as

$$\rho_{t+1}(x) = \frac{1}{2}\rho_t\left(\frac{x}{2}\right) + \frac{1}{2}\rho_t\left(\frac{x}{2} + \frac{1}{2}\right) , \tag{4.8}$$

meaning that $x/2$ and $x/2+1/2$ are the preimages of x (see Fig. 4.3). From Eq. (4.8) it easily follows that if $\rho_0 = 1$ then $\rho_t = 1$ for all $t \geq 0$, in other terms the uniform distribution is an invariant density for the Bernoulli map, $\rho^{inv}(x) = 1$. By numerical studies similar to those represented in Fig. 4.1b, one can see that, for any generic $\rho_0(x)$, $\rho_t(x)$ evolves for $t \to \infty$ toward $\rho^{inv}(x) = 1$. This can be explicitly shown with the choice

$$\rho_0(x) = 1 + \alpha\left(x - \frac{1}{2}\right) \quad \text{with} \quad |\alpha| \leq 2 ,$$

for which Eq. (4.8) implies that

$$\rho_t(x) = 1 + \frac{1}{2^t}\alpha\left(x - \frac{1}{2}\right) = \rho^{inv}(x) + \mathcal{O}(2^{-t}) , \tag{4.9}$$

i.e. $\rho_t(x)$ converges to $\rho^{inv}(x) = 1$ exponentially fast.

For generic maps, $x(t+1) = f(x(t))$, Eq. (4.8) straightforwardly generalizes to:

$$\rho_{t+1}(x) = \int \mathrm{d}y\, \rho_t(y)\, \delta(x - f(y)) = \sum_k \frac{\rho_t(y_k)}{|f'(y_k)|} = \mathcal{L}_{\text{PF}}\rho_t(x) , \tag{4.10}$$

where the first equality is just the request that y is a preimage of x as made explicit in the second expression where y_k's are the solutions of $f(y_k) = x$ and f' indicates the derivative of f with respect to its argument. The last expression defines the Perron-Frobenius (PF) operator $\mathcal{L}_{PF}$ (see, e.g., Ruelle (1978b); Lasota and Mackey (1985); Beck and Schlögl (1997)), which is the linear[5] operator ruling the evolution of the probability density. The invariant density satisfies the equation

$$\mathcal{L}_{\rm PF}\,\rho^{inv}(x) = \rho^{inv}(x)\,, \tag{4.11}$$

meaning that $\rho^{inv}(x)$ is the eigenfunction with eigenvalue equal to 1 of the Perron-Frobenius operator. In general, $\mathcal{L}_{\rm PF}$ admits infinite eigenfunctions $\psi^{(k)}(x)$,

$$\mathcal{L}_{\rm PF}\,\psi^{(k)}(x) = \alpha_k \psi^{(k)}(x)\,,$$

with eigenvalues α_k, that can be complex. The generalization of the Perron-Frobenius theorem, originally formulated in the context of matrices,[6] asserts the existence of a real eigenvalue equal to unity, $\alpha_1 = 1$, associated to the invariant density, $\psi^{(1)}(x) = \rho^{inv}(x)$, and the other eigenvalues are such that $|\alpha_k| \leq 1$ for $k \geq 2$. Thus all eigenvalues belong to the unitary circle of the complex plane.[7]

For the case of PF-operators with a non-degenerate and discrete spectrum, it is rather easy to understand how the invariant density is approached. Assume that the eigenfunctions $\{\psi^{(k)}\}_{k=1}^{\infty}$, ordered according to the eigenvalues, form a complete basis, we can then express any initial density as a linear combination of them $\rho_0(x) = \rho^{inv}(x) + \sum_{k=2}^{\infty} A_k \psi^{(k)}(x)$ with the coefficients A_k such that $\rho_0(x)$ is real and non-negative for any x. The density at time t can thus be related to that at time $t = 0$ by

$$\rho_t(x) = \mathcal{L}^t_{PF}\rho_0(x) = \rho^{inv}(x) + \sum_{k=2}^{\infty} A_k \alpha_k^t \psi^{(k)}(x) = \rho^{inv}(x) + \mathcal{O}\left(\mathrm{e}^{-t\ln\left|\frac{1}{\alpha_2}\right|}\right)\,, \tag{4.12}$$

where $\mathcal{L}^t_{PF}$ indicates t successive applications of the operator. Such an expression conveys two important pieces of information: (i) independently of the initial condition $\rho_t \to \rho^{inv}$ and (ii) the convergence is exponentially fast with the rate $-\ln|1/\alpha_2|$. From Eq. (4.9) and Eq. (4.12), one recognizes that $\alpha_2 = 1/2$ for the Bernoulli map.

What does happen when the dynamics of the map is regular? In this case, for typical initial conditions, the Perron-Frobenius dynamics may be either attracted by a unique invariant density or may never converge to a limiting distribution, exhibiting a periodic or quasiperiodic behavior. For instance, this can be understood by considering the logistic map for $r < r_\infty$, where period-2^k orbits are stable. Recalling the results of Sec. 3.1, the following scenario arises. For $r < 3$, there is a unique attracting fixed point x^* and thus, for large times

$$\rho_t(x) \to \delta(x - x^*)\,,$$

[5]One can easily see that $\mathcal{L}_{\rm PF}(a\rho_1 + b\rho_2) = a\mathcal{L}_{\rm PF}\rho_1 + b\mathcal{L}_{\rm PF}\rho_2$.

[6]The matrix formulation naturally appear in the context of random processes known as Markov Chains, whose properties are very similar (but in the stochastic world) to those of deterministic dynamical systems, see Box B.6 for a brief discussion highlighting these similarities.

[7]Under some conditions it is possible to prove that, for $k \geq 2$, $|\alpha_k| < 1$ strictly, which is a very useful and important result as we will see below.

independently of $\rho_0(x)$. For $r_{n-1} < r < r_n$, the trajectories are attracted by a period-2^n orbit $x^{(1)}, x^{(2)}, \cdots x^{(2^n)}$, so that after a transient

$$\rho_t(x) = \sum_{k=1}^{2^n} c_k(t)\delta(x - x^{(k)}) \,,$$

where $c_1(t), c_2(t), \cdots, c_{2^n}(t)$ evolve in a cyclic way, i.e.: $c_1(t+1) = c_{2^n}(t)$; $c_2(t+1) = c_1(t)$; $c_3(t+1) = c_2(t)$; $\cdots$ and depend on $\rho_0(x)$. Clearly, for $n \to \infty$, i.e. in the case of the Feigenbaum attractor, the PF-operator is not even periodic as the orbit has an infinite period.

We can summarize the results as follows: regular dynamics entails $\rho_t(x)$ not forgetting the initial density $\rho_0(x)$ while chaotic dynamics are characterized by densities relaxing to a well-defined and unique invariant density $\rho^{inv}(x)$, moreover typically the convergence is exponentially fast.

We conclude this section by explicitly deriving the invariant density for the logistic map at $r = 4$. The idea is to exploit its topological conjugation with the tent map (Sec. 3.1). The PF-operator takes a simple form also for the tent map

$$y(t+1) = g(y(t)) = 1 - 2|y(t) - 1/2| \,.$$

A construction similar to that of Fig. 4.3 shows that the equivalent of (4.8) reads

$$\rho_{t+1}(y) = \frac{1}{2}\rho_t\left(\frac{y}{2}\right) + \frac{1}{2}\rho_t\left(1 - \frac{y}{2}\right) \,,$$

for which $\rho^{inv}(y) = 1$. We should now recall that tent and logistic map at the Ulam point $x(t+1) = f(x(t)) = 4x(t)(1 - x(t))$ are topologically conjugated (Box B.3) through the change of variables $y = h(x)$ whose inverse is (see Sec. 3.1)

$$x = h^{(-1)}(y) = \frac{1 - \cos(\pi y)}{2} \,. \tag{4.13}$$

As discussed in the Box B.3, the dynamical properties of the two maps are not independent. In particular, the invariant densities are related to each other through the change of variable, namely: if $y = h(x)$, from $\rho^{inv}_{(x)}(x)\mathrm{d}x = \rho^{inv}_{(y)}(y)\mathrm{d}y$ then

$$\rho^{inv}_{(y)}(y) = \left|\frac{\mathrm{d}h}{\mathrm{d}x}\right|^{-1} \rho^{inv}_{(x)}(x = h^{(-1)}(y))$$

where $\mathrm{d}h/\mathrm{d}x$ is evaluated at $x = h^{(-1)}(y)$. For the tent map $\rho^{inv}_{(y)}(y) = 1$ so that, from the above formula and using (4.13), after some simple algebra, one finds

$$\rho^{inv}_{(x)}(x) = \frac{1}{\pi\sqrt{x(1-x)}} \,, \tag{4.14}$$

which is exactly the density we found numerically as a limiting distribution in Fig. 4.1b. Moreover, we can analytically study how the initial density $\rho_0(x) = 1$ approaches the invariant one, as in Fig. 4.1b. Solving Eq. (4.10) for $t = 1, 2$ the density is given by

$$\rho_1(x) = \frac{1}{2\sqrt{1-x}}$$

$$\rho_2(x) = \frac{\sqrt{2}}{8\sqrt{1-x}}\left[\frac{1}{\sqrt{1+\sqrt{1-x}}} + \frac{1}{\sqrt{1-\sqrt{1-x}}}\right] \,,$$

these two steps describe the evolution obtained numerically in Fig. 4.1b. For $t = 2$, $\rho_2 \approx \rho^{inv}$ apart from very small deviations. Actually, we know from Eq. (4.12) that the invariant density is approached exponentially fast.

General formulation of the problem

The generalization of the Perron-Frobenius formalism to d-dimensional maps

$$\boldsymbol{x}(t+1) = \boldsymbol{g}(\boldsymbol{x}(t)) \ ,$$

straightforwardly gives

$$\rho_{t+1}(\boldsymbol{x}) = \mathcal{L}_{PF}\,\rho_t(\boldsymbol{x}) = \int \mathrm{d}\boldsymbol{y}\ \rho_t(\boldsymbol{y})\delta(\boldsymbol{x} - \boldsymbol{g}(\boldsymbol{y})) = \sum_k \frac{\rho_t(\boldsymbol{y}_k)}{|\det[\mathbb{L}(\boldsymbol{y}_k)]|} \tag{4.15}$$

where $\boldsymbol{g}(\boldsymbol{y}_k) = \boldsymbol{x}$, and $\mathbb{L}_{ij} = \partial g_i/\partial x_j$ is the stability matrix (Sec. 2.4).

For time continuous dynamical systems described by a set of ODEs

$$\frac{\mathrm{d}\boldsymbol{x}}{\mathrm{d}t} = \boldsymbol{f}(\boldsymbol{x}) \ , \tag{4.16}$$

the evolution of a density $\rho(\boldsymbol{x}, t)$ is given by Eq. (2.4), which we rewrite here as

$$\frac{\partial \rho}{\partial t} = \mathcal{L}_L\,\rho(\boldsymbol{x}, t) = -\nabla \cdot (\boldsymbol{f}\rho(\boldsymbol{x}, t)) \tag{4.17}$$

where $\mathcal{L}_L$ is the Liouville operator, see e.g. Lasota and Mackey (1985). In this case the invariant density can be found solving by

$$\mathcal{L}_L\,\rho^{inv}(\boldsymbol{x}, t) = 0 \, .$$

Equations (4.15) and (4.17) rule the evolution of probability densities of a generic deterministic time-discrete or -continuous dynamical systems, respectively. As for the logistic map, the behavior of $\rho_t(\boldsymbol{x})$ (or $\rho(\boldsymbol{x}, t)$) depends on the specific dynamics, in particular, on whether the system is chaotic or not.

We conclude by noticing that for the evolution of densities, but not only, chaotic systems share many formal similarities with stochastic processes known as Markov Processes [Feller (1968)], see Box B.6 and Sec. 4.5.

Box B.6: Markov Processes

A: Finite states Markov Chains

A Markov chain (MC), after the Russian mathematician A.A. Markov, is one of the simplest example of nontrivial, discrete-time and discrete-state stochastic processes. We consider a random variable x_t which, at any discrete time t, may assume S possible values (states) $X_1, ..., X_S$. In the sequel, to ease the notation, we shall indicate with i the state

X_i. Such a process is a Markov chain if it verifies the Markov property: every future state is conditionally independent of every prior state but the present one, in formulae

$$\mathrm{Prob}(x_n = i_n | x_{n-1} = i_{n-1}, \ldots, x_{n-k} = i_{n-k}, \ldots) = \mathrm{Prob}(x_n = i_n | x_{n-1} = i_{n-1}) \,, \qquad \text{(B.6.1)}$$

for any n, where $i_n = 1, \ldots, S$. In other words the jump from the state $x_t = X_i$ to $x_{t+1} = X_j$ takes place with probability $\mathrm{Prob}(x_{t+1} = j | x_t = i) = p(j|i)$ independently of the previous history. At this level $p(j|i)$ may depend on the time t. We restrict the discussion to time-homogeneous Markov chains which, as we will see, are completely characterized by the time-independent, single-step transition matrix $\mathbb{W}$ with elements[8]

$$\mathbb{W}_{jk} = p(j|k) = \mathrm{Prob}(x_{t+1} = j | x_t = k) \,,$$

such that $\mathbb{W}_{ij} \geq 0$ and $\sum_{i=1}^{S} \mathbb{W}_{ij} = 1$. For instance, consider the two states MC defined by the transition matrix:

$$\mathbb{W} = \begin{pmatrix} p & 1-q \\ 1-p & q \end{pmatrix} \qquad \text{(B.6.2)}$$

with $p, q \in [0:1]$. Any MC admits a weighted graph representation (see, e.g., Fig. B6.1), often very useful to visualize the properties of Markov chains.

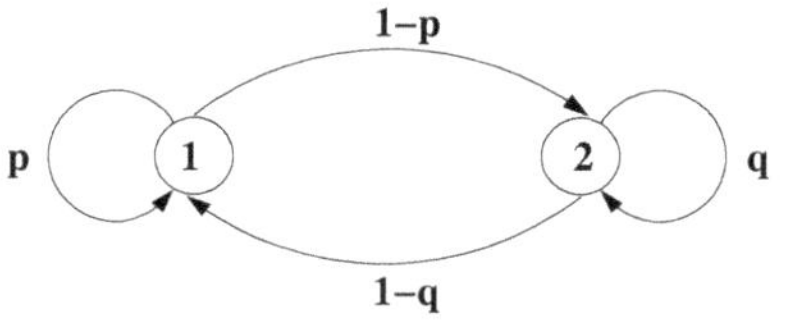

Fig. B6.1 Graph representation of the MC (B.6.2). The states are the nodes and the links between nodes, when present, are weighted with the transition probabilities.

Thanks to the Markov property (B.6.1), the knowledge of $\mathbb{W}$ (i.e. of the probabilities $\mathbb{W}_{ij}$ to jump from state j to state i in one-step) is sufficient to determine the n-step transition probability, which is given by the so-called Chapman-Kolmogorov equation

$$\mathrm{Prob}(x_n = j | x_0 = i) = \sum_{r=1}^{S} \mathbb{W}^k_{jr} \mathbb{W}^{n-k}_{ri} = \mathbb{W}^n_{ji} \qquad \text{for any} \quad 0 \leq k \leq n$$

where $\mathbb{W}^n$ denotes the n-power of the matrix. It is useful to briefly review the basic classification of Markov Chains. According to the structure of the transition matrix, the states of a Markov Chain can be classified in *transient* if a finite probability exists that a given state, once visited by the random process, will never be visited again, or *recurrent* if with probability one it is visited again. The latter class is then divided in *null* or *non-null* depending on whether the mean recurrence time is infinite or finite, respectively. Recurrent non-null states can be either *periodic* or *aperiodic*. The state is said periodic if the probability to come back to it in k-steps is null unless k is multiple of a given value T, which is the period of such a state, otherwise it is said aperiodic. A recurrent, non-null aperiodic state is called *ergodic*. Then we distinguish between *irreducible* (indecomposable)

[8] Usually in books of probability theory, such as Feller (1968), $\mathbb{W}_{ij}$ is the transpose of what is called transition matrix.

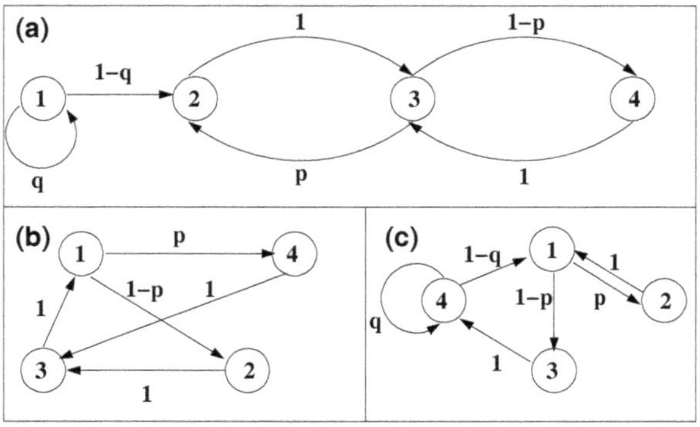

Fig. B6.2 Three examples of MC with 4 states. (a) Reducible MC where state 1 is transient and $2, 3, 4$ are recurrent and periodic with period 2. (b) Period-3 irreducible MC. (c) Ergodic irreducible MC. In all examples $p, q \neq 0, 1$.

and *reducible* (decomposable) Markov Chains according to the fact that each state is accessible from any other or not. The property of being *accessible*, in practice, means that there exists a $k \geq 1$ such that $\mathbb{W}^k_{ij} > 0$ for each i, j. The notion of irreducibility is important in virtue of a theorem (see, e.g., Feller, 1968) stating that the states of an irreducible chain are all of the same kind. Therefore, we shall call a MC ergodic if it is irreducible and its states are ergodic. Figure B6.1 is an example of ergodic irreducible MC with two states, other examples of MC are shown in Fig. B6.2.

Consider now an ensemble of random variables all evolving with the same transition matrix, analogously to what has been done for the logistic map, we can investigate the evolution of the probability $P_j(t) = \text{Prob}(X_t = j)$ to find the random variable in state j at time t. The time-evolution for such a probability is obtained from Eq. (B.6.1):

$$P_j(t) = \sum_{k=1}^{S} \mathbb{W}_{jk} P_k(t-1)\,, \tag{B.6.3}$$

i.e. the probability to be in j at time t is equal to the probability to have been in k at $t-1$ times the probability to jump from k to j summed over all the possible previous states k. Equation (B.6.3) takes a particularly simple form introducing the column vector $\boldsymbol{P}(t) = (P_1(t), .., P_S(t))$, and using the matrix notation

$$\boldsymbol{P}(t) = \mathbb{W}\boldsymbol{P}(t-1) \quad \Longrightarrow \quad \boldsymbol{P}(t) = \mathbb{W}^t \boldsymbol{P}(0)\,. \tag{B.6.4}$$

A question of obvious relevance concerns the convergence of the probability vector $\boldsymbol{P}(t)$ to a certain limit and, if so, whether such a limit is unique. Of course, if such limit exists, it is the invariant (or equilibrium) probability $\boldsymbol{P}^{inv}$ that satisfies the equation

$$\boldsymbol{P}^{inv} = \mathbb{W}\boldsymbol{P}^{inv}\,, \tag{B.6.5}$$

i.e. it is the eigenvector of the matrix $\mathbb{W}$ with eigenvalue equal to unity.
The following important theorem holds:

For an irreducible ergodic Markov Chain, the limit

$$\boldsymbol{P}(t) = \mathbb{W}^t \boldsymbol{P}(0) \to \boldsymbol{P}(\infty) \quad \textit{for} \quad t \to \infty\,,$$

exists and is unique – independent of the initial distribution. Moreover, $\boldsymbol{P}(\infty) = \boldsymbol{P}^{inv}$ *and satisfies Eq. (B.6.5), i.e.* $\boldsymbol{P}^{inv} = \mathbb{W}\boldsymbol{P}^{inv}$ *, meaning that the limit probability is invariant (stationary). [Notice that for irreducible periodic MC the invariant distribution exists and is unique, but it does not exist the limit* $\boldsymbol{P}(\infty)$*.]*

The convergence of $\boldsymbol{P}(t)$ towards $\boldsymbol{P}^{inv}$ is exponentially fast:

$$\boldsymbol{P}(t) = \mathbb{W}^t\boldsymbol{P}(0) = \boldsymbol{P}^{inv} + \mathcal{O}(|\alpha_2|^t) \text{ and } \mathbb{W}^t_{ij} = P^{inv}_i + \mathcal{O}(|\alpha_2|^t) \qquad \text{(B.6.6)}$$

where[9] α_2 is the second eigenvalue of $\mathbb{W}$. Equation (B.6.6) can be derived following step by step the procedure which lead to Eq. (4.12).

The above results can be extended to understand the behavior of the correlation function between two generic functions g and h defined on the states of the Markov Chain, $C_{gh}(t) = \langle g(x_{(t_0+t)})h(x_{t_0})\rangle = \langle g(x_t)h(x_0)\rangle$, which for stationary MC only depends on the time lapse t. The average $\langle[\ldots]\rangle$ is performed over the realizations of the Markov Chain, that is on the equilibrium probability $\boldsymbol{P}^{inv}$. The correlation function $C_{gh}(t)$ can be written in terms of $\mathbb{W}^n$ and $\boldsymbol{P}^{inv}$ and, moreover, can be shown to decay exponentially

$$C_{gh}(t) = \langle g(x)\rangle\langle h(x)\rangle + \mathcal{O}\left(\mathrm{e}^{-\frac{t}{\tau_c}}\right) , \qquad \text{(B.6.7)}$$

where in analogy to Eq. (B.6.6) $\tau_c = 1/\ln(1/|\alpha_2|)$ as we show in the following. By denoting $g_i = g(x_t = i)$ and $h_i = h(x_t = i)$, the correlation function can be explicitly written as

$$\langle g(x_t)h(x_0)\rangle = \sum_{i,j} P^{inv}_j \, h_j \mathbb{W}^t_{ij} \, g_i \, ,$$

so that from Eq. (B.6.6)

$$\langle g(x_t)h(x_0)\rangle = \sum_{i,j} P^{inv}_i P^{inv}_j g_j h_i + \mathcal{O}(|\alpha_2|^t) \, ,$$

and finally Eq. (B.6.7) follows, noting that $\sum_{i,j} P^{inv}_i P^{inv}_j g_j h_i = \langle g(x)\rangle\langle h(x)\rangle$.

B: Continuous Markov processes

The Markov property (B.6.1) can be generalized to a N-dimensional continuous stochastic process $\boldsymbol{x}(t) = (x_1(t), \ldots, x_N(t))$, where the variables $\{x_j\}$'s and time t are continuous valued. In particular, Eq. (B.6.1) can be stated as follows. For any sequence of times $t_1, \ldots t_n$ such that $t_1 < t_2 < \ldots < t_n$, and given the values of the random variable $\boldsymbol{x}^{(1)}$, $\ldots$, $\boldsymbol{x}^{(n-1)}$ at times $t_1, \ldots, t_{n-1}$, the probability $w_n(\boldsymbol{x}^{(n)}, t_n|\boldsymbol{x}^{(1)}, t_1, ..., \boldsymbol{x}^{(n-1)}, t_{n-1})\,\mathrm{d}\boldsymbol{x}$ that at time t_n $x_j(t_n) \in [x_j : x_j + \mathrm{d}x_j]$ (for each j) is only determined by the present $\boldsymbol{x}^{(n)}$ and the previous state $\boldsymbol{x}^{(n-1)}$, i.e. it reduces to $w_2(\boldsymbol{x}^{(n)}, t_n|\boldsymbol{x}^{(n-1)}, t_{n-1})$ in formulae

$$w_n(\boldsymbol{x}^{(n)}, t_n|\boldsymbol{x}^{(1)}, t_1, ..., \boldsymbol{x}^{(n-1)}, t_{n-1}) = w_2(\boldsymbol{x}^{(n)}, t_n|\boldsymbol{x}^{(n-1)}, t_{n-1}) \, . \qquad \text{(B.6.8)}$$

[9]We ordered the eigenvalues α_k as follows: $\alpha_1 = 1 > |\alpha_2| \geq |\alpha_3| \ldots$. We remind that in an ergodic MC $|\alpha_2| < 1$, as consequence of the the Perron-Frobenius theorem on the non degeneracy of the first (in absolute value) eigenvalue of a matrix with real positive elements [Grimmett and Stirzaker (2001)].

For time stationary processes the conditional probability $w_2(\boldsymbol{x}^{(n)}, t_n|\boldsymbol{x}^{(n-1)}, t_{n-1})$ only depends on the time difference $t_n - t_{n-1}$ so that, in the following, we will use the notation $w_2(\boldsymbol{x}, t|\boldsymbol{y})$ for $w_2(\boldsymbol{x}, t|\boldsymbol{y}, 0)$.

Analogously to finite state MC, the probability density function $\rho(\boldsymbol{x}, t)$ at time t can be expressed in terms of its initial condition $\rho(\boldsymbol{x}, 0)$ and the transition probability $w_2(\boldsymbol{x}, t|\boldsymbol{y})$:

$$\rho(\boldsymbol{x}, t) = \int \mathrm{d}\boldsymbol{y}\, w_2(\boldsymbol{x}, t|\boldsymbol{y})\, \rho(\boldsymbol{y}, 0)\,, \tag{B.6.9}$$

and from Eq. (B.6.8) it follows the Chapman-Kolmogorov equation

$$w_2(\boldsymbol{x}, t|\boldsymbol{y}) = \int \mathrm{d}\boldsymbol{z}\, w_2(\boldsymbol{x}, t - t_0|\boldsymbol{z}) w_2(\boldsymbol{z}, t_0|\boldsymbol{y}) \tag{B.6.10}$$

stating that the probability to have a transition from state $\boldsymbol{y}$ at time 0 to $\boldsymbol{x}$ at time t can be obtained integrating over all possible intermediate transitions $\boldsymbol{y} \to \boldsymbol{z} \to \boldsymbol{x}$ at any time $0 < t_0 < t$.

An important class of Markov processes is represented by those processes in which to an infinitesimal time interval Δt corresponds a infinitesimal displacement $\boldsymbol{x} - \boldsymbol{y}$ having the following properties

$$a_j(\boldsymbol{x}, \Delta t) = \int \mathrm{d}\boldsymbol{y}\, (y_j - x_j) w_2(\boldsymbol{y}, \Delta t|\boldsymbol{x}) = \mathcal{O}(\Delta t) \tag{B.6.11}$$

$$b_{ij}(\boldsymbol{x}, \Delta t) = \int \mathrm{d}\boldsymbol{y}\, (y_j - x_j)(y_i - x_i) w_2(\boldsymbol{y}, \Delta t|\boldsymbol{x}) = \mathcal{O}(\Delta t)\ , \tag{B.6.12}$$

while higher order terms are negligible

$$\int \mathrm{d}\boldsymbol{y}\, (y_j - x_j)^n w_2(\boldsymbol{y}, \Delta t|\boldsymbol{x}) = \mathcal{O}(\Delta t^k) \text{ with } k > 1 \text{ for } n \geq 3\,. \tag{B.6.13}$$

As the functions a_j and b_{ij} are both proportional to Δt, it is convenient to introduce

$$f_j(\boldsymbol{x}) = \lim_{\Delta t \to 0} \frac{1}{\Delta t} a_j(\boldsymbol{x}, \Delta t) \quad \text{and} \quad Q_{ij}(\boldsymbol{x}) = \lim_{\Delta t \to 0} \frac{1}{\Delta t} b_{ij}(\boldsymbol{x}, \Delta t)\ . \tag{B.6.14}$$

Then, from a Taylor expansion in $\boldsymbol{x} - \boldsymbol{y}$ of Eq. (B.6.10) with $t_0 = \Delta t$ and using Eqs. (B.6.11)–(B.6.14) we obtain the Fokker-Planck equation

$$\frac{\partial w_2}{\partial t} = -\sum_j \frac{\partial}{\partial x_j}\left[f_j w_2\right] + \frac{1}{2}\sum_{ij} \frac{\partial^2}{\partial x_j \partial x_i}\left[Q_{ij} w_2\right]\,, \tag{B.6.15}$$

which also rules the evolution of $\rho(\boldsymbol{x}, t)$, as follows from Eq. (B.6.9).

The Fokker-Planck equation can be linked to a stochastic differential equation — the Langevin equation. In particular, for the case in which Q_{ij} does not depend on $\boldsymbol{x}$, one can easily verify that Eq. (B.6.15) rules the evolution of the density associated to stochastic process

$$x_j(t + \Delta t) = x_j(t) + f_j(x(t))\Delta t + \sqrt{\Delta t}\, \eta_j(t)\,,$$

where $\eta_j(t)$'s are Gaussian distributed with $\langle \eta_j(t) \rangle = 0$ and $\langle \eta_j(t + n\Delta t)\eta_i(t + m\Delta t) \rangle = Q_{ij}\delta_{nm}$. Formally, we can perform the limit $\Delta t \to 0$, leading to the Langevin equation

$$\frac{\mathrm{d}x_j}{\mathrm{d}t} = f_j(\boldsymbol{x}) + \eta_j(t)\,, \tag{B.6.16}$$

where $j = 1, \cdots, N$ and $\eta_j(t)$ is a multi-variate Gaussian white noise, i.e. $\langle \eta_j(t) \rangle = 0$ and $\langle \eta_j(t)\eta_i(t') \rangle = Q_{ij}\delta(t - t')$, where the covariance matrix $\{Q_{ij}\}$ is positive definite [Chandrasekhar (1943)].

C: Dynamical systems with additive noise

The connection between Markov processes and dynamical systems is evident if we consider Eq. (4.16) with the addition of a white noise term $\{\eta_j\}$, so that it becomes a Langevin equation as Eq. (B.6.16). In this case, for the evolution of the probability density Eq. (4.17) is replaced by [Gardiner (1982)]

$$\frac{\partial \rho}{\partial t} = \mathcal{L}_L \rho + \frac{1}{2}\sum_{ij} Q_{ij}\frac{\partial^2 \rho}{\partial x_j \partial x_j}\,,$$

where the symmetric matrix $\{Q_{ij}\}$, as discussed above, depends on the correlations among the $\{\eta_i\}$'s. In other terms the Liouville operator is replaced by the Fokker-Planck operator:

$$\mathcal{L}_{FP} = \mathcal{L}_L + \frac{1}{2}\sum_{ij} Q_{ij}\frac{\partial^2}{\partial x_j \partial x_j}\,.$$

Physically speaking, one can think about the noise $\{\eta_j(t)\}$ as a way to emulate the effects of fast internal dynamics, as in Brownian motion or in noisy electric circuits.

For the sake of completeness, we briefly discuss the modification of the Perron-Frobenius operator for noisy maps $x(t + 1) = g(x(t)) + \eta(t)$, being $\{\eta(t)\}$ a stationary stochastic process with zero average and pdf $P_\eta(\eta)$. Equation (4.7) modifies in

$$\mathcal{L}_{PF}\rho_t(x) = \int \mathrm{d}y\mathrm{d}\eta\, \rho_t(y)P_\eta(\eta)\delta(x - g(y) - \eta) = \sum_k \int \mathrm{d}\eta\, \frac{\rho_t(y_k(\eta))}{|g'(y_k(\eta))|}P_\eta(\eta)\,,$$

where $y_k(\eta)$ are the points such that $g(y_k(\eta)) = x - \eta$.

In Sec. 4.5 we shall see that the connection between chaotic maps and Markov processes goes much further than the mere formal similarity.

4.3 Ergodicity

In Section 4.1 we left unexplained the coincidence of the invariant density obtained by following a generic trajectory of the logistic map at $r = 4$ with the limit distribution Eq. (4.14), obtained iterating the Perron-Frobenius operator (see Fig. 4.1). This is a generic and important property shared by a very large class of chaotic systems, standing at the core of the *ergodic and mixing problems*, which we explore in this Section.

4.3.1 *An historical interlude on ergodic theory*

Ergodic theory began with Boltzmann's attempt, in kinetic theory, at justifying the equivalence of theoretical expected values (ensemble or phase averages) and experimentally measured ones, computed as "infinite" time averages. Modern ergodic theory can be viewed as a branch of abstract theory of measure and integration, and its aim goes far beyond the original formulation of Boltzmann. In a nutshell Boltzmann's program was to derive thermodynamics from the knowledge of the microscopic laws ruling the huge number of degrees of freedom composing a macroscopic system as, e.g. a gas with $N \approx \mathcal{O}(10^{23})$ molecules (particles).

In the dynamical system framework, we can formulate the problem as follows. Let $\boldsymbol{q}_i$ and $\boldsymbol{p}_i$ be the position and momentum vectors of the i-th particle, the microscopic state of a N-particle system, at time t, is given by the vector $\boldsymbol{x}(t) \equiv (\boldsymbol{q}_1(t), \ldots, \boldsymbol{q}_N(t); \boldsymbol{p}_1(t), \ldots, \boldsymbol{p}_N(t))$ in a $6\,N$-dimensional phase space Γ (we assume that the gas is in the three-dimensional Euclidean space). Then, microscopic evolution follows from Hamilton's equations (Chap. 2). Thermodynamics consists in passing from $6N$ degrees of freedom to a few macroscopic parameters such as, for instance, the temperature or the pressure, which can be experimentally accessed through time averages. Such averages are typically performed on a macroscopic time scale $\mathcal{T}$ (the observation time window) much larger than the microscopic time scale characterizing fast molecular motions. This means that an experimental measurement is actually the result of a single observation during which the system explores a huge number of microscopic states. Formally, given a macroscopic observable Φ, depending on the microscopic state $\boldsymbol{x}$, we have to compute

$$\overline{\Phi}^{\mathcal{T}}(\boldsymbol{x}(0)) = \frac{1}{\mathcal{T}} \int_{t_0}^{t_0+\mathcal{T}} \mathrm{d}t\, \Phi(\boldsymbol{x}(t)) \,.$$

For example, the temperature of a gas corresponds to choosing $\Phi = \frac{1}{N}\sum_{i=1}^{N} p_i^2/m$. In principle, computing $\overline{\Phi}^{\mathcal{T}}$ requires both the knowledge of the complete microscopic state of the system at a given time and the determination of its trajectory. It is evident that this an impossible task. Moreover, even if such an integration could be possible, the outcome $\overline{\Phi}^{\mathcal{T}}$ may presumably depend on the initial condition, making meaningless even statistical predictions.

The *ergodic hypothesis* allows this obstacle to be overcome. The trajectories of the energy conserving Hamiltonian system constituted by the N molecules evolve onto the $(6N-1)$-dimensional hypersurface $H = E$. The invariant measure for the microstates $\boldsymbol{x}$ can be written as $\mathrm{d}^{6N}\boldsymbol{x}\, \delta(E - H(\boldsymbol{x}))$, that is the microcanonical measure $\mathrm{d}\mu_{mc}$ which, by deriving the δ-function, can be equivalently written as

$$\mathrm{d}\mu_{mc}(\boldsymbol{x}) = \frac{\mathrm{d}\Sigma(\boldsymbol{x})}{|\nabla H|} \,,$$

where $d\Sigma$ is the energy-constant hypersurface element and the vector $\boldsymbol{\nabla} H = (\partial_{\boldsymbol{q}_1} H, ..., \partial_{\boldsymbol{q}_N} H; \partial_{\boldsymbol{p}_1} H, ..., \partial_{\boldsymbol{p}_N} H)$. The microcanonical is the invariant measure for any Hamiltonian system. The ergodic hypothesis consists in assuming that

$$\overline{\Phi} \equiv \lim_{\mathcal{T}\to\infty} \frac{1}{\mathcal{T}} \int_{t_0}^{t_0+\mathcal{T}} dt\ \Phi(\boldsymbol{x}(t)) = \int_{\Gamma} d\mu_{mc}(\boldsymbol{x})\ \Phi(\boldsymbol{x}) \equiv \langle \Phi \rangle\,, \tag{4.18}$$

i.e. that the time average is independent of the initial condition and coincides with the ensemble average. Whether (4.18) is valid or not, i.e. if it is possible to substitute the temporal average with an average performed in terms of the microcanonical measure, stays at the core of the ergodic problem in statistical mechanics.

From a physical point of view, it is important to understand how long the time $\mathcal{T}$ must be to ensure the convergence of the time average. In general, this is a rather difficult issue depending on several factors (see also Chapter 14) among which the number of degrees of freedom and the observable Φ. For instance, if we choose as observable the characteristic function of a certain set A of the phase space, in order to observe the expected result

$$\frac{1}{\mathcal{T}} \int_{t_0}^{t_0+\mathcal{T}} dt\ \Phi(\boldsymbol{x}(t)) \simeq \mu(A)$$

$\mathcal{T}$ must be much larger than $1/\mu(A)$, which is exponentially large in the number of degrees of freedom, as a consequence of the statistics of Poincaré recurrence times (Box B.7).

Box B.7: Poincaré recurrence theorem

Poincaré recurrence theorem states that

> *Given a Hamiltonian system with a bounded phase space Γ, and a set $A \subset \Gamma$, all the trajectories starting from $\boldsymbol{x} \in A$ will return back to A after some time repeatedly and infinitely many times, except for some of them in a set of zero measure.*

The proof in rather simple by *reductio ad absurdum*. Indicate with $B_0 \subseteq A$ the set of points that never return to A. There exists a time t_1 such that $B_1 = \mathcal{S}^{t_1} B_0$ does not overlap A and therefore $B_0 \bigcap B_1 = \emptyset$. In a similar way there should be times $t_N > t_{N-1} > .... > t_2 > t_1$ such that $B_n \bigcap B_k = \emptyset$ for $n \neq k$ where $B_n = \mathcal{S}^{(t_n - t_{n-1})} B_{n-1} = \mathcal{S}^{t_n} B_0$. This can be understood noting that if $C = B_n \bigcap B_k \neq \emptyset$, for instance for $n > k$, one has a contradiction with the hypothesis that the points in B_0 do not return in A. The sets $D_1 = \mathcal{S}^{-t_n} C$ and $D_2 = \mathcal{S}^{-t_k} C$ are both contained in B_0, and D_2 can be written as $D_2 = \mathcal{S}^{(t_n - t_k)} \mathcal{S}^{-t_n} C = \mathcal{S}^{(t_n - t_k)} D_1$, therefore the points in D_1 are recurrent in B_0 after a time $t_n - t_k$, in disagreement with the hypothesis. Consider now the set $\bigcup_{n=1}^{N} B_n$, using the fact that the sets $\{B_n\}$ are non overlapping and, because of the Liouville theorem $\mu(B_n) = \mu(B_0)$, one has

$$\mu\left(\bigcup_{n=1}^{N} B_n\right) = \sum_{n=1}^{N} \mu(B_n) = N\mu(B_0)\,.$$

Since $\mu(\bigcup_{n=1}^{N} B_n)$ must be smaller than 1, and N can be arbitrarily large, the unique possibility is that $\mu(B_0) = 0$. Applying the result after any return to A one realizes that any trajectory, up to zero measure exclusions, returns infinitely many times to A. Let us note that the proof requires just Liouville theorem, so Poincaré recurrence theorem holds not only for Hamiltonian systems but also for any conservative dynamics. This theorem was at the core of the objection raised by Zermelo against Boltzmann's view on irreversibility. Zermelo indeed argued that due to the recurrence theorem the neighborhood of any microscopic state will be visited an infinite number of times, making meaningless the explanation of irreversibility given by Boltzmann in terms of the H-theorem [Cercignani (1998)]. However, Zermelo overlooked the fact that Poincaré theorem does not give information about the time of Poincaré recurrences which, as argued by Boltzmann in his reply, can be astronomically long. Recently, the statistics of recurrence times gained a renewed interest in the context of statistical properties of weakly chaotic systems [Buric *et al.* (2003); Zaslavsky (2005)]. Let us briefly discuss this important aspect. For the sake of notation simplicity we consider discrete time systems defined by the evolution law $\mathcal{S}^t$ the phase space Γ and the invariant measure μ. Given a measurable set $A \subset \Gamma$, define the recurrence time $\tau_A(\boldsymbol{x})$ as:

$$\tau_A(\boldsymbol{x}) = \inf_{k \geq 1} \{\boldsymbol{x} \in A : \mathcal{S}^k \boldsymbol{x} \in A\}$$

and the average recurrence time:

$$\langle \tau_A \rangle = \frac{1}{\mu(A)} \int_A \mathrm{d}\mu(\boldsymbol{x})\, \tau_A(\mathbf{x}) \ .$$

For an ergodic system a classical result (Kac's lemma) gives [Kac (1959)]:

$$\langle \tau_A \rangle = \frac{1}{\mu(A)} \ . \tag{B.7.1}$$

This lemma tells us that the average return time to a set is inversely proportional to its measure, we notice that instead the residence time (i.e. the total time spent in the set) is proportional to the measure of the set. In a system with N degrees of freedom, if A is a hypercube of linear size $\varepsilon < 1$ one has $\langle \tau_A \rangle = \varepsilon^{-N}$, i.e. an exponentially long average return time. This simple result has been at the basis of Boltzmann reply to Zermelo and, with little changes, it is technically relevant in the data analysis problem, see Chap. 10. More interesting is the knowledge of the distribution function $\rho_A(t)\mathrm{d}t = \mathrm{Prob}[\tau_A(\boldsymbol{x}) \in [t\!:\!t+\mathrm{d}t]]$. The shape of $\rho_A(t)$ depends on the underlying dynamics. For instance, for Anosov systems (see Box B.10 for a definition), the following exact result holds [Liverani and Wojtkowski (1995)]:

$$\rho_A(t) = \frac{1}{\langle \tau_A \rangle} \mathrm{e}^{-t/\langle \tau_A \rangle} \ .$$

Numerical simulations show that the above relation is basically verified also in systems with strong chaos, i.e. with a dominance of chaotic regions, e.g. in the standard map (2.18) with $K \gg 1$. On the contrary, for weak chaos (e.g. close to integrability, as the standard map for small value of K) at large t, $\rho_A(t)$ shows a power law decay [Buric *et al.* (2003)]. The difference between weak and strong chaos will become clearer in Chap. 7.

4.3.2 Abstract formulation of the Ergodic theory

In abstract terms, a generic continuous or discrete time dynamical system can be defined through the triad $(\Omega, \mathcal{U}^t, \mu)$ where $\mathcal{U}^t$ is a time evolution operator acting in phase space Ω:

$$\boldsymbol{x}(0) \to \boldsymbol{x}(t) = \mathcal{U}^t \boldsymbol{x}(0)$$

(e.g. for maps $\mathcal{U}^t \boldsymbol{x}(0) = \boldsymbol{f}^{(t)}(\boldsymbol{x}(0)))$, and μ a measure invariant under the evolution $\mathcal{U}^t$ i.e., generalizing Eq. (4.6), for any measurable set $B \subset \Omega$

$$\mu(B) = \mu(\mathcal{U}^{-t} B) \,.$$

We used μ and not the density ρ, because in dissipative systems the invariant measure is typically singular with respect to the Lebesgue measure (Fig. 4.2).

The dynamical system $(\Omega, \mathcal{U}^t, \mu)$ is *ergodic*, with respect the invariant measure μ, if for every integrable (measurable) function $\Phi(\boldsymbol{x})$

$$\overline{\Phi} \equiv \lim_{\mathcal{T} \to \infty} \frac{1}{\mathcal{T}} \int_{t_0}^{t_0+\mathcal{T}} \mathrm{d}t \; \Phi(\boldsymbol{x}(t)) = \int_{\Gamma} \mathrm{d}\mu(\boldsymbol{x}) \, \Phi(\boldsymbol{x}) \equiv \langle \Phi \rangle \,,$$

where $\boldsymbol{x}(t) = \mathcal{U}^{t-t_0} \boldsymbol{x}(t_0)$, for almost all (respect to the measure μ) the initial conditions $\boldsymbol{x}(t_0)$. Of course, in the case of maps the integral must be replaced by a sum. We can say that if a system is ergodic, a very long trajectory gives the same statistical information of the measure $\mu(\boldsymbol{x})$. Ergodicity is then at the origin of the physical relevance of the density defined by Eq. (4.2).[10]

The definition of ergodicity is more subtle than it may look and requires a few remarks.

First, notice that all statements of ergodic theory hold only *with respect to the measure* μ, meaning that they may fail for sets of zero μ-measure, which however can be non-zero with respect to another invariant measure.

Second, ergodicity is not a distinguishing property of chaos, as the next example stresses once more. Consider the rotation on the torus $[0:1] \times [0:1]$

$$\begin{cases} x_1(t) = x_1(0) + \omega_1 t & \mod 1 \\ x_2(t) = x_2(0) + \omega_2 t & \mod 1 \,, \end{cases} \tag{4.19}$$

for which the Lebesgue measure $\mathrm{d}\mu(\boldsymbol{x}) = \mathrm{d}x_1 \mathrm{d}x_2$ is invariant. If ω_1/ω_2 is rational, the evolution (4.19) is periodic and non-ergodic with respect to the Lebesgue measure; while if ω_1/ω_2 is irrational the motion is quasiperiodic and ergodic with respect to the Lebesgue measure (Fig. B1.1b). It is instructive to illustrate this point by explicitly computing the temporal and ensemble averages. Let $\Phi(\boldsymbol{x})$ be a smooth function, as e.g.

$$\Phi(x_1, x_2) = \Phi_{0,0} + \sum_{(n,m) \neq (0,0)} \Phi_{n,m} \mathrm{e}^{i2\pi(nx_1 + mx_2)} \tag{4.20}$$

[10]To explain the coincidence of the density defined by Eq. (4.2) with the limiting density of the Perron-Frobenius evolution, we need one more ingredient which is the mixing property, discussed in the following.

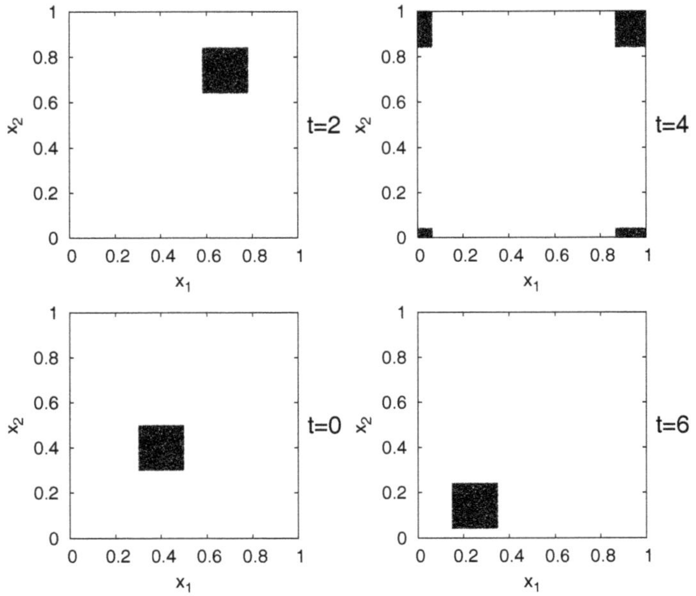

Fig. 4.4 Evolution of an ensemble of 10^4 points for the rotation on the torus (4.19), with $\omega_1 = \pi, \omega_2 = 0.6$ at $t = 0, 2, 4, 6$.

where n and m are integers $0, \pm 1, \pm 2, \ldots$. The ensemble average over the Lebesgue measure on the torus yields

$$\langle \Phi \rangle = \Phi_{0,0} \ .$$

The time average can be obtained plugging the evolution Eq. (4.19) into the definition of Φ (4.20) and integrating in $[0 : T]$. If ω_1/ω_2 is irrational, it is impossible to find $(n, m) \neq (0, 0)$ such that $n\omega_1 + m\omega_2 = 0$, and thus for $T \to \infty$

$$\overline{\Phi}^T = \Phi_{0,0} + \frac{1}{T} \sum_{(n,m)\neq(0,0)} \Phi_{n,m} \frac{\mathrm{e}^{i2\pi(n\omega_1+m\omega_2)T} - 1}{i2\pi(n\omega_1 + m\omega_2)} \mathrm{e}^{i2\pi[nx_1(0)+mx_2(0)]} \to \Phi_{0,0} = \langle \Phi \rangle \ ,$$

i.e. the system is ergodic. On the contrary, if ω_1/ω_2 is rational, the time average $\overline{\Phi}$ depends on the initial condition $(x(0), y(0))$ and, therefore, the system is not ergodic:

$$\overline{\Phi}^T \to \Phi_{0,0} + \sum_{\omega_1 n+\omega_2 m=0} \Phi_{n,m} \mathrm{e}^{i2\pi[nx(0)+my(0)]} \neq \langle \Phi \rangle \ .$$

The rotation on the torus example (4.19) also shows that ergodicity does not imply relaxation to the invariant density. This can be appreciated by looking at Fig. 4.4, where the evolution of a localized distribution of points is shown. As one can see such a distribution is merely translated by the transformation and remains localized, instead of uniformly spreading on the torus.

Both from a mathematical and a physical point of view, it is natural to wonder under which conditions a dynamical system is ergodic. At an abstract level, this problem had been tackled by Birkhoff (1931) and von Neumann (1932), who proved the following fundamental theorems:

Theorem I. *For almost every initial condition $\boldsymbol{x}_0$ the infinite time average*

$$\overline{\Phi}(\boldsymbol{x}_0) \equiv \lim_{\mathcal{T}\to\infty} \frac{1}{\mathcal{T}} \int_0^{\mathcal{T}} \mathrm{d}t\, \Phi(\mathcal{U}^t \boldsymbol{x}_0)$$

exists.

Theorem II. *Necessary and sufficient condition for the system to be ergodic, i.e. the time average $\overline{\Phi}(\boldsymbol{x}_0)$ does not depend on the initial condition (for almost all $\boldsymbol{x}_0$), is that the phase space Ω is metrically indecomposable, meaning that Ω cannot be split into two invariant sets, say A and B, (i.e. $\mathcal{U}^t A = A$ and $\mathcal{U}^t B = B$) having both positive measure. In other terms, if A is an invariant set either $\mu(A) = 1$ or $\mu(A) = 0$. [Sometimes, instead of metrically indecomposable the equivalent term metrically transitive is used.]*

The first statement **I** is rather general and not very stringent: the existence of the time average $\overline{\Phi}(\boldsymbol{x}_0)$ does not rule out its dependence on the initial condition. The second statement **II** is more interesting, although often of little practical usefulness as, in general, deciding whether a system satisfies the metrical indecomposability condition is impossible.

The concept of metric indecomposability or transitivity can be illustrated with the following example. Suppose that a given system admits two unstable fixed points x_1^* and x_2^*, clearly both $\mathrm{d}\mu_1 = \delta(x - x_1^*)\mathrm{d}x$ and $\mathrm{d}\mu_2 = \delta(x - x_2^*)\mathrm{d}x$ are invariant measures and the system is ergodic with respect to μ_1 and μ_2, respectively. The measure $\mu = p\mu_1 + (1-p)\mu_2$ with $0 < p < 1$ is, of course, also an invariant measure but it is not ergodic.[11]

We conclude by noticing that ergodicity is somehow the analogous in the dynamical system context of the law of large numbers in probability theory. If $X_1, X_2, X_3, \ldots$ is an infinite sequence of random variables such that they are independent and identically distributed with a probability density function $p(X)$, characterized by an expected value $\langle X \rangle = \int \mathrm{d}X\, p(X) X$ and variance $\sigma^2 = \langle X^2 \rangle - \langle X \rangle^2$, which are both finite, then the sample average (which corresponds to the time average)

$$\overline{X}^N = \frac{1}{N} \sum_{n=1}^{N} X_n$$

converges to the expected value $\langle X \rangle$ (which, in dynamical systems theory, is the equivalent of the ensemble average). More formally, for any positive number ϵ we have

$$\mathrm{Prob}\Big[|\overline{X}^N - \langle X \rangle| \geq \epsilon\Big] \to 0 \text{ as } N \to \infty\,.$$

[11] With probability $p > 0$ $(1 - p > 0)$ one picks the point x_1^* (x_2^*) and the time averages do not coincide with the ensemble average. The phase space is indeed parted into two invariant sets.

The difficulty with dynamical systems is that we cannot assume the independence of the successive states along a given trajectory so that ergodicity should be demonstrated without invoking the law of large numbers.

4.4 Mixing

The example of rotation on a torus (Fig. 4.4) shows that ergodicity is not sufficient to ensure the relaxation to an invariant measure which is, however, often realized in chaotic systems. In order to figure out the conditions for such a relaxation, it is necessary to introduce the important concept of *mixing*.

A dynamical system $(\Omega, \mathcal{U}^t, \mu)$ is mixing if for all sets $A, B \subset \Omega$

$$\lim_{t\to\infty} \mu(A \cap \mathcal{U}^t B) = \mu(A)\mu(B) \,, \tag{4.21}$$

whose interpretation is rather transparent: $\boldsymbol{x} \in A \cap \mathcal{U}^t B$ means that $\boldsymbol{x} \in A$ and $\mathcal{U}^t \boldsymbol{x} \in B$, Eq. (4.21) implies that the fraction of points starting from B and landing in A, after a (large) time t, is nothing but the product of the measures of A and B, for any $A, B \subset \Omega$.

The Arnold cat map (2.11)-(2.12) introduced in Chapter 2

$$\begin{cases} x_1(t+1) = x_1(t) + x_2(t) & \mod 1 \\ x_2(t+1) = x_1(t) + 2x_2(t) & \mod 1 \end{cases} \tag{4.22}$$

is an example of two-dimensional area preserving map which is mixing. As shown in Fig. 4.5, the action of the map on a cloud of points recalls the stirring of a spoon over the cream in a cup of coffee (where physical space coincides with the phase space). The interested reader may find a brief survey on other relevant properties of the cat map in Box B.10 at the end of the next Chapter.

It is worth remarking that mixing is a stronger condition than ergodicity, indeed mixing implies ergodicity. Consider a mixing system and let A be an invariant set of Ω, that is $\mathcal{U}^t A = A$ which implies $A \cap \mathcal{U}^t A = A$. From the latter expression and taking $B = A$ in Eq. (4.21) we have $\mu(A) = \mu(A)^2$ and thus $\mu(A) = 1$ or $\mu(A) = 0$. From theorem **II**, this is nothing but the condition for the ergodicity. As clear from the torus map (4.19) example, the opposite is not generically true.

The mixing condition ensures convergence to an invariant measure which, as mixing implies ergodicity, is also ergodic. Therefore, assuming a discrete time dynamics and the existence of a density ρ, if a system is mixing then for large t

$$\rho_t(\boldsymbol{x}) \to \rho^{inv}(\boldsymbol{x}) \,,$$

regardless of the initial density ρ_0. Moreover, as from Eq. (4.12) (see, also Lasota and Mackey, 1985; Ruelle, 1989), similarly to Markov chains (Box B.6), such a relaxation to the invariant density is typically[12] exponential

$$\rho_t(\boldsymbol{x}) = \rho^{inv}(\boldsymbol{x}) + \mathcal{O}\left(e^{-\frac{t}{\tau_c}}\right) \,,$$

[12] At least if the spectrum of the PF-operator is not degenerate.

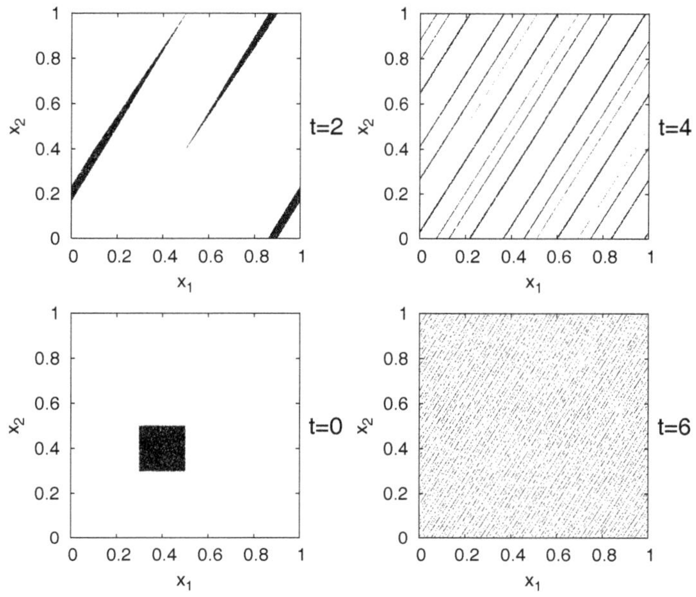

Fig. 4.5 Same as Fig. 4.4 for the cat map Eq. (4.22).

with the decay time τ_c related to the second eigenvalue of the Perron-Frobenius operator (4.12).

Mixing can be regarded as the capacity of the system to rapidly lose memory of the initial conditions, which can be characterized by the correlation function

$$C_{gh}(t) = \langle g(\boldsymbol{x}(t))h(\boldsymbol{x}(0))\rangle = \int_\Omega \mathrm{d}x\, \rho^{inv}(\boldsymbol{x})g(\mathcal{U}^t\boldsymbol{x})h(\boldsymbol{x})\,,$$

where g and h are two generic functions, and we assumed time stationarity. It is not difficult to show (e.g. one can repeat the procedure discussed in Box B.6 for the case of Markov Chains) that the relaxation time τ_c also describes the decay of the correlation functions:

$$C_{gh}(t) = \langle g(\boldsymbol{x})\rangle\langle h(\boldsymbol{x})\rangle + \mathcal{O}\left(\mathrm{e}^{-\frac{t}{\tau_c}}\right)\,. \tag{4.23}$$

The connection with the mixing condition becomes transparent by choosing g and h as the characteristic functions of the set A and B, respectively, i.e. $g(\boldsymbol{x}) = \mathcal{X}_B(\boldsymbol{x})$ and $h(\boldsymbol{x}) = \mathcal{X}_A(\boldsymbol{x})$ with $\mathcal{X}_E(\boldsymbol{x}) = 1$ if $\boldsymbol{x} \in E$ and 0 otherwise. In this case Eq. (4.23) becomes

$$C_{\mathcal{X}_A,\mathcal{X}_B}(t) = \int_\Omega \mathrm{d}\boldsymbol{x}\; \rho^{inv}(\boldsymbol{x})\, \mathcal{X}_B(\mathcal{U}^t\boldsymbol{x})\mathcal{X}_A(\boldsymbol{x}) = \mu(A\cap\mathcal{U}^tB) = \mu(A)\mu(B) + \mathcal{O}\left(\mathrm{e}^{-\frac{t}{\tau_c}}\right)$$

which is the mixing condition (4.21).

4.5 Markov chains and chaotic maps

The fast memory loss of mixing systems may suggest an analogy with Markov processes (Box B.6). Under certain conditions, this parallel can be made tight for a specific class of chaotic maps.

In general, it is not clear how and why a deterministic system can give rise to an evolution characterized by the Markov property (B.6.1), i.e. the probability of the future state of the system only depends on the current state and not on the entire history. In order to illustrate how this can be realized, let us proceed heuristically. Consider, for simplicity, a one-dimensional map $x(t+1) = g(x(t))$ of the unit interval, $x \in [0:1]$, and assume that the invariant measure is absolute continuous with respect to the Lebesgue measure, $d\mu^{inv}(x) = \rho^{inv}(x)dx$. Then, suppose to search for a coarse-grained description of the system evolution, which may be desired either for providing a compact description of the system or, more interestingly, to discretize the Perron-Frobenius operator and thus reduce it to a matrix. To this aim we can introduce a partition of $[0:1]$ into N non overlapping intervals (cells) B_j, $j = 1, \ldots, N$ such that $\cup_{j=1}^{N} B_j = [0:1]$. Each interval will be of the form $B_j = [b_{j-1} : b_j[$ with $b_0 = 0$, $b_N = 1$, and $b_{j+1} > b_j$. In this way we can construct a coarse-grained (symbolic) description of the system evolution by mapping a trajectory $x(0), x(1), x(2), \ldots x(t) \ldots$ into a sequence of symbols $i(0), i(1), i(2), \ldots, i(t), \ldots$, belonging to a finite alphabet $\{1, \ldots, N\}$, where $i(t) = k$ if $x(t) \in B_k$. Now let's introduce the $(N \times N)$-matrix

$$\mathbb{W}_{ij} = \frac{\mu^L(g^{-1}(B_i) \cap B_j)}{\mu^L(B_j)} \qquad i, j = 1, \ldots N\,, \tag{4.24}$$

where μ^L indicates the Lebesgue measure. In order to work out the analogy with MC, we can interpret $p_j = \mu^L(B_j)$ as the probability that $x(t) \in B_j$, and $p(i,j) = \mu^L(g^{-1}(B_i) \cap B_j)$ as the joint probability that $x(t\text{–}1) \in B_j$ and $x(t) \in B_i$. Therefore, $\mathbb{W}_{ij} = p(i|j) = p(i,j)/p(j)$ is the probability to find $x(t) \in B_i$ under the condition that $x(t-1) \in B_j$. The definition is consistent as $\sum_{i=0}^{N} \mu^L(g^{-1}(B_i) \cap B_j) = \mu^L(B_j)$ and hence $\sum_{i=1}^{N} \mathbb{W}_{ij} = 1$.

Recalling the basic notions of finite state Markov Chains (Box B.6A, see also Feller (1968)), we can now wonder about the connection between the MC generated by the transition matrix $\mathbb{W}$ and the original map. In particular, we can ask whether the invariant probability $\boldsymbol{P}^{inv} = \mathbb{W}\boldsymbol{P}^{inv}$ of the Markov chain has some relation with the invariant density $\rho^{inv}(x) = \mathcal{L}_{PF}\,\rho^{inv}(x)$ of the original map.

A rigorous answer exists in some cases: Li (1976) proved the so-called Ulam conjecture stating that if the map is expanding, i.e. $|dg(x)/dx| > 1$ everywhere, then $\boldsymbol{P}^{inv}$ defined by (4.24) approaches the invariant density of the original problem, $P_j^{inv} \to \int_{B_j} dx\, \rho^{inv}(x)$, when the partition becomes more and more refined ($N \to \infty$). Although the approximation can be good for N not too large [Ding and Li (1991)], this is somehow not very satisfying because the limit $N \to \infty$ prevents us from any true coarse-grained description.

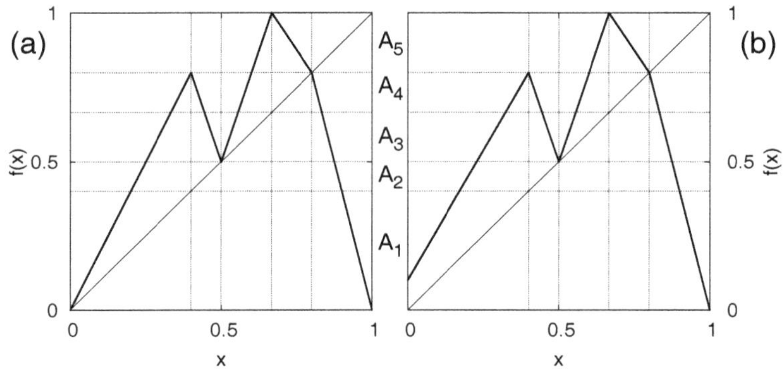

Fig. 4.6 Two examples of piecewise linear map (a) with a Markov partition (here coinciding with the intervals of definition of the map, i.e. $B_i = A_i$ for any i) and (b) with a non-Markov partition, indeed $f(0)$ is not an endpoint of any sub-interval.

Remarkably, there exists a class of maps — *piecewise linear, expanding maps* [Collet and Eckmann (1980)] — and of partitions — *Markov partitions* [Cornfeld *et al.* (1982)] — such that the MC defined by (4.24) provides the exact invariant density even for finite N.

A Markov partition $\{B_i\}_{i=1}^N$ is defined by the property

$$f(B_j) \cap B_i \neq \emptyset \quad \text{if and only if} \quad B_i \subset f(B_j)\,,$$

which, in $d = 1$, is equivalent to require that endpoints b_k of the partition get mapped onto other endpoints (in case the same one), i.e. $f(b_k) \in \{b_0, b_1, \ldots, b_N\}$ for any k, and the interval contained between two endpoints get mapped onto a single or a union of sub-intervals of the partition (to compare Markov and non-Markov partition see Fig. 4.6a and b).

Piecewise linear expanding maps have constant derivative in sub-intervals of $[0:1]$. For example, let $\{A_i\}_{i=1}^N$ be a finite non-overlapping partition of the unit interval, a generic piecewise linear expanding map $f(x)$ is such that

$$|f'(x)| = c_i > 1 \text{ for } x \in A_i\ ,$$

moreover $0 \leq f(x) \leq 1$ for any x. The expansivity condition $c_i > 1$ ensures that any fixed point is unstable making the map chaotic. For such maps the invariant measure is absolute continuous with respect to the Lebesgue measure [Lasota and Yorke (1982); Lasota and Mackey (1985); Beck and Schlögl (1997)]. Actually, it is rather easy to realize that the invariant density should be piecewise constant. We already encountered examples of piecewise linear maps as the Bernoulli shift map or the tent map, for a generic one see Fig. 4.6.

Note that in principle the Markov partition $\{B_i\}_{i=1}^N$ of a piecewise linear map may be different from the partition $\{A_i\}_{i=1}^N$ defining the map either in the position of the endpoints or in the number of sub-intervals (see for example two possible Markov partitions for the tent map Fig. 4.7b and c).

Piecewise maps represent analytically treatable cases showing, in a rather transparent way, the connection between chaos and Markov chains. To see how the connection establishes let's first consider the example in Fig. 4.6a, which is particularly simple as the Markov partition coincides with the intervals where the map has constant derivative. The five intervals of the Markov partition are mapped by the dynamics as follows: $A_1 \to A_1 \cup A_2 \cup A_3 \cup A_4$, $A_2 \to A_3 \cup A_4$, $A_3 \to A_3 \cup A_4 \cup A_5$, $A_4 \to A_5$, $A_5 \to A_1 \cup A_2 \cup A_3 \cup A_4$. Then it is easy to see that the equation defining the invariant density (4.11) reduces to a linear system of five algebraic equations for the probabilities P_i^{inv}:

$$P_i^{inv} = \sum_j \mathbb{W}_{ij} P_j^{inv} , \tag{4.25}$$

where the matrix elements $\mathbb{W}_{ij}$ are either zero, when the transition from j to i is impossible (as e.g. $0 = \mathbb{W}_{51} = \mathbb{W}_{12} = \mathbb{W}_{22} = \ldots = \mathbb{W}_{55}$) or equal to

$$\mathbb{W}_{ij} = \frac{\mu^L(B_i)}{c_j \mu^L(B_j)} , \tag{4.26}$$

as easily derived from Eq. (4.24). The invariant density for the map is constant in each interval A_i and equal to

$$\rho^{inv}(x) = P_i^{inv} / \mu^L(A_i) \quad \text{for} \quad x \in A_i .$$

In the case of the tent map one can see that the two Markov partitions (Fig. 4.7a and b) are equivalent. Indeed, labeling with (a) and (b) as in the figure, it is straightforward to derive[13]

$$\mathbb{W}^{(a)} = \begin{pmatrix} \frac{1}{2} & \frac{1}{2} \\ \frac{1}{2} & \frac{1}{2} \end{pmatrix} \qquad \mathbb{W}^{(b)} = \begin{pmatrix} \frac{1}{2} & 1 \\ \frac{1}{2} & 0 \end{pmatrix} .$$

Equation (4.25) is solved by $\boldsymbol{P}_{(a)}^{inv} = (1/2, 1/2)$ and $\boldsymbol{P}_{(b)}^{inv} = (2/3, 1/3)$, respectively which, since $\mu^L(B_1^{(a)}) = \mu^L(B_2^{(a)}) = 1/2$ and $\mu^L(B_1^{(b)}) = 2/3$, $\mu^L(B_2^{(b)}) = 1/3$, correspond to the same invariant density $\rho^{inv}(x) = 1$.

However, although the two partitions lead to the same invariant density, the second one has an extra remarkable property.[14] The second eigenvalue of $\mathbb{W}^{(b)}$, which is equal to $1/2$, is exactly equal to the second eigenvalue of the Perron-Frobenius operator associated with the tent map. In particular, this means that $\boldsymbol{P}(t) = \mathbb{W}^{(b)} \boldsymbol{P}(t-1)$ is an exact coarse-grained description of the Perron-Frobenius evolution, provided that the initial density $\rho_0(x)$ is chosen constant in the two interval $B_1^{(a)}$ and $B_2^{(b)}$, and $\boldsymbol{P}(0)$ accordingly (see Nicolis and Nicolis (1988) for details).

[13] Note that, in general, Eq. (4.26) cannot be used if the partitions $\{B_i\}$ does not coincide with the intervals of definition of the map $\{A_i\}$, as in the example (b).

[14] Although, the first partition is more "fundamental" than the second one, being a *generating partition* as discussed in Chap. 8.

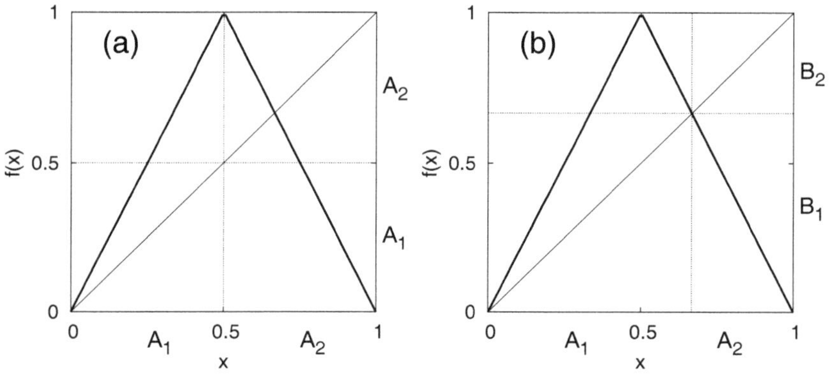

Fig. 4.7 Two Markov partitions for the tent map $f(x) = 1/2 - 2|x - 1/2|$ in (a) the Markov partition $\{B_i\}_{i=1}^2$ coincides with the one which defines the map $\{A_i\}_{i=1}^N$, in (b) they are different.

We conclude this section by quoting that MC or higher order MC,[15] can be often used to obtain reasonable approximations for some properties of a system [Cecconi and Vulpiani (1995); Cencini *et al.* (1999b)], even if the used partition does not constitute a Markov partition.

4.6 Natural measure

As the reader may have noticed, unlike other parts of the book, in this Chapter we have been a little bit careful in adopting a mathematically oriented notation for a dynamical systems as $(\Omega, \mathcal{U}^t, \mu)$. Typically, in the physical literature the invariant measure does not need to be specified. This is an important and delicate point deserving a short discussion. When the measure is not indicated, implicitly it is assumed to be the one "selected by the dynamics", i.e. the *natural measure*.

As there are a lot of ergodic measures associated with a generic dynamical system, a criterion to select the physically meaningful measure is needed. Let's consider once again the logistic map (4.1). Although for $r = 4$ the map is chaotic, we have seen that there exists an infinite number of unstable periodic trajectories $(x^{(1)}, x^{(2)}, \cdots, x^{(2^n)})$ of period 2^n, with $n = 1, 2, \cdots$. Therefore, besides the ergodic density (4.14), there is an infinite number of ergodic measures of the form

$$\rho^{(n)}(x) = \sum_{k=1}^{2^n} 2^{-n}\delta(x - x^{(k)}) \,. \tag{4.27}$$

Is there a reason to prefer $\rho^{inv}(x)$ of (4.14) instead of one of the $\rho^{(n)}(x)$ (4.27)?

[15] The idea is to assume that the state at time $t+1$ is determined by the previous k-states only, in formulae Eq. (B.6.1) becomes

$$\mathrm{Prob}(x_n = i_n | x_{n-1} = i_{n-1}, \ldots, x_{n-m} = i_{n-m} \ldots) = \mathrm{Prob}(x_n = i_n | x_{n-1} = i_{n-1}, \ldots, x_{n-k} = i_{n-k}) \,.$$

In the physical world, it makes sense to assume that the system under investigation is inherently noisy (e.g. due to the influence of the environment, not accounted for in the system description). This suggests to consider a stochastic modification of the logistic map

$$x(t+1) = rx(t)(1 - x(t)) + \epsilon\,\eta(t)$$

where $\eta(t)$ is a random and time-uncorrelated variable[16] with zero mean and unit variance. Changing ϵ tunes the relative weight of the stochastic/deterministic component of the dynamics. Clearly, for $\epsilon = 0$ the measures $\rho^{(n)}(x)$ in (4.27) are invariant, but as soon as $\epsilon \neq 0$ the small amount of noise drives the system away from the unstable periodic orbit. As a consequence, the measures $\rho^{(n)}(x)$'s are no more invariant and play no longer a physical role. On the contrary, the density (4.14), slightly modified by the presence of noise, remains a well defined invariant density for the noisy system.[17]

We can thus assume that the "correct" measure is the one obtained by adding a noisy term of intensity ϵ to the dynamical system, and then performing the limit $\epsilon \to 0$. Such a measure is the natural (or *physical*) measure and is, by construction, "dynamically robust". We notice that in any numerical simulation both the computer processor and the algorithm in use are not "perfect", so that there are unavoidable "errors" (see Chap. 10) due to truncations, round-off, etc., which play the role of noise. Similarly, noisy interactions with the environment cannot be removed in laboratory experiments. Therefore, it is self-evident (at least from a physical point of view) that numerical simulations and experiments provide access to an approximation of the natural measure.

Eckmann and Ruelle (1985), according to whom the above idea dates back to Kolmogorov, stress that such a definition of natural measure may give rise to some difficulties in general, because the added noise may induce jumps among different asymptotic states of motion (i.e. different attractors, see next Chapter). To overcome this ambiguity they suggest the use of an alternative definition of physical measure based on the request that the measure defined by

$$\rho(x; x(0)) = \lim_{T\to\infty} \frac{1}{T} \sum_{t=1}^{T} \delta(x - x(t))$$

exists and is independent of the initial condition, for almost all $x(0)$ with respect to the *Lebesgue measure*,[18] i.e. for almost all $x(0)$ randomly chosen in suitable set. This idea makes use of the concept of Sinai-Ruelle-Bowen measure that will be briefly discussed in Box B.10, for further details see Eckmann and Ruelle (1985).

[16] One should be careful to exclude those realization which bring $x(t)$ outside of the unit interval.

[17] Notice that in the presence of noise the Perron-Frobenius operator is modified (see Box B.6C).

[18] Note that the ergodic theorem would require such a property with respect to the invariant measure, which is typically different from the Lebesgue one. This is not a mere technical point indeed, as emphasized by Eckmann and Ruelle, *"Lebesgue measure corresponds to a more natural notion of sampling that the invariant measure ρ, which is carried by an attractor and usually singular"*.

4.7 Exercises

Exercise 4.1: Numerically study the time evolution of $\rho_t(x)$ for the logistic map $x(t+1) = r\,x(t)\,(1-x(t))$ with $r = 4$. Use as initial condition

$$\rho_0(x) = \begin{cases} 1/\Delta & \text{if} \quad x \in [x_0 : x_0 + \Delta] \\ 0 & \text{elsewhere} \end{cases},$$

with $\Delta = 10^{-2}$ and $x_0 = 0.1$ or $x_0 = 0.45$. Look at the evolution and compare with the invariant density $\rho^{inv}(x) = (\pi\sqrt{x(1-x)})^{-1}$.

Exercise 4.2: Consider the map $x(t+1) = x(t) + \omega \mod 1$ and show that
(1) the Lebesgue measure in $[0:1]$ is invariant;
(2) the map is periodic if ω is rational;
(3) the map is ergodic if ω is irrational.

Exercise 4.3: Consider the two-state Markov Chain defined by the transition matrix

$$\mathbb{W} = \begin{pmatrix} p & 1-p \\ 1-p & p \end{pmatrix}:$$

provide a graphical representation; find the invariant probabilities; show that a generic initial probability relax to the invariant one as $\boldsymbol{P}(t) \approx \boldsymbol{P}^{inv} + \mathcal{O}(\mathrm{e}^{-t/\tau})$ and determine τ; explicitly compute the correlation function $C(t) = \langle x(t)x(0)\rangle$ with $x(t) = 1, 0$ if the process is in the state 1 or 2.

Exercise 4.4: Consider the Markov Chains defined by the transition probabilities

$$\mathbb{F} = \begin{pmatrix} 0 & 1/2 & 1/2 & 0 \\ 1/2 & 0 & 0 & 1/2 \\ 1/2 & 0 & 0 & 1/2 \\ 0 & 1/2 & 1/2 & 0 \end{pmatrix} \qquad \mathbb{T} = \begin{pmatrix} 0 & 1/2 & 1/2 \\ 1/2 & 0 & 1/2 \\ 1/2 & 1/2 & 0 \end{pmatrix}$$

which describe a random walk within a ring of 4 and 3 states, respectively.
(1) provide a graphical representation of the two Markov Chains;
(2) find the invariant probabilities in both cases;
(3) is the invariant probability asymptotically reached from any initial condition?
(4) after a long time what is the probability of visiting each state?
(5) generalize the problem to the case with $2n$ or $2n+1$ states, respectively.
Hint: *What does happen if one starts with the first state, e.g. if* $\boldsymbol{P}(t=0) = (1, 0, 0, 0)$*?*

Exercise 4.5: Consider the standard map

$$\mathrm{I}(t+1) = \mathrm{I}(t) + K\sin(\phi(t)) \mod 2\pi\,, \qquad \phi(t+1) = \phi(t) + \mathrm{I}(t+1) \mod 2\pi\,,$$

and numerically compute the pdf of the time return function in the set $A = \{(\phi, \mathrm{I}) : (\phi-\phi_0)^2 + (\mathrm{I}-\mathrm{I}_0)^2 < 10^{-2}\}$ for $K = 10$, with $(\phi_0, \mathrm{I}_0) = (1.0, 1.0)$ and $K = 0.9$, with $(\phi_0, \mathrm{I}_0) = (0, 0)$. Compare the results with the expectation for ergodic systems (Box B.7).

Exercise 4.6: Consider the Gauss map defined in the interval $[0:1]$ by $F(x) = x^{-1} - [x^{-1}]$ if $x \neq 0$ and $F(x=0) = 0$, where $[\ldots]$ denotes the integer part. Verify that $\rho(x) = \frac{1}{\ln 2}\frac{1}{1+x}$ is an invariant measure for the map.

Exercise 4.7:
Show that the one-dimensional map defined by the equation (see figure on the right)

$$x(t+1) = \begin{cases} x(t) + 3/4 & 0 \le x(t) < 1/4 \\ x(t) + 1/4 & 1/4 \le x(t) < 1/2 \\ x(t) - 1/4 & 1/2 \le x(t) < 3/4 \\ x(t) - 3/4 & 3/4 \le x(t) \le 1 \end{cases}$$

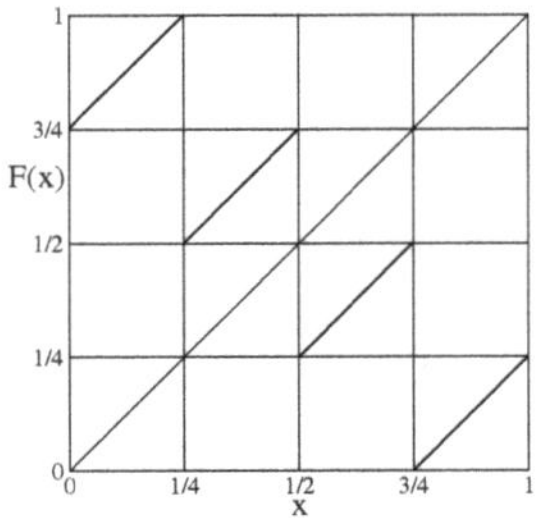

is not ergodic with respect to the Lebesgue measure which is invariant.

Hint: *Use of the Birkhoff's second theorem (Sec. 4.3.2).*

Exercise 4.8: Numerically investigate the Arnold cat map and reproduce Fig. 4.5, compute also the auto-correlation function of x and y.

Exercise 4.9: Consider the map defined by $F(x) = 3x \mod 1$ and show that the Lebesgue measure is invariant. Then consider the characteristic function $\chi(x) = 1$ if $x \in [0:1/2]$ and zero elsewhere. Numerically verify the ergodicity of the system for a set of generic initial conditions, in particular study how the time average $1/T\sum_{t=0}^{T}\chi(x(t))$ converges to the expected value 1/2 for generic initial conditions and, in particular for $x(0) = 7/8$, what's special in this point? Compute also the correlation function $\langle\chi(x(t+\tau))\chi(x(t))\rangle - \langle\chi(x(t))\rangle^2$ for generic initial conditions.

Exercise 4.10:
Consider the roof map defined by

$$F(x) = \begin{cases} F_l(x) = a + 2(1-a)x & 0 \le x < 1/2 \\ F_r(x) = 2(1-x) & 1/2 \le x < 1 \end{cases}$$

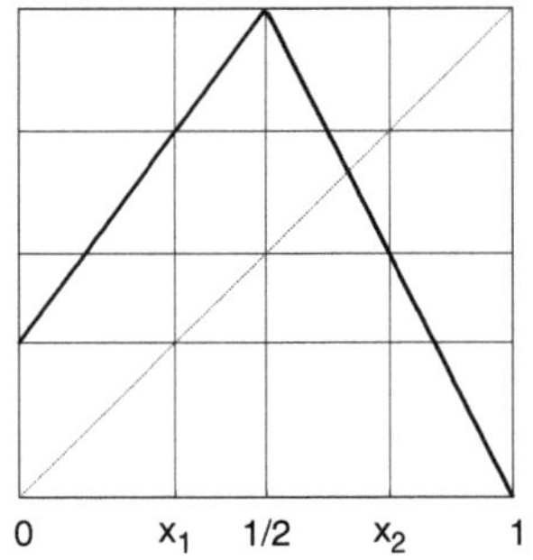

with $a = (3-\sqrt{3})/4$. Consider the points $x_1 = F_l^{-1}(x_2)$ and $x_2 = F_r^{-1}(1/2) = 3/4$ where $F_{l,r}^{-1}$ is the inverse of the $F_{l,r}$ map show that

(1) $[0:1/2[$ $[1/2:1]$ is not a Markov partitions;
(2) $[0:x_1[$ $[x_1:1/2[$ $[1/2:x_2[$ $[x_2:1]$ is a Markov partition and compute the transition matrix;
(3) compute the invariant density.

Hint: *Use the definition of Markov partition and use the Markov partition to compute the invariant probability, hence the density.*

Chapter 5

Characterization of Chaotic Dynamical Systems

> Geometry is nothing more than a branch of physics; the geometrical truths are not essentially different from physical ones in any aspect and are established in the same way.
>
> *David Hilbert (1862–1943)*

> The farther you go, the less you know.
>
> *Lao Tzu (6th century BC)*

In this Chapter, we first review the basic mathematical concepts and tools of fractal geometry, which are useful to characterize strange attractors. Then, we give a precise mathematical meaning to the sensitive dependence on initial conditions introducing the Lyapunov exponents.

5.1 Strange attractors

The concept of attractor as "geometrical locus" where the motion asymptotically converges is strictly related to the presence of dissipative mechanisms, leading to a contraction of phase-space volumes (see Sec. 2.1.1). In typical systems, the attractor emerges as an asymptotic stationary regime after a transient behavior. In Chapter 2 and 3, we saw the basic types of attractor: regular attractors such as stable fixed points, limit cycles and tori, and irregular or strange ones, such as the chaotic Lorenz (Fig. 3.6) and the non-chaotic Feigenbaum attractors (Fig. 3.12).

In general, a system may possess several attractors and the one selected by the dynamics depends on the initial condition. The ensemble of all initial conditions converging to a given attractor defines its *basin of attraction*. For example, the attractor of the damped pendulum (1.4) is a fixed point, representing the pendulum at rest, and the basin of attraction is the full phase space. Nevertheless, basins of attraction may also be objects with very complex (fractal) geometries [McDonald *et al.* (1985); Ott (1993)] as, for example, the Mandelbrot and Julia sets [Mandelbrot (1977); Falconer (2003)]. All points in a given basin of attraction asymptotically

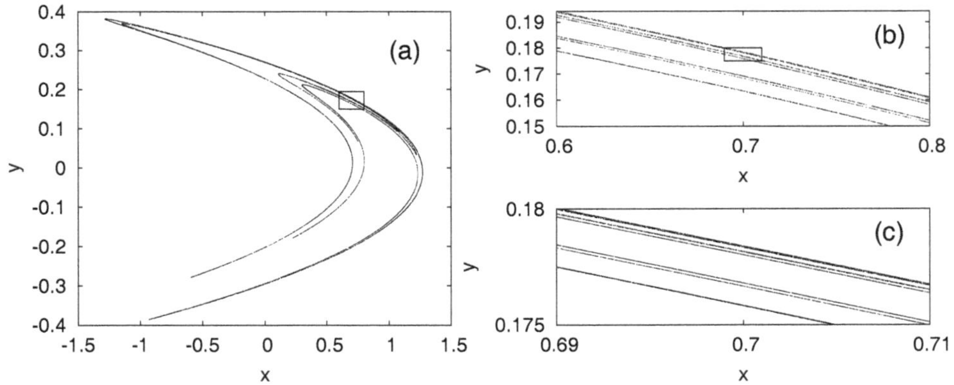

Fig. 5.1 (a) The Hénon attractor generated by the iteration of Eqs. (5.1) with parameters $a = 1.4$ and $b = 0.3$. (b) Zoom of the rectangles in (a). (c) Zoom of the rectangle in (b).

evolve toward an attractor A, which is invariant under the dynamics: if a point belongs to A, its evolution also belongs to A. We can thus define the attractor A as the smallest invariant set which cannot be decomposed into two or more subsets with distinct basins of attraction (see, e.g. Jost (2005)).

Strange attractors, unlike regular ones, are geometrically very complicated, as revealed by the evolution of a small phase-space volume. For instance, if the attractor is a limit cycle, a small two-dimensional volume does not change too much its shape: in a direction it maintains its size, while in the other it shrinks till becoming a "very thin strand" with an almost constant length. In chaotic systems, instead, the dynamics continuously stretches and folds an initial small volume transforming it into a thinner and thinner "ribbon" with an exponentially increasing length. The visualization of the stretching and folding process is very transparent in discrete time systems as, for example, the Hénon map (1976) (Sec. 2.2.1)

$$\begin{aligned} x(t+1) &= 1 - ax(t)^2 + y(t) \\ y(t+1) &= bx(t)\,. \end{aligned} \tag{5.1}$$

After many iterations the initial points will set onto the Hénon attractor shown in Fig. 5.1a. Consecutive zooms (Fig. 5.1b,c) highlight the complicated geometry of the Hénon attractor: at each blow-up, a series of stripes emerges which appear to self-similarly reproduce themselves on finer and finer length-scales, analogously to the Feigenbaum attractor (Fig. 3.12).

Strange attractors are usually characterized by a non-smooth geometry, as it is easily realized by considering a generic three-dimensional dissipative ODE. On the one hand, due to the dissipative nature of the system, the attractor cannot occupy a portion of non-zero volume in $\mathbb{R}^3$. On the other hand, a non-regular attractor cannot lie on a regular two-dimensional surface, because of the Poincarè-Bendixon theorem (Sec. 2.3) which prevents motions from being irregular on a two-dimensional surface. As a consequence, the strange attractor of a dissipative dynamical system should be a set of vanishing volume in $\mathbb{R}^3$ and, at the same time, it cannot be a smooth curve, so that it should necessarily have a rough and irregular geometrical structure.

The next section introduces the basic mathematical concepts and numerical tools to analyze such irregular geometrical entities.

5.2 Fractals and multifractals

Likely, the most intuitive concept to characterize a geometrical shape is its dimension: why do we say that in a three-dimensional space curves and surfaces have dimension 1 and 2, respectively? The classical answer is that a curve can be set in biunivocal and continuous correspondence with an interval of the real axes, so that at each point P of the curve corresponds a unique real number x and *viceversa*. Moreover, close points on the curve identify close real numbers on the segment (continuity). Analogously, a biunivocal correspondence can be established between a point P of a surface and a couple of real numbers (x, y) in a domain of $\mathbb{R}^2$. For example, a point on Earth is determined by two coordinates: the latitude and the longitude. In general, a geometrical object has a dimension d, when points belonging to it are in biunivocal and continuous correspondence with a set of $\mathbb{R}^d$, whose elements are arrays $(x_1, x_2, ..., x_d)$ of d real numbers.

The above introduced geometrical dimension d coincides with the number of independent directions accessible to a point sampling the object. This is said *topological dimension* which, by definition, is a non-negative integer lower than or equal to the dimension of the space in which the object is embedded. This integer number d, however, might be insufficient to fully quantify the dimensionality of a generic set of points, characterized by a "bizarre" arrangement of segmentation, voids or discontinuities such as the Hénon or Feigenbaum attractors. It is then useful to introduce an alternative definition of dimension based on the "measure" of the considered object, a transparent example of this procedure is as follows. Let's approximate a smooth curve of length L_0 with a polygonal of length

$$L(\ell) = N(\ell)\ell$$

where $N(\ell)$ represents the number of segments of length ℓ needed to approximate the whole curve. In the limit $\ell \to 0$, of course, $L(\ell) \to L_0$ and so $N(l) \to \infty$ as:

$$N(\ell) \sim \ell^{-1}, \tag{5.2}$$

i.e. with an exponent $d = -\lim_{\ell \to 0} \ln N(\ell)/\ln \ell = 1$ equal to the topological dimension. In order to understand why this new procedure can be helpful in coping with more complex objects consider now the von Kock curve shown in Fig. 5.2. Such a curve is obtained recursively starting from the unitary segment $[0\!:\!1]$ which is divided in three equal parts of length $1/3$. The central element is removed and

Fig. 5.2 Iterative procedure to construct the fractal von Koch curve, from top to bottom.

replaced by two segments of equal length 1/3 (Fig. 5.2). The construction is then repeated for each of the four edges so that, after many steps, the outcome is the weird line shown in Fig. 5.2. Of course, the curve has topological dimension $d = 1$. However, let's repeat the procedure which lead to Eq. (5.2). At each step, the number of segments increases as $N(k+1) = 4N(k)$ with $N(0) = 1$, and their length decreases as $\ell(k) = (1/3)^k$. Therefore, at the n-th generation, the curve has length

$$L(n) = \left(\frac{4}{3}\right)^n$$

is composed by $N(n) = 4^n$ segments of length $\ell(n) = (1/3)^n$. By eliminating n between $\ell(n)$ and $N(n)$, we obtain the scaling law

$$N(\ell) = \ell^{-\frac{\ln 4}{\ln 3}} \, ,$$

so that the exponent

$$D_F = -\lim_{\ell \to 0} \frac{\ln N(\ell)}{\ln \ell} = \frac{\ln 4}{\ln 3} = 1.2618 \ldots$$

is now actually larger than the topological dimension and, moreover, is not integer. The index D_F is the *fractal dimension* of the von Kock curve. In general, we call fractal any object characterized by $D_F \neq d$ [Falconer (2003)].

One of the peculiar properties of fractals is the self-similarity (or scale invariance) under scale deformation, dilatation or contraction. Self-similarity means that a part of a fractal reproduces the same complex structure of the whole object. This feature is present by construction in the von Kock curve, but can also be found, at least approximately, in the Hénon (Fig. 5.1a-c) and Feigenbaum (Fig. 3.12a-c) attractors. Another interesting example is to consider the set obtained by removing, at each generation, the central interval (instead of replacing it with two segments) the resulting fractal object is the *Cantor set* which has dimension $D_F = \ln 2 / \ln 3 =$

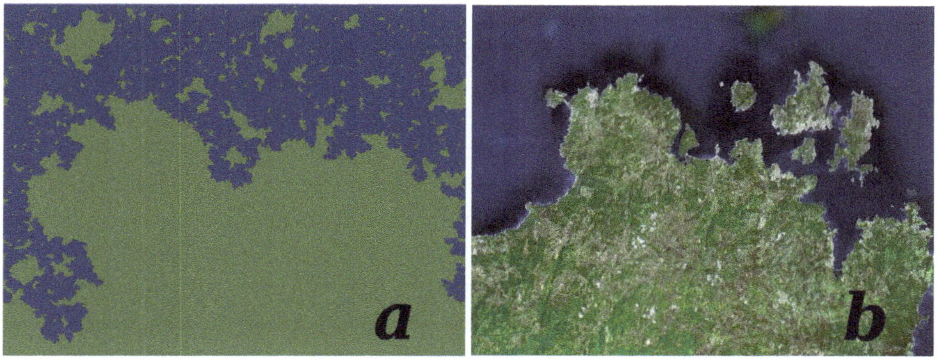

Fig. 5.3 Fractal-like nature of the coastline of Sardinia Island, Italy. (a) The fractal profile obtained by simulating the erosion model proposed by Sapoval *et al.* (2004), (b) the true coastline is on the right. Typical rocky coastlines have $D_F \approx 4/3$. [Courtesy of A. Baldassarri]

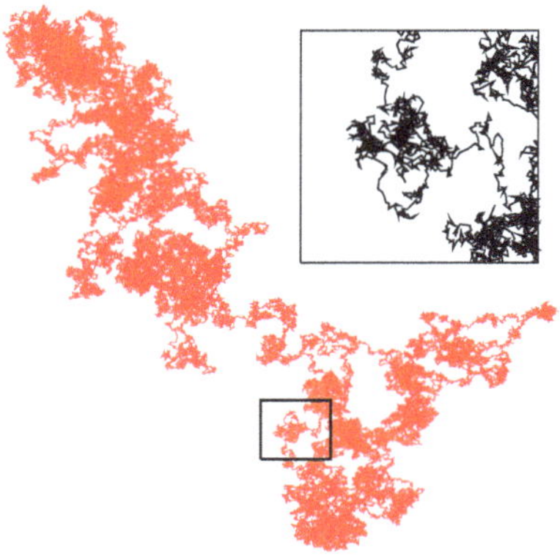

Fig. 5.4 Typical trajectory of a two-dimensional Brownian motion. The inset shows a zoom of the small box in the main figure, notice the self-similarity. The figure represents only a small portion of the trajectory, as it would densely fill the whole plane because its fractal dimension is $D_F = 2$, although the topological one is $d = 1$.

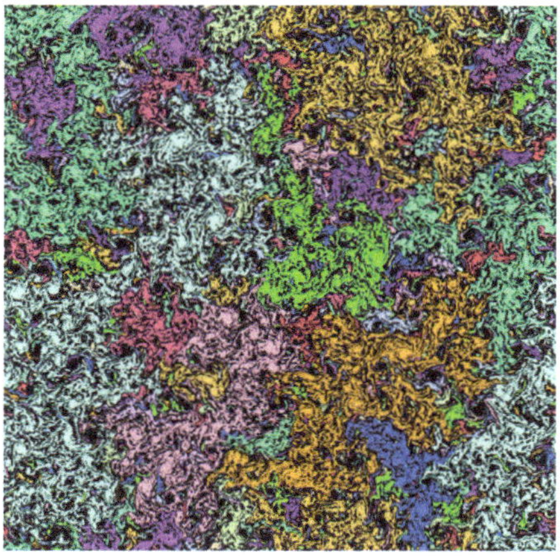

Fig. 5.5 Isolines of zero-vorticity in two-dimensional turbulence in the inverse cascade regime (Chap. 13). Colors identify different vorticity clusters, i.e. regions with equal sign of the vorticity. The boundaries of such clusters are fractals with $D_F = 4/3$ as shown by Bernard *et al.* (2006). [Courtesy of G. Boffetta]

0.63092..., i.e. less than the topological dimension (to visualize such a set retain only segments of the von Koch curve which lie on the horizontal axis).

The value D_F provides a measure of the roughness degree of the geometrical object it refers: the rougher the shape, the larger the deviation of D_F from the topological dimension.

Fractals are not mere mathematical curiosities or exceptions from usual geometry, but represent typical non-smooth geometrical structures ubiquitous in Nature [Mandelbrot (1977); Falconer (2003)]. Many natural processes such as growth, sedimentation or erosion may generate rough landscapes and profiles rich of discontinuities and fragmentation [Erzan *et al.* (1995)]. Although the self-similarity in natural fractals is only approximated and, sometimes, hidden by elements of randomness, fractal geometry represents better the variety of natural shapes than Euclidean geometry. A beautiful example of naturally occurring fractal is provided by rocky coastlines (Fig. 5.3) which, according to Sapoval *et al.* (2004), undergo a process similar to erosion, leading to $D_F \approx 4/3$. Another interesting example is the trajectory drawn by the motion of a small impurity (as pollen) suspended on the surface of a liquid, which moves under the effect of collisions with fluid molecules. It is very well known, after Brown at the beginning of 19-th century, that such motion is so irregular that exhibits fractal properties. A Brownian motion on the plane has $D_F = 2$ (Fig. 5.4) [Falconer (2003)]. Fully developed turbulence is another generous source of natural fractals. For instance, the energy dissipated is known to concentrate on small scale fractal structures [Paladin and Vulpiani (1987)]. Figure 5.5 shows the patterns emerging by considering the zero-vorticity (the vorticity is the curl of the velocity) lines of a two dimensional turbulent flow. These isolines separating regions of the fluid with vorticity of opposite sign exhibit a fractal geometry [Bernard *et al.* (2006)].

5.2.1 *Box counting dimension*

We now introduce an intuitive definition of fractal dimension which is also operational: the *box counting dimension* [Mandelbrot (1985); Falconer (2003)], which can be obtained by the procedure sketched in Fig. 5.6. Let A be a set of points embedded in a d-dimensional space, then construct a covering of A by d-dimensional hypercubes of side ℓ. Analogously to Eq. (5.2), the number $N(\ell)$ of occupied boxes, i.e. the cells that contain at least one point of A, is expected to scale as

$$N(\ell) \sim \ell^{-D_F} \,. \tag{5.3}$$

Therefore, the fractal or *capacity* dimension of a set A can be defined through the exponent

$$D_F = -\lim_{\ell \to 0} \frac{\ln N(\ell)}{\ln \ell} \,. \tag{5.4}$$

Whenever the set A is regular D_F, coincides with the topological dimension.

In practice, after computing $N(\ell)$ for several ℓ, one looks at the plot of $\ln N(\ell)$ versus $\ln \ell$, which is typically linear in a well defined region of scales $\ell_1 \ll \ell \ll \ell_2$, the slope of the plot estimates the fractal dimension D_F. The upper cut-off ℓ_2 reflects the finite extension of the set A, while the lower one ℓ_1 critically depends on the number of points used to sample the set A. Roughly, below ℓ_1, each cell contains a single point, so that $N(\ell)$ saturates to the number of points for any $\ell < \ell_1$.

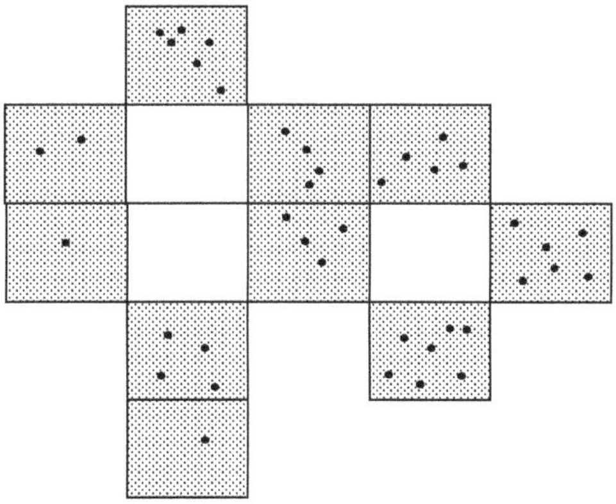

Fig. 5.6 Sketch of the box counting procedure. Shadowed boxes have occupation number greater than zero and contribute to the box counting.

For instance, the box counting method estimates a fractal dimension $D_F \simeq 1.26$ for the Hénon attractor with parameters $a = 1.4, b = 0.3$ (Fig. 5.1a), as shown in Fig. 5.7. In the figure one can also see that at reducing the number M of points representative of the attractor, the scaling region shrinks due to the shift of the lower cut-off ℓ_1 towards higher values. The same procedure can be applied to the Lorenz system obtaining $D_F \simeq 2.05$, meaning that Lorenz attractor is something slightly more complex than a surface.

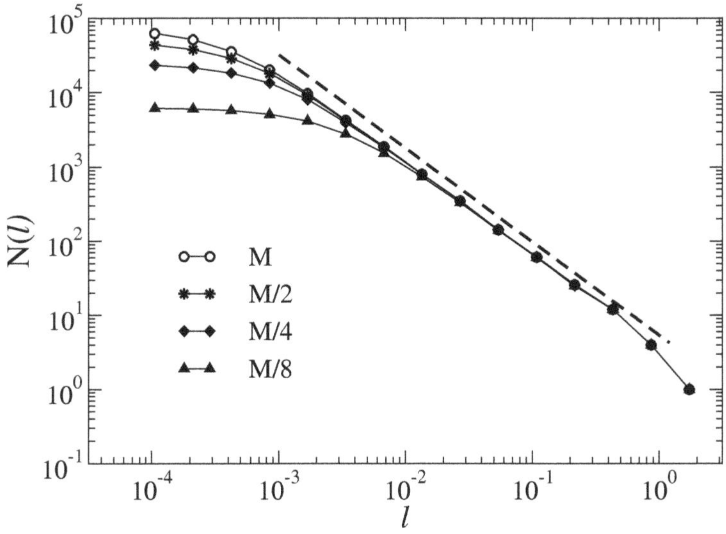

Fig. 5.7 $N(\ell)$ vs ℓ from box counting method applied to Hénon attractor (Fig. 5.1a). The slope of the dashed straight line gives $D_F = 1.26$. The computation is performed using a different number of points, as in label where $M = 10^5$. Notice how scaling at small scales is spoiled by decreasing the number of points. The presence of the large scale cutoff is also evident.

In dynamical systems, the dimensions D_F provides not only a geometrical characterization of strange attractors but also indicates the number of effective degrees of freedom, meant as the independent coordinates of dynamical relevance. It can be argued that if the fractal dimension is D_F, then the dynamics on the attractor can be described by $[D_F] + 1$ coordinates, where the symbol $[\ldots]$ denotes the integer part of a real number. In general, finding the right coordinates, which faithfully describe the motion on the attractor, is a task of paramount difficulty. Nevertheless, knowing that D_F is reasonably small would suggest the possibility of modeling a given phenomenon with a low dimensional deterministic system.

In principle, the computation of the fractal dimension by using Eq. (5.4) does not present conceptual difficulties. As discussed below, the greatest limitation of box counting method actually lies in the finite memory storage capacity of computers.

5.2.2 *The stretching and folding mechanism*

Stretching and folding mechanisms, typical of chaotic systems, are tightly related to sensitive dependence on initial conditions and the fractal character of strange attractors. In order to understand this link, take a small set A of close initial conditions in phase space and let them evolve according to a chaotic evolution law. As close trajectories quickly separate, the set A will be stretched. However, dissipation entails attractors of finite extension, so that the divergence of trajectories cannot take place indefinitely and will saturate to the natural bound imposed by the actual size of the attractor (see e.g. Fig. 3.7b). Therefore, sooner or later, the set A during its evolution has to fold onto itself. The chaotic evolution at each step continuously reiterates the process of stretching and folding which, in dissipative systems, is also responsible for the fractal nature of the attractors.

Stretching and folding can be geometrically represented by a mapping of the plane onto itself proposed by Smale (1965), known as *horseshoe* transformation. The basic idea is to start with the rectangle ABCD of Fig.5.8 with edges L_1 and L_2 and to transform it by the composition of the following two consecutive operations:

(a) The rectangle ABCD is stretched by a factor 2 in the horizontal direction and contracted in the vertical direction by the amount 2η (with $\eta > 1$), thus ABCD becomes a stripe with $L_1 \to 2L_1$ and $L_2 \to L_2/(2\eta)$;
(b) The stripe obtained in (a) is then bent, without changing its area, in a horseshoe manner so to bring it back to the region occupied by the original rectangle ABCD.

The transformation is dissipative because the area reduces by a factor $1/\eta$ at each iteration. By repeating the procedure a) and b), the area is further reduced by a factor $1/\eta^2$ while its length becomes $4L_1$. At the end of the n-th iteration, the thickness will be $L_2/(2\eta)^n$, the length $2^n L_1$, the area $L_1 L_2/\eta^n$ and the stripe will be refolded 2^n times. In the limit $n \to \infty$, the original rectangle is transformed

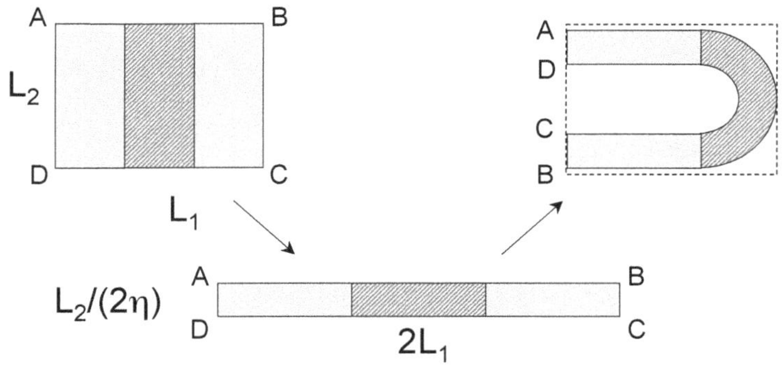

Fig. 5.8 Elementary steps of Smale's horseshoe transformation. The rectangle ABCD is first horizontally stretched and vertically squeezed, then it is bent over in a horseshoe shape so to fit into the original area.

into a fractal set of zero volume and infinite length. The resulting object can be visualized by considering the line which vertically cuts the rectangle ABCD in two identical halves. After the first application of the horseshoe transformation, such a line will intercept the image of the rectangle in two intervals of length $L_2/(4\eta^2)$. At the second application, the intervals will be 4 with size $L_2/(2\eta)^3$. At the k-step, we have 2^k intervals of length $L_2/(2\eta)^{k+1}$. It is easy to realize that the outcome of this construction is a vertical Cantor set with fractal dimension $\ln 2/\ln(2\eta)$. Therefore, the whole Smale's attractor can be regarded as the Cartesian product of a Cantor set with dimension $\ln 2/\ln(2\eta)$ and a one-dimensional continuum in the expanding direction so that its fractal dimension is

$$D_F = 1 + \frac{\ln 2}{\ln(2\eta)}$$

intermediate between 1 and 2. In particular, for $\eta = 1$, Smale's transformation becomes area preserving. Clearly by such a procedure two trajectories (initially very close) double their distance at each stretching operation, i.e. they separate exponentially in time with rate $\ln 2$, as we shall see in Sec. 5.3 this is the Lyapunov exponent of the horseshoe transformation.

Somehow, the action of Smale's horseshoe recalls the operations that a baker executes to the dough when preparing the bread. For sure, the image of bread preparation has been a source inspiration also for other scientists who proposed the so-called baker's map [Aizawa and Murakami (1983)]. Here, in particular we focus on a generalization of the baker's map [Shtern (1983)] transforming the unit square $Q = [0\!:\!1] \times [0\!:\!1]$ onto itself according to the following equations

$$(x(t+1), y(t+1)) = \begin{cases} \left(a\, x(t), \dfrac{y(t)}{h}\right) & \text{if} \quad 0 < y(t) \le h \\ \left(b\,(x(t)-1)+1, \dfrac{y(t)-h}{1-h}\right) & \text{if} \quad h < y(t) \le 1\,, \end{cases} \tag{5.5}$$

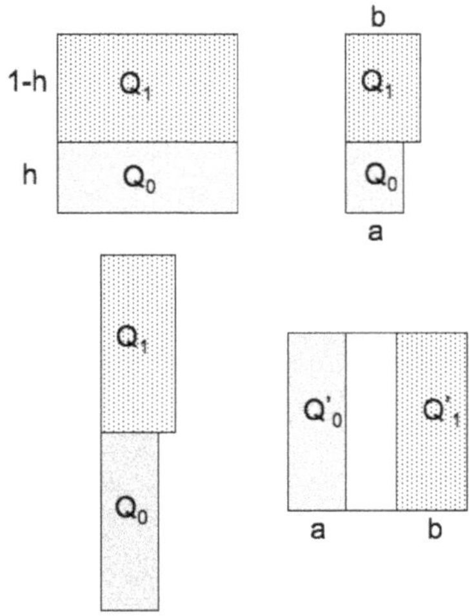

Fig. 5.9 Geometrical transformation induced on the square $Q = [0:1] \times [0:1]$ by the first step of the generalized baker's map (5.5). Q is horizontally cut in two subsets Q_0, Q_1, that are, at the same time, squeezed on x-direction and vertically dilated. Finally the two sets are rearranged in the original area Q.

with $0 < h < 1$ and $a + b \leq 1$. With reference to Fig. 5.9, the map cuts horizontally the square Q into two rectangles $Q_0 = \{(x, y) \in Q |\ y < h\}$ and $Q_1 = \{(x, y) \in Q |\ y > h\}$, and contracts them along the $x-$direction by a factor a and b, respectively (see Fig. 5.9). The two new sets are then vertically magnified by a factor $1/h$ and $1/(1 - h)$ to retrieve both unit height. Since the attractor must be bounded, finally, the upper rectangle is placed back into the rightmost part of Q and the lower one into the leftmost part of Q. Therefore, in the first step, the map (5.5) transforms the unit square Q into the two vertical stripes of Q: $Q'_0 = \{(x, y) \in Q |\ 0 < x < a\}$ and $Q'_0 = \{(x, y) \in Q |\ 1-b < x < 1\}$ with area equal to a and b, respectively.

The successive application of the map generates four vertical stripes on Q, two of area a^2, b^2 and two of area ab each, by recursion the n-th iteration results in a series of 2^n parallel vertical strips of width $a^m b^{n-m}$, with $m = 0, \ldots, n$. In the limit $n \to \infty$, the attractor of the baker's map becomes a fractal set consisting in vertical parallel segments of unit height located on a Cantor set. In other words, the asymptotic attractor is the Cartesian product of a continuum (along y-axis) with dimension 1 and a Cantor set (along x-axis) of dimension D_F, so that the whole attractor has dimension $1 + D_F$. For $a = b$ and h arbitrary, the Cantor set generated by the baker's map can be shown, via the same argument applied to the horseshoe map, to have fractal dimension

$$D_F = \frac{\ln 2}{\ln(1/a)}, \tag{5.6}$$

which is independent of h. Fig. 5.10 shows the set corresponding to $h = 1/2$.

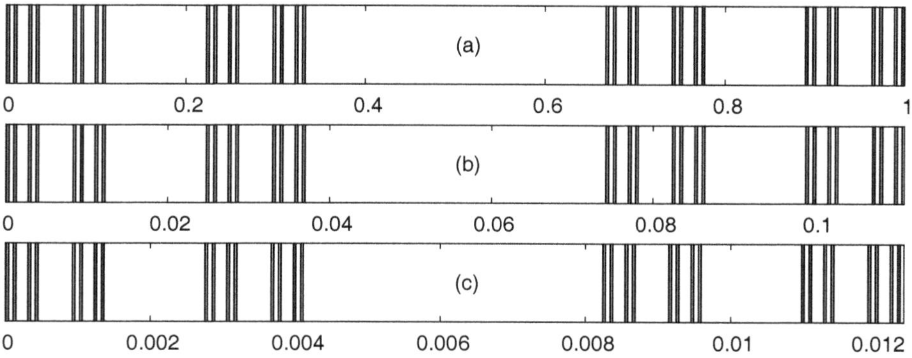

Fig. 5.10 (a) Attractor of the baker's map (5.5) for $h = 1/2$ and $a = b = 1/3$. (b) Close up of the leftmost block in (a). (c) Close up of the leftmost block in (b). Note the perfect self-similarity of this fractal set.

5.2.3 *Multifractals*

Fractals observed in Nature, including strange attractors, typically have more complex self-similar properties than, e.g., those of von Koch's curve (Fig. 5.2). The latter is characterized by geometrical properties (summarized by a unique index D_F) which are invariant under a generic scale transformation: by construction a magnification of any portion of the curve would be equivalent to the whole curve–perfect self-similarity. The same holds true for the attractor of the baker's map for $h = 1/2$ and $a = b = 1/3$ (Fig. 5.10). However, there are other geometrical sets for which a unique index D_F is insufficient to fully characterize their properties. This is particularly evident if we look at the set shown in Fig. 5.11 that was generated by the baker's map for $h = 0.2$ and $a = b = 1/3$. According to Eq. (5.6) this set shares the same fractal dimension of that shown in Fig. 5.10, but differs in the self-similarity properties as evident by comparing Fig. 5.10 with Fig. 5.11. In the former, we can see that vertical bars are dense in the same way (eyes do not distinguish one region from the other). On the contrary, in the latter eyes clearly resolve darker from lighter regions, corresponding to portions where bars are denser. Accounting for such non-homogeneity naturally call for introducing the concept of *multifractal*, in which the self-similar properties become locally depending on the position on the set. In a nutshell the idea is to imagine that, instead of a single fractal dimension globally characterizing the set, a spectrum of fractal dimensions differing from point to point has to be introduced.

This idea can be better formalized by introducing the generalized fractal dimensions (see, e.g, Paladin and Vulpiani, 1987; Grassberger *et al.*, 1988). In particular, we need a statistical description of the fractal capable of weighting inhomogeneities. In the box counting approach, the inhomogeneities manifest through the fluctuations of the occupation number from one box to another (see, e.g., Fig. 5.6). Notice that the box counting dimension D_F (5.4) is blind to these fluctuations as it only

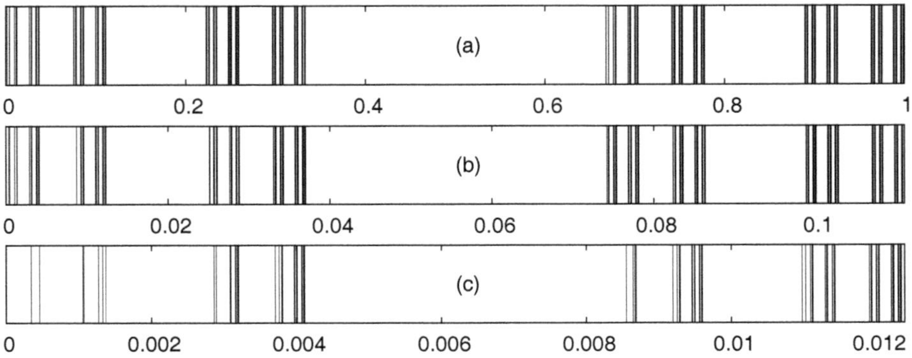

Fig. 5.11 Same as Fig. 5.10 for $h = 0.2$ and $a = b = 1/3$. Note that despite Eq. (5.6) implies that the fractal dimension of this set is the same as that of Fig. 5.10, in this case self-similarity appears to be broken.

discriminates occupied from empty cells regardless the actual — *crowding* — number of points. The different crowding can be quantified by assigning a weight $p_n(\ell)$ to the n-th box according to the fraction of points it contains. When $\ell \to 0$, for simple homogeneous fractals (Fig. 5.10) $p_n(\ell) \sim \ell^\alpha$ with $\alpha = D_F$ independently of n, while for multifractals (Fig. 5.11) α depends on the considered cell, $\alpha = \alpha_n$, and is said the crowding or *singularity* index.

Standard multifractal analysis studies the behavior of the function

$$\mathcal{M}_q(\ell) = \sum_{n=1}^{N(\ell)} p_n^q(\ell) = \langle p^{q-1}(\ell) \rangle \,, \tag{5.7}$$

where $N(\ell)$ indicates the number of non-empty boxes of the covering at scale ℓ. The function $\mathcal{M}_q(\ell)$ represents the moments of order $q-1$ of the probabilities p_n's. Changing q selects certain contributions to become dominant, allowing the scaling properties of a certain class of subsets to be sampled. When the covering is sufficiently fine that a scaling regime occurs, in analogy with box counting, we expect

$$\mathcal{M}_q(\ell) \sim \ell^{(q-1)D(q)} \,.$$

In particular, for $q = 0$ we have $\mathcal{M}_0(\ell) = N(\ell)$ and Eq. (5.7) reduces to Eq. (5.3), meaning that $D(0) = D_F$. The exponent

$$D(q) = \frac{1}{q-1} \lim_{\ell \to 0} \frac{\ln \mathcal{M}_q(\ell)}{\ln \ell} \tag{5.8}$$

is called the generalized fractal dimension of order q (or Rényi dimension) and characterizes the multifractal properties of the measure. As already said $D(0) = D_F$ is nothing but the box counting dimension. Other relevant values are: the information dimension

$$\lim_{q \to 1} D(q) = D(1) = \lim_{\ell \to 0} \sum_{n=1}^{N(\ell)} \frac{p_n(\ell) \ln p_n(\ell)}{\ln \ell}$$

and the correlation dimension $D(2)$. The physical interpretation of these two indexes is as follows. Consider the attractor of a chaotic dissipative system. Picking at random a point on the attractor with a probability given by the natural measure, and looking in a sphere of radius ℓ around it one would find that the local fractal dimension is given by $D(1)$. While picking two points at random with probabilities given by the natural measure, the probability to find them at a distance not larger than ℓ scales as $\ell^{D(2)}$.

An alternative procedure to perform the multifractal analysis consists in grouping all the boxes having the same singularity index α, i.e. all n's such that $p_n(\ell) \sim \ell^\alpha$. Let $N(\alpha, \ell)$ be the number of such boxes, by definition we can rewrite the sum (5.7) as a sum over the indexes

$$\mathcal{M}_q(\ell) = \sum_\alpha N(\alpha, \ell)\ell^{\alpha q} ,$$

where we have used the scaling relation $p_n(\ell) \sim \ell^\alpha$. We can then introduce the multifractal spectrum of singularities as the fractal dimension, $f(\alpha)$, of the subset with singularity α. In the limit $\ell \to 0$, the number of boxes with crowding index in the infinitesimal interval $[\alpha : \alpha + \mathrm{d}\alpha]$ is

$$\mathrm{d}N(\alpha, \ell) \sim \ell^{-f(\alpha)}\mathrm{d}\alpha,$$

thus we can write $\mathcal{M}_q(\ell)$ as an integral

$$\mathcal{M}_q(\ell) \simeq \int_{\alpha_{min}}^{\alpha_{max}} \mathrm{d}\alpha \; \rho(\alpha)\, \ell^{[\alpha q - f(\alpha)]} , \tag{5.9}$$

where $\rho(\alpha)$ is a smooth function independent of ℓ, for ℓ small enough, and $\alpha_{\mathrm{min}/\mathrm{max}}$ is the smallest/largest point-wise dimension of the set. In the limit $\ell \to 0$, the above integral receives the leading contribution from $\min_\alpha\{q\alpha - f(\alpha)\}$, corresponding to the solution α^* of

$$\frac{\mathrm{d}}{\mathrm{d}\alpha}[\alpha q - f(\alpha)] = q - f'(\alpha) = 0 \tag{5.10}$$

with $f''(\alpha^*) < 0$. Therefore, asymptotically we have

$$\mathcal{M}_q(\ell) \sim \ell^{[q\alpha^* - f(\alpha^*)]}$$

that inserted into Eq. (5.8) determines the relationship between $f(\alpha)$ and $D(q)$

$$D(q) = \frac{1}{q-1}[q\alpha^* - f(\alpha^*)] , \tag{5.11}$$

amounting to say that the singularity spectrum $f(\alpha)$ is the Legendre transform of the generalized dimension $D(q)$. In Equation (5.11), α^* is parametrized by q upon inverting the equation $f'(\alpha^*) = q$, which is nothing but Eq. (5.10). Therefore, when $f(\alpha)$ is known, we can determine $D(q)$ as well. Conversely, from $D(q)$, the Legendre transformation can be inverted to obtain $f(\alpha)$ as follows. Multiply Eq. (5.11) by $q - 1$ and differentiate both members with respect to q to get

$$\frac{\mathrm{d}}{\mathrm{d}q}[(q-1)D(q)] = \alpha(q) , \tag{5.12}$$

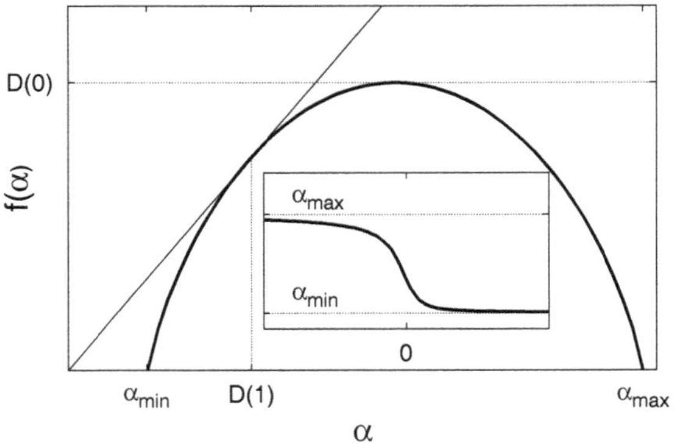

Fig. 5.12 Typical shape of the multifractal spectrum $f(\alpha)$ vs α, where noteworthy points are indicated explicitly. Inset: the corresponding $D(q)$.

where we used the condition Eq. (5.10). Thus, the singularity spectrum reads

$$f(\alpha) = q\alpha - (q-1)D(q) \tag{5.13}$$

where q is now a function of α upon inverting Eq. (5.12). The dimension spectrum $f(\alpha)$ is a concave function of α (i.e. $f''(\alpha) < 0$). A typical graph of $f(\alpha)$ is shown in Fig. 5.12, where we can identify some special features. Setting $q = 0$ in Eq. (5.13), it is easy to realize that $f(\alpha)$ reaches its maximum D_F, at the box counting dimension. While, setting $q = 1$, from Eqs. (5.12)-(5.13) we have that for $\alpha = D(1)$ the graph is tangent to bisecting line, $f(\alpha) = \alpha$. Around the value $\alpha = D(1)$, the multifractal spectrum can be typically approximated by a parabola of width σ

$$f(\alpha) \approx \alpha - \frac{[\alpha - D(1)]^2}{2\sigma^2}$$

so that by solving Eq. (5.12) an explicit expression of the generalized dimension close to $q = 1$ can be given as:

$$D(q) \approx D(1) - \frac{\sigma^2}{2}(q-1) .$$

Furthermore, from the integral (5.9) and Eq. (5.11) it is easy to obtain $\lim_{q\to\infty} D(q) = \alpha_{min}$ while $\lim_{q\to-\infty} D(q) = \alpha_{max}$.

We conclude by discussing a simple example of multifractal. In particular, we consider the two scale Cantor set that can also be obtained by horizontally sectioning the baker-map attractor (e.g. Fig. 5.11). As from previous section, at the n-th iteration, the action of the map generates 2^n stripes of width $a^m b^{n-m}$ each of weight (the darkness of vertical bars of Fig. 5.11)

$$p_i(n) = h^m(1-h)^{n-m} ,$$

where $m = 0, \dots, n$. For fixed n, the number of stripes with the same area $a^m b^{n-m}$ is provided by the binomial coefficient,

$$\binom{n}{m} = \frac{n!}{m!(n-m)!} .$$

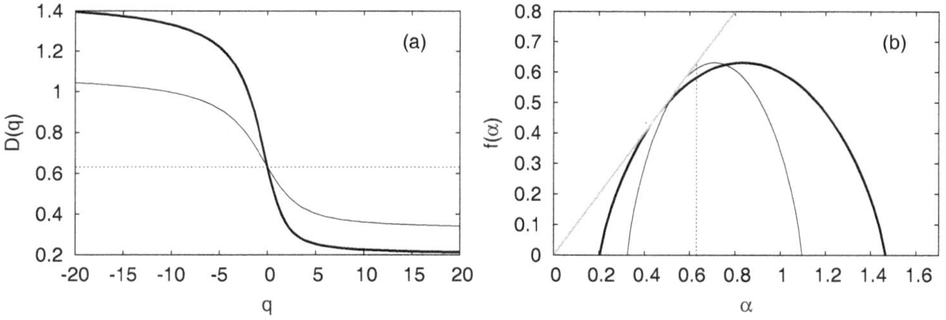

Fig. 5.13 (a) $D(q)$ vs q for the two scale Cantor set obtained from the baker's map (5.5) with $a = b = 1/3$ and $h = 1/2$ (dotted line), 0.3 (solid line) and 0.2 (thick black line). Note that $D(0)$ is independent of h. (b) The corresponding spectrum $f(\alpha)$ vs α. In gray we show the line $f(\alpha)=\alpha$. Note that for $h=1/2$ the spectrum is defined only at $\alpha = D(0) = D_F$ and $D(q)=D(0)=D_F$, i.e. it is a homogeneous fractal.

We can now compute the $(q-1)$-moments of the distribution $p_i(n)$

$$\mathcal{M}_n(q) = \sum_{i=1}^{2^n} p_i^q(n) = \sum_{m=0}^{n} \binom{n}{m} [h^m(1-h)^{n-m}]^q = [h^q + (1-h)^q]^n \,,$$

where the second equality stems from the fact that binomial coefficient takes into account the multiplicity of same-length segments, and the third equality from the expression perfect binomial. In the case $a = b$, i.e. equal length segments,[1] the limit in Eq. (5.8) corresponds to $n \to \infty$ with $\ell = a^n$, and the generalized dimension $D(q)$ reads

$$D(q) = \frac{1}{q-1} \frac{\ln[h^q + (1-h)^q]}{\ln a} \,,$$

and is shown in Fig. 5.13 together with the corresponding dimension spectrum $f(\alpha)$. The generalized dimension of the whole baker-map attractor is $1 + D(q)$ because in the vertical direction we have a one-dimensional continuum.

Two observations are in order. First, setting $q = 0$ recovers Eq. (5.6), meaning that the box counting dimension does not depend on h. Second if $h = 1/2$, we have the homogeneous fractal of Fig. 5.10 with $D(q) = D(0)$, where $f(\alpha)$ is defined only for $\alpha = D_F$ with $f(D_F) = D_F$ (Fig. 5.13b). It is now clear that only knowing the whole $D(q)$ or, equivalently, $f(\alpha)$ we can characterize the richness of the set represented in Fig. 5.11.

Usually $D(q)$ of a strange attractor is not amenable to analytical computation and it has to be estimated numerically. Next section presents one of the most efficient and widely employed algorithm for $D(q)$ estimation.

From a mathematical point of view, the multifractal formalism here presented belongs the more general framework of Large Deviation Theory, which is briefly reviewed in Box B.8.

[1]The case $a \neq b$ can also be considered at the price of a slight more complicated derivation of the limit, involving a covering of the set with cells of variable sizes.

Box B.8: Brief excursion on Large Deviation Theory

Large deviation theory (LDT) studies rare events, related to the tails of distributions [Varadhan (1987)] (see also Ellis (1999) for a physical introduction). The limit theorems of probability theory (law of large numbers and central limit [Feller (1968); Gnedenko and Ushakov (1997)]) guarantee the convergence toward determined distribution laws in a limited interval around the mean value. Large deviation theory, instead, addresses the problem of the statistical properties outside this region. The simplest way to approach LDT consists in considering the distribution of the sample average

$$X_N = \frac{1}{N}\sum_i^N x_i \,,$$

of N independent random variables $\{x_1, \ldots, x_N\}$ that, for simplicity, are assumed equally distributed with expected value $\mu = \langle x \rangle$ and variance $\sigma^2 = \langle (x-\mu)^2 \rangle < \infty$. The issue is how much the empirical value X_N deviates from its mathematical expectation μ, for N finite but sufficiently large. The Central Limit Theorem (CLT) states that, for large N, the distribution of X_N becomes

$$P_N(X) \sim \exp[-N(X-\mu)^2/2\sigma^2] \,,$$

and thus typical fluctuations of X_N around μ are of order $\mathcal{O}(N^{-1/2})$. However, CLT does not concern non-typical fluctuations of X_N larger than a certain value $f \gg \sigma/\sqrt{N}$, which instead are the subject of LDT. In particular, LDT states that, under suitable hypotheses, the probability to observe such large deviations is exponentially small

$$\Pr\left(|\mu - X_N| \simeq f\right) \sim e^{-N\mathcal{C}(f)} \,, \tag{B.8.1}$$

where $\mathcal{C}(f)$ is called *Cramer's function* or *rate function* [Varadhan (1987); Ellis (1999)]. The Bernoulli process provides a simple example of how LDT works. Let $x_n = 1$ and $x_n = 0$ be the entries of a Bernoulli process with probability p and $1-p$, respectively. A simple calculus gives that X_N has average p and variance $p(1-p)/N$. The distribution of X_N is

$$P(X_N = k/N) = \frac{N!}{k!(N-k)!}p^k(1-p)^{N-k} \,.$$

If $P(X_N)$ is written in exponential form, via Stirling approximation $\ln s! \simeq s\ln s - s$, for large N we obtain

$$P_N(X \simeq x) \sim e^{-N\mathcal{C}(x)} \tag{B.8.2}$$

where we set $x = k/N$ and

$$\mathcal{C}(x) = (1-x)\ln\left(\frac{1-x}{1-p}\right) + x\ln\left(\frac{x}{p}\right) , \tag{B.8.3}$$

which is defined for $0 < x < 1$, i.e. the bounds of X_N. Expression (B.8.2) is formally identical to Eq. (B.8.1) and represents the main result of LDT which goes beyond the central limit theorem as it allows the statistical feature of exponentially small (in N) tails to be estimated. The Cramer function (B.8.3) is minimal in $x = p$ where it also vanishes, $\mathcal{C}(x = p) = 0$, and a Taylor expansion of Eq. (B.8.3) around its minimum provides

$$\mathcal{C}(x) \simeq \frac{1}{2}\frac{(x-p)^2}{p(1-p)} - \frac{1-2p}{6}\frac{(x-p)^3}{p^2(1-p)^2} + \dots .$$

The quadratic term recovers the CLT once plugged into Eq. (B.8.2), while for $|x - p| > \mathcal{O}(N^{-1/2})$, higher order terms are relevant and thus tails lose the Gaussian character. We notice that the Cramer function cannot have an arbitrary shape, but possesses the following properties:

a- $\mathcal{C}(x)$ must be a convex function;

b- $\mathcal{C}(x) > 0$ for $x \neq \langle x \rangle$ and $\mathcal{C}(\langle x \rangle) = 0$ as a consequence of the law of large numbers;

c- further, whenever the central limit theorem hypothesis are verified, in a neighborhood of $\langle x \rangle$, $\mathcal{C}(x)$ has a parabolic shape: $\mathcal{C}(x) \simeq (x - \langle x \rangle)^2/(2\sigma^2)$.

5.2.4 *Grassberger-Procaccia algorithm*

The box counting method, despite its simplicity, is severely limited by memory capacity of computers which prevents from the direct use of Eq. (5.3). This problem dramatically occurs in high dimensional systems, where the number of cells needed of the covering exponentially grows with the dimension d, i.e. $N(\ell) \sim (L/\ell)^d$, L being the linear size of the object. For example, if the computer has 1Gb of memory and $d = 5$ the smallest scale which can be investigated is $\ell/L \simeq 1/64$, typically too large to properly probe the scaling region.

Such limitation can be overcome, by using the procedure introduced by Grassberger and Procaccia (1983c) (GP). Given a d-dimensional dynamical system, the basic point of the techniques is to compute the correlation sum

$$C(\ell, M) = \frac{2}{M(M-1)} \sum_{i,\, j>i} \Theta(\ell - ||\boldsymbol{x}_i - \boldsymbol{x}_j||) \tag{5.14}$$

from a sequence of M points $\{\boldsymbol{x}_1, \dots, \boldsymbol{x}_M\}$ sampled, at each time step τ, from a trajectory exploring the attractor, i.e. $\boldsymbol{x}_i = \boldsymbol{x}(i\tau)$, with $i = 1, \dots, M$. The sum (5.14) is an unbiased estimator of the correlation integral

$$C(\ell) = \int \mathrm{d}\mu(\boldsymbol{x}) \int \mathrm{d}\mu(\boldsymbol{y})\, \Theta(\ell - ||\boldsymbol{x} - \boldsymbol{y}||)\,, \tag{5.15}$$

where μ is the natural measure (Sec. 4.6) of the dynamics. In principle, the choice of the sampling time τ is irrelevant, however it may matter in practice as we shall see in Chapter 10. The symbol $||\dots||$, in Eq. (5.14), denotes the distance in some norm and $\Theta(s)$ is the unitary step function: $\Theta(s) = 1$ for $s \geq 0$ and $\Theta(s) = 0$ when $s < 0$. The function $C(\ell, M)$ represents the fraction of pairs of points with mutual distance less than or equal to ℓ. For $M \to \infty$, $C(\ell)$ can be interpreted as the

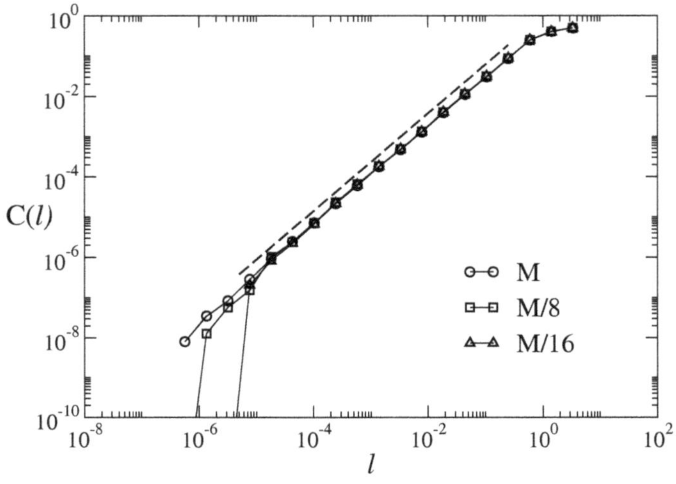

Fig. 5.14 Hénon attractor: scaling behavior of the correlation integral $C(\ell)$ vs ℓ at varying the number of points, as in label with $M = 10^5$. The dashed line has slope $D(2) \approx 1.2$, slightly less than the box counting dimension D_F (Fig. 5.7), this is consistent with the inequality $D_F \geq D(2)$ and provides evidence for the multifractal nature of Hénon attractor.

probability that two points randomly chosen on the attractor lie within a distance ℓ from each other. When ℓ is of the order of the attractor size, $C(\ell)$ saturates to a plateau, while it decreases monotonically to zero as $\ell \to 0$. At scales small enough, $C(\ell, M)$ is expected to decrease like a power law, $C(\ell) \sim \ell^\nu$, where the exponent

$$\nu = \lim_{\ell \to 0} \frac{\ln C(\ell, M)}{\ln \ell}$$

is a good estimate to the *correlation dimension* $D(2)$ of the attractor which is lower bound for D_F.

The advantage of GP algorithm with respect to box counting can be read from Eq. (5.14): it does require to store M data point only, greatly reducing the memory occupation. However, computing the correlation integral becomes quite demanding at increasing M, as the number of operations grows as $\mathcal{O}(M^2)$. Nevertheless, a clever use of the neighbor listing makes the computation much more efficient (see, e.g., Kantz and Schreiber (1997) for an updated review of all possible tricks to fasten the computation of $C(\ell, M)$).

A slight modification of GP algorithm also allows the generalized dimensions $D(q)$ to be estimated by avoiding the partition in boxes. The idea is to estimate the occupation probabilities $p_k(\ell)$ of the k-th box without using the box counting. Assume that a hypothetical covering in boxes $\mathcal{B}_k(\ell)$ of side ℓ was performed and that $\boldsymbol{x}_i \in \mathcal{B}_k(\ell)$. Then instead of counting all points which fall into $\mathcal{B}_k(\ell)$, we compute

$$n_i(\ell) = \frac{1}{M-1} \sum_{j \neq i} \Theta(\ell - ||\boldsymbol{x}_i - \boldsymbol{x}_j||) ,$$

which, if the points are distributed according to the natural measure, estimates the occupation probability, i.e. $n_i(l) \sim p_k(\ell)$ with $\boldsymbol{x}_i \in \mathcal{B}_k(\ell)$. Now let $f(\boldsymbol{x})$ be a generic function, its average on the natural measure may be computed as

$$\frac{1}{M}\sum_{i=1}^{M} f(\boldsymbol{x}_i) = \frac{1}{M}\sum_{k}\sum_{\boldsymbol{x}_i\in\mathcal{B}_k(\ell)} f(\boldsymbol{x}_i) \sim \sum_k f(\boldsymbol{x}_{i(k)})p_k(\ell)\,,$$

where the first equality stems from a trivial regrouping of the points, the last one from estimating the number of points in box $\mathcal{B}_k(\ell)$ with $Mp_k(\ell) \simeq Mn_i(\ell)$ and the function evaluated at the center $\boldsymbol{x}_{i(k)}$ of the cell $\mathcal{B}_k(\ell)$. By choosing for f the probability itself, we have:

$$C_q(\ell, M) = \frac{1}{M}\sum_i n_i^q(\ell) \sim \sum_k p_k^{q+1}(\ell) \sim \ell^{\,q\,D(q+1)}$$

which allows the generalized dimensions $D(q)$ to be estimated from a power law fitting. It is now also clear why $\nu = D(2)$.

Similarly to box counting, GP algorithm estimates dimensions from the small-ℓ scaling behavior of $C_q(\ell, M)$, involving an extrapolation to the limit $\ell \to 0$. The direct extrapolation to $\ell \to 0$ is practically impossible because if M is finite $C_q(\ell, M)$ drops abruptly to zero at scales $\ell \leq \ell_c = \min_{ij}\{||\boldsymbol{x}_i - \boldsymbol{x}_j||\}$, where no pairs are present. Even if, a paramount collection of data is stored to get l_c very small, near this bound the pair statistics becomes so poor that any meaningful attempt to reach the limit $\ell \to 0$ is hopeless. Therefore, the practical way to estimate the $D(q)$'s amounts to plotting C_q against ℓ on a log-log scale. In a proper range of small ℓ, the points adjust on a straight line (see e.g. Fig. 5.14) whose linear fit provides the slope corresponding to $D(q)$. See Kantz and Schreiber (1997) for a thorough insight on the use and abuse of the GP method.

5.3 Characteristic Lyapunov exponents

This section aims to provide the mathematical framework for characterizing sensitive dependence on initial conditions. This leads us to introduce a set of parameters associated to each trajectory $\boldsymbol{x}(t)$, called Characteristic Lyapunov exponents (CLE or simply LE), providing a measure of the degree of its instability. They quantify the mean rate of divergence of trajectories which start infinitesimally close to a reference one, generalizing the concept of linear stability (Sec. 2.4) to aperiodic motions.

We introduce CLE considering a generic d-dimensional map

$$\boldsymbol{x}(t+1) = \boldsymbol{f}(\boldsymbol{x}(t))\,, \tag{5.16}$$

nevertheless all the results can be straightforwardly extended to flows. The stability of a single trajectory $\boldsymbol{x}(t)$ can be studied by looking at the evolution of its nearby trajectories $\boldsymbol{x}'(t)$, obtained from initial conditions $\boldsymbol{x}'(0)$ displaced from $\boldsymbol{x}(0)$ by an infinitesimal vector: $\boldsymbol{x}'(0) = \boldsymbol{x}(0) + \delta\boldsymbol{x}(0)$ with $\Delta(0) = |\delta\boldsymbol{x}(0)| \ll 1$. In non-chaotic systems, the distance $\Delta(t)$ between the reference trajectory and perturbed ones either remains bounded or increases algebraically. In chaotic systems it grows exponentially with time

$$\Delta(t) \sim \Delta(0)\mathrm{e}^{\gamma t}\,,$$

where γ is the local exponential rate of expansion. As shown in Fig. 3.7b for the Lorenz model, the exponential growth is observable until $\Delta(t)$ remains much smaller than the attractor size while, at large times, $\Delta(t)$ erratically fluctuates around a finite value. A non-fluctuating parameter characterizing trajectory instability can be defined through the double limit

$$\lambda_{max} = \lim_{t\to\infty} \lim_{\Delta(0)\to 0} \frac{1}{t} \ln\left(\frac{\Delta(t)}{\Delta(0)}\right), \tag{5.17}$$

which is the mean exponential rate of divergence and is called the maximum Lyapunov exponent. Notice that the two limits cannot be exchanged, otherwise, in bounded attractors, the result would be trivially 0. When the limit λ exists positive, the trajectory shows sensitivity to initial conditions and thus the system is chaotic.

The maximum LE alone does not fully characterize the instability of a d-dimensional dynamical system. Actually, there exist d LEs defining the Lyapunov spectrum, which can be computed by studying the time-growth of d independent infinitesimal perturbations $\{\boldsymbol{w}^{(i)}\}_{i=1}^{d}$ with respect to a reference trajectory. In mathematical language, the vectors $\boldsymbol{w}^{(i)}$ span a linear space: the *tangent space*.[2] The evolution of a generic tangent vector is obtained by linearizing Eq. (5.16):

$$\boldsymbol{w}(t+1) = \mathbb{L}[\boldsymbol{x}(t)]\boldsymbol{w}(t), \tag{5.18}$$

where $\mathbb{L}_{ij}[\boldsymbol{x}(t)] = \partial f_i(\boldsymbol{x})/\partial x_j|_{\boldsymbol{x}(t)}$ is the linear stability matrix (Sec. 2.4). Equation (5.18) shows that the stability problem reduces to study the asymptotic properties of products of matrices, indeed the iteration of Eq. (5.18) from the initial condition $\boldsymbol{x}(0)$ and $\boldsymbol{w}(0)$ can be written as $\boldsymbol{w}(t) = \mathbb{P}_t[\boldsymbol{x}(0)]\boldsymbol{w}(0)$, where

$$\mathbb{P}_t[\boldsymbol{x}(0)] = \prod_{k=0}^{t-1} \mathbb{L}[\boldsymbol{x}(k)]\,.$$

In this context, a result of particular relevance is provided by Oseledec (1968) multiplicative theorem (see also Raghunathan (1979)) which we enunciate without proof.

Let $\{\mathbb{L}(1), \mathbb{L}(2), \ldots, \mathbb{L}(k), \ldots\}$ *be a sequence of* $d \times d$ *stability matrices referring to the evolution rule (5.16), assumed to be an application of the compact manifold* A *onto itself, with continuous derivatives. Moreover, let*

[2] The use of tangent vectors implies the limit of infinitesimal distance as in Eq. (5.17).

μ be an invariant measure on A under the evolution (5.16). The matrix product $\mathbb{P}_t[\boldsymbol{x}(0)]$ is such that, the limit

$$\lim_{t\to\infty}\left[\mathbb{P}_t^T[\boldsymbol{x}(0)]\mathbb{P}_t[\boldsymbol{x}(0)]\right]^{\frac{1}{2t}} = \mathbb{V}[\boldsymbol{x}(0)]$$

exists with the exception of a subset of initial conditions of zero measure. Where $\mathbb{P}^T$ denotes the transpose of $\mathbb{P}$.

The symmetric matrix $\mathbb{V}[\boldsymbol{x}(0)]$ has d real and positive eigenvalues $\nu_i[\boldsymbol{x}(0)]$ whose logarithm defines the Lyapunov exponents

$$\lambda_i(\boldsymbol{x}(0)) = \ln(\nu_i[\boldsymbol{x}(0)]).$$

Customarily, they are listed in descending order $\lambda_{max} = \lambda_1 \geq \lambda_2 .... \geq \lambda_d$, equal sign accounts for multiplicity due to a possible eigenvalue degeneracy. Oseledec theorem guarantees the existence of LEs for a wide class of dynamical systems, under very general conditions.

However, it is worth remarking that CLE are associated to a single trajectory, so that we are not allowed to drop out the dependence on the initial condition $\boldsymbol{x}(0)$ unless the dynamics is ergodic. In that case Lyapunov spectrum is independent of the initial condition becoming a global property of the system. Nevertheless, mostly in low dimensional symplectic systems, the phase space can be parted in disconnected ergodic components with a different LE each. For instance, this occurs in planar billiards [Benettin and Strelcyn (1978)].

An important consequence of the Oseledec theorem concerns the expansion rate of k-dimensional oriented volumes $\text{Vol}_k(t) = \text{Vol}[\boldsymbol{w}^{(1)}(t), \boldsymbol{w}^{(2)}(t), \ldots, \boldsymbol{w}^{(k)}(t)]$ delimited by k independent tangent vectors $\boldsymbol{w}^{(1)}, \boldsymbol{w}^{(2)}, \ldots, \boldsymbol{w}^{(k)}$. Under the effect of the dynamics, the k-parallelepiped is distorted and its volume-rate of expansion/contraction is given by the sum of the first k Lyapunov exponents:

$$\sum_{i=1}^{k} \lambda_i = \lim_{t\to\infty} \frac{1}{t} \ln\left[\frac{\text{Vol}_k(t)}{\text{Vol}_k(0)}\right] . \tag{5.19}$$

For $k = 1$ this result recovers Eq. (5.17), notice that here the limit $\text{Vol}_k(0) \to 0$ is not necessary as we are directly working in tangent space. Equation (5.19) also enables to devise an algorithm for numerically computing the whole Lyapunov spectrum, by monitoring the evolution of k tangent vectors (see Box B.9).

When we consider k-volumes with $k = d$, d being the phase-space dimensionality, the sum (5.19) gives the phase-space contraction rate,

$$\sum_{i=1}^{d} \lambda_i = \langle \ln|\det[\mathbb{L}(\boldsymbol{x})]|\rangle ,$$

which for continuous time dynamical systems reads

$$\sum_{i=1}^{d} \lambda_i = \langle \boldsymbol{\nabla} \cdot \boldsymbol{f}(\boldsymbol{x})\rangle, \tag{5.20}$$

angular brackets indicates time average. Therefore, recalling the distinction between conservative and dissipative dynamical systems (Sec. 2.1.1), we have that for the former the Lyapunov spectrum sums to zero. Moreover, for Hamiltonian system or symplectic maps, the Lyapunov spectrum enjoys a remarkable symmetry referred to as *pairing rule* in the literature [Benettin *et al.* (1980)]. This symmetry is a straightforward consequence of the symplectic structure and, for a system with N degrees of freedom (having $2N$ Lyapunov exponents), it consists in the relationship

$$\lambda_i = -\lambda_{2N-i+1} \quad i = 1, \dots, N\,, \tag{5.21}$$

so that only half of the spectrum needs to be computed. The reader may guess that pairing stems from the property discussed in Box B.2.

In autonomous continuous time systems without stable fixed points at least one Lyapunov exponent is vanishing. Indeed there cannot be expansion or contraction along the direction tangent to the trajectory. For instance, consider a reference trajectory $\boldsymbol{x}(t)$ originating from $\boldsymbol{x}(0)$ and take as a perturbed trajectory that originating from $\boldsymbol{x}'(0) = \boldsymbol{x}(\tau)$ with $\tau \ll 1$, clearly if the system is autonomous $|\boldsymbol{x}(t) - \boldsymbol{x}'(t)|$ remains constant. Of course, in autonomous continuous time Hamiltonian system, Eq. (5.21) implies that a couple of vanishing exponents occur.

In particular cases, the phase-space contraction rate is constant, $\det[\mathbb{L}(\boldsymbol{x})] =$ const or $\boldsymbol{\nabla} \cdot \boldsymbol{f}(\boldsymbol{x}) =$ const. For instance, for the Lorenz model $\boldsymbol{\nabla} \cdot \boldsymbol{f}(\boldsymbol{x}) = -(\sigma+b+1)$ (see Eq. (3.12)) and thus, through Eq. (5.20), we know that $\lambda_1 + \lambda_2 + \lambda_3 = -(\sigma + b + 1)$. Moreover, one exponent has to be zero, as Lorenz model is an autonomous set of ODEs. Therefore, to know the full spectrum we simply need to compute λ_1 because $\lambda_3 = -(\sigma + b + 1) - \lambda_1$ (λ_2 being zero).

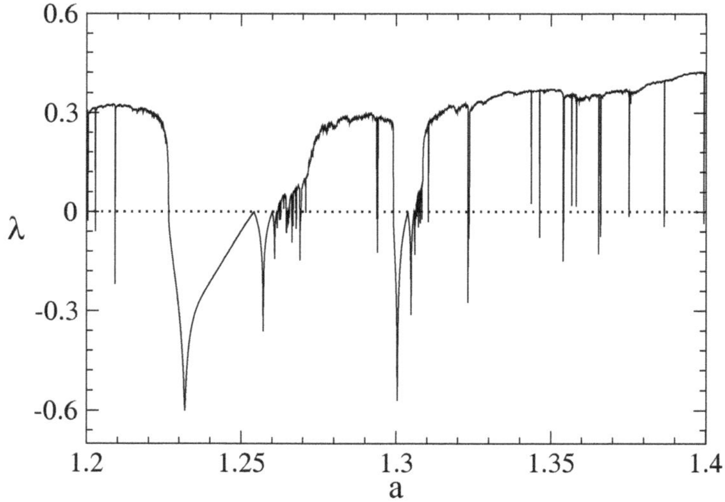

Fig. 5.15 Maximal Lyapunov exponent λ_1 for the Hénon map as a function of the parameter a with $b = 0.3$. The horizontal line separates parameter regions with chaotic ($\lambda_1 > 0$) non-chaotic ($\lambda_1 < 0$) behaviors.

As seen in the case of the logistic map (Fig. 3.5), sometimes, chaotic and non-chaotic motions may alternate in a complicated fashion when the control parameter is varied. Under these circumstances, the LE displays an irregular alternation between positive and negative values, as for instance in the Hénon map (Fig. 5.15).

In the case of dissipative systems, the set of LEs is informative about qualitative features of the attractor. For example, if the attractor reduces to

(a) stable fixed point, all the exponents are negative;
(b) limit cycle, an exponent is zero and the remaining ones are all negative;
(c) k-dimensional stable torus, the first k LEs vanish and the remaining ones are negative;
(d) for strange attractor generated by a chaotic dynamics at least one exponent is positive.

Box B.9: Algorithm for computing Lyapunov Spectrum

A simple and efficient numerical technique for calculating the Lyapunov spectrum has been proposed by Benettin *et al.* (1978b, 1980). The idea is to employ Eq. (5.19) and thus to evolve a set of d linearly independent tangent vectors $\{\boldsymbol{w}^{(1)}, \ldots, \boldsymbol{w}^{(d)}\}$ forming a d-dimensional parallelepiped of volume Vol_d. Equation (5.19) allows us to compute $\Lambda_k = \sum_{i=1}^{k} \lambda_i$. For $k = 1$ we have the maximal LE $\lambda_1 = \Lambda_1$ and then the k-th LE is simply obtained from the recursion $\lambda_k = \Lambda_k - \Lambda_{k-1}$.

We start describing the first necessary step, i.e. the computation of λ_1. Choose an arbitrary tangent vector $\boldsymbol{w}^{(1)}(0)$ of unitary modulus, and evolve it up to a time t by means of Eq. (5.18) (or the equivalent one for ODEs) so to obtain $\boldsymbol{w}^{(1)}(t)$. When λ_1 is positive, $\boldsymbol{w}^{(1)}$ exponentially grows without any bound and its direction identifies the direction of maximal expansion. Therefore, to prevent computer overflow, $\boldsymbol{w}^{(1)}(t)$ must be periodically renormalized to unitary amplitude, at each interval τ of time. In practice, τ should be neither too small, to avoid wasting of computational time, nor too large, to maintain $\boldsymbol{w}^{(1)}(\tau)$ far from the computer overflow limit. Thus, $\boldsymbol{w}^{(1)}(0)$ is evolved to $\boldsymbol{w}^{(1)}(\tau)$, and its length $\alpha_1(1) = |\boldsymbol{w}^{(1)}(\tau)|$ computed; then $\boldsymbol{w}^{(1)}(\tau)$ is rescaled as $\boldsymbol{w}^{(1)}(\tau) \to \boldsymbol{w}^{(1)}(\tau)/|\boldsymbol{w}^{(1)}(\tau)|$ and evolved again up to time 2τ. During the evolution, we repeat the renormalization and store all the amplitudes $\alpha_1(n) = |\boldsymbol{w}^{(1)}(n\tau)|$, obtaining the largest Lyapunov exponent as:

$$\lambda_1 = \lim_{n\to\infty} \frac{1}{n\tau} \sum_{m=1}^{n} \ln \alpha_1(m) \,. \tag{B.9.1}$$

It is worth noticing that, as the tangent vector evolution (5.18) is linear, the above result is not affected by the renormalization procedure.

To compute λ_2, we need two initially orthogonal unitary tangent vectors $\{\boldsymbol{w}^{(1)}(0), \boldsymbol{w}^{(2)}(0)\}$. They identify a parallelogram of area $\mathrm{Vol}_2(0) = |\boldsymbol{w}^{(1)} \times \boldsymbol{w}^{(2)}|$ (where $\times$ denotes the cross product). The evolution deforms the parallelogram and changes its area because both $\boldsymbol{w}^{(1)}(t)$ and $\boldsymbol{w}^{(2)}(t)$ tend to align along the direction of maximal expansion, as shown in Fig. B9.1. Therefore, at each time interval τ, we rescale $\boldsymbol{w}^{(1)}$ as before

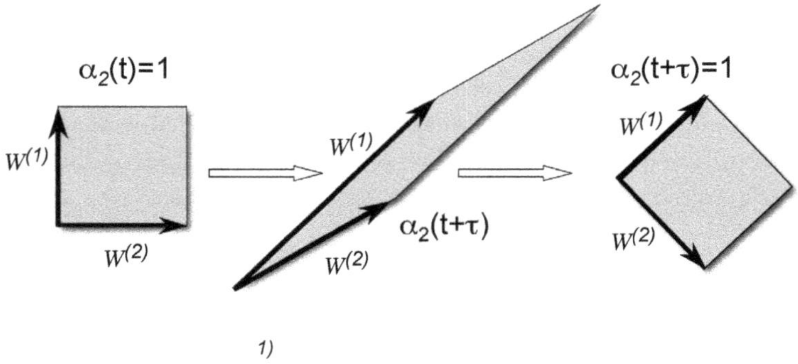

Fig. B9.1 Pictorial representation of the basic step of the algorithm for computing the Lyapunov exponents. The orthonormal basis at time $t = j\tau$ is evolved till $t = (j+1)\tau$ and then it is again orthonormalized. Here $k = 2$.

and replace $\boldsymbol{w}^{(2)}$ with a unitary vector orthogonal to $\boldsymbol{w}^{(1)}$. In practice we can use the Gram-Schmidt orthonormalization method. In analogy with Eq. (B.9.1) we have

$$\Lambda_2 = \lambda_1 + \lambda_2 = \lim_{n\to\infty} \frac{1}{n\tau} \sum_{m=1}^{n} \ln \alpha_2(m)$$

where α_2 is the area of the parallelogram before each re-orthonormalization.

The procedure can be iterated for a k-volume formed by k independent tangent vectors to compute all the Lyapunov spectrum, via the relation

$$\Lambda_k = \lambda_1 + \lambda_2 + \ldots + \lambda_k = \lim_{n\to\infty} \frac{1}{n\tau} \sum_{m=1}^{n} \ln \alpha_k(s) \,,$$

α_k being the volume of the k-parallelepiped before re-orthonormalization.

5.3.1 *Oseledec theorem and the law of large numbers*

Oseledec theorem constitutes the main mathematical result of Lyapunov analysis, the basic difficulty relies on the fact that it deals with product of matrices, generally a non-commutative operation. The essence of this theorem becomes clear when considering the one dimensional case, for which the stability matrix reduces to a scalar multiplier $a(t)$ and the tangent vectors are real numbers obeying the multiplicative process $w(t+1) = a(t)w(t)$, which is solved by

$$w(t) = \prod_{k=0}^{t-1} a(k)\, w(0) \,. \tag{5.22}$$

As we are interested in the asymptotic growth of $|w(t)|$ for large t, it is convenient to transform the product (5.22) into the sum

$$\ln|w(t)| = \sum_{k=0}^{t-1} \ln|a(k)| + \ln|w(0)| \,.$$

From the above expression it is possible to realize that Oseledec's theorem reduces to the law of large numbers for the variable $\ln|a(k)|$ [Gnedenko and Ushakov (1997)], and for the average exponential growth, we have

$$\lambda = \lim_{t\to\infty} \frac{1}{t} \ln\left|\frac{w(t)}{w(0)}\right| = \lim_{t\to\infty} \frac{1}{t} \sum_{k=0}^{t-1} \ln|a(k)| = \langle \ln|a| \rangle \tag{5.23}$$

where λ is the LE. In other words, with probability 1 as $t \to \infty$, an infinitesimal displacement w expands with the law

$$|w(t)| \sim \exp(\langle \ln|a| \rangle\, t) \,.$$

Oseledec's theorem is the equivalent of the law of large numbers for the product of non-commuting matrices.

To elucidate the link between Lyapunov exponents, invariant measure and ergodicity, it is instructive to apply the above computation to a one-dimensional map. Consider the map $x(t+1) = g(x(t))$ with initial condition $x(0)$, for which the tangent vector $w(t)$ evolves as $w(t+1) = g'(x(t))w(t)$. Identifying $a(t) = |g'(x(t))|$, from Eq. (5.23) we have that the LE can be written as

$$\lambda = \lim_{T\to\infty} \frac{1}{T} \sum_{t=1}^{T} \ln|g'(x(t))| \,.$$

If the system is ergodic, λ does not depend on $x(0)$ and can be obtained as an average over the invariant measure $\rho^{inv}(x)$ of the map:

$$\lambda = \int \mathrm{d}x\, \rho^{inv}(x)\, \ln|g'(x)| \,, \tag{5.24}$$

In order to be specific, consider the generalized tent map (or *skew tent map*) defined by

$$x(t+1) = g(x(t)) = \begin{cases} \dfrac{x(t)}{p} & 0 \le x(t) < p \\[2ex] \dfrac{1-x(t)}{1-p} & p \le x(t) \le 1 \,. \end{cases} \tag{5.25}$$

with $p \in [0:1]$. It is easy to show that $\rho^{inv}(x) = 1$ for any p, moreover, the multiplicative process describing the tangent evolution is particularly simple as $|g'(x)|$ takes only two values, $1/p$ and $1/(1-p)$. Thus the LE is given by

$$\lambda = -p\ln p - (1-p)\ln(1-p) \,,$$

maximal chaoticity is thus obtained for the usual tent map ($p = 1/2$).

The above discussed connection among Lyapunov exponents, law of large numbers and ergodicity essentially tells us that the LEs are self-averaging objects.[3] In concluding this section, it is useful to wonder about the rate of convergence of the limit $t \to \infty$ that, though mathematically clear, cannot be practically (numerically) realized. For reasons which will become much clearer reading the next two chapters, we anticipate here that very different convergence behaviors are typically observed when considering dissipative or Hamiltonian systems. This is exemplified in Fig. 5.16 where we compare the convergence to the maximal LE by numerically following a single trajectory of the standard and Hénon maps. As a matter of facts, the convergence is much slower in Hamiltonian systems, due to presence of "regular" islands, around which the trajectory may stay for long times, a drawback rarely encountered in dissipative systems.

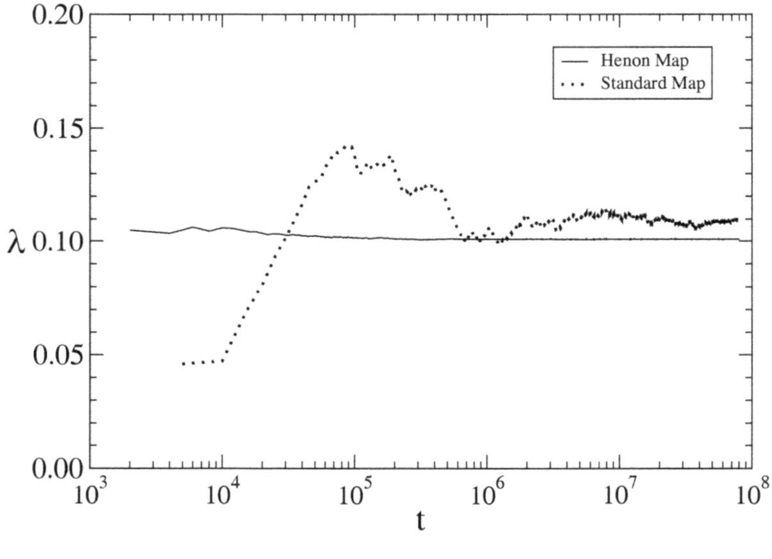

Fig. 5.16 Convergence to the maximal LE in the standard map (2.18) with $K = 0.97$ and Hénon map (5.1) with $a = 1.271$ and $b = 0.3$, as obtained by using Benettin *et al.* algorithm (Box B.9).

5.3.2 *Remarks on the Lyapunov exponents*

5.3.2.1 *Lyapunov exponents are topological invariant*

As anticipated in Box B.3, Lyapunov exponents of topologically conjugated dynamical systems as, for instance the logistic map at $r = 4$ and the tent map, are

[3]Readers accustomed to statistical mechanics of disordered systems, use the term self-averaging to mean that in the thermodynamic limit it is not necessary to perform an average over samples with different realizations of the disorder. In this context, the self-averaging property indicates that it is not necessary an average over many initial conditions.

identical. We show the result for a one-dimensional map

$$x(t+1) = g(x(t))\,, \tag{5.26}$$

which is assumed to be ergodic with Lyapunov exponent

$$\lambda^{(x)} = \lim_{T\to\infty} \frac{1}{T} \sum_{t=1}^{T} \ln |g'(x(t))|\,. \tag{5.27}$$

Under the invertible change of variable $y = h(x)$ with $h' \neq 0$, Eq. (5.26) becomes

$$y(t+1) = f(y(t)) = h(g(h^{-1}(y(t))))\,,$$

and the corresponding Lyapunov exponent is

$$\lambda^{(y)} = \lim_{T\to\infty} \frac{1}{T} \sum_{t=1}^{T} \ln |f'(y(t))|\,. \tag{5.28}$$

Equations (5.27) and (5.28) can be, equivalently, rewritten as:

$$\lambda^{(x)} = \lim_{T\to\infty} \frac{1}{T} \sum_{t=1}^{T} \ln \left| \frac{z^{(x)}(t)}{z^{(x)}(t-1)} \right|\,,$$

$$\lambda^{(y)} = \lim_{T\to\infty} \frac{1}{T} \sum_{t=1}^{T} \ln \left| \frac{z^{(y)}(t)}{z^{(y)}(t-1)} \right|\,,$$

where the tangent vector $z^{(x)}$ associated to Eq. (5.26) evolves according to $z^{(x)}(t+1) = g'(x(t))z^{(x)}(t)$, and analogously $z^{(y)}(t+1) = f'(y(t))z^{(y)}(t)$. From the chain rule of differentiation we have $z^{(y)} = h'(x)z^{(x)}$ so that

$$\lambda^{(y)} = \lim_{T\to\infty} \frac{1}{T} \sum_{t=1}^{T} \ln \left| \frac{z^{(x)}(t)}{z^{(x)}(t-1)} \right| + \lim_{T\to\infty} \frac{1}{T} \sum_{t=1}^{T} \ln \left| \frac{h'(x(t))}{h'(x(t-1))} \right|\,.$$

Noticing that the second term of the right hand side of the above expression is $\lim_{T\to\infty}(1/T)(\ln|h'(x(T))| - \ln|h'(x(0))|) = 0$, it follows

$$\lambda^{(x)} = \lambda^{(y)}\,.$$

5.3.2.2 *Relationship between Lyapunov exponents of flows and Poincaré maps*

In Section 2.1.2 we saw that a Poincaré map

$$\boldsymbol{P}_{n+1} = \boldsymbol{G}(\boldsymbol{P}_n) \quad \text{with} \quad \boldsymbol{P}_k \in \mathrm{I\!R}^{d-1} \tag{5.29}$$

can always be associated to a d dimensional flow

$$\frac{d\boldsymbol{x}}{dt} = \boldsymbol{f}(\boldsymbol{x}) \quad \text{with} \quad \boldsymbol{x} \in \mathrm{I\!R}^{d}\,. \tag{5.30}$$

It is quite natural to wonder about the relation between the CLE spectrum of the flow (5.30) and that of the corresponding Poincaré section (5.29). Such a relation can written as

$$\lambda_k = \frac{\widetilde{\lambda}_{k'}}{\langle \tau \rangle}, \tag{5.31}$$

where the tilde indicates the LE of the Poincaré map. As for the correspondence between k and k', one should notice that any chaotic autonomous ODE, as Eq. (5.30), always admits a zero-Lyapunov exponent and, therefore, except for this one (which is absent in the discrete time description) Eq. (5.31) always applies with $k' = k$ or $k' = k - 1$.

The average $\langle \tau \rangle$ corresponds to the mean return time on the Poincaré section, i.e. $\langle \tau \rangle = \langle t_n - t_{n-1} \rangle$, t_n being the time at which the trajectory $\boldsymbol{x}(t)$ cross the Poincaré surface for the n-th time. Such a relation confirms once again that there is no missing of information in the Poincaré construction.

We show how relation (5.31) stems by discussing the case of the maximum LE. From the definition of Lyapunov exponent we have that for infinitesimal perturbations

$$|\delta \boldsymbol{P}_n| \sim e^{\widetilde{\lambda}_1 n} \quad \text{and} \quad |\delta \boldsymbol{x}(t)| \sim e^{\lambda_1 t}$$

for the flow and map, respectively. Clearly, $|\delta \boldsymbol{P}_n| \sim |\delta \boldsymbol{x}(t_n)|$ and if $n \gg 1$ then $t_n \approx n\langle \tau \rangle$, so that relation (5.31) follows.

We conclude with an example. Lorenz model seen in Sec. 3.2 possesses three LEs. The first λ_1 is positive, the second λ_2 is zero and the third λ_3 must be negative. Its Poincaré map is two-dimensional with one, $\widetilde{\lambda}_1$, positive and one, $\widetilde{\lambda}_2$, negative Lyapunov exponent. From Eq. (5.31): $\lambda_1 = \widetilde{\lambda}_1/\langle \tau \rangle$ and $\lambda_3 = \widetilde{\lambda}_2/\langle \tau \rangle$.

5.3.3 *Fluctuation statistics of finite time Lyapunov exponents*

Lyapunov exponents are related to the "typical" or "average behavior" of the expansion rates of nearby trajectories, and do not take into account finite time fluctuations of these rates. In some systems such fluctuations must be characterized as they represent the relevant aspect of the dynamics as, e.g., in *intermittent* chaotic systems [Fujisaka and Inoue (1987); Crisanti *et al.* (1993a); Brandenburg *et al.* (1995); Contopoulos *et al.* (1997)] (see also Sec. 6.3).

The fluctuations of the expansion rate can be accounted for by introducing the so-called Finite Time Lyapunov Exponent (FTLE) [Fujisaka (1983); Benzi *et al.* (1985)] in a way similar to what has been done in Sec. 5.2.3 for multifractals, i.e. by exploiting the large deviation formalism (Box B.8). The FTLE, hereafter indicated by γ, is the fluctuating quantity defined as

$$\gamma(\tau, t) = \frac{1}{t} \ln \left[\frac{|\boldsymbol{w}(\tau + t)|}{|\boldsymbol{w}(\tau)|} \right] = \frac{1}{t} \ln R(\tau, t) ,$$

indicating the partial, or local, growth rate of the tangent vectors within the time interval $[\tau, \tau{+}t]$. The knowledge of the distribution of the so-called *response function*

$R(\tau, t)$ allows a complete characterization of local expansion rates. By definition, the LE is recovered by the limit

$$\lambda = \lim_{t\to\infty} \langle \gamma(\tau, t) \rangle_\tau = \lim_{t\to\infty} \frac{1}{t} \langle \ln R(\tau, t) \rangle_\tau \,,$$

where $\langle [\ldots] \rangle_\tau$ has the meaning of time-average over τ, in ergodic systems it can be replaced by phase-average.

Fluctuations can be characterized by studying the q-moments of the response function

$$\mathcal{R}_q(t) = \langle R^q(\tau, t) \rangle_\tau = \langle \mathrm{e}^{q\gamma(t,\tau)\, t} \rangle_\tau$$

which, due to trajectory instability, for finite but long enough times are expected to scale asymptotically as

$$\mathcal{R}_q(t) \sim \mathrm{e}^{t\, L(q)} \,,$$

with

$$L(q) = \lim_{t\to\infty} \frac{1}{t} \ln \langle R^q(\tau, t) \rangle_\tau = \lim_{t\to\infty} \frac{1}{t} \ln \mathcal{R}_q(t) \tag{5.32}$$

is called generalized Lyapunov exponent, characterizing the fluctuations of the FTLE $\gamma(t)$. The generalized LE $L(q)$ (5.32) plays exactly the same role of the $D(q)$ in Eq. (5.8).[4] The maximal LE is nothing but the limit

$$\lambda_1 = \lim_{q\to 0} \frac{L(q)}{q} = \left. \frac{\mathrm{d}L(q)}{\mathrm{d}q} \right|_{q=0} \,,$$

and is the counterpart of the information dimension $D(1)$ in the multifractal analysis. In the absence of fluctuations $L(q) = \lambda_1 q$. In general, the higher the moment, the more important is the contribution to the average coming from trajectories with a growth rate largely different from λ. In particular, the limits $\lim_{q\to\pm\infty} L(q)/q = \gamma_{max/min}$ select the maximal and minimal expanding rate, respectively.

For large times, Oseledec's theorem ensures that values of γ largely deviating from the most probable value λ_1 are rare, so that the distribution of γ will be peaked around λ_1 and, according to large deviation theory (Box B.8), we can make the *ansatz*

$$\mathrm{d}P_t(\gamma) = \rho(\gamma) \mathrm{e}^{-S(\gamma) t} \mathrm{d}\gamma \,,$$

where $\rho(\gamma)$ is a regular density in the limit $t \to \infty$ and $S(\gamma)$ is the rate or Cramer function (for its properties see Box B.8), which vanishes for $\gamma = \lambda_1$ and is positive for $\gamma \neq \lambda_1$.

Clearly $S(\gamma)$ is the equivalent of the multifractal spectrum of dimensions $f(\alpha)$. Thus, following the same algebraic manipulations of Sec. 5.2.3, we can connect $S(\gamma)$ to $L(q)$. In particular, the moment $\mathcal{R}_q$ can be rewritten as

$$\mathcal{R}_q(t) = \int \mathrm{d}\gamma\, \rho(\gamma) \mathrm{e}^{t\,[q\gamma - S(\gamma)]} \,, \tag{5.33}$$

[4] In particular, the properties of $L(q)$ are the same as those of the function $(q-1)D(q)$.

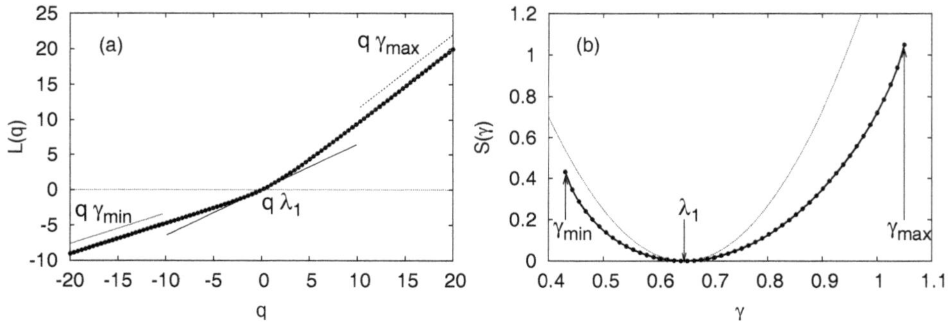

Fig. 5.17 (a) $L(q)$ vs q as from Eq. (5.34) for $p = 0.35$. The asymptotic $q \to \pm\infty$ behaviors are shown as dotted lines while in solid lines we depict the behavior close to the origin. (b) The rate function $S(\gamma)$ vs γ corresponding to (a). Critical points are indicated by arrows. The parabolic approximation of $S(\gamma)$ corresponding to(5.35) is also shown, see text for details.

where we used the asymptotic expression $R(t) \sim \exp(\gamma t)$. In the limit $t \to \infty$, the asymptotic value of the integral (5.33) is dominated by the leading contribution (saddle point) coming from those γ-values which maximize the exponent, so that

$$L(q) = \max_{\gamma}\{q\gamma - S(\gamma)\}\,.$$

As for $D(q)$ and $f(\alpha)$, this expression establishes that $L(q)$ and $S(\gamma)$ are linked by a Legendre transformation.

As an example we can reconsider the skew tent map (5.25), for which an easy computation shows that

$$\langle R^q(t,\tau)\rangle_\tau = \left[p\left(\frac{1}{p}\right)^q + (1-p)\left(\frac{1}{1-p}\right)^q\right]^t \tag{5.34}$$

and thus

$$L(q) = \ln[p^{1-q} + (1-p)^{1-q}]\,,$$

whose behavior is illustrated in Fig. 5.17a. Note that asymptotically, for $q \to \pm\infty$, $L(q) \sim q\gamma_{max,min}$, while, in $q = 0$, the tangent to $L(q)$ has slope $\lambda_1 = L'(q) = -p\ln p - (1-p)\ln(1-p)$. Through the inverse Legendre transformation we can obtain the Cramer function $S(\gamma)$ associated to $L(q)$ (shown in Fig. 5.17b). Here, for brevity, we omit the algebra which is a straightforward repetition of that discussed in Sec. 5.2.3.

In general, the distribution $P_t(\gamma)$ is not known a priori and should be sampled via numerical simulations. However, its shape can be guessed and often well approximated around the peak by assuming that, due to the randomness and decorrelation induced by the chaotic motion, $\gamma(t)$ behaves as a random variable. In particular, assuming the validity of central limit theorem (CLT) for $\gamma(t)$ [Gnedenko and Ushakov (1997)], for large times P_t converges to the Gaussian

$$P_t(\gamma) \sim \exp\left[-\frac{t(\gamma-\lambda_1)^2}{2\sigma^2}\right] \tag{5.35}$$

characterized by two parameters, namely $\lambda_1 = L'(0)$ and $\sigma^2 = \lim_{t\to\infty} t\,\langle(\gamma(t) - \lambda_1)^2\rangle = L''(0)$. Note that the variance of γ behaves as σ^2/t, i.e. the probability distribution shrinks to a δ-function for $t \to \infty$ (another way to say that the law of large numbers is asymptotically verified). Equation (5.35) corresponds to approximate the Cramer function as the parabola $S(\gamma) \approx (\gamma - \lambda_1)^2/(2\sigma^2)$ (see Fig. 5.17b). In this approximation we have that the generalized Lyapunov exponent reads:

$$L(q) = \lambda_1 q + \frac{\sigma^2 q^2}{2} \,.$$

We may wonder how well the approximation (5.35) performs in reproducing the true behavior of $P_t(\gamma)$. Due to dynamical correlations, the tails of the distribution are typically non-Gaussian and sometimes $\gamma(t)$ violates so hardly the CLT that even the bulk deviates from (5.35). Therefore, in general, the distribution of finite time Lyapunov exponent $\gamma(t)$ cannot be characterized in terms of λ and σ^2 only.

5.3.4 *Lyapunov dimension*

In dissipative systems, the Lyapunov spectrum $\{\lambda_1, \lambda_2, ..., \lambda_d\}$ can be used also to extract important quantitative information concerning the fractal dimension.

Simple arguments show that for two dimensional dissipative chaotic maps

$$D_F \approx D_L = 1 + \frac{\lambda_1}{|\lambda_2|} \,, \tag{5.36}$$

where D_L is usually called Lyapunov or Kaplan-Yorke dimension. The above relation can be derived by observing that a small circle of radius ϵ is deformed by the dynamics into an ellipsoid of linear dimensions $L_1 = \epsilon \exp(\lambda_1 t)$ and $L_2 = \epsilon \exp(-|\lambda_2| t)$. Therefore, the number of square boxes of side $\ell = L_2$ needed to cover the ellipsoid is proportional to

$$N(\ell) = \frac{L_1}{L_2} = \frac{\exp(\lambda_1 t)}{\exp(-|\lambda_2| t)} \sim \ell^{-\left(1+\frac{\lambda_1}{|\lambda_2|}\right)}$$

that via Eq. (5.4) supports the relation (5.36). Notice that this result is the same we obtained for the horseshoe map (Sec. 5.2.2), since in that case $\lambda_1 = \ln 2$ and $\lambda_2 = -\ln(2\eta)$.

The relationship between fractal dimension and Lyapunov spectrum also extends to higher dimensions and is known as the Kaplan and Yorke (1979) formula, which is actually a conjecture however verified in several cases:

$$D_F \approx D_L = j + \frac{\sum_{i=1}^{j} \lambda_i}{|\lambda_{j+1}|} \tag{5.37}$$

where j is the largest index such that $\sum_{i=1}^{j} \lambda_j \geq 0$, once LEs are ranked in decreasing order. The j-dimensional hyper-volumes should either increase or remain constant, while the $(j+1)$-dimensional ones should contract to zero. Notice that formula (5.37) is a simple linear interpolation between j and $j+1$, see Fig. 5.18.

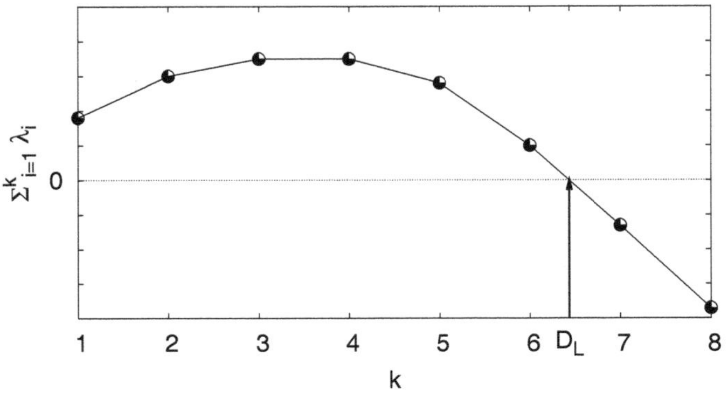

Fig. 5.18 Sketch of the construction for deriving the Lyapunov dimension. In this example $d = 8$ and the CLE spectrum is such that $6 < D_L < 7$. Actually D_L is just the intercept with the x-axis of the segment joining the point $(6, \sum_{i=1}^{6} \lambda_i)$ with $(7, \sum_{i=1}^{7} \lambda_i)$.

For N-degree of freedom Hamiltonian systems, the pairing symmetry (5.21) implies that $D_L = d$, where $d = 2 \times N$ is the phase-space dimension. This is another way to see that in such systems no attractors exist.

Although the Kaplan-York conjecture has been rigorously proved for a certain class of dynamical systems [Ledrappier (1981); Young (1982)] (this is the case, for instance, of systems possessing an SRB measure, see Box B.10 and also Eckmann and Ruelle (1985)), there is no proof for its general validity. Numerical simulations suggest the formula to hold approximately quite in general. We remark that due to a practical impossibility to directly measure fractal dimensions larger than 4, formula (5.37) practically represents the only viable estimate of the fractal dimension of high dimensional attractors and, for this reason, it assumes a capital importance in the theory of systems with many degrees of freedom.

We conclude by a numerical example concerning the Hénon map (5.1) for $a = 1.4$ and $b = 0.3$. A direct computation of the maximal Lyapunov exponent gives $\lambda_1 \approx 0.419$ which, being $\lambda_1 + \lambda_2 = \ln|\det(\mathbb{L})| = \ln b = -1.20397$, implies $\lambda_2 \approx -1.623$ and thus $D_L = 1 + \lambda_1/|\lambda_2| \approx 1.258$. As seen in Figure 5.7 the box counting and correlation dimension of Hénon attractor are $D_F \approx 1.26$ and $\nu = D(2) \approx 1.2$. These three values are very close each other because the multifractality is weak.

Box B.10: Mathematical chaos

Many results and assumptions that have been presented for chaotic systems, such as e.g. the existence of ergodic measures, the equivalence between Lyapunov and fractal dimension or, as we will see in Chapter 8, the Pesin relation between the sum of positive Lyapunov exponents and the Kolmogorov-Sinai entropy, cannot be proved unless imposing some restriction on the mathematical properties of the considered systems [Eckmann and Ruelle

(1985)]. This box aims to give a flavor of the rigorous approaches to chaos by providing hints on some important mathematical aspects. The reader may find a detailed treatment in more mathematical oriented monographs [Ruelle (1989); Katok and Hasselblatt (1995); Collet and Eckmann (2006)] or in surveys as Eckmann and Ruelle (1985).

A: Hyperbolic sets and Anosov systems

Consider a system evolving according to a discrete time map or a ODE, and a compact set Ω invariant under the time evolution $\mathcal{S}^t$, a point $\boldsymbol{x} \in \Omega$ is *hyperbolic* if its associated tangent space $T_{\boldsymbol{x}}$ can be decomposed into the direct sum of the stable ($E^s_{\boldsymbol{x}}$), unstable ($E^u_{\boldsymbol{x}}$) and neutral ($E^0_{\boldsymbol{x}}$) subspaces (i.e. $T_{\boldsymbol{x}} = E^s_{\boldsymbol{x}} \oplus E^u_{\boldsymbol{x}} \oplus E^0_{\boldsymbol{x}}$), defined as follows:
if $\boldsymbol{z}(0) \in E^s_{\boldsymbol{x}}$ there exist $K > 0$ and $0 < \alpha < 1$ such that

$$|\boldsymbol{z}(t)| \leq K\alpha^t |\boldsymbol{z}(0)|$$

while if $\boldsymbol{z}(0) \in E^u_{\boldsymbol{x}}$

$$|\boldsymbol{z}(-t)| \leq K\alpha^t |\boldsymbol{z}(0)| \, ,$$

where $\boldsymbol{z}(t)$ and $\boldsymbol{z}(-t)$ denote the forward and backward time evolution of the tangent vector, respectively. Finally, if $\boldsymbol{z}(0) \in E^0_{\boldsymbol{x}}$ then $|\boldsymbol{z}(\pm t)|$ remains bounded and finite at any time t. Note that $E^0_{\boldsymbol{x}}$ must be one dimensional for ODE and it reduces to a single point in case of maps. The set Ω is said hyperbolic if all its points are hyperbolic. In a hyperbolic set all tangent vectors, except those directed along the neutral space, grow or decrease at exponential rates, which are everywhere bounded away from zero.

The concept of hyperbolicity allows us to define two classes of systems.
Anosov systems are smooth (differentiable) maps of a compact smooth manifold with the property that the entire space is a hyperbolic set.
Axiom A systems are dissipative smooth maps whose attractor Ω is a hyperbolic set and periodic orbits are dense in Ω.[5] Axiom A attractors are structurally stable, i.e. their structure survive a small perturbation of the map.

Systems which are Anosov or Axiom A possess nice properties which allows the rigorous derivation of many results [Eckmann and Ruelle (1985); Ruelle (1989)]. However, apart from special cases, attractors of chaotic systems are typically not hyperbolic. For instance, the Hénon attractor (Fig. 5.1) contains points $\boldsymbol{x}$ where the *stable and unstable manifolds*[6] are tangent to one another in some locations and, as a consequence, $E^{u,s}_{\boldsymbol{x}}$ cannot be defined, and the attractor is not a hyperbolic set. On the contrary, the baker's map (5.5) is hyperbolic but, since it is not differentiable, is not Axiom A.

B: SRB measure

For conservative systems, we have seen in Chap. 4 that the Lebesgue measure (i.e. uniform distribution) is invariant under the time evolution and, in the presence of chaos, is

[5]Note that an Anosov system is always also Axiom A.

[6]Stable and unstable manifolds generalize the concept of stable and unstable directions outside the tangent space. Given a point $\boldsymbol{x}$, its stable $W^s_{\boldsymbol{x}}$ and unstable $W^u_{\boldsymbol{x}}$ manifold are defined by

$$W^{s,u}_{\boldsymbol{x}} = \{\boldsymbol{y} : \lim_{t\to\pm\infty} \boldsymbol{y}(t) = \boldsymbol{x}\} \, ,$$

namely these are the set of all points in phase space converge forwardly or backwardly in time to $\boldsymbol{x}$, respectively. Of course, infinitesimally close to $\boldsymbol{x}$ $W^{s,u}_{\boldsymbol{x}}$ coincides with $E^{s,u}_{\boldsymbol{x}}$.

the obvious candidate for being the ergodic and mixing measure of the system. Such an assumption, although not completely correct, is often reasonable (e.g., the standard map for high values of the parameter controlling the nonlinearity, see Sec. 7.2). In chaotic dissipative systems, on the contrary, the non trivial invariant ergodic measures are usually singular with respect to the Lebesgue one. Indeed, attracting sets are typically characterized by discontinuous (fractal) structures, transversal to the stretching directions, produced by the folding of unstable manifolds, think of the Smale's Horseshoe (Sec. 5.2.2). This suggests thus that invariant measures may be very rough transversely to the unstable manifolds, making them non-absolute continuous with respect to the Lebesgue measure. It is reasonable, however, to expect the measure to be smooth along the unstable directions, where stretching is acting.

This consideration leads to the concept of *SRB measures* from Sinai, Bowen and Ruelle [Ruelle (1989)]. Given a smooth dynamical system (diffeomorphism)[7] and an invariant measure μ, we call μ a SRB measure if *the conditional measure of* μ *on the unstable manifold is absolutely continuous with respect to the Lebesgue measure on the unstable manifold* (i.e. is uniform on it) [Eckmann and Ruelle (1985)]. Thus, in a sense the SRB measures generalize to dissipative systems the notion of smooth invariant measures for conservative systems.

SRB measures are relevant in physics because they are good candidates to describe natural measures (Sec. 4.6) [Eckmann and Ruelle (1985); Ruelle (1989)].

It is possible to prove that Axiom A attractors always admit SRB measures, and very few rigorous results can be proved relaxing the Axiom A hypothesis, even though recently the existence of SRB measures for the Hénon map has been shown by Benedicks and Young (1993), notwithstanding its non-hyperbolicity.

C: The Arnold cat map

A famous example of Anosov system is Arnold cat map

$$\begin{pmatrix} x(t+1) \\ y(t+1) \end{pmatrix} = \begin{pmatrix} 1 & 1 \\ 1 & 2 \end{pmatrix} \begin{pmatrix} x(t) \\ y(t) \end{pmatrix} \quad \bmod 1\,,$$

that we already encountered in Sec. 4.4 while studying the mixing property. This system, although conservative, illustrates the meaning of the above discussed concepts.

The Arnold map, being a diffeomorphism, has no neutral directions, and its tangent space at any point $\boldsymbol{x}$ is the real plane $\mathbb{R}^2$. The eigenvalues of the associated stability matrix are $l^{u,s} = \left(3 \pm \sqrt{5}\right)/2$ with eigenvectors

$$\boldsymbol{v}^u = \begin{pmatrix} 1 \\ \mathcal{G} \end{pmatrix}, \qquad \boldsymbol{v}^s = \begin{pmatrix} 1 \\ -\mathcal{G}^{-1} \end{pmatrix},$$

$\mathcal{G} = \left(1+\sqrt{5}\right)/2$ being the golden ratio. Since both eigenvalues and eigenvectors are independent of $\boldsymbol{x}$, the stable and unstable directions are given by $\boldsymbol{v}^s$ and $\boldsymbol{v}^u$, respectively. Then, thanks to the irrationality of ϕ and the modulus operation wrapping any line into the unitary square, it is straightforward to figure out that the stable and unstable manifolds,

[7] Given two manifolds A and B, a bijective map f from A to B is called a diffeomorphism if both f and its inverse f^{-1} are differentiable.

associated to any point $\boldsymbol{x}$, consist of lines with slope $\mathcal{G}$ or $-\mathcal{G}^{-1}$, respectively densely filling the unitary square. The exponential rates of growth and decrease of the tangent vectors are given by l^u and l^s, because any tangent vector is a linear combinations of $\boldsymbol{v}^u$ and $\boldsymbol{v}^s$. In other words, if one thinks of such manifolds as the trajectory of a point particle, which moves at constant velocity, exits the square at given instants of times, and re-enters the square form the opposite side, one realizes that it can never re-enter at a point which has been previously visited. In other words, this trajectory, i.e. the unstable manifold, wraps around densely exploring all the square $[0:1] \times [0:1]$, and the invariant SRB measure is the Lebesgue measure $d\mu = dxdy$.

5.4 Exercises

Exercise 5.1: Consider the subset A, of the interval $[0:1]$, whose elements are the infinite sequence of points: $A = \left\{1, \frac{1}{2^\alpha}, \frac{1}{3^\alpha}, \frac{1}{4^\alpha}, \ldots, \frac{1}{n^\alpha}, \ldots\right\}$ with $\alpha > 0$. Show that the Box-counting dimension D_F of set A is $D_F = 1/(1+\alpha)$.

Exercise 5.2: Show that the invariant set (repeller) of the map

$$x(t+1) = \begin{cases} 3x(t) & 0 \leq x(t) < 1/2; \\ 3(1-x(t)) & 1/2 \leq x(t) < 0\,, \end{cases}.$$

is the Cantor set discussed in Sec. 5.2 with fractal dimension $D_F = \ln 2/\ln 3$.

Exercise 5.3: Numerically compute the Grassberger-Procaccia dimension for:
(1) Hénon attractor obtained with $a = 1.4, b = 0.3$;
(2) Feigenbaum attractor obtained with logistic map at $r = r_\infty = 3.569945...$

Exercise 5.4: Consider the following two-dimensional map

$$\begin{aligned} x(t+1) &= \lambda_x x(t) \qquad \mod 1 \\ y(t+1) &= \lambda_y y(t) + \cos(2\pi x(t)) \end{aligned}$$

λ_x and λ_y being positive integers with $\lambda_x > \lambda_y$. This map has no attractors with finite y, as almost every initial condition generates an orbit escaping to $y = \pm\infty$. Show that:

(1) the basin of attraction boundary is given by the Weierstrass' curve [Falconer (2003)] defined by

$$y = -\sum_{n=1}^{\infty} \lambda_y^{-n} \cos(2\pi \lambda_x^{n-1}\, x)\,;$$

(2) the fractal dimension of such a curve is $D_F = 2 - \frac{\ln \lambda_y}{\ln \lambda_x}$ with $1 < D_F < 2\,.$

Hint: *Use the property that curves/surfaces separating two basins of attractions are invariant under the dynamics.*

Exercise 5.5:
Consider the fractal set A generated by infinite iteration of the geometrical rule of basic step as in the figure on the right. We define a measure on this fractal as follows: let $\alpha_1, \ldots, \alpha_5$ be positive numbers such that $\sum_{i=1}^{5} \alpha_i = 1$. At the first stage of the construction, we assign to the upper-left box the measure α_1, α_2 to the upper-right box and so on, as shown in the figure. Compute the dimension $D(q)$.

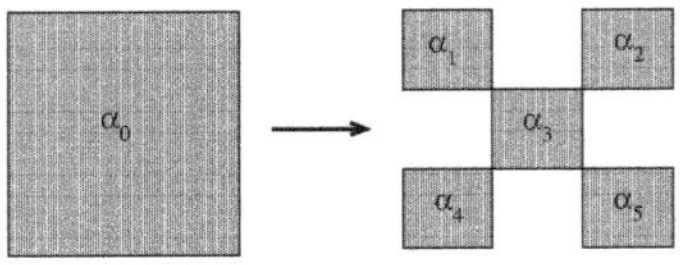

Hint: *Consider the covering with appropriate boxes and compute the number of such boxes.*

Exercise 5.6: Compute the Lyapunov exponents of the two-dimensional map:

$$
\begin{aligned}
x(t+1) &= \lambda_x x(t+1) + \sin^2(2\pi y(t+1)) \quad \mod 1 \\
y(t+1) &= 4y(t)(1-y(t)) .
\end{aligned}
$$

Hint: *Linearize the map and observe the properties of the Jacobian matrix.*

Exercise 5.7: Consider the two-dimensional map

$$
\begin{aligned}
x(t+1) &= 2x(t) \quad \mod 1 \\
y(t+1) &= ay(t) + 2\cos(2\pi x(t)) .
\end{aligned}
$$

(1) Show that if $|a| < 1$ there exists a finite attractor.
(2) Compute Lyapunov exponents $\{\lambda_1, \lambda_2\}$.

Exercise 5.8: Numerically compute the Lyapunov exponents $\{\lambda_1, \lambda_2\}$ of the Hénon map for $a = 1.4, b = 0.3$, check that $\lambda_1 + \lambda_2 = \ln b$; and test the Kaplan-Yorke conjecture with the fractal dimension computed in Ex. 5.3
Hint: *Evolve the map together with the tangent map, use Gram-Schmidt orthonormalization trying different values for the number of steps between two successive orthonormalization.*

Exercise 5.9: Numerically compute the Lyapunov exponents for the Lorenz model. Compute the whole spectrum $\{\lambda_1, \lambda_2, \lambda_3\}$ for $r = 28, \sigma = 10, b = 8/3$ and verify that: $\lambda_2 = 0$ and $\lambda_3 = -(\sigma + b + 1) - \lambda_1$.
Hint: *Solve first Ex.5.8. Check the dependence on the time and orthonormalization step.*

Exercise 5.10: Numerically compute the Lyapunov exponents for the Hénon-Heiles system. Compute the whole spectrum $\{\lambda_1, \lambda_2, \lambda_3, \lambda_4\}$, for trajectory starting from an initial condition in "chaotic sea" on the energy surface $E = 1/6$. Check that: $\lambda_2 = \lambda_3 = 0$; $\lambda_4 = -\lambda_1$.
Hint: *Do not forget that the system is conservative, check the conservation of energy during the simulation.*

Exercise 5.11: Consider the one-dimensional map defined as follows

$$x(t+1) = \begin{cases} 4x(t) & 0 \le x(t) < \frac{1}{4} \\ \frac{4}{3}(x(t) - 1/4) & \frac{1}{4} \le x(t) \le 1 \,. \end{cases}$$

Compute the generalized Lyapunov exponent $L(q)$ and show that:
(1) $\lambda_1 = \lim_{q \to 0} L(q)/q = \ln 4/4 + 3/4 \ln(4/3)$;
(2) $\lim_{q \to \infty} L(q)/q = \ln 4$;
(3) $\lim_{q \to -\infty} L(q)/q = \ln(4/3)$.
Finally, compute the Cramer function $S(\gamma)$ for the effective Lyapunov exponent.
Hint: *Consider the quantity* $\langle |\delta x^q(t)| \rangle$, *where* $\delta x(t)$ *is the infinitesimal perturbation evolving according the linearized map.*

Exercise 5.12:
Consider the one-dimensional map

$$x(t+1) = \begin{cases} 3x(t) & 0 \le x(t) < 1/3 \\ 1 - 2(x(t) - 1/3) & 1/3 \le x(t) < 2/3 \\ 1 - x(t) & 2/3 \le x(t) \le 1 \end{cases}$$

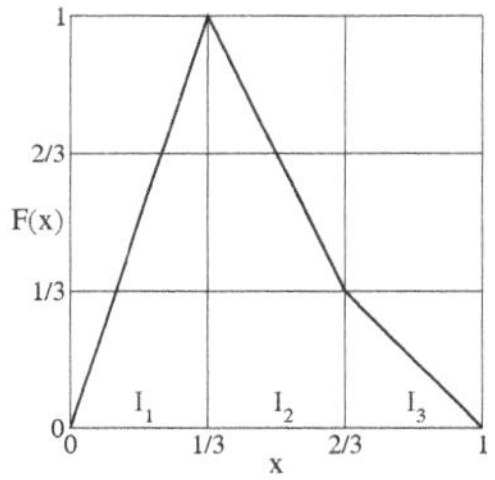

illustrated on the right. Compute the LE and the generalized LE.

Chapter 6

From Order to Chaos in Dissipative Systems

> It is not at all natural that "laws of nature" exist, much less that man is able to discover them. The miracle of the appropriateness of the language of mathematics for the formulation of the laws of physics is a wonderful gift for which we neither understand nor deserve.
>
> *Eugene Paul Wigner (1902–1995)*

We have seen that the qualitative behavior of a dynamical system dramatically changes as a nonlinearity control parameter, r, is varied. At varying r, the system dynamics changes from regular (such as stable fixed points, periodic or quasiperiodic motion) to chaotic motions, characterized by a high degree of irregularity and by sensitive dependence on the initial conditions.

The study of the qualitative changes in the behavior of dynamical systems goes under the name of bifurcation theory or theory of the transition to chaos. Entire books have been dedicated to it, where all the possible mechanisms are discussed in details, see Bergé *et al.* (1987). Here, mostly illustrating specific examples, we deal with the different routes from order to chaos in dissipative systems.

6.1 The scenarios for the transition to turbulence

We start by reviewing the problem of the transition to turbulence, which has both a pedagogical and conceptual importance.

The existence of qualitative changes in the dynamical behavior of a fluid in motion is part of every day experience. A familiar example is the behavior of water flowing through a faucet (Fig. 6.1). Everyone should have noticed that when the faucet is partially open the water flows in a regular way as a jet stream, whose shape is preserved in time: this is the so-called *laminar regime*. Such a kind of motion is analogous to a fixed point because water velocity stays constant in time. When the faucet is opened by a larger amount, water discharge increases and the flow qualitatively changes: the jet stream becomes thicker and variations in time can be seen by looking at a specific location, moreover different points of the jet behave in

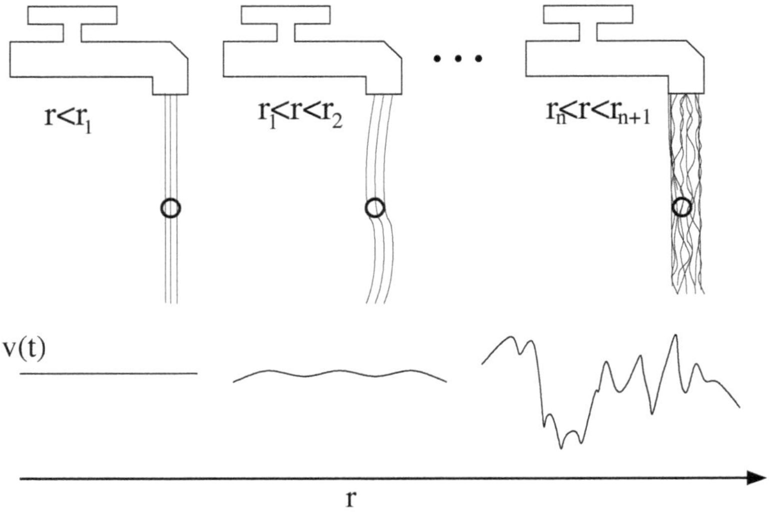

Fig. 6.1 Sketch of the transition to irregular motion in the faucet. Circles indicate the location where the velocity component $v(t)$ (bottom) is measured.

slightly different ways. As a result the size and shape of the water jet irregularly varies in time. This is the *turbulent regime*, which is characterized by complicated, irregular variations of all the kinematic and dynamical quantities.[1] For a cartoon of this transition see Fig. 6.1.

In this specific case, nonlinearity is controlled by the Reynolds number (Re), a dimensionless number proportional to the average velocity of the water U, to the size L of the open hole in the faucet, and to the inverse of the viscosity ν measuring fluid internal resistance:

$$\mathrm{Re} = \frac{LU}{\nu} \, .$$

What is the mechanism ruling the transition from laminar to turbulent motion?

This is the problem of the transition to turbulence, which is indeed a remarkable example for historical and conceptual reasons. It is thus interesting to think back the history of the proposed "scenarios" for the transition to turbulence so to appreciate the conceptual changes occurred in the course of the seventies.

6.1.1 *Landau-Hopf*

The first mechanism for the onset of turbulence was proposed by the soviet physicist Landau (1944). In a nutshell, the idea is the following: the irregular (chaotic in modern language) behavior characterizing fluids with high Reynolds numbers results from the superposition of a growing with Re (hereafter denoted with r) number of regular oscillations with different frequencies.

[1] In Chapter 13 we shall come back to the problem of turbulence.

With reference to Fig. 6.1, we focus now on the behavior of one velocity component $v(t)$ in a specific point of the water jet (e.g. measuring it in the circles shown in Fig. 6.1). At varying r, Landau theory can be summarized as follows. Below a critical Reynolds number (say r_1) the velocity is constant $v(t) = U$, this is the thin and regular water jet stream observed for low water discharge. As soon as $r > r_1$, an oscillation with frequency ω_1 superimposes to the mean flow U. Another oscillation with frequency ω_2 appears further opening the faucet till r raises up to a second critical value r_2, and so forth. In formulae:

$$
\begin{aligned}
&v(t) = U && \text{for} \quad r < r_1 \\
&v(t) = U + A_1 \sin(\omega_1 t + \phi_1) && \text{for} \quad r_1 < r < r_2 \\
&v(t) = U + A_1 \sin(\omega_1 t + \phi_1) + A_2 \sin(\omega_2 t + \phi_2) && \text{for} \quad r_2 < r < r_3 \\
&\quad \vdots \\
&v(t) = U + \textstyle\sum_{k=1}^{N} A_k \sin(\omega_k t + \phi_k) && \text{for} \quad r_N < r < r_{(N+1)} \,,
\end{aligned}
\tag{6.1}
$$

or in a more compact notation

$$
v(t) = U + \sum_{k=1}^{\infty} A_k(r) \sin(\omega_k t + \phi_k) \quad \text{with} \quad A_k(r) = 0 \quad \text{for} \quad r < r_k \,, \tag{6.2}
$$

where the phases $\phi_1, \dots, \phi_N$ are determined by the initial conditions. When r is sufficiently high that the number N of frequencies is large enough the resulting velocity $v(t)$ can be very irregular, provided the frequencies $\omega_1, \dots, \omega_N$ are rationally independent (i.e. no vanishing linear combination with integer coefficients can be formed).

About in the same years, the German mathematician Hopf (1943) proved that the asymptotic solutions of a wide class of differential equations change, by varying the control parameter, from stable fixed points to periodic orbits via a rather generic mechanism (see Box B.11 for details).[2] Therefore, at least the first step (from the first to the second line in Eq. (6.1)) of Landau theory is mathematically well based. Further support to the first step of Landau theory comes from the onset of limit cycles in the van der Pol oscillator (see Box B.12), although here the mechanism is different from that of Hopf.

The proposal outlined by Landau was thus in agreement, at least partially, with some pieces of rigorous mathematics, and with the common believe of that time that irregular and hence complicated behaviors were the result of the superposition of many simple (regular) causes. This mechanism for the transition to turbulence was generally accepted as correct until the seventies. However, it should be said that such a believe was not supported by systematic experimental investigations aimed to check the validity of the proposed theory.

[2]Actually this result often goes under the name of Poincaré-Andronov-Hopf theorem as it was independently obtained by Andronov in 1929 and Poincaré in 1882.

The attitude of the scientific community towards the problem of turbulence and, actually, towards most of classical physics problems at that time can be understood by considering that their attention was captured by Quantum Mechanics and other branches of science. Only after the seventies the interest on turbulence raised up again. Nowadays, both in physics and in mathematics, turbulence still stays at the frontiers of our understanding (see Chapter 13).

Box B.11: Hopf bifurcation

A bifurcation occurs when a dynamical system qualitatively modifies its behavior upon varying a control parameter. In particular, we consider here the case of a fixed point that loses stability giving rise to a *limit cycle* (Sec. 2.4.2.1). In autonomous nonlinear systems, one of the most common bifurcation of such kind has been theoretically studied by Hopf (1943), who showed that oscillations near an equilibrium point can be understood by looking at the eigenvalues of the linearized equations.

Consider the autonomous dynamical system of d degrees of freedom described by the ODE

$$\frac{\mathrm{d}\boldsymbol{x}}{\mathrm{d}t} = \boldsymbol{f}_\mu(\boldsymbol{x}) \tag{B.11.1}$$

depending on the control parameter μ. As seen in Chap. 2, a fixed point for the system (B.11.1) is the solution $\boldsymbol{x}_c(\mu)$ such that $\boldsymbol{f}_\mu(\boldsymbol{x}_c) = 0$, and its linear stability is characterized by the eigenvalues $\{\lambda_1, ..., \lambda_d\}$ of the stability matrix $\mathbb{L}_{ij}(\mu) = \partial f_i/\partial x_j|_{\boldsymbol{x}_c}$. With reference to Table 2.1, $\boldsymbol{x}_c$ is stable, for a given value of μ, if the eigenvalues $\lambda_k(\mu) = \alpha_k(\mu) + i\omega_k(\mu)$ have a negative real part, $\alpha_k(\mu) < 0$ for any $k = 1, \dots, d$. Generally, a stable fixed point $\boldsymbol{x}_c(\mu)$ is said to undergo a bifurcation for $\mu = \mu_c$ when at least one of the eigenvalues has a vanishing real part at μ_c, i.e. it exists at least a $\bar{k}$ such that $\alpha_{\bar{k}}(\mu_c) = 0$.

The Hopf bifurcation occurs under the following conditions. First of all, the fixed point should be stable for $\mu < \mu_c$, i.e. with $\alpha_k(\mu) < 0$ for any k, and should lose stability because a pair of eigenvalues acquire a zero real part $\alpha = 0$, with $\mathrm{d}\alpha/\mathrm{d}\mu|_{\mu_c} > 0$, this additional condition implies a non-tangent crossing to the zero. As a final requirement the fixed point should be, for $\mu = \mu_c$, a *vague attractor* [Ruelle and Takens (1971); Gallavotti (1983)], meaning that any trajectory in a neighborhood of $\boldsymbol{x}_c$ should be attracted toward it. Notice that the validity of the latter request depends upon the nonlinear terms in the expansion around the fixed point. Therefore, the knowledge of the stability matrix is not enough to determine if the bifurcation is Hopf-like.

If all the above conditions are fulfilled, we have a Hopf bifurcation where, as soon as μ is slightly larger than μ_c, the asymptotic dynamics passes from a fixed point to a limit cycle (Fig. B11.1), whose radius can be shown to grow as $\sqrt{\mu - \mu_c}$, for $\mu - \mu_c \ll 1$. In the presence of symmetries, more than a pair of eigenvalues may have real parts crossing zero, however here we shall not discuss these non-generic cases.

Instead of proving the theorem, we show how Hopf's bifurcation works in practice, resorting to the following example

$$\begin{aligned} \frac{\mathrm{d}x}{\mathrm{d}t} &= \mu x - \omega y + a(x^2 + y^2)x \\ \frac{\mathrm{d}y}{\mathrm{d}t} &= \omega x + \mu y + a(x^2 + y^2)y\,, \end{aligned} \tag{B.11.2}$$

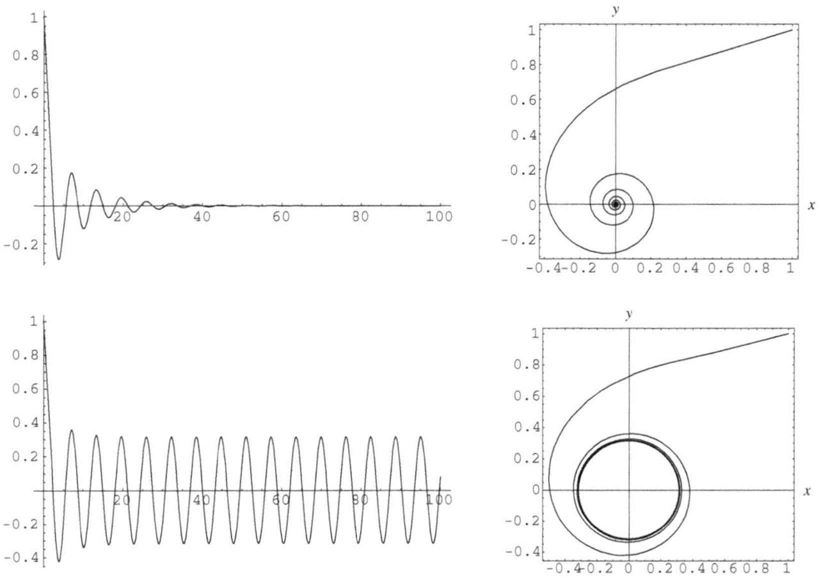

Fig. B11.1 Phase portrait (right) and time course of the x-coordinate (left), illustrating the Hopf mechanism for the system (B.11.2) with $a=-1$ and $\omega=1$. The two upper panels refer to $\mu=-0.1$ (stable fixed point) while the two bottom panels show the onset of the limit cycle for $\mu=0.1$.

which catches the basic features. The origin $(0,0)$ is a fixed point with eigenvalues $\mu \pm i\omega$. For $\mu < 0$ and $\omega \neq 0$ the fixed point is stable and all the hypothesis of the theorem hold, provided that a is negative so to have a vague attractor. It is particularly instructive to look at (B.11.2) in polar coordinate (r,θ)

$$\frac{\mathrm{d}r}{\mathrm{d}t} = (\mu + ar^2)r$$
$$\frac{\mathrm{d}\theta}{\mathrm{d}t} = \omega\,.$$

It is now evident that as μ passes through zero, a Hopf bifurcation occurs and a stable limit cycle appears with radius $\sqrt{\mu/|a|}$ and period $2\pi/\omega$.

In discrete-time dynamical systems Hopf's bifurcation corresponds to the exit from the unitary circle of a pair of complex conjugate eigenvalues $\lambda_{\bar{k}}(\mu) = \rho_{\bar{k}}(\mu)\mathrm{e}^{\pm i\theta_{\bar{k}}(\mu)}$ associated to a fixed point, i.e. $\rho_{\bar{k}}(\mu)$ becomes greater then 1 as $\mu > \mu_c$.

Box B.12: The Van der Pol oscillator and the averaging technique

The van der Pol equation was introduced to model self-sustained current oscillations in a triode circuit employed in early electronic devices [van der Pol (1927)]. Nowadays, although it is just an historical curiosity from a technological point of view, it remains an interesting example of limit cycle generation with a mechanism different from Hopf's bifurcation. The equation that describes the system is

$$\frac{\mathrm{d}^2 x}{\mathrm{d}t^2} - \mu(1-x^2)\frac{\mathrm{d}x}{\mathrm{d}t} + \omega^2 x = 0\,, \tag{B.12.1}$$

which is the second order differential equation corresponding to first order ODEs (2.26).

It is easy to see that, when $\mu < 0$, the stable fixed point $(x, \mathrm{d}x/\mathrm{d}t) = (0,0)$ attracts the motion. For $\mu = 0$, Eq. (B.12.1) reduces to the standard harmonic oscillator with frequency ω. For $\mu > 0$, the fixed point becomes unstable and the motion sets onto a limit cycle, shown in Fig. B12.1. This behavior can be understood using the averaging technique, originally introduced in mechanics (see e.g. Arnold (1989)), which deserves a brief discussion due to its common applicability. To illustrate the method, consider the Hamilton equations written in the action-angle variables:

$$\begin{aligned}\frac{\mathrm{d}\phi}{\mathrm{d}t} &= \frac{1}{\epsilon}[\omega(\mathrm{I}) + f(\phi, \mathrm{I})] \\ \frac{\mathrm{dI}}{\mathrm{d}t} &= g(\phi, \mathrm{I})\,,\end{aligned}$$

where the functions f and g are 2π-periodic in the angle ϕ. Assuming $\epsilon \ll 1$, ϕ and I can be identified as the fast and slow variables, respectively with time scale ratio $\mathcal{O}(\epsilon)$. The averaging method consists in introducing a "smoothed" action J describing the "slow motion" obtained by filtering out the fast $\mathcal{O}(\epsilon)$ oscillations. The dynamics of J is ruled by the "force" acting on I averaged over the fast variable ϕ

$$\frac{\mathrm{dJ}}{\mathrm{d}t} = G(\mathrm{J}) = \frac{1}{2\pi}\int_0^{2\pi} \mathrm{d}\phi\, g(\phi, \mathrm{J})\ .$$

The evolution of J gives the leading order behavior of I [Arnold (1989)].

Let us now apply the above procedure to Eq. (B.12.1). The non Hamiltonian character of the van der Pol equation is not a limitation for the use of averaging method. We thus introduce ϕ and I as

$$\phi = \arctan\left(\frac{1}{x}\frac{\mathrm{d}x}{\mathrm{d}t}\right)\,, \quad \mathrm{I} = \frac{1}{2}\left[x^2 + \left(\frac{\mathrm{d}x}{\mathrm{d}t}\right)^2\right]\,. \tag{B.12.2}$$

Equation (B.12.1) and (B.12.2), with $\omega = 1$, yield

$$\frac{\mathrm{dI}}{\mathrm{d}t} = \mu(1-x^2)\left(\frac{\mathrm{d}x}{\mathrm{d}t}\right)^2 = 2\mu\mathrm{I}(1 - 2\mathrm{I}\cos^2\phi)\sin^2\phi\,.$$

The time scales of ϕ and I are $\mathcal{O}(1)$ and $\mathcal{O}(\mu^{-1})$, respectively. Therefore, for $\mu \ll 1$, time scale separation occurs and the averaging method can be applied. In particular, averaging over ϕ we obtain

$$\frac{\mathrm{dJ}}{\mathrm{d}t} = G(\mathrm{J}) = \mu\mathrm{J}(1 - \frac{\mathrm{J}}{2})\,,$$

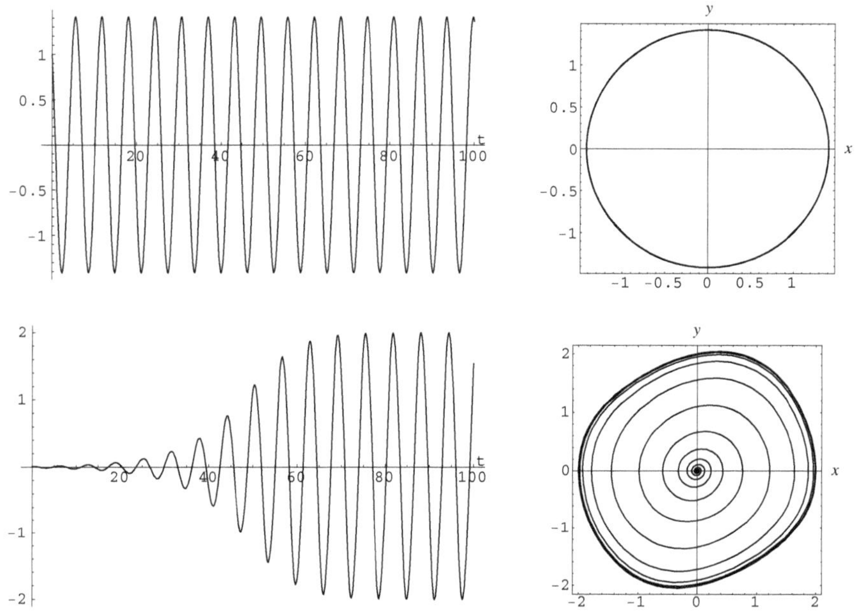

Fig. B12.1 Phase portrait (right) and time course of the x-coordinate (left), illustrating the bifurcation occurring in the van der Pol equation (B.12.1) with $\omega = 1$. For $\mu = 0$, a simple harmonic motion is observed (top) while for $\mu = 0.1$ a nontrivial limit cycle sets in (bottom).

which admits two fixed points: $\mathrm{J} = 0$ and $\mathrm{J} = 2$. For $\mu < 0$, $\mathrm{J} = 0$ is stable and $\mathrm{J} = 2$ unstable, while the reverse if true for $\mu > 0$. Note that $\mathrm{J} = 2$ corresponds to a circle of radius $R = 2$, so for small positive values of μ an attractive limit cycle exists.

We conclude by noticing that, notwithstanding the system (B.12.1) and (B.11.2) have a similar linear structure, unlike Hopf's bifurcation (see Box B.11) here the limit-cycle radius is finite, $R = 2$, independently of the value of μ. It is important to stress that such a difference has its roots in the form of the nonlinear terms. Technically speaking, in the van der Pol equation, the original fixed point does not constitute a vague attractor for the dynamics.

6.1.2 *Ruelle-Takens*

Nowadays, we know from experiments (see Sect. 6.5) and rigorous mathematical studies that Landau's scenario is inconsistent. In particular, Ruelle and Takens (1971) (see also Newhouse, Ruelle and Takens (1978)) proved that the Landau-Hopf mechanism cannot be valid beyond the transition from one to two frequencies, the quasiperiodic motion with three frequencies being *structurally unstable.*

Let us open a brief digression on structural stability. Consider a generic differential equation

$$\frac{\mathrm{d}\boldsymbol{x}}{\mathrm{d}t} = \boldsymbol{f}_r(\boldsymbol{x}) \,, \tag{6.3}$$

and the same equation with a "small" modification in its r.h.s.

$$\frac{\mathrm{d}\boldsymbol{x}}{\mathrm{d}t} = \tilde{\boldsymbol{f}}_r(\boldsymbol{x}) = \boldsymbol{f}_r(\boldsymbol{x}) + \delta\boldsymbol{f}_r(\boldsymbol{x}) \,, \tag{6.4}$$

where $\tilde{\boldsymbol{f}}_r(\boldsymbol{x})$ is "close" to $\boldsymbol{f}_r(\boldsymbol{x})$, in the sense that the symbol $\delta\boldsymbol{f}_r(\boldsymbol{x})$ denotes a very "small" perturbation. Given the dynamical system (6.3), one of its property is said to be *structurally stable* if that property still holds in Eq. (6.4) for any — non *ad hoc* — choice of the perturbation $\delta\boldsymbol{f}_r(\boldsymbol{x})$, provided this is small enough in some *norm*. We stress that in any rigorous treatment the norm needs to be specified [Berkooz (1994)]. Here, for the sake of simplicity, we remain at general level and leave the norm unspecified.

In simple words, Ruelle and Takens have rigorously shown that even if there exists a certain dynamical system (say described by Eq. (6.3)) that exhibits a Landau-Hopf scenario, the same mechanism is not preserved for generic small perturbations such as (6.4), unless *ad hoc* choices of $\delta\boldsymbol{f}_r$ are adopted.

This result is not a mere technical point and has a major conceptual importance. In general, it is impossible to know with arbitrary precision the "true" equation describing the evolution of a system or ruling a certain phenomenon (for example, the precise values of the control parameters). Therefore, an explanation or theory based through a mechanism which, although proved to work in specific conditions, disappears as soon as the laws of motion are changed by a very tiny amount should be seen with suspect.

After Ruelle and Takens, we known that Landau-Hopf theory for the transition to chaos is meaningful for the first two steps only: from a stable fixed point to a limit cycle and from a limit cycle to a motion characterized by two frequencies. The third step was thus substituted by a transition to a strange attractor with sensitive dependence on the initial conditions.

It is important to underline that while Landau-Hopf mechanism to explain complicated behaviors requires a large number of degrees of freedom, Ruelle-Takens predicted that for chaos to appear three degrees of freedom ODE is enough, which explains the ubiquity of chaos in nonlinear low dimensional systems.

We conclude this section by stressing another pivotal consequence of the scenario proposed by Ruelle and Takens. This was the first mechanism able to interpret a physical phenomenon, such as the transition to turbulence in fluids, in terms of chaotic dynamical systems, which till that moment were mostly considered as mathematical toys. Nevertheless, it is important to recall that Ruelle-Takens scenario is not the only mechanism for the transition to turbulence. In the following we describe other two quite common possibilities for the transition to chaos that have been identified in low dimensional dynamical systems.

6.2 The period doubling transition

In Sec. 3.1 we have seen that the logistic map,

$$x(t+1) = f_r(x(t)) = r\,x(t)(1-x(t))\,,$$

follows a peculiar route from order to chaos — the *period doubling* transition — characterized by an infinite series of control parameter values $r_1, r_2, \dots, r_n, \dots$ such that if $r_n < r < r_{n+1}$ the dynamics is periodic with period 2^n. The first few steps of this transition are shown in Fig. 6.2. The series $\{r_n\}$ accumulates to the finite limiting value

$$r_\infty = \lim_{n\to\infty} r_n = 3.569945\dots$$

beyond which the dynamics passes from periodic (though with a very high, diverging, period) to chaotic.

This bifurcation scheme is actually common to many different systems, e.g. we saw in Chap. 1 that also the motion of a vertically driven pendulum becomes chaotic through period doubling [Bartuccelli *et al.* (2001)], and may also be present (though with slightly different characteristics) in conservative systems [Lichtenberg and Lieberman (1992)]. Period doubling is remarkable also, and perhaps more importantly, because it is characterized by a certain degree of universality, as recognized by Feigenbaum (1978).

Before illustrating and explaining this property, however, it is convenient to introduce the concept of *superstable orbits.* A periodic orbit $x_1^*, x_2^*, \dots, x_T^*$ of period T is said superstable if

$$\left.\frac{\mathrm{d}f_r^{(T)}(x)}{\mathrm{d}x}\right|_{x_1^*} = \prod_{k=1}^{T} \left.\frac{\mathrm{d}f_r(x)}{\mathrm{d}x}\right|_{x_k^*} = 0\,,$$

the second equality, obtained by applying the chain rule of differentiation, implies that for the orbit to be superstable it is enough that in at least one point of the orbit, say x_1^*, the derivative of the map vanishes. Therefore, for the logistic map, superstable orbits contain $x = 1/2$ and are realized for specific values of the control parameter R_n, defined by

$$\left.\frac{\mathrm{d}f_{R_n}^{(2^n)}(x)}{\mathrm{d}x}\right|_{x_1^*=1/2} = 0\,, \tag{6.5}$$

such values are identified by vertical lines in Fig. 6.2. It is interesting to note that the series $R_0, R_1, \dots, R_n, \dots$ is also infinite and that $R_\infty = r_\infty$.

Pioneering numerical investigations by Feigenbaum in 1975 have highlighted some intriguing properties:

-1- At each r_n the number of branches doubles (Fig. 6.2), and the distance between two consecutive branchings, $r_{n+1} - r_n$, is in constant ratio with the distance of the branching of the previous generation $r_n - r_{n-1}$ i.e.

$$\frac{r_n - r_{n-1}}{r_{n+1} - r_n} \approx \delta = 4.6692\dots\,, \tag{6.6}$$

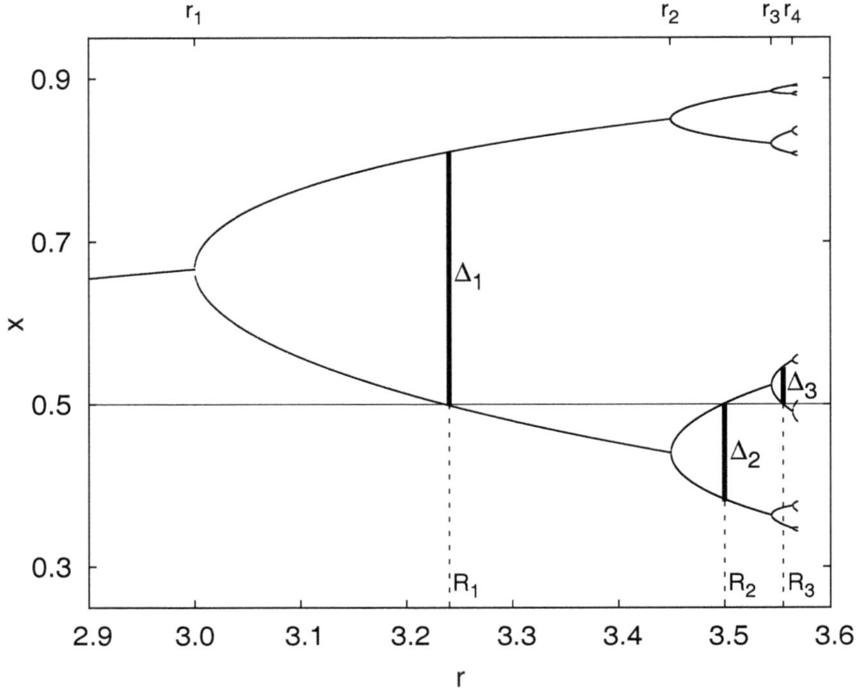

Fig. 6.2 Blow up of the bifurcation diagram shown in Fig. 3.5 in the interval $r \in [2.9, 3.569]$, range in which the orbits pass from having period 1 to 16. The depicted doubling transitions happen at $r_1 = 3$, $r_2 \approx 3.449\ldots$, $r_3 \approx 3.544\ldots$ and $r_4 \approx 3.5687\ldots$, respectively. The vertical dashed lines locate the values of r at which one finds superstable periodic orbits of period 2 (at R_1), 4 (at R_2) and 8 (at R_3). Thick segments indicate the distance between the points of superstable orbits which are the closest to $x = 1/2$. See text for explanation.

thus by plotting the bifurcation diagram against $\ln(r_\infty - r)$ one would obtain that the branching points will appear as equally spaced. The same relation holds true for the series $\{R_n\}$ characterizing the superstable orbits.

-2- As clear from Fig. 6.2, the bifurcation tree possesses remarkable geometrical similarities, each branching reproduces the global scheme on a reduced scale. For instance, the four upper points at $r = r_4$ (Fig. 6.2) are a rescaled version of the four points of the previous generation (at $r = r_3$). We can give a more precise mathematical definition of such a property considering the superstable orbits at R_1, R_2 .... Denoting with Δ_n the signed distance between the two points of period-2^n superstable orbits which are closer to $1/2$ (see Fig. 6.2), we have that

$$\frac{\Delta_n}{\Delta_{n+1}} \approx -\alpha = -2.5029\ldots\,, \tag{6.7}$$

the minus sign indicates that Δ_n and $\Delta n + 1$ lie on opposite sides of the line $x = 1/2$, see Fig. 6.2.

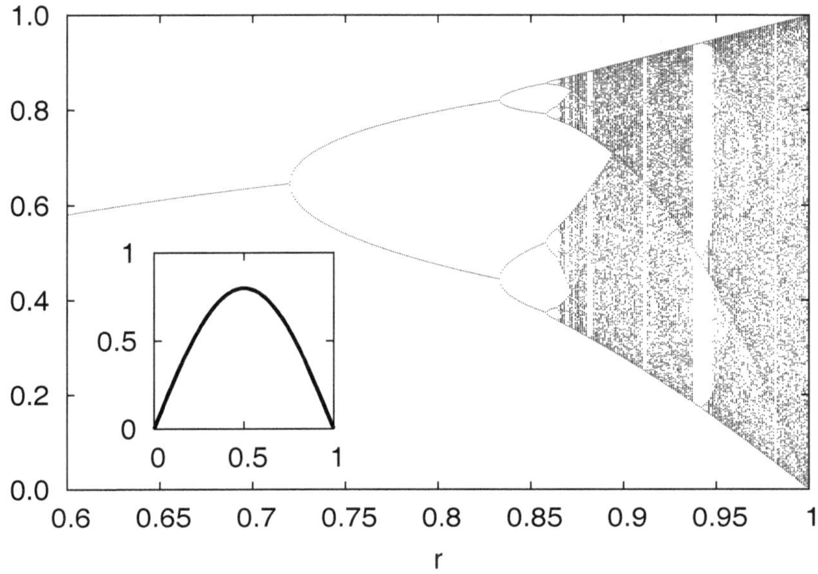

Fig. 6.3 Bifurcation diagram of the sin map Eq. (6.8) (in the inset), generated in the same way as that of the logistic map (Fig. 3.5).

Equations (6.6) and (6.7) becomes more and more well verified as n increases. Moreover, and very interestingly, the values of α and δ, called Feigenbaum's constants, are not specific to the logistic map but are universal, as they characterize the period doubling transition of all maps with a unique quadratic maximum (so-called quadratic unimodal maps). For example, notice the similarity of the bifurcation diagram of the sin map:

$$x(t+1) = r\sin(\pi x(t)), \tag{6.8}$$

shown in Fig. 6.3, with that of the logistic map (Fig. 3.5). The correspondence of the doubling bifurcations in the two maps is perfect.

Actually, also continuous time differential equations can display a period doubling transition to chaos with the same α and δ, and it is rather natural to conjecture that hidden in the system it should be a suitable return map (as the Lorenz map shown in Fig. 3.8, see Sec. 3.2) characterized by a single quadratic maximum.

We thus have that, for a large class of evolution laws, the mechanism for the transition to chaos is universal. For unimodal maps with non-quadratic maximum, universality applies too. For instance, if the function behaves as $|x - x_c|^z$ (with $z > 1$) close to the maximum [Feigenbaum (1978); Derrida *et al.* (1979); Feigenbaum (1979)], the universality class is selected by the exponent z, meaning that α and δ are universal constants which only depends upon z.

6.2.1 *Feigenbaum renormalization group*

The existence of universal constants and the presence of self-similarity (e.g. in the organization of the bifurcation diagram or in the appearance of the transition values r_n or equivalently R_n) closely recalls critical phenomena [Kadanoff (1999)], whose unifying understanding in terms of the Renormalization Group (RG) [Wilson (1975)] came about in the same years of the discovery by Feigenbaum of properties (6.6) and (6.7). Feigenbaum himself recognized that such a formal similarity could be used to analytically predict the values of α and δ and to explain their universality in terms of the RG approach of critical phenomena.

The fact that scaling laws such as (6.6) are present indicates an underlying self-similar structure: a blow up of a portion of the bifurcation diagram is similar to the entire diagram. This property is not only aesthetically nice, but also strengthens the contact with phase transitions, the physics of which, close to the critical point, is characterized by scale invariance.

For its conceptual importance, here we shall discuss in some details how RG can be applied to derive α in maps with quadratic maximum. A complete treatment can be found in Feigenbaum (1978, 1979) or, for a more compact description, the reader may refer to Schuster and Just (2005).

To better illustrate the idea of Feigenbaum's RG, we consider superstable orbits of the logistic maps defined by Eq. (6.5). Fig. 6.4a shows the logistic map at R_0 where the first superstable orbit of period $2^0 = 1$ appears. Then, consider the 2-nd iterate of the map at R_1 (Fig. 6.4b), where the superstable orbit has period $2^1 = 2$, and the 4-th iterate at R_2 (Fig. 6.4c), where it has period $2^2 = 4$. If we focus on the boxed area around the point $(x, f(x)) = (1/2, 1/2)$ in Fig. 6.4b–c, we can realize that the graph of the first superstable map $f_{R_0}(x)$ is reproduced, though on smaller scales. Actually, in Fig. 6.4b the graph is not only reduced in scale but also reflected with respect to $(1/2, 1/2)$. Now imagine to rescale the x-axis and the y-axis in the neighborhood of $(1/2, 1/2)$, and to operate a reflection when necessary, so that the graph of Fig. 6.4b-c around $(1/2, 1/2)$ superimposes to that of Fig. 6.4a. Such an operation can be obtained by performing the following steps: first shift the origin such that the maximum of the first iterate of the map is obtained in $x = 0$ and call $\tilde{f}_r(x)$ the resulting map; then draw

$$(-\alpha)^n \tilde{f}^{(2^n)}_{R_n}\left[\frac{x}{(-\alpha)^n}\right]. \tag{6.9}$$

The result of these two steps is shown in Fig. 6.4d, the similarity between the graphs of these curves suggests that the limit

$$g_0(x) = \lim_{n\to\infty} (-\alpha)^n \tilde{f}^{(2^n)}_{R_n}\left[\frac{x}{(-\alpha)^n}\right]$$

exists and well characterizes the behavior of the 2^n-th iterate of the map close to the critical point $(1/2, 1/2)$. In analogy with the above equation, we can introduce the functions

$$g_k(x) = \lim_{n\to\infty} (-\alpha)^n \tilde{f}^{(2^n)}_{R_{n+k}}\left[\frac{x}{(-\alpha)^n}\right],$$

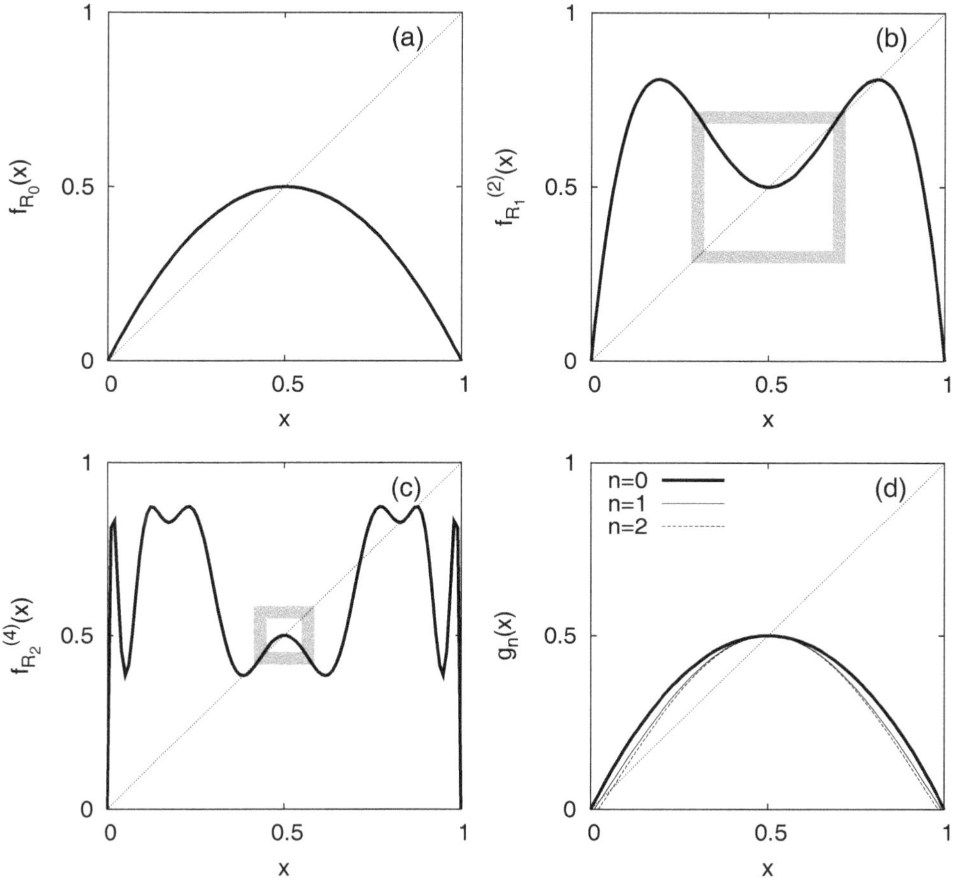

Fig. 6.4 Illustration of the renormalization group scheme for computing Feigenbaum's constant α. (a) Plot of $f_{R_0}(x)$ vs x with $R_0 = 2$ being the superstable orbit of period-1. (b) Second iterate at the superstable orbit of period-2, i.e. $f^{(2)}_{R_1}(x)$ vs x. (c) Fourth iterate at the superstable orbit of period 2, i.e. $f^{(4)}_{R_2}(x)$ vs x. (d) Superposition of first, second and fourth iterates of the map under the doubling transformation (6.9). This corresponds to superimposing (a) with the gray boxed area in (b) and in (c).

which are related among each other by the so-called doubling transformation $\mathcal{D}$

$$g_{k-1}(x) = \mathcal{D}[g_k(x)] \equiv (-\alpha) g_k \left[g_k \left[\frac{x}{(-\alpha)} \right] \right] ,$$

as can be derived noticing that

$$\begin{aligned} g_{k-1}(x) &= \lim_{n \to \infty} (-\alpha)^n \tilde{f}^{(2^n)}_{R_{n+k-1}} \left[\frac{x}{(-\alpha)^n} \right] \\ &= \lim_{n \to \infty} (-\alpha)(-\alpha)^{n-1} \tilde{f}^{(2^{n-1+1})}_{R_{n-1+k}} \left[\frac{1}{(-\alpha)} \frac{x}{(-\alpha)^{n-1}} \right] \end{aligned}$$

then by posing $i = n - 1$ we have

$$\begin{aligned} g_{k-1}(x) &= \lim_{i\to\infty} (-\alpha)(-\alpha)^i \tilde{f}^{(2^{i+1})}_{R_{i+k}} \left[\frac{1}{(-\alpha)} \frac{x}{(-\alpha)^i} \right] \\ &= \lim_{i\to\infty} (-\alpha)(-\alpha)^i \tilde{f}^{(2^i)}_{R_{i+k}} \left[\frac{1}{(-\alpha)^i} (-\alpha)^i \tilde{f}^{(2^i)}_{R_{i+k}} \left[\frac{1}{-\alpha} \frac{x}{(-\alpha)^i} \right] \right] \\ &= (-\alpha) g_k \left[g_k \left(\frac{x}{-\alpha} \right) \right] . \end{aligned}$$

The limiting function $g(x) = \lim_{n\to\infty} g_n(x)$ solves the "fixed point" equation

$$g(x) = \mathcal{D}[g(x)] = (-\alpha) g \left[g \left[\frac{x}{(-\alpha)} \right] \right] , \tag{6.10}$$

from which we can determine α after fixing a "scale", indeed we notice that if $g(x)$ solves Eq. (6.10) also $\nu g(x/\nu)$ (with arbitrary $\nu \neq 0$) is a solution. Therefore, we have the freedom to set $g(0) = 1$. The final step consists in using Eq. (6.10) by searching for better and better approximations of $g(x)$. The lowest nontrivial approximation can be obtained assuming a simple quadratic maximum

$$g(x) = 1 + c_2 x^2$$

and plugging it in the fixed point equation (6.10)

$$1 + c_2 x^2 = -\alpha(1 + c_2) - \frac{2c_2^2}{\alpha} x^2 + o(x^4)$$

from which we obtain $\alpha = -2c_2$ and $c_2 = -(1 + \sqrt{3})/2$ and thus

$$\alpha = 1 + \sqrt{3} = 2.73 \ldots$$

which is only 10% wrong. Next step would consist in choosing a quartic approximation $g(x) = 1 + c_2 x^2 + c_4 x^4$ and to determine the three constants c_2, c_4 and α. Proceeding this way one obtains

$$g(x) = 1 - 1.52763x^2 + 0.104815x^4 + 0.0267057x^6 - \ldots \Longrightarrow \alpha = 2.502907875 \ldots .$$

Universality of α follows from the fact that we never specified the form of the map in this derivation, the period doubling transformation can be defined for any map; we only used that the quadratic shape (plus corrections) around its maximum.

A straightforward generalization allows us to compute α for maps behaving as x^z around the maximum.

Determining δ is slightly more complicated and requires to linearize the doubling transformation $\mathcal{D}$ around r_∞. The interested reader may find the details of such a procedure in Schuster and Just (2005) or in Briggs (1997) where α and δ are reported up to about one hundred digits.

6.3 Transition to chaos through intermittency: Pomeau-Manneville scenario

Another important mechanism of transition to chaos was discovered by Pomeau and Manneville (1980). Their theory originates from the observation of a particular behavior called *intermittency* in some chemical and fluid mechanical systems: long intervals of time characterized by laminar/regular behavior interrupted by abrupt and short periods of very irregular motion. This phenomenon is observed in several systems when the control parameter r exceeds a critical value r_c. Here, we will mainly follow the original work of Pomeau and Manneville (1980) to describe the way it appears.

In Figure 6.5, a typical example of intermittent behavior is exemplified. Three time series are represented as obtained from the time evolution of the z variable of the Lorenz system (see Sec. 3.2)

$$\frac{\mathrm{d}x}{\mathrm{d}t} = -\sigma x + \sigma y$$
$$\frac{\mathrm{d}y}{\mathrm{d}t} = -y + rx - xz$$
$$\frac{\mathrm{d}z}{\mathrm{d}t} = -bz + xy$$

with the usual choice $\sigma = 10$ and $b = 8/3$ but for r close to 166. As clear from the figure, at $r = 166$ one has periodic oscillations, for $r > r_c = 166.05\ldots$ the regular

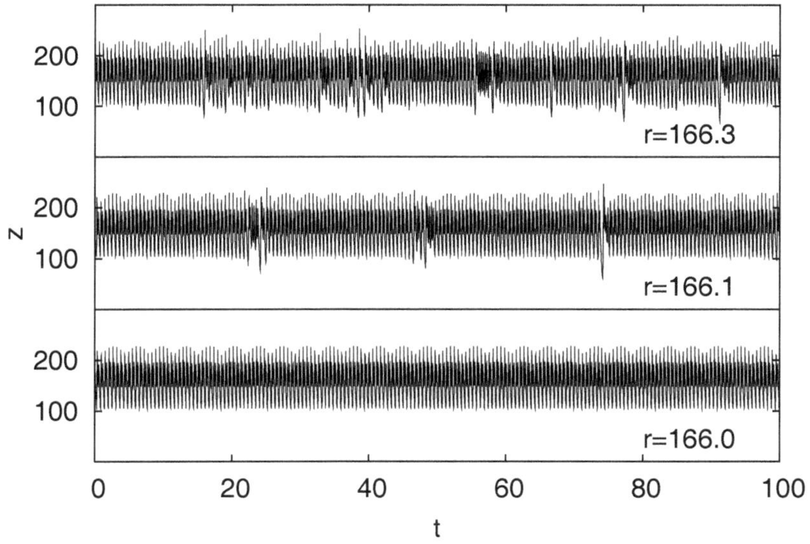

Fig. 6.5 Typical evolution of a system which becomes chaotic through intermittency. The three series represent the evolution of z in the Lorenz systems for $\sigma = 10$, $b = 8/3$ and for three different values of r as in the legend.

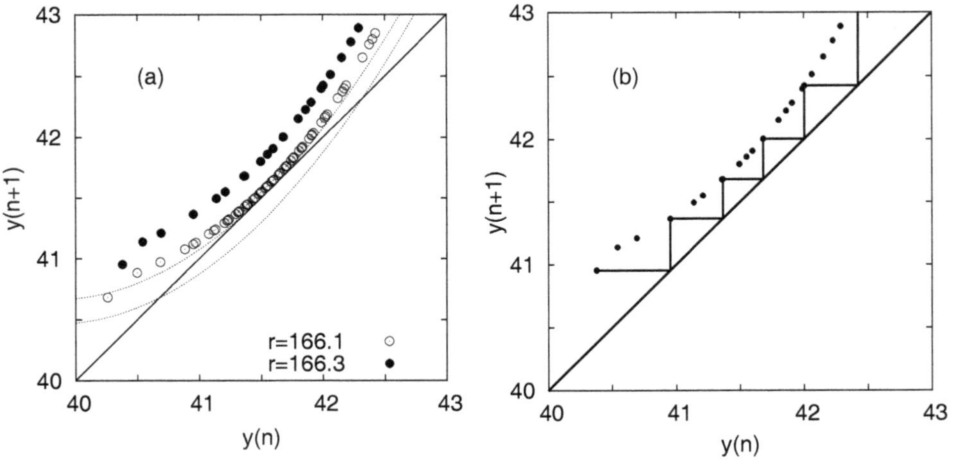

Fig. 6.6 (a) First return map $y(n+1)$ vs $y(n)$ for $r = 166.1$ (open circles) and $r = 166.3$ (filled circles) obtained by recording the intersections with the plane $x = 0$ for the $y > 0$ values (see text). The two dotted curves pictorially represent the expected behavior of such a map for $r = r_c \approx 166.05$ (upper curve) and $r < r_c$ lower curve. (b) Again the first return map for $r = 166.3$ with representation of the evolution, clarifying the mechanism for the long permanence in the channel.

oscillations are interrupted by irregular oscillations, which becomes more and more frequent as $r - r_c$ becomes larger and larger.

Similarly to the Lorenz return map (Fig. 3.8) discussed in Sec. 3.2, an insight into the mechanism of this transition to chaos can be obtained by constructing a return map associated to the dynamics. In particular, consider the map

$$y(k+1) = f_r(y(k)),$$

where $y(k)$ is the (positive) y-coordinate of the k-th intersection of trajectory with the $x = 0$ plane. For the same values of r of Fig. 6.5, the map is shown in Fig. 6.6a. At increasing $\epsilon = r - r_c$, a channel of growing width appears in between the graph of the map and the bisectrix. At $r = r_c$ the map is tangent to the bisectrix (see the dotted curves in the figure) and, for $r > r_c$, it detaches from the line opening a channel. This occurrence is usually termed *tangent bifurcation.*

The graphical representation of the iteration of discrete time maps shown in Fig. 6.6b provides a rather intuitive understanding of the origin of intermittency. For $r < r_c$, a fast convergence toward the stable periodic orbit occurs. For $r = r_c + \epsilon$ $(0 < \epsilon \ll 1)$ $y(k)$ gets trapped in the channel for a very long time, proceeding by very small steps, the narrower the channel the smaller the steps. Then it escapes, performing a rapid irregular excursion, after which it re-enters the channel for another long period. The duration of the "quiescent" periods will be generally different each time, being strongly dependent on the point of injection into the channel. Pomeau and Manneville have shown that the average quiescent time is proportional to $1/\sqrt{\epsilon}$.

In dynamical-systems jargon, the above described transition is usually called *intermittency transition of kind I* which, in the discrete-time domain, is generally represented by the map

$$x(n+1) = r + x(n) + x^2(n) \qquad \text{mod } 1\,, \tag{6.11}$$

which for $r = 0$ is tangent to the bisecting line at the origin, while for $0 = r_c < r \ll 1$ a narrow channel opens. Interestingly this type of transition can also be observed in the logistic map close $r = 1 + \sqrt{8}$ where period-3 orbits appears [Hirsch *et al.* (1982)].

Several other types of transition to chaos through intermittency have been identified so far. The interested reader may refer to more focused monographs as, e.g. Bergé *et al.* (1987).

6.4 A mathematical remark

Dissipative systems, as seen in the previous sections, exhibit several different scenarios for the transition to chaos. The reader may thus have reached the wrong conclusion that there is a sort of zoology of possibilities without any connections among them. Actually, this is not the case. For example, the different transitions encountered above can be understood as the generic ways a fixed point or limit cycle[3] loses stability, see e.g. Eckmann (1981). This issue can be appreciated, without loss of generality, considering time discrete maps

$$\boldsymbol{x}(t+1) = \boldsymbol{f}_\mu(\boldsymbol{x}(t))\,.$$

Assume that the fixed point $\boldsymbol{x}^* = \boldsymbol{f}_\mu(\boldsymbol{x}^*)$ is stable for $\mu < \mu_c$ and unstable for $\mu > \mu_c$. According to linear stability theory (Sec. 2.4), this means that for $\mu < \mu_c$ the stability eigenvalues $\lambda_k = \rho_k e^{i\theta_k}$ are all inside the unit circle ($\rho_k < 1$). Whilst for $\mu = \mu_c$, stability is lost because at least one or a pair of complex conjugate eigenvalues touch the unitary circle.

The exit of the eigenvalues from the unitary circle may, in general, happen in three distinct ways as sketched in the left panel of Fig. 6.7:

(a) one real eigenvalue equal to 1 ($\rho = 1$, $\theta = 0$);
(b) one real eigenvalue equal to -1 ($\rho = 1$, $\theta = \pi$);
(c) a pair of complex conjugate eigenvalues with modulo equal to 1 ($\rho = 1$, $\theta \neq n\pi$ for n integer).

Case (a) refers to Pomeau-Manneville scenario, i.e. intermittency of kind I. Technically speaking, this is an inverse saddle-node bifurcation as sketched in the right panel of Fig. 6.7: for $\mu < \mu_c$ a stable and an unstable fixed points coexist and merge at $\mu = \mu_c$; both disappear for $\mu > \mu_c$. For instance, this happens for the map in

[3]We recall that limit cycle or period orbits can be always thought as fixed point for an appropriate mapping. For instance, a period-2 orbit of a map $f(x)$ corresponds to a fixed point of the second iterate of the map, i.e. $f(f(x))$. So we can speak about fixed points without loss of generality.

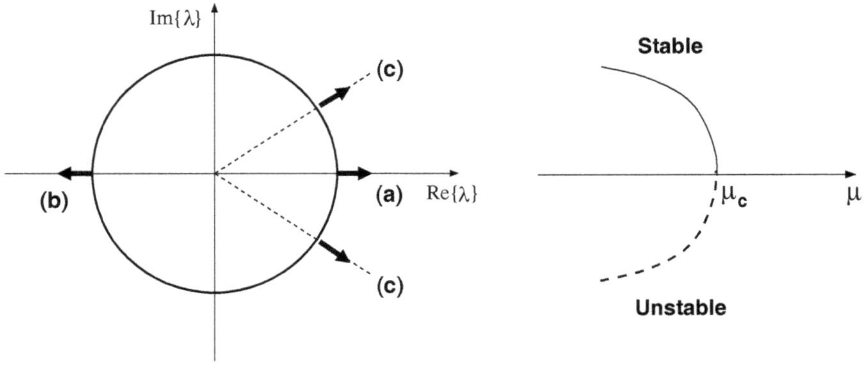

Fig. 6.7 (left) Sketch of the possible routes of exit of the eigenvalue from the unitary circle, see text for explanation of the different labels. (right) Sketch of the inverse saddle-node bifurcation, see text for further details.

Fig. 6.6a. Case (b) characterizes two different kinds of transition: period doubling and the so-called intermittency transition of kind III. Finally case (c) pertains to Hopf's bifurcation (first step of the Ruelle-Takens scenario) and the intermittency transition of kind II. We do not detail here the intermittency transitions of kind II and III, they are for some aspects similar to that of kind I encountered in Sect. 6.3. Most of the differences lie indeed in the statistics of the duration of laminar periods. The reader can find an exhaustive discussion of these kinds of route to chaos in Bergé *et al.* (1987).

6.5 Transition to turbulence in real systems

Several mechanisms have been identified for the transition from fixed points (f.p.) to periodic orbits (p.o.) and finally to chaos when the control parameter r is varied. They can be schematically summarized as follows:

Landau-Hopf for $r = r_1, r_2, \ldots, r_n, r_{n+1}, \ldots$ (the sequence being unbounded and ordered, $r_j < r_{j+1}$) the following transitions occur:

> f.p. $\to$ p.o. with 1 frequency $\to$ p.o. with 2 frequencies $\to$ p.o. with 3 frequencies $\to \ldots \to$ p.o. with n frequencies $\to$ p.o. with $n+1$ frequencies $\to \ldots$ (after Ruelle and Takens we know that only the first two steps are structurally stable).

Ruelle-Takens there are three critical values $r = r_1, r_2, r_3$ marking the transitions:

> f.p. $\to$ p.o. with 1 frequency $\to$ p.o. with 2 frequencies $\to$ chaos with aperiodic solutions and the trajectories settling onto a strange attractor.

Feigenbaum infinite critical values $r_1, \ldots, r_n, r_{n+1}, \ldots$ ordered ($r_j < r_{j+1}$) with a finite limit $r_\infty = \lim_{n\to\infty} r_n < \infty$ for which:

> p.o. with period-1 $\to$ p.o. with period-2 $\to$ p.o. with period-4 $\to \ldots \to$ p.o. with period-2^n $\to \ldots \to$ chaos for $r > r_\infty$.

Pomeau-Manneville there is a single critical parameter r_c :

f.p. or p.o. $\to$ chaos characterized by intermittency.

It is important to stress that the mechanisms above listed do not only work in abstract mathematical examples.

Time discreteness is not an indispensable requirement. This should be clear from the discussion of the Pomeau-Manneville transition, which can be found also in ordinary differential equations such as the Lorenz model. Time discrete representation is anyway very useful because it provides an easy visualization of the structural changes induced by variations of the control parameter r.

As a further demonstration of the generality of the kind of transitions found in maps, we mention another example taken by fluid dynamics. Franceschini and Tebaldi (1979) studied the transition to turbulence in two-dimensional fluids, using a set of five nonlinear ordinary differential equations obtained from Navier-Stokes equation with the Galerkin truncation (Chap. 13), similarly to Lorenz derivation (Box B.4). Here the the control parameter r is the Reynolds number. At varying r, they observed a period doubling transition to chaos: steady dynamics for $r < r_1$, periodic motion of period T_0 for $r_1 < r < r_2$, periodic motion of period $2T_0$ for $r_2 < r < r_3$ and so forth. Moreover, the sequence of critical numbers r_n was characterized by the same universal properties of the logistic map. The period doubling transition has been observed also in the Hénon map in some parameter range.

6.5.1 *A visit to laboratory*

Experimentalists have been very active during the '70s and '80s and studied the transition to chaos in different physical contexts. In this respect, it is worth mentioning the experiments by Arecchi *et al.* (1982); Arecchi (1988); Ciliberto and Rubio (1987); Giglio *et al.* (1981); Libchaber *et al.* (1983); Gollub and Swinney (1975); Gollub and Benson (1980); Maurer and Libchaber (1979, 1980); Jeffries and Perez (1982), see also Eckmann (1981) and references therein. In particular, various works devoted their attention to two hydrodynamic problems: the convective instability for fluids heated from below — the Rayleigh-Bénard convection — and the motion of a fluid in counterotating cylinders — the circular Taylor-Couette flow. In the former laboratory experience, the parameter controlling the nonlinearity is the Rayleigh number Ra (see Box B.4) while, in the latter, nonlinearity is tuned by the difference between the angular velocities of the inner and external rotating cylinders. Laser Doppler techniques [Albrecht *et al.* (2002)] allows a single component $v(t)$ of the fluid velocity and/or the temperature in a point to be measured for different values of the control parameter r in order to verify, e.g. that Landau-Hopf mechanism never occurs. In practice, given the signal $v(t)$ in a time period $0 < t < T_{\max}$, the power spectrum $S(\omega)$ can be computed by Fourier transform

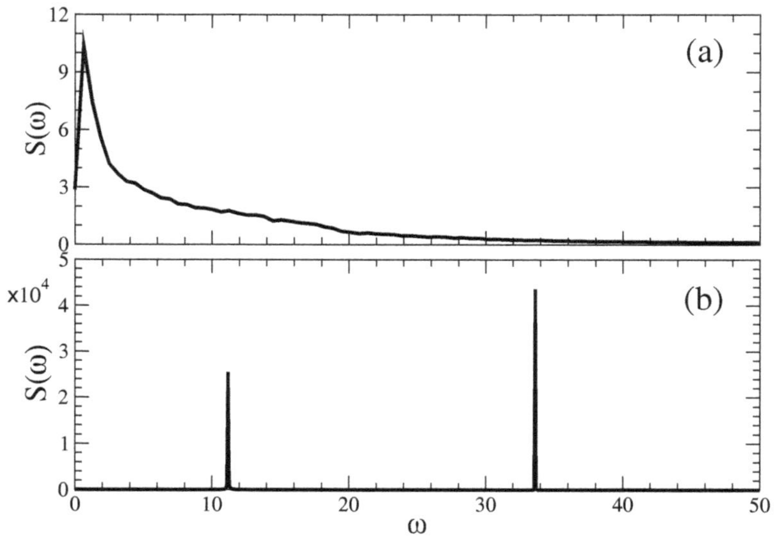

Fig. 6.8 Power spectrum $S(\omega)$ vs ω associated to the Lorenz system with $b = 8/3$ and $\sigma = 10$ for the chaotic case $r = 28$ (a) and the periodic one $r = 166$ (b). The power spectrum is obtained by Fourier transforming the corresponding correlation functions (Fig. 3.11).

(see, e.g. Monin and Yaglom (1975)):

$$S(\omega) = \left| \frac{1}{T_{\max}} \int_0^{T_{\max}} \mathrm{d}t \, v(t) \mathrm{e}^{i\,\omega t} \right|^2 .$$

The power spectrum $S(\omega)$ quantifies the contribution of the frequency ω to the signal $v(t)$. If $v(t)$ results from a process like (6.2), $S(\omega)$ would simply be a sum of δ-function at the frequencies $\omega_1, \ldots \omega_n$ present in the signal i.e. :

$$S(\omega) = \sum_{k=0}^{n} B_k \, \delta(\omega - \omega_k) \, . \tag{6.12}$$

In such a situation the power spectrum would appear as separated spikes in a spectrum analyzer, while chaotic trajectories generate broad band continuous spectra. This difference is exemplified in Figures 6.8a and b, where $S(\omega)$ is shown for the Lorenz model in chaotic and non-chaotic regimes, respectively.

However, in experiments a sequence of transitions described by a power spectrum such as (6.12) has never been observed, while all the other scenarios we have described above (along with several others not discussed here) are possible, just to mention a few examples:

- Ruelle-Takens scenario has been observed in Rayleigh-Bénard convection at high Prandtl number fluids (Pr $= \nu/\kappa$ measures the ratio between viscosity and thermal diffusivity of the fluid) [Maurer and Libchaber (1979); Gollub and Benson (1980)], and in the Taylor-Couette flow [Gollub and Swinney (1975)].

- Feigenbaum period doubling transition is very common, and it can be found in lasers, plasmas, or in the Belousov-Zhabotinsky chemical reaction [Zhang *et al.* (1993)] (see also Sec. 11.3.3 for a discussion on chaos in chemical reactions) for certain values of the concentration of chemicals. Period doubling has been found also in Rayleigh-Bénard convection for low Pr number fluids, such as in mercury or liquid helium (see Maurer and Libchaber (1979); Giglio *et al.* (1981); Gollub and Benson (1980) and references therein).
- Pomeau-Manneville transition to chaos through intermittency has been observed in Rayleigh-Bénard system under particular conditions and in Belousov-Zhabotinsky reaction [Zhang *et al.* (1993)]. It has been also found in driven nonlinear semiconductors [Jeffries and Perez (1982)]

All the above mentioned examples might suggest non universal mechanisms for the transition to chaos. Moreover, even in the same system, disparate mechanisms can coexist in different ranges of the control parameters. However, the number of possible scenarios is not infinite, actually it is rather limited, so that we can at least speak about different classes of universality for such kind of transitions, similarly to what happen in phase transitions of statistical physics [Kadanoff (1999)]. It is also clear that Landau-Hopf mechanism is never observed and the passage from order to chaos always happens through a low dimensional strange attractor. This is evident from numerical and laboratory experiments. Although in the latter the evidences are less direct than in computer simulation, as rather sophisticated concepts and tools are needed to extract the low dimensional strange attractor from measurements based on a scalar signal (Chap. 10).

6.6 Exercises

Exercise 6.1: Consider the system

$$\frac{\mathrm{d}x}{\mathrm{d}t} = y\,, \quad \frac{\mathrm{d}y}{\mathrm{d}t} = z^2 \sin x \cos x - \sin x - \mu y\,, \quad \frac{\mathrm{d}z}{\mathrm{d}t} = k(\cos x - \rho)$$

with μ as control parameter. Assume that $\mu > 0$, $k = 1$, $\rho = 1/2$. Describe the bifurcation of the fixed points at varying μ.

Exercise 6.2: Consider the set of ODEs

$$\frac{\mathrm{d}x}{\mathrm{d}t} = 1 - (b+1)x + ax^2 y\,, \qquad \frac{\mathrm{d}y}{\mathrm{d}t} = bx - ax^2 y$$

known as *Brusselator* which describes a simple chemical reaction.
(1) Find the fixed points and study their stability.
(2) Fix a and vary b. Show that at $b_c = a + 1$ there is a Hopf bifurcation and the appearance of a limit cycle.
(3) Estimate the dependence of the period of the limit cycle as a function of a close to b_c.

Hint: *You need to see that the eigenvalues of the stability matrix are pure imaginary at* b_c*. Note that the imaginary part of a complex eigenvalue is related to the period.*

Exercise 6.3: Estimate the Feigenbaum constants of the sin map (Ex.3.3) from the first, say 4, 6 period doubling bifurcations and see how they approach the known universal values.

Exercise 6.4: Consider the logistic map at $r = r_c - \epsilon$ with $r_c = 1 + \sqrt{8}$ (see also Eq. (3.2)). Graphically study the evolution of the third iterate of the map for small ϵ and, specifically, investigate the region close to $x = 1/2$. Is it similar to the Lorenz map for $r = 166.3$? Why? Expand the third iterate of the map close to its fixed point and compare the result with Eq. (6.11). Study the behavior of the correlation function at decreasing ϵ. Do you have any explanation for its behavior?
Hint: *It may be useful to plot the absolute value of the correlation function each* 3 *iterates.*

Exercise 6.5: Consider the one-dimensional map defined by

$$F(x) = x_c - (1+\epsilon)(x - x_c) + \alpha(x - x_c)^2 + \beta(x - x_c)^3 \mod 1$$

(1) Study the change of stability of the fixed point x_c at varying ϵ, in particular perform the graphical analysis using the second iterate $F(F(x))$ for $x_c = 2/3$, $\alpha = 0.3$ and $\beta = \pm 1.1$ at increasing ϵ, what is the difference between the $\beta > 0$ and $\beta < 0$ case?
(2) Consider the case with negative β and iterate the map comparing the evolution with that of the map Eq. (6.11).
The kind of behavior displayed by this map has been termed intermittency of III-type (see Sec. 6.4).

Chapter 7

Chaos in Hamiltonian Systems

> At any given time there is only a thin layer separating what is trivial from what is impossibly difficult. It is in that layer that mathematical discoveries are made.
>
> *Andrei Nikolaevich Kolmogorov (1903–1987)*

Hamiltonian systems constitute a special class of dynamical systems. A generic perturbation indeed destroys their Hamiltonian/symplectic structure. Their peculiar properties reflect on the routes such systems follow from order (integrability) to chaos (non-integrability), which are very different from those occurring in dissipative systems. Discussing in details the problem of the appearance of chaos in Hamiltonian systems would require several Chapters or, perhaps, a book by itself. Here we shall therefore remain very much qualitative by stressing what are the main problems and results. The demanding reader may deepen the subject by referring to dedicated monographs such as Berry (1978); Lichtenberg and Lieberman (1992); Benettin *et al.* (1999).

7.1 The integrability problem

A Hamiltonian system is integrable when its trajectories are periodic or quasiperiodic. More technically, a given Hamiltonian $H(\boldsymbol{q}, \boldsymbol{p})$ with $\boldsymbol{q}, \boldsymbol{p} \in \mathbb{R}^N$ is said integrable if there exists N independent conserved quantities, including energy. Proving integrability is equivalent to provide the explicit time evolution of the system (see Box B.1). In practice, one has to find a canonical transformation from coordinates $(\boldsymbol{q}, \boldsymbol{p})$ to action-angle variables $(\mathbf{I}, \boldsymbol{\phi})$ such that the new Hamiltonian depends on the actions $\mathbf{I}$ only:

$$H = H(\mathbf{I}) \,. \tag{7.1}$$

Notice that for this to be possible, the conserved quantities (the actions) should be in *involution*. In other terms the Poisson brackets between any two conserved quantities should vanish

$$\{I_i, I_j\} = 0 \quad \text{for all} \quad i, j\,. \tag{7.2}$$

When the conditions for integrability are fulfilled, the time evolution is trivially given by

$$\begin{cases} I_i(t) = I_i(0) \\ \phi_i(t) = \phi_i(0) + \omega_i(\mathbf{I}(0))\, t\,, \end{cases} \qquad i = 1, \cdots, N \tag{7.3}$$

where $\omega_i = \partial H_0/\partial I_i$ are the frequencies. It is rather easy to see that the motion obtained by Eq. (7.3) evolves on N-dimensional tori. The periodicity or quasiperiodicity of the motions depends upon the commensurability or not of the frequencies $\{\omega_i\}$'s (see Fig. B1.1 in Box B.1).

The Solar system provides an important example of Hamiltonian system. When planetary interactions are neglected, the system reduces to the two-body problem Sun-Planet, whose integrability can be easily proved. This means that if in the Solar system we had only Earth and Sun, Earth motion would be completely regular and fully predictable. Unfortunately, Earth is gravitationally influenced by other astronomical bodies, the Moon above all, so that we have to consider, at least, a three-body problem for which integrability is not granted (see also Sec. 11.1).

It is thus natural to wonder about the effect of perturbations on an integrable Hamiltonian system H_0, i.e. to study the near-integrable Hamiltonian

$$H(\mathbf{I}, \boldsymbol{\phi}) = H_0(\mathbf{I}) + \epsilon H_1(\mathbf{I}, \boldsymbol{\phi})\,, \tag{7.4}$$

where ϵ is assumed to be small. The main questions to be asked are:
i) Will the trajectories of the perturbed Hamiltonian system (7.4) be "close" to those of the integrable one H_0?
ii) Does integrals of motion, besides energy, exist when the perturbation term $\epsilon H_1(\mathbf{I}, \boldsymbol{\phi})$ is present?

7.1.1 *Poincaré and the non-existence of integrals of motion*

The second question was answered by Poincaré (1892, 1893, 1899) (see also Poincaré (1890)), who showed that, as soon as $\epsilon \neq 0$, a system of the form (7.4) does not generally admit analytic first integrals, besides energy. This result can be understood as follows. If $F_0(\mathbf{I})$ is a conserved quantity of H_0, for small ϵ, it is natural to seek for a new integral of motion of the form

$$F(\mathbf{I}, \boldsymbol{\phi}) = F_0(\mathbf{I}) + \epsilon F_1(\mathbf{I}, \boldsymbol{\phi}) + \epsilon^2 F_2(\mathbf{I}, \boldsymbol{\phi}) + \ldots \quad . \tag{7.5}$$

The perturbative strategy can be exemplified considering the first order term F_1 which, as the angular variables $\boldsymbol{\phi}$ are cyclic, can be expressed via the Fourier series

$$F_1(\mathbf{I}, \boldsymbol{\phi}) = \sum_{m_1=-\infty}^{+\infty} \cdots \sum_{m_N=-\infty}^{+\infty} f^{(1)}_{\boldsymbol{m}}(\mathbf{I}) e^{i(m_1\phi_1 + \cdots + m_N\phi_N)} = \sum_{\boldsymbol{m}} f^{(1)}_{\boldsymbol{m}}(\mathbf{I}) e^{i\boldsymbol{m}\cdot\boldsymbol{\phi}} \tag{7.6}$$

where $\boldsymbol{m} = (m_1, \ldots, m_N)$ is an N-component vector of integers. The definition of conserved quantity implies the condition $\{H, F\} = 0$, which by using (7.5) leads to the equation for F_1:

$$\{H_0, F_1\} = -\{H_1, F_0\} \,. \tag{7.7}$$

The perturbation H_1 is assumed to be a smooth function which can also be expanded in Fourier series

$$H_1 = \sum_{\boldsymbol{m}} h^{(1)}_{\boldsymbol{m}}(\mathbf{I}) e^{i\boldsymbol{m}\cdot\boldsymbol{\phi}} \,. \tag{7.8}$$

Substituting the expressions (7.6) and (7.8) in Eq. (7.7), for $F_0 = \mathrm{I}_j$, yields

$$F_1 = \sum_{\boldsymbol{m}} \frac{m_j h^{(1)}_{\boldsymbol{m}}(\mathbf{I})}{\boldsymbol{m}\cdot\boldsymbol{\omega}_0(\mathbf{I})} e^{i\boldsymbol{m}\cdot\boldsymbol{\phi}} \,, \tag{7.9}$$

$\boldsymbol{\omega}_0(\mathbf{I}) = \boldsymbol{\nabla}_{\mathbf{I}} H_0(\mathbf{I})$ being the unperturbed N-dimensional frequency vector for the torus corresponding to action $\mathbf{I}$. The reason of the nonexistence of first integrals can be directly read from Eq. (7.9): for any $\boldsymbol{\omega}_0$ there will be some $\boldsymbol{m}$ such that $\boldsymbol{m}\cdot\boldsymbol{\omega}_0$ becomes arbitrarily small, posing problems for the meaning of the series (7.9) — this is the *small denominators problem* , see e.g. Arnold (1963b); Gallavotti (1983).

The series (7.9) may fail to exist in two situations. The obvious one is when the torus is *resonant* meaning that the frequencies $\boldsymbol{\omega}_0 = (\omega_1, \omega_2, \ldots, \omega_N)$ are rationally dependent, so that $\boldsymbol{m}\cdot\boldsymbol{\omega}_0(\mathbf{I}) = 0$ for some $\boldsymbol{m}$. Resonant tori are destroyed by the perturbation as a consequence of the Poincaré-Birkhoff theorem, that will be discussed in Sec. 7.3. The second reason is that, also in the case of rationally independent frequencies, the denominator $\boldsymbol{m}\cdot\boldsymbol{\omega}_0(\mathbf{I})$ can be arbitrarily small, making the series not convergent.

Already on the basis of these observations the reader may conclude that analytic first integrals (besides energy) cannot exist and, therefore, any perturbation of an integrable system should lead to chaotic orbits. Consequently, also the question *i)* about the "closeness" of perturbed trajectories to integrable ones is expected to have a negative answer. However, this negative conclusion contradicts intuition as well as many results obtained with analytical approximations or numerical simulations. For example, in Chapter 3 we saw that Hénon-Heiles system for small nonlinearity exhibits rather regular behaviors (Fig. 3.10a). Worse than this, the presumed overwhelming presence of chaotic trajectories in a perturbed system leaves us with the unpleasant feeling to live in a completely chaotic Solar system with an uncertain fate, although, so far, this does not seem to be the case.

7.2 Kolmogorov-Arnold-Moser theorem and the survival of tori

Kolmogorov (1954) was able to reconcile the mathematics with the "intuition" and laid the basis of an important theorem, sketching the essential lines of the proof, which was subsequently completed by Arnold (1963a) and Moser (1962), whence the name KAM for the theorem which reads:

Given a Hamiltonian $H(\mathbf{I}, \boldsymbol{\phi}) = H_0(\boldsymbol{I}) + \epsilon H_1(\mathbf{I}, \boldsymbol{\phi})$, *with* $H_0(\mathbf{I})$ *sufficiently regular and such that* $\det |\partial^2 H_0(\mathbf{I})/\partial \mathrm{I}_i \partial \mathrm{I}_j| = \det |\partial \omega_i / \partial \mathrm{I}_j| \neq 0$, *if* ϵ *is small enough, then on the constant-energy surface, invariant tori survive in a region whose measure tends to* 1 *as* $\epsilon \to 0$.

These tori, called KAM tori, result from a small deformation of those of the integrable system ($\epsilon = 0$).

At first glance, KAM theorem might seem obvious, while in the light of the small denominator problem, the existence of KAM tori constitutes a rather subtle result.

In order to appreciate such subtleties we need to recall some elementary notions of number theory. Resonant tori, those destroyed as soon as the perturbation is present, correspond to motions with frequencies that are rationally dependent, whilst non-resonant tori relate to rationally independent ones. Rationals are dense[1] in $\mathbb{R}$ and this is enough to forbid analytic first integrals besides energy. However, there are immeasurably more, with respect to the Lebesgue measure, irrationals than rationals. Therefore, KAM theorem implies that, even in the absence of global analytic integrals of motion, the measure of non-resonant tori, which are not destroyed but only slightly deformed, tend to 1 for $\epsilon \to 0$. As a consequence, the perturbed system behaves similarly to the integrable one, at least for generic initial conditions. In conclusion, the absence of conserved quantities does not imply that all the perturbed trajectories will be far from the unperturbed ones, meaning that a negative answer to question *ii*) does not imply a negative answer to question *i*).

We do not enter the technical details of KAM theorem, here we just sketch the basic ideas. The small denominator problem prevents us from finding integrals of motion other than energy. However, relaxing the request of global constant of motions, i.e. valid in the whole phase space, we may look for the weaker condition of "local" integrals of motions, i.e. existing in a portion of non-zero measure of the constant energy surface. This is possible if the Fourier terms of F_1 in (7.9) are small enough. Assuming that H_1 is an analytic function, the coefficients $h^{(1)}_{\boldsymbol{m}}$'s exponentially decrease with $m = |m_1| + |m_2| + \cdots + |m_N|$. Nevertheless, there will exist tori with frequencies $\boldsymbol{\omega}_0(\mathbf{I})$ such that the denominator is *not too small*, specifically

$$|\boldsymbol{m} \cdot \boldsymbol{\omega}_0(\mathbf{I})| > \alpha(\boldsymbol{\omega}_0) m^{-\tau} , \tag{7.10}$$

for all integer vectors $\boldsymbol{m}$ (except the zero vector), α and $\tau \geq N - 1$ being positive constants — this is the so-called *Diophantine* inequality [Arnold (1963b); Berry (1978)]. Tori fulfilling condition (7.10) are strongly non-resonating and are infinitely many, as the set of frequencies $\boldsymbol{\omega}_0$ for which inequality (7.10) holds has a non-zero measure. Thus, the function F_1 can be built locally, in a suitable non-zero measure region, excluding a small neighborhood around non-resonant tori. Afterwords, the procedure should be iterated for $F_2, F_3, ...$ and the convergence of the series controlled. For a given $\epsilon > 0$, however, not all the non-resonant tori fulfilling condition (7.10) survive: this is true only for those such that $\alpha \gg \sqrt{\epsilon}$ (see Pöschel (2001) for a rigorous but gentle discussion of KAM theorem).

[1]For any real number x and every $\delta > 0$ there is a rational number q such that $|x - q| < \delta$.

The strong irrationality degree on the torus frequencies, set by inequality (7.10), is crucial for the theorem, as it implies that the more irrational the frequencies the larger the perturbation has to be to destroy the torus. To appreciate this point we open a brief digression following Berry (1978) (see also Livi *et al.* (2003)). Consider a two-dimensional torus with frequencies ω_1 and ω_2. If $\omega_1/\omega_2 = r/s$ with r and s coprime integers, we have a resonant torus which is destroyed. Now suppose that $\omega_1/\omega_2 = \sigma$ is irrational, it is always possible to find a rational approximation, e.g.

$$\sigma = \pi = 3.14159265\cdots \approx \frac{r}{s} = \frac{3}{1}, \frac{31}{10}, \frac{314}{100}, \frac{3141}{1000}, \frac{31415}{10000}, \ldots .$$

Such kind of naive approximation can be proved to converge as

$$\left| \sigma - \frac{r}{s} \right| < \frac{1}{s} .$$

Actually, a faster convergence rate can be obtained by means of continued fractions [Khinchin (1997)]:

$$\sigma = \lim_{n\to\infty} \frac{r_n}{s_n} \quad \text{with} \quad \frac{r_n}{s_n} = [a_0;\, a_1, \ldots, a_n]$$

where

$$[a_0;\, a_1] = a_0 + \frac{1}{a_1}, \quad [a_0;\, a_1, a_2] = a_0 + \cfrac{1}{a_1 + \cfrac{1}{a_2}}$$

for which it is possible to prove that

$$\left| \sigma - \frac{r_n}{s_n} \right| < \frac{r_n}{s_n\, s_{n-1}} .$$

A theorem ensures that continued fractions provide the best, in the sense of faster converging, approximation to a real number [Khinchin (1997)]. Clearly the sequence r_n/s_n converges faster the faster the sequence a_n diverges, so we have now a criterion to define the degree of "irrationality" of a number in terms of the rate of convergence (divergence) of the sequence σ_n (a_n, respectively). For example, the Golden Ratio $\mathcal{G} = (\sqrt{5}+1)/2$ is the more irrational number, indeed its continued fraction representation is $\mathcal{G} = [1;\, 1, 1, 1, \ldots]$ meaning that the sequence $\{a_n\}$ does not diverge. Tori associated to $\mathcal{G} \pm k$, with k integer, will be thus the last tori to be destroyed.

The above considerations are nicely illustrated by the standard map

$$\begin{aligned} \mathrm{I}(t+1) &= \mathrm{I}(t) + K \sin(\phi(t)) \\ \phi(t+1) &= \phi(t) + \mathrm{I}(t+1) \qquad \mod 2\pi . \end{aligned} \tag{7.11}$$

For $K = 0$ this map is integrable, so that K plays the role of ϵ, while the *winding* or rotation number

$$\sigma = \lim_{t\to\infty} \frac{\phi(t) - \phi(0)}{t}$$

defines the nature of the tori.

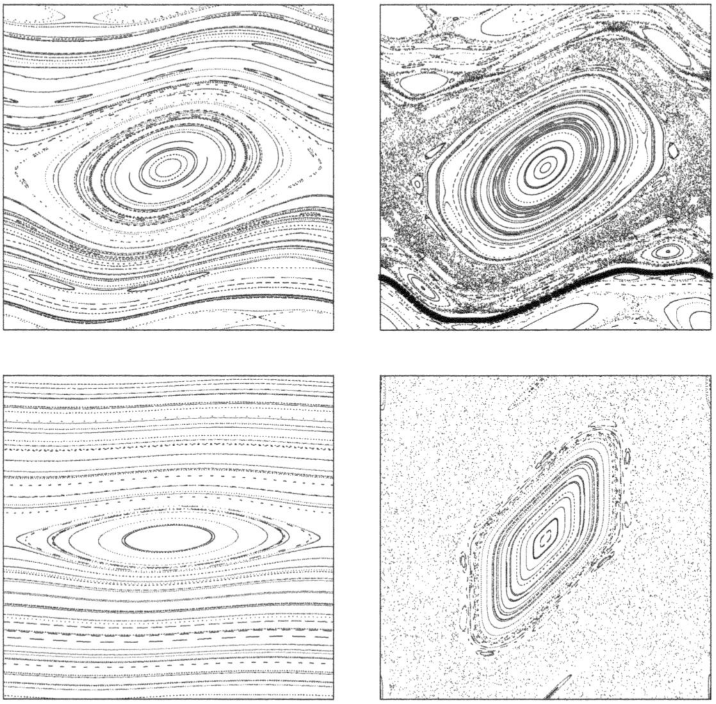

Fig. 7.1 Phase-portrait of the standard map (7.11) for $K = 0.1, 0.5, 0.9716, 2.0$ (turning clockwise from the bottom left panel). The thick black curve in the top-right panel is a quasiperiodic orbit with winding number very close to the golden ratio $\mathcal{G}$, actually to $\mathcal{G}-1$. The portion of phase space represented is a square $2\pi \times 2\pi$, chosen by symmetry considerations to represent the elementary cell, indeed the motions are by construction spatially periodic with respect to such a cell.

We have to distinguish two different kinds of KAM tori: "separating" ones, which cut the phase space horizontally acting as a barrier to the trajectories, and "non-separating" ones, as those of regular islands which derive from resonant tori and which survive also for very large values of the perturbation. Examples of these two classes of KAM tori can be seen in Fig. 7.1, where we show the phase-space portrait for different values of K. The invariant curves identified by the value of the action I, filling the phase space at $K = 0$, are only slightly perturbed for $K = 0.1$ and $K = 0.5$. Indeed for $K = 0$, independently of irrationality or rationality of the winding number, tori fill densely the phase space, and appear as horizontal straight lines. For small K, the presence of a chaotic orbits, forming a thin layer in between surviving tori, can be hardly detected. However, for $K = K_c$, portion of phase space covered by chaotic orbits gets larger. The critical value K_c is associated to the "death" of the last "separating" KAM torus, corresponding to the orbit with winding number equal to $\mathcal{G}$ (thick curve in the figure). For $K > K_c$, the barrier constituted by the last separating KAM torus is eliminated and no more separated regions exist: now the action I(t) can wander in the entire phase space giving rise to a diffusive behavior (see Box B.14 for further details). However, the phase portrait is still characterized by the presence of regular islands of quasi-periodic motion — the "non-separating" KAM tori — embedded in a chaotic sea which gets larger as K increases. Similar features has been observed while studying Hénon-Heiles system in Sec. 3.3. We emphasize that in non-Hamiltonian, conservative systems (or non-symplectic, volume-preserving maps) the transition to chaos is very similar to that described above for Hamiltonian systems and, in particular cases, invariant surfaces survive a nonlinear perturbation in a KAM-like way [Feingold *et al.* (1988)].

It is worth observing that the behavior of two degrees of freedom systems ($N = 2$) is rather peculiar and different from that of $N > 2$ degrees of freedom systems. For $N = 2$, KAM tori are bi-dimensional and thus can separate regions of the three-dimensional surface of constant energy. Then disjoint chaotic regions, separated by invariant surfaces (KAM tori), can coexist, at least until the last tori are destroyed, e.g. for $K < K_c$ in the standard map example. The situation changes for $N > 2$, as KAM tori have dimension N while the energy hypersurface has dimension $2N - 1$. Therefore, for $N \geq 3$, the complement of the set of invariant tori is connected allowing, in principle, the wandering of chaotic orbits. This gives rise to the so-called Arnold diffusion [Arnold (1964); Lichtenberg and Lieberman (1992)]: trajectories can move on the whole surface of constant energy, by diffusing among the unperturbed tori (see Box B.13).

The existence of invariant tori prescribed by KAM theorem is a result "local" in space but "global" in time: those tori lasting forever live only in a portion of phase space. If we are interested to times smaller than a given (large) T_{max} and to generic initial conditions (i.e. globally in phase space), KAM theorem is somehow too restrictive because of the infinite time requirement and not completely satisfactory due to its "local" validity. An important theorem by Nekhoroshev (1977) provides some bounds valid globally in phase space but for finite time intervals. In particular, it states that the actions remain close to their initial values for a very long time, more formally

Given a Hamiltonian $H(\mathrm{I}, \phi) = H_0(\mathrm{I}) + \epsilon H_1(\mathrm{I}, \phi)$, *with* $H_0(\mathrm{I})$, *under the same assumptions of the KAM theorem. Then there exist positive constants* A, B, C, α, β, *such that*

$$|\mathrm{I}_n(t) - \mathrm{I}_n(0)| \leq A\epsilon^{\alpha} \qquad n = 1, \cdots, N \tag{7.12}$$

for times such that

$$t \leq B \exp(C\epsilon^{-\beta}) \,. \tag{7.13}$$

KAM and Nekhoroshev theorems show clearly that both ergodicity and integrability are non-generic properties of Hamiltonian systems obtained as perturbation of integrable ones. We end this section observing that, despite the importance of these two theorems, it is extremely difficult to have a precise control, even at a qualitative level, of important aspects as, for instance, how the measure of KAM tori varies as function of both ϵ and the number of degrees of freedom N or how the constants in Eqs. (7.12) and (7.13) depend on N.

Box B.13: Arnold diffusion

There is a sharp qualitative difference between the behavior of Hamiltonian systems with two degrees of freedom, and those with $N \geq 3$ because in the latter case the N-dimensional KAM tori cannot separate the $(2N-1)$-dimensional phase space in disjoint regions, able to confine trajectories. Therefore, even for arbitrary small ϵ, there is the possibility that any trajectory initially close to a KAM torus may invade any region of phase space compatible with the constant-energy constraint. Arnold (1964) was the first to show the existence of such a phenomenon, resembling diffusion, in a specific system, whence the name of Arnold diffusion. Roughly speaking the wandering of chaotic trajectories occurs in the set of the energy hypersurface complementary to the union of the KAM tori, or more precisely in the so-called Arnold web (AW), which can be defined as a suitable neighborhood of resonant orbits,

$$\sum_{i=1}^{N} k_i \omega_i = 0$$

with some integers $(k_1, ..., k_N)$. The size δ of the AW depends both on perturbation strength ϵ and on order k of the resonance, $k = |k_1| + |k_2| + \cdots + |k_N|$: typically $\delta \sim \sqrt{\epsilon}/k$ [Guzzo *et al.* (2002, 2005)]. Of course, trajectories in the AW can be chaotic and the simplest assumption is that at large times the action $\mathrm{I}(t)$ performs a sort of random walk on AW so that

$$\langle |\mathrm{I}(t) - \mathrm{I}(0)|^2 \rangle = \langle \Delta \mathrm{I}(t)^2 \rangle \simeq 2Dt \qquad \text{(B.13.1)}$$

where $\langle\ \rangle$ denotes the average over initial conditions. If Eq. (B.13.1) holds true, Nekhoroshev theorem can be used to set an upper bound for the diffusion coefficient D, in particular from (7.13) we have

$$D < \frac{A^2}{B} \epsilon^{2\alpha} \exp(-C\epsilon^{-\beta}) .$$

Benettin *et al.* (1985) and Lochak and Neishtadt (1992) have shown that generically $\beta \sim 1/N$ implying that, for large N, the exponential factor can be $\mathcal{O}(1)$ so that the values of A and B (which are not easy to be determined) play the major role. Strong numerical evidence shows that standard diffusion (B.13.1) occurs on the AW and $D \to 0$ faster than any power as $\epsilon \to 0$. This result was found by Guzzo *et al.* (2005) studying some quasi-integrable Hamiltonian system (or symplectic maps) with $N = 3$, where both KAM and Nekhoroshev theorems apply. For systems with $N = 4$, obtained coupling two standard maps, some theoretical arguments give $\beta = 1/2$ in agreement with numerical simulations [Lichtenberg and Aswani (1998)]. Actually, the term "diffusion" can be misleading, as behaviors different from standard diffusion (B.13.1) can be present. For instance, Kaneko and Konishi (1989), in numerical simulations of high dimensional symplectic maps, observed a sub-diffusive behavior

$$\langle \Delta \mathrm{I}^2(t) \rangle \sim t^{\nu} \qquad \text{with} \qquad \nu < 1 ,$$

at least for finite but long times. We conclude with a brief discussion of the numerical results for high dimensional symplectic maps of the form

$$\begin{aligned} \phi_n(t+1) &= \phi_n(t) + \mathrm{I}_n(t) && \mod 2\pi \\ \mathrm{I}_n(t+1) &= \mathrm{I}_n(t) + \epsilon \frac{\partial F(\phi(t+1))}{\partial \phi_n(t+1)} && \mod 2\pi\,, \end{aligned}$$

where $n = 1, \ldots, N$. The above symplectic map is nothing but a canonical transformation from the "old" variables (I, ϕ), i.e. those at time t, to the "new" variables (I', ϕ'), at time $t+1$ [Arnold (1989)]. When the coupling constant ϵ vanishes the system is integrable, and the term $\epsilon F(\phi)$ plays the role of the non-integrable perturbation. Numerical studies by Falcioni *et al.* (1991) and Hurd *et al.* (1994) have shown that: on the one hand, irregular behaviors becomes dominant at increasing N, specifically the volume of phase space occupied by KAM tori decreases exponentially with N; on the other hand individual trajectories forget their initial conditions, invading a non-negligible part of phase space, only after extremely long times (see also Chap. 14). Therefore, we can say that usually Arnold diffusion is very weak and different trajectories, although with a high value of the first Lyapunov exponent, maintain memory of their initial conditions for considerable long times.

7.3 Poincaré-Birkhoff theorem and the fate of resonant tori

KAM theorem determines the conditions for a torus to survive a perturbation: KAM tori resist a weak perturbation, being only slightly deformed, while resonant tori, for which a linear combination of the frequencies with integer coefficients $\{k\}_{i=1}^{N}$ exists such that $\sum_{i=1}^{N} \omega_i k_i = 0$, are destroyed. Poincaré-Birkhoff [Birkhoff (1927)] theorem concerns the "fate" of these resonant tori.

The presentation of this theorem is conveniently done by considering the twist map [Tabor (1989); Lichtenberg and Lieberman (1992); Ott (1993)] which is the transformation obtained by a Poincaré section of a two-degree of freedom integrable Hamiltonian system, whose equation of motion in action-angle variables reads

$$\begin{aligned} \mathrm{I}_k(t) &= \mathrm{I}_k(0) \\ \theta_k(t) &= \theta_k(0) + \omega_k t\,, \end{aligned}$$

where $\omega_k = \partial H/\partial \mathrm{I}_k$ and $k = 1, 2$. The initial value of the actions $\mathrm{I}(0)$ selects a trajectory which lies in a 2-dimensional torus. Its Poincaré section with the plane $\Pi \equiv \{\mathrm{I}_2 = \text{const and } \theta_2 = \text{const}\}$ identifies a set of points forming a smooth closed curve for irrational rotation number $\alpha = \omega_1/\omega_2$ or a finite set of points for α rational. The time $T_2 = 2\pi/\omega_2$ is the period for the occurrence of two consecutive intersections of the trajectory with the plane Π. During the interval of time T_2, θ_1 changes as $\theta_1(t + T_2) = \theta_1(t) + 2\pi\omega_1/\omega_2$. Thus, the intersections with the plane Π

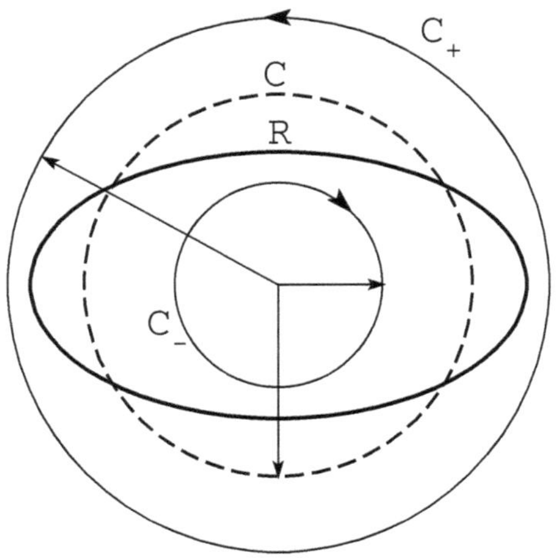

Fig. 7.2 The circles $\mathcal{C}_-, \mathcal{C}, \mathcal{C}_+$ and the non-rotating set $\mathcal{R}$ used to sketch the Poincaré-Birkhoff theorem. [After Ott (1993)]

define the twist map T_0

$$\mathrm{T}_0 \;:\; \begin{cases} \mathrm{I}(t+1) = \mathrm{I}(t) \\ \theta(t+1) = \theta(t) + 2\pi\alpha(\mathrm{I}(t+1)) \mod 2\pi \,, \end{cases} \tag{7.14}$$

where I and θ are now used instead of I_1 and θ_1, respectively, and time is measured in units of T_2.[2] The orbits generated by T_0 depend on the value of the action I and, without loss of generality, can be considered as a family of concentric circles parametrized by the polar coordinates $\{\mathrm{I}, \theta\}$. Consider a specific circle $\mathcal{C}$ corresponding to a resonant torus with $\alpha(\mathrm{I}) = p/q$ (where p, q are coprime integers). Each point of the circle $\mathcal{C}$ is a fixed point of T_0^q, because after q iterations of map (7.14) we have $\mathrm{T}_0^q\theta = \theta + 2\pi q(p/q) \mod 2\pi = \theta$. We now consider a weak perturbation of T_0

$$\mathrm{T}_\epsilon \;:\; \begin{cases} \mathrm{I}(t+1) = \mathrm{I}(t) + \epsilon f(\mathrm{I}(t+1), \theta(t)) \\ \theta(t+1) = \theta(t) + 2\pi\alpha(\mathrm{I}(t+1)) + \epsilon g(\mathrm{I}(t+1), \theta(t)) \mod 2\pi \,, \end{cases}$$

which must be interpreted again as the Poincaré section of the perturbed Hamiltonian, so that f and g cannot be arbitrary but must preserve the symplectic structure (see Lichtenberg and Lieberman (1992)). The issue is to understand what happens to the circle $\mathcal{C}$ of fixed points of T_0^q under the action of the perturbed map.

Consider the following construction. Without loss of generality, α can be considered a smooth increasing function of I. We can thus choose two values of the

[2] In the second line of Eq. (7.14) for convenience we used $\mathrm{I}(t+1)$ instead of $\mathrm{I}(t)$. In this case it makes no difference as $I(t)$ is constant, but in general the use of $\mathrm{I}(t+1)$ helps in writing the map in a symplectic form (see Sec. 2.2.1.2).

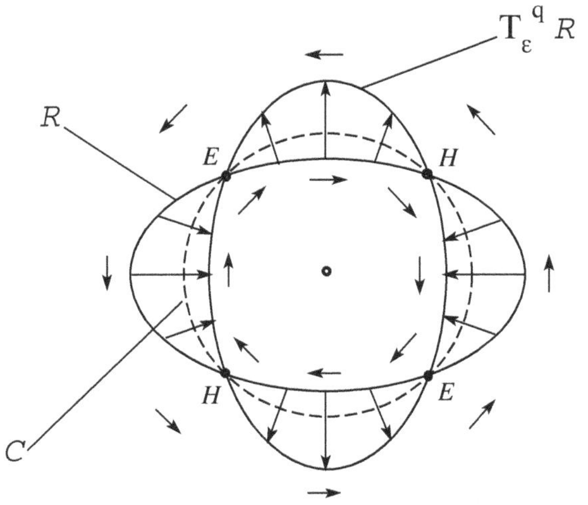

Fig. 7.3 Poincaré-Birkhoff theorem: geometrical construction illustrating the effect of a perturbation on the resonant circle $\mathcal{C}$ of the unperturbed twist map. The curve $\mathcal{R}$ is modified in the radial direction under the action of T_ϵ^q. The original $\mathcal{R}$ and evolved $T_\epsilon^q\mathcal{R}$ curves intersect in an even number of points which form an alternate sequence of elliptic (E) and hyperbolic (H) fixed point for the perturbed map T_ϵ^q. The radial arrows indicate the action of T_ϵ^q on $\mathcal{R}$ while the other arrows the action of the map on the interior or exterior of $\mathcal{R}$. Following the arrow directions the identification of hyperbolic and elliptic fixed points is straightforward. [After Ott (1993)]

action $I_\pm$ such that $I_- < I < I_+$ and thus $\alpha(I_-) < p/q < \alpha(I_+)$ with $\alpha(I_-)$ and $\alpha(I_+)$ irrational, selecting two KAM circles $\mathcal{C}_-$ and $\mathcal{C}_+$, respectively. The two circles $\mathcal{C}_-$ and $\mathcal{C}_+$ are on the interior and exterior of $\mathcal{C}$, respectively. The map T_0^q leaves $\mathcal{C}$ unchanged while rotates $\mathcal{C}_-$ and $\mathcal{C}_+$ clockwise and counterclockwise with respect to $\mathcal{C}$, as shown in Fig. 7.2.

For ϵ small enough, KAM theorem ensures that $\mathcal{C}_\pm$ survive the perturbation, even if slightly distorted and hence $T_\epsilon^q\mathcal{C}_+$ and $T_\epsilon^q\mathcal{C}_-$ still remain rotated anticlockwise and clockwise with respect to the original $\mathcal{C}$. Then by continuity it should be possible to construct a closed curve $\mathcal{R}$ between $\mathcal{C}_-$ and $\mathcal{C}_+$ such that T_ϵ^q acts on $\mathcal{R}$ as a deformation in the radial direction only, the transformation from $\mathcal{R}$ to $T_\epsilon^q\mathcal{R}$ is illustrated in Fig 7.3. Since T_ϵ^q is area preserving, the areas enclosed by $\mathcal{R}$ and $T_\epsilon^q\mathcal{R}$ are equal and thus the two curves must intersect in an even number of points (under the simplifying assumption that generically the tangency condition of such curves does not occur). Such intersections determine the fixed points of the perturbed map T_ϵ^q. Hence, the whole curve $\mathcal{C}$ of fixed points of the unperturbed twist map T_0^q is replaced by a finite (even) number of fixed points when the perturbation is active. More precisely, the theorem states that the number of fixed points is an even multiple of q, $2kq$ (with k integer), but it does not specify the value of k (for example Fig. 7.3 refers to the case $q = 2$ and $k = 1$). The theorem also determines the nature of the new fixed points. In Figure 7.3, the arrows depict the displace-

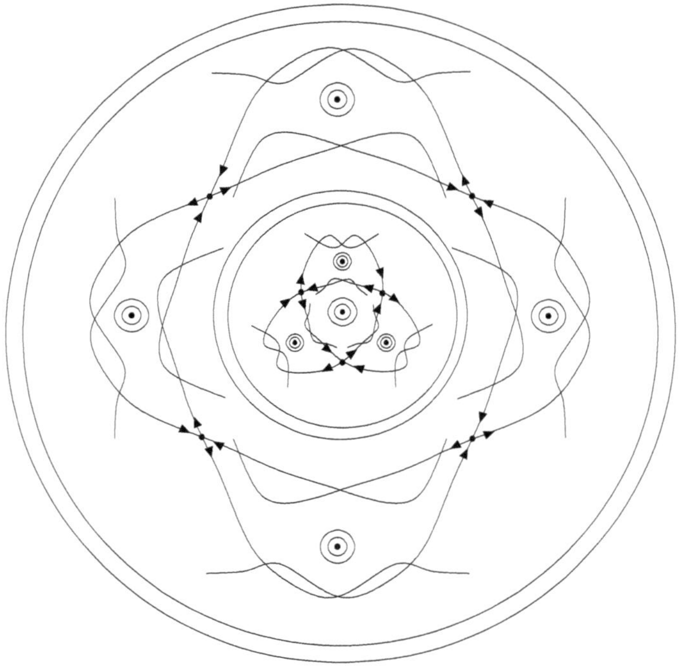

Fig. 7.4 Self-similar structure off-springing from the "explosion" of a resonant torus. [After Ott (1993)]

ments produced by T_ϵ^q. The elliptic/hyperbolic character of the fixed points can be clearly identified by looking at the direction of rotations and the flow lines.

In summary, Poincaré-Birkhoff theorem states that a generic perturbation destroys a resonant torus $\mathcal{C}$ with winding number p/q, giving rise to $2kq$ fixed points, half of which are hyperbolic and the other half elliptic in alternating sequence.

Around each elliptic fixed point, we can find again resonant tori which undergo Poincaré-Birkhoff theorem when perturbed, generating a new alternating sequence of elliptic and hyperbolic fixed points. Thus by iterating the Poincaré-Birkhoff theorem, a remarkable structure of fixed points that repeats self-similarly at all scales must arise around each elliptic fixed point, as sketched in Fig. 7.4. These are the *regular islands* we described for the Hénon-Heiles Hamiltonian (Fig. 3.10).

7.4 Chaos around separatrices

In Hamiltonian systems the mechanism at the origin of chaos can be understood looking at the behavior of trajectories close to fixed points, which are either hyperbolic or elliptic. In the previous section we saw that Poincaré-Birkhoff theorem predicts resonant tori to "explode" in a sequence of alternating (stable) elliptic and (unstable) hyperbolic couples of fixed points. Elliptic fixed points thus become the

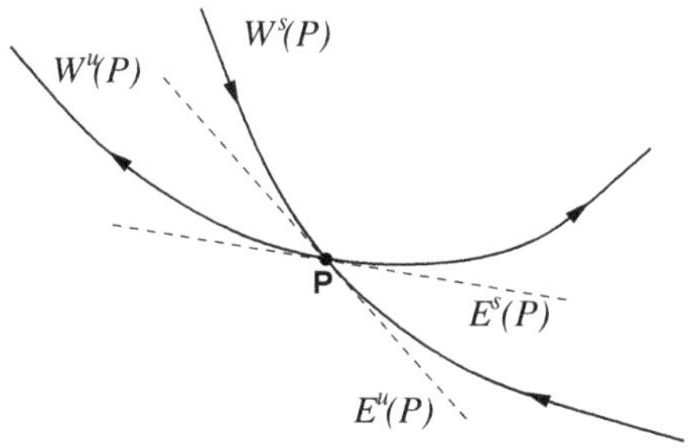

Fig. 7.5 Sketch of the stable $\mathcal{W}^s(P)$ and unstable $\mathcal{W}^u(P)$ manifolds of the point P, which are tangent to the stable $E^s(P)$ and unstable $E^u(P)$ linear spaces.

center of stable regions, called nonlinear resonance islands sketched in Fig. 7.4 (and which are well visible in Fig. 7.1 also for large perturbations), embedded into a sea of chaotic orbits. Unstable hyperbolic fixed points instead play a crucial role in originating chaotic trajectories.

We focus now on trajectories close to a hyperbolic point P.[3] The linearization of the dynamics identifies the stable and unstable spaces $E^s(P)$ and $E^u(P)$, respectively. Such notions can be generalized out of the tangent space (i.e. beyond linear theory) by introducing the *stable and unstable manifolds*, respectively (see Fig. 7.5). We start describing the latter. Consider the set of all points converging to P under the application of the time reversed dynamics of a system. Very close to P, the points of this set should identify the unstable direction given by the linearized dynamics $E^u(P)$, while the entire set constitutes the unstable manifold $W^u(P)$ associated to point P, formally

$$W^u(P) = \{\boldsymbol{x} : \lim_{t\to-\infty} \boldsymbol{x}(t) = P\},$$

where $\boldsymbol{x}$ is a generic point in phase space generating the trajectory $\boldsymbol{x}(t)$. Clearly from its definition $W^u(P)$ is an invariant set that, moreover, cannot have self-intersections for the theorem of existence and uniqueness. By reverting the direction of time, we can define the stable manifold $W^s(P)$ as

$$W^s(P) = \{\boldsymbol{x} : \lim_{t\to\infty} \boldsymbol{x}(t) = P\},$$

identifying the set of all points in phase space that converge to P forward in time. This is also an invariant set and cannot cross itself.

For an integrable Hamiltonian system, stable and unstable manifolds smoothly connect to each other either onto the same fixed point (homoclinic orbits) or in a different one (heteroclinic orbits), forming the separatrix (Fig. 7.6). We recall that these orbits usually separate regions of phase space characterized by different kinds of trajectories (e.g. oscillations from rotations as in the nonlinear pendulum

[3] Fixed points in a Poincaré section corresponds to periodic orbits of the original system, therefore the considerations of this section extend also to hyperbolic periodic orbits.

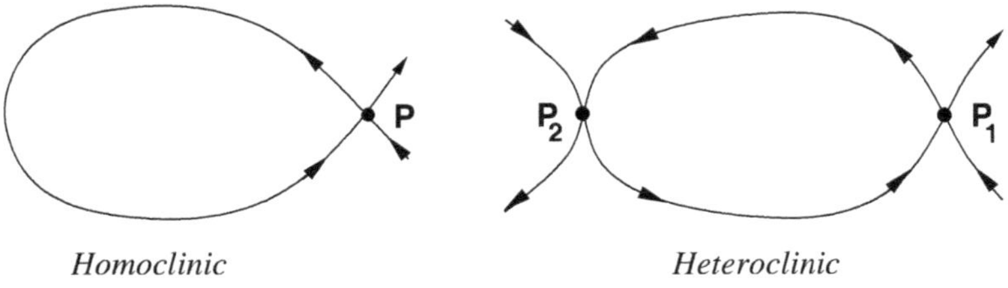

Fig. 7.6 Sketch of homoclinic and heteroclinic orbits.

of Fig. 1.1c). Notice that separatrices are periodic orbits with an infinite period. What does it happen in the presence of a perturbation?

Typically the smooth connection breaks. If the stable manifold W^s intersects the unstable one W^u in at least one other point (homoclinic point when the two manifold originate from the same fixed point or heteroclinic if from different ones), chaotic motion occurs around the region of these intersections. The underlying mechanism can be easily illustrated for non tangent contact between stable and unstable manifolds. First of all notice that a single intersection between W^s and W^u implies an infinite number of intersections (Figs. 7.7a,b,c). Indeed being the two manifold invariant, each point should be mapped by the forward or backward iteration onto another point of the unstable or stable manifold, respectively. This is true, of course, also for the intersection point, and thus there should be infinite intersections (homoclinic points), although both W^s and W^u cannot have self-intersections. Poincaré wrote:

> *The intersections form a kind of trellis, a tissue, an infinite tight lattice; each of curves must never self-intersect, but it must fold itself in a very complex way, so as to return and cut the lattice an infinite numbers of times.*

Such a complex structure depicted in Fig. 7.7 for the standard map is called *homoclinic tangle* (analogously there exist *heteroclinic tangles*). The existence of one, and therefore infinite, homoclinic intersection entails chaos. In virtue of the conservative nature of the system, the successive loops formed between homoclinic intersections must have the same area (see Fig. 7.7d). At the same time the distance between successive homoclinic intersections should decrease exponentially as the fixed point is approached. These two requirements imply a concomitant exponential growth of the loop lengths and a strong bending of the invariant manifolds near the fixed point. As a result a small region around the fixed point will be stretched and folded and close points will separate exponentially fast. These features are illustrated in Fig. 7.7 showing the homoclinic tangle of the standard map (7.11) around one of its hyperbolic fixed points for $K = 1.5$.

The existence of homoclinic tangles is rather common and constitute the generic mechanism for the appearance of chaos. This is further exemplified by considering

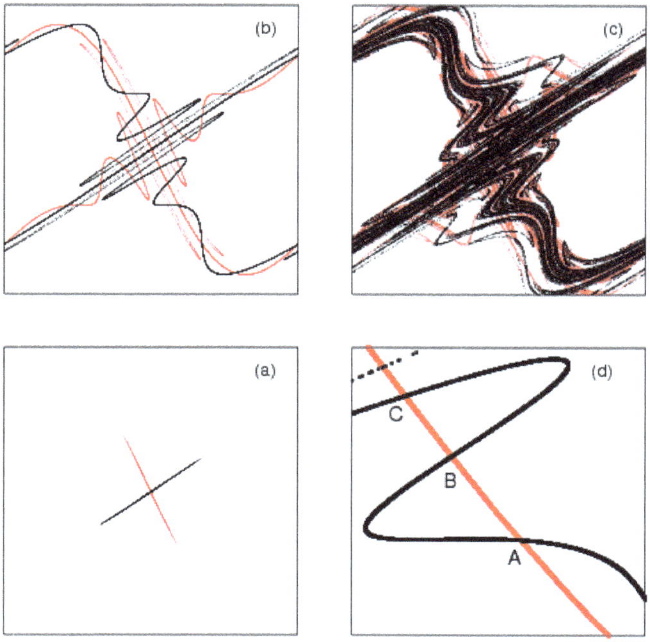

Fig. 7.7 (a)-(c) Typical example of homoclinic tangle originating from an unstable hyperbolic point. The three figures has been obtained by evolving an initially very small clouds of about 10^4 points around the fixed point $(\mathrm{I}, \phi) = (0,0)$ of the standard map. The black curve represents the unstable manifold and is obtained by forward iterating the map (7.11) for $5, 10, 22$ steps (a), (b) and (c), respectively. The stable manifold in red is obtained by iterating backward in time the map. Note that at early times (a) one finds what expected by the linearized theory, while as times goes on the tangle of intersections becomes increasingly complex. (d) Enlargement of a portion of (b). A,B and C are homoclinic points, the area enclosed by the black and red arcs AB and that enclosed by the black and red arcs BC are equal. [After Timberlake (2004)]

a typical Hamiltonian system obtained as a perturbation of an integrable one as, for instance, the (frictionless) Duffing oscillator

$$H(q,p,t) = H_0(q,p) + \epsilon H_1(q,p,t) = \frac{p^2}{2} - \frac{q^2}{2} + \frac{q^4}{4} + \epsilon q \cos(\omega t) . \qquad (7.15)$$

where the perturbation H_1 is a periodic function of time with period $T = 2\pi/\omega$. By recording the motion of the perturbed system at every $t_n = t_0 + nT$, we can construct the stroboscopic map in (q,p)-phase space

$$\boldsymbol{x}(t_0) \to \boldsymbol{x}(t_0 + T) = \boldsymbol{S}_\epsilon[\boldsymbol{x}(t_0)] ,$$

where $\boldsymbol{x}$ denotes the canonical coordinates (q,p), and $t_0 \in [0\!:\!T]$ plays the role of a phase and can be seen as a parameter of the area-preserving map $\boldsymbol{S}_\epsilon$.

In the absence of the perturbation ($\epsilon = 0$), a hyperbolic fixed point $\tilde{\boldsymbol{x}}_0$ is located in $(0,0)$ and the separatrix $\boldsymbol{x}_0(t)$ corresponds to the orbit with energy $H = 0$, in

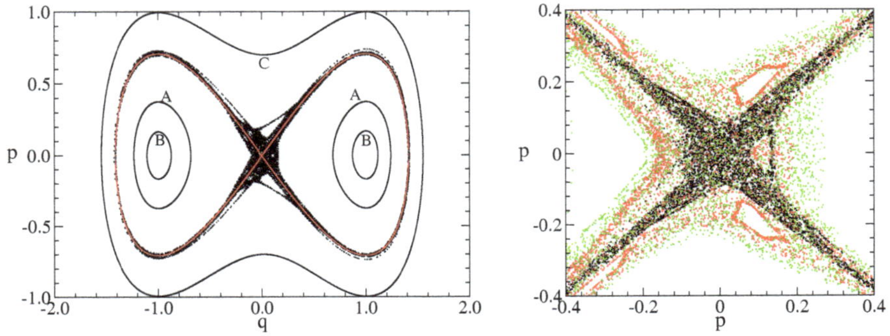

Fig. 7.8 (left) Phase-space portrait of the Hamiltonian system (7.15), The points indicate the Poincarè section obtained by a stroboscopic sampling of the orbit at every period $T = 2\pi/\omega$. The separatrix of the unperturbed system ($\epsilon = 0$) is shown in red. The sets A and B are the regular orbits around the two stable fixed points $(\pm 1, 0)$ of the unperturbed system; C is the regular orbit that originates from an initial condition far from the separatrix. Dots indicate the chaotic behavior around the separatrix. (right) Detail of the chaotic behavior near the separatrix for different values of ϵ showing the growth of the chaotic layer when ϵ increases from 0.01 (black) to 0.04 (red) and 0.06 (green).

red in Fig. 7.8 left. Moreover, there are two elliptic fixed points in $\boldsymbol{x}_{\pm}(t) = (\pm 1, 0)$, also shown in the figure.

For small positive ϵ, the unstable fixed point $\tilde{\boldsymbol{x}}_\epsilon$ of $\boldsymbol{S}_\epsilon$ is close to the unperturbed one $\tilde{\boldsymbol{x}}_0$ and a homoclinic tangle forms, so that chaotic trajectories appear around the unperturbed separatrix (Fig. 7.8 left). As long as ϵ remains very small, chaos is confined to a very thin layer around the separatrix: this sort of "stochastic layer" corresponds to a situation of bounded chaos, because far from the separatrix, orbits remain regular. The thickness of the chaotic layer increases with ϵ (Fig 7.8 right). The same features have been observed in Hénon-Heiles model (Fig. 3.10).

So far, we saw what happens around one separatrix. What does change when two or more separatrices are present? Typically the following scenario is observed. For small ϵ, bounded chaos appears around each separatrix, and regular motion occurs far from them. For perturbation large enough $\epsilon > \epsilon_c$ (ϵ_c being a system dependent critical value), the stochastic layers can overlap so that chaotic trajectories may diffuse in the system. This is the so-called phenomenon of the *overlap of resonances*, see Box B.14. In Sec. 11.2.1 we shall come back to this problem in the context of transport properties in fluids.

Box B.14: The resonance-overlap criterion

This box presents a simple but powerful method to determine the transition from "local chaos" — chaotic trajectories localized around separatrices — to "large scale chaos" — chaotic trajectories spanning larger and larger portions of phase space — in Hamiltonian systems. This method, called *resonance-overlap criterion*, has been introduced by Chirikov (1979) and, although not rigorous, it is one of the few valuable analytical techniques which can successfully be used in Hamiltonian systems.

The basic idea can be illustrated considering the Chirikov-Taylor (standard) map

$$\begin{aligned} \mathrm{I}(t+1) &= \mathrm{I}(t) + K \sin\theta(t) \\ \theta(t+1) &= \theta(t) + \mathrm{I}(t+1) \mod 2\pi\,, \end{aligned}$$

which can be derived from the Hamiltonian of the kicked rotator

$$H(\theta,\mathrm{I},t) = \frac{\mathrm{I}^2}{2} + K\cos\theta \sum_{m=-\infty}^{\infty} \delta(t-m) = \frac{\mathrm{I}^2}{2} + K \sum_{m=-\infty}^{\infty} \cos(\theta - 2\pi m t)\,,$$

describing a pendulum without gravity and driven by periodic Dirac-δ shaped impulses [Ott (1993)]. From the second form of H we can identify the presence of resonances $\mathrm{I} = d\theta/dt = 2\pi m$, corresponding to actions equal to one of the external driving frequencies. If the perturbation is small, $K \ll 1$, around each resonance $\mathrm{I}_m = 2\pi m$, the dynamics is approximately described by the pendulum Hamiltonian

$$H \approx \frac{(\mathrm{I}-\mathrm{I}_m)^2}{2} + K\cos\psi \qquad \text{with} \qquad \psi = \theta - 2\pi m t\,.$$

In (ψ, I)-phase space one can identify two qualitatively different kinds of motion (phase oscillations for $H < K$ and phase rotations for $H > K$) distinguished by the separatrix

$$\mathrm{I} - \mathrm{I_m} = \pm 2\sqrt{K} \sin\left(\frac{\psi}{2}\right)\,.$$

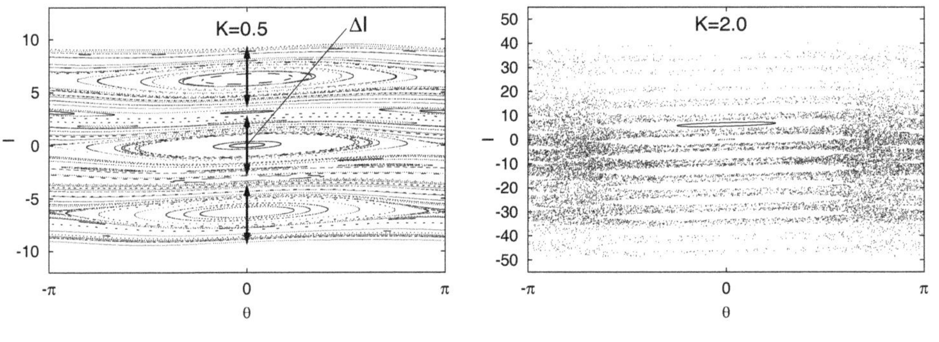

Fig. B14.1 Phase portrait of the standard map for $K = 0.5 < K_c$ (left) for $K = 2 > K_c$ (right).

For $H = K$, the separatrix starts from the unstable fixed point $(\psi = 0, \mathrm{I} = \mathrm{I}_m)$ and has width

$$\Delta \mathrm{I} = 4\sqrt{K}\,. \tag{B.14.1}$$

In the left panel of Figure B14.1 we show the resonances $m = 0, \pm 1$ whose widths are indicated by arrows. If K is small enough, the separatrix labeled by m does not overlap the adjacent ones $m \pm 1$ and, as a consequence, when the initial action is close to m-th resonance, $I(0) \approx I_m$, its evolution $I(t)$ remains bounded, i.e. $|I(t) - I(0)| < \mathcal{O}(\sqrt{K})$. On the contrary, if K is large enough, ΔI becomes larger than 2π (the distance between I_m and $I_{m\pm 1}$) and the separatrix of m-th resonance overlaps the nearest neighbor ones ($m\pm 1$). An approximate estimate based on Eq. (B.14.1) for the overlap to occur is

$$K > K_{ovlp} = \frac{\pi^2}{4} \simeq 2.5\,.$$

When $K > K_{ovlp}$, it is rather natural to conjecture that the action $I(t)$ may jump from one resonance to another performing a sort of random walk among the separatrices (Fig. B14.1 right panel), which can give rise to a diffusive behavior (Fig. B14.2)

$$\langle (I(t) - I(0))^2 \rangle = 2Dt\,,$$

D being the diffusion constant. Let us note that the above diffusive behavior is rather different from Arnold diffusion (Box B.13). This is clear for two-degrees of freedom systems, where Arnold diffusion is impossible while diffusion by resonances overlap is often encountered. For systems with three or more degrees of freedom both mechanisms are present, and their distinction requires careful numerical analysis [Guzzo *et al.* (2002)].

As discussed in Sec. 7.2, the last "separating" KAM torus of the standard map disappears for $K_c \simeq 0.971\ldots$, beyond which action diffusion is actually observed. Therefore, Chirikov's resonance-overlap criterion $K_{ovlp} = \pi^2/4$ overestimates K_c. This difference stems from both the presence of secondary order resonances and the finite size of the chaotic layer around the separatrices. A more elaborated version of the resonance-overlap criterion provides $K_{ovlp} \simeq 1$ much closer to the actual value [Chirikov (1988)].

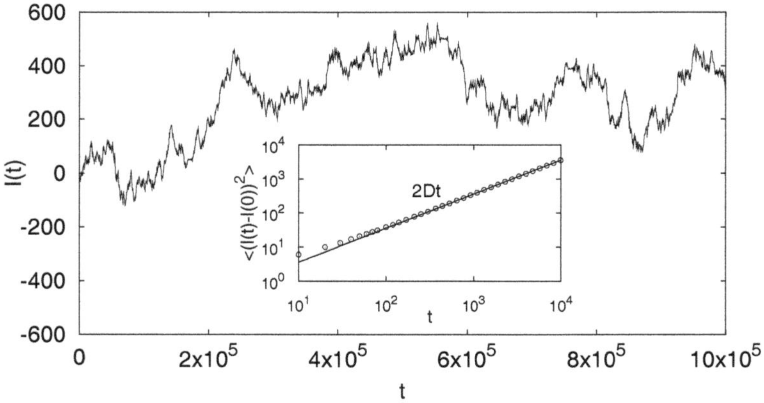

Fig. B14.2 Diffusion behavior of action $I(t)$ for the standard map above the threshold, i.e. $K = 2.0 > K_c$. The inset shows the linear growth of mean square displacement $\langle (I(t) - I(0))^2 \rangle$ with time, D being the diffusion coefficient.

For a generic system, the resonance overlap criterion amount to identify the resonances and perform a local pendulum approximation of the Hamiltonian around each resonance, from which one computes $\Delta I(K)$ and finds K_{ovlp} as the minimum value of K such that two separatrices overlap.

Although up to now a rigorous justification of the method is absent[4] and sometimes it fails, as for the Toda lattice, this criterion remains the only physical approach to determine the transition from "local" to "large scale" chaos in Hamiltonian systems.

The difficulty of finding a mathematical basis for the resonance-overlap criterion relies on the need of an analytical approach to heteroclinic crossings, i.e. the intersection of the stable and unstable manifolds of two distinct resonances. Unlike homoclinic intersections, which can be treated in the framework of perturbation of the integrable case (Melnikov method, see Sec. 7.5), the phenomenon of heteroclinic intersection is not perturbative. The resonance-overlap criterion had been applied to systems such as particles in magnetic traps [Chirikov (1988)] and highly excited hydrogen atoms in microwave fields [Casati *et al.* (1988)].

7.5 Melnikov's theory

When a perturbation causes homoclinic intersections, chaotic motion is expected to appear in proximity of the separatrix (homoclinic orbit), it is then important to determine whether and at which strength of the perturbation such intersections occur. To this purpose, we now describe an elegant perturbative approach to determine whether homoclinic intersections happen or not [Melnikov (1963)].

The essence of this method can be explained by considering a one-degree of freedom Hamiltonian system driven by a small periodic perturbation $\boldsymbol{g}(q,p,t) = (g_1(q,p,t), g_2(q,p,t))$ of period T

$$\frac{\mathrm{d}q}{\mathrm{d}t} = \frac{\partial H(q,p)}{\partial p} + \epsilon g_1(q,p,t)$$
$$\frac{\mathrm{d}p}{\mathrm{d}t} = -\frac{\partial H(q,p)}{\partial q} + \epsilon g_2(q,p,t)\,.$$

Suppose that the unperturbed system admits a single homoclinic orbit associated to a hyperbolic fixed point P_0 (Fig. 7.9). The perturbed system is non autonomous requiring to consider the enlarged phase space $\{q,p,t\}$. However, time periodicity enables to get rid of time dependence by taking the (stroboscopic) Poincaré section recording the motion every period T (Sec. 2.1.2), $(q_n(t_0), p_n(t_0)) = (q(t_0+nT), p(t_0+nT))$ where t_0 is any reference time in the interval $[0:T]$ and parametrically defines the stroboscopic map. The perturbation shifts the position of the hyperbolic fixed point P_0 to $P_\epsilon = P_0 + \mathcal{O}(\epsilon)$ and splits the homoclinic orbit into a stable $W^s(P_\epsilon)$ and unstable manifolds $W^u(P_\epsilon)$ associated to P_ϵ, as in Fig. 7.9. We have now to determine whether these two manifolds cross each other with possible onset of chaos by homoclinic tangle. The perturbation $\boldsymbol{g}$ can be, in principle, either Hamiltonian or dissipative. The former generates surely a homoclinic tangle, while the latter not always leads to a homoclinic tangle [Lichtenberg

---

[4]When Chirikov presented this criterion to Kolmogorov, the latter said *one should be a very brave young man to claim such things.*

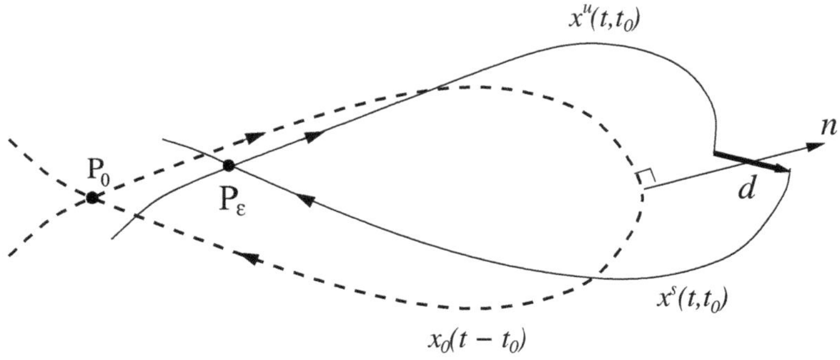

Fig. 7.9 Melnikov's construction applied to the homoclinic separatrix of hyperbolic fixed point P_0 (dashes loop). The full lines represent the stable and unstable manifolds of the perturbed fixed point P_ϵ. Vector $\boldsymbol{d}$ is the displacement at time t of the two manifolds whose projection along the normal $\boldsymbol{n}(t)$ to the unperturbed orbits is the basic element of Melnikov's method.

and Lieberman (1992)]. Thus, Melnikov's theory proves particularly useful when applied to dissipative perturbations.

It is now convenient to introduce the compact notation for the Hamiltonian flow

$$\frac{\mathrm{d}\boldsymbol{x}}{\mathrm{d}t} = \boldsymbol{f}(\boldsymbol{x}) + \epsilon \boldsymbol{g}(\boldsymbol{x}, t) \qquad \boldsymbol{x} = (\boldsymbol{q}, \boldsymbol{p}) \,. \tag{7.16}$$

To detect the crossing between $W^u(P_\epsilon)$ and $W^u(P_\epsilon)$, we need to construct a function quantifying the "displacement" between them,

$$\boldsymbol{d}(t, t_0) = \boldsymbol{x}^s(t, t_0) - \boldsymbol{x}^u(t, t_0) \,,$$

where $\boldsymbol{x}^{s,u}(t, t_0)$ is the orbit corresponding to $W^{s,u}(P_\epsilon)$ (Fig. 7.9). In a perturbative approach, the two manifolds remain close to each other and to the unperturbed homoclinic orbit $\boldsymbol{x}_0(t-t_0)$, thus they can be expressed as a series in power of ϵ, which to first order reads

$$\boldsymbol{x}^{s,u}(t, t_0) = \boldsymbol{x}_0(t - t_0) + \epsilon \boldsymbol{x}_1^{s,u}(t, t_0) + \mathcal{O}(\epsilon^2) \,. \tag{7.17}$$

A direct substitution of expansion (7.17) into Eq. (7.16) yields the differential equation for the lowest order term $\boldsymbol{x}_1^{u,s}(t, t_0)$

$$\frac{\mathrm{d}\boldsymbol{x}_1^{s,u}}{\mathrm{d}t} = \mathbb{L}(\boldsymbol{x}_0(t - t_0))\boldsymbol{x}_1^{s,u} + \boldsymbol{g}(\boldsymbol{x}_0(t - t_0), t) \,, \tag{7.18}$$

where $\mathbb{L}_{ij} = \partial f_i / \partial x_j$ is the stability matrix. A meaningful function characterizing the distance between W^s and W^u is the scalar product

$$d_n(t, t_0) = \boldsymbol{d}(t, t_0) \cdot \boldsymbol{n}(t, t_0)$$

projecting the displacement $\boldsymbol{d}(t, t_0)$ along the normal $\boldsymbol{n}(t, t_0)$ to the unperturbed separatrix $\boldsymbol{x}_0(t - t_0)$ at time t (Fig. 7.9). The function d_n can be computed as

$$d_n(t, t_0) = \frac{\boldsymbol{f}^{\perp}[\boldsymbol{x}_0(t - t_0)] \cdot \boldsymbol{d}(t, t_0)}{|\boldsymbol{f}[\boldsymbol{x}_0(t - t_0)]|} \,,$$

where the vector $\boldsymbol{f}^{\perp} = (-f_2, f_1)$ is orthogonal to the unperturbed flow $\boldsymbol{f} = (f_1, f_2)$ and everywhere normal to unperturbed trajectory $\boldsymbol{x}_0(t - t_0)$, i.e.

$$\boldsymbol{n}(t, t_0) = \frac{\boldsymbol{f}^{\perp}[\boldsymbol{x}_0(t - t_0)]}{|\boldsymbol{f}[\boldsymbol{x}_0(t - t_0)]|} .$$

Notice that in two dimensions $\boldsymbol{a} \cdot \boldsymbol{b}^{\perp} = \boldsymbol{a} \times \boldsymbol{b}$ (where $\times$ denotes cross product) for any vector $\boldsymbol{a}$ and $\boldsymbol{b}$, so that

$$d_n(t, t_0) = \frac{\boldsymbol{f}[\boldsymbol{x}_0(t - t_0)] \times \boldsymbol{d}(t, t_0)}{|\boldsymbol{f}[\boldsymbol{x}_0(t - t_0)]|} . \tag{7.19}$$

Melnikov realized that there is no need to solve Eq. (7.18) for $\boldsymbol{x}_1^u(t, t_0)$ and $\boldsymbol{x}_1^s(t, t_0)$ to obtain an explicit expression of $d_n(t, t_0)$ at reference time t_0 and at the first order in ϵ. Actually, as $\boldsymbol{d}(t, t_0) \simeq \epsilon\, [\boldsymbol{x}_1^u(t, t_0) - \boldsymbol{x}_1^s(t, t_0)]$, we have to evaluate the functions

$$\Delta^{s,u}(t, t_0) = \boldsymbol{f}[\boldsymbol{x}_0(t - t_0)] \times \boldsymbol{x}_1^{s,u}(t, t_0) \tag{7.20}$$

at the numerator of Eq. (7.19). Differentiation of $\Delta^{s,u}$ with respect to time yields

$$\frac{\mathrm{d}\Delta^{s,u}}{\mathrm{d}t} = \frac{\mathrm{d}\boldsymbol{f}(\boldsymbol{x}_0)}{\mathrm{d}t} \times \boldsymbol{x}_1^{s,u} + \boldsymbol{f}(\boldsymbol{x}_0) \times \frac{\mathrm{d}\boldsymbol{x}_1^{s,u}}{\mathrm{d}t}$$

which, by means of the chain rule in the first term, becomes

$$\frac{\mathrm{d}\Delta^{s,u}}{\mathrm{d}t} = \mathbb{L}(\boldsymbol{x}_0)\frac{\mathrm{d}\boldsymbol{x}_0}{\mathrm{d}t} \times \boldsymbol{x}_1^{s,u} + \boldsymbol{f}(\boldsymbol{x}_0) \times \frac{\mathrm{d}\boldsymbol{x}_1^{s,u}}{\mathrm{d}t} .$$

Substituting Eqs. (7.16) and (7.18) in the above expression, we obtain

$$\frac{\mathrm{d}\Delta^{s,u}}{\mathrm{d}t} = \mathbb{L}(\boldsymbol{x}_0)\boldsymbol{f}(\boldsymbol{x}_0) \times \boldsymbol{x}_1^{s,u} + \boldsymbol{f}(\boldsymbol{x}_0) \times [\mathbb{L}(\boldsymbol{x}_0)\boldsymbol{x}_1^{s,u} + \boldsymbol{g}(\boldsymbol{x}_0, t)]$$

that, via the vector identity $\mathbb{A}\boldsymbol{a} \times \boldsymbol{b} + \boldsymbol{a} \times \mathbb{A}\boldsymbol{b} = \mathrm{Tr}(\mathbb{A})\, \boldsymbol{a} \times \boldsymbol{b}$ (Tr indicating the trace operation), can be recast as

$$\frac{\mathrm{d}\Delta^{s,u}(t, t_0)}{\mathrm{d}t} = \mathrm{Tr}[\mathbb{L}(\boldsymbol{x}_0)]\, \boldsymbol{f}(\boldsymbol{x}_0) \times \boldsymbol{x}_1^{s,u} + \boldsymbol{f}(\boldsymbol{x}_0) \times \boldsymbol{g}(\boldsymbol{x}_0, t) .$$

Finally, recalling the definition of $\Delta^{s,u}$ (7.20), the last equation takes the form

$$\frac{\mathrm{d}\Delta^{s,u}}{\mathrm{d}t} = \mathrm{Tr}[\mathbb{L}(\boldsymbol{x}_0)]\Delta^{s,u} + \boldsymbol{f}(\boldsymbol{x}_0) \times \boldsymbol{g}(\boldsymbol{x}_0, t) , \tag{7.21}$$

which, as $\mathrm{Tr}(\mathbb{L})=0$ for Hamiltonian systems,[5] further simplifies to

$$\frac{\mathrm{d}\Delta^{s,u}(t, t_0)}{\mathrm{d}t} = \boldsymbol{f}(\boldsymbol{x}_0) \times \boldsymbol{g}(\boldsymbol{x}_0, t) .$$

The last step of Melnikov's method requires to integrate the above equation forward in time for the stable manifold

$$\Delta^s(\infty, t_0) - \Delta^s(t_0, t_0) = \int_{t_0}^{\infty} \mathrm{d}t\; \boldsymbol{f}[\boldsymbol{x}_0(t - t_0)] \times \boldsymbol{g}[\boldsymbol{x}_0(t - t_0), t] .$$

[5]Note that Eq. (7.21) holds also for non Hamiltonian, dissipative systems.

and backward for the unstable

$$\Delta^u(t_0, t_0) - \Delta^u(-\infty, t_0) = \int_{-\infty}^{t_0} \mathrm{d}t\; \boldsymbol{f}[\boldsymbol{x}_0(t - t_0)] \times \boldsymbol{g}[\boldsymbol{x}_0(t - t_0), t] \,.$$

Since the stable and unstable manifolds share the fixed point P_ϵ (Fig. 7.9), then $\Delta^u(-\infty, t_0) = \Delta^s(\infty, t_0) = 0$, and by summing the two above equations we have

$$\Delta^u(t_0, t_0) - \Delta^s(t_0, t_0) = \int_{-\infty}^{\infty} \mathrm{d}t\; \boldsymbol{f}[\boldsymbol{x}_0(t - t_0)] \times \boldsymbol{g}[\boldsymbol{x}_0(t - t_0), t] \,.$$

The Melnikov function or integral

$$M(t_0) = \int_{-\infty}^{\infty} \mathrm{d}t\; \boldsymbol{f}[\boldsymbol{x}_0(t)] \times \boldsymbol{g}[\boldsymbol{x}_0(t), t + t_0] \tag{7.22}$$

is the crucial quantity of the method: whenever $M(t_0)$ changes sign at varying t_0, the perturbed stable $W^s(P_\epsilon)$ and unstable $W^u(P_\epsilon)$ manifolds cross each other transversely, inducing chaos around the separatrix.

Two remarks are in order:
(1) the method is purely perturbative;
(2) the method works also for dissipative perturbations $\boldsymbol{g}$, providing that the flow for $\epsilon = 0$ is Hamiltonian [Holmes (1990)].

The original formulation of Melnikov refers to time-periodic perturbations, see [Wiggins and Holmes (1987)] for an extension of the method to more general kinds of perturbation.

7.5.1 *An application to the Duffing's equation*

As an example, following Lichtenberg and Lieberman (1992); Nayfeh and Balachandran (1995), we apply Melnikov's theory to the forced and damped Duffing oscillator

$$\begin{aligned} \frac{\mathrm{d}q}{\mathrm{d}t} &= p \\ \frac{\mathrm{d}p}{\mathrm{d}t} &= q - q^3 + \epsilon[F\cos(\omega t) - 2\mu p] \,, \end{aligned}$$

which, for $\mu = 0$, was discussed in Sec. 7.4.

For $\epsilon = 0$, this system is Hamiltonian, with

$$H(q, p) = \frac{p^2}{2} - \frac{q^2}{2} + \frac{q^4}{4} \,,$$

and it has two elliptic and one hyperbolic fixed points in $(\pm 1, 0)$ and $(0, 0)$, respectively. The equation for the separatrix, formed by two homoclinic loops (red curve in the left panel of Fig. 7.8), is obtained by solving the algebraic equation $H = 0$ with respect to p,

$$p = \pm\sqrt{q^2\left(1 - \frac{q^2}{2}\right)} \,. \tag{7.23}$$

The time parametrization of the two homoclinic orbits is obtained by integrating Eq. (7.23) with $p = \mathrm{d}q/\mathrm{d}t$ and initial conditions $q(0) = \pm\sqrt{2}$ and $p(0) = 0$, so that

$$\begin{aligned} q(t) &= \pm\,\sqrt{2}\,\mathrm{sech}(t) \\ p(t) &= \mp\,\mathrm{sech}(t)\tanh(t)\,. \end{aligned} \tag{7.24}$$

With the above expressions, Melnikov's integral (7.22) reads

$$M(t_0) = -\sqrt{2}\int_{-\infty}^{\infty} \mathrm{d}t\ \mathrm{sech}(t)\tanh(t)\Big\{F\cos[\omega(t+t_0)] + 2\sqrt{2}\,\mu\,\mathrm{sech}(t)\tanh(t)\Big\}$$

where we have considered

$$\boldsymbol{f} = [p(t), q(t) - q^3(t)] \qquad \boldsymbol{g} = [0, F\cos(\omega t) - 2\mu p(t)].$$

The exact integration yields the result

$$M(t_0) = -\frac{8}{3}\,\mu + 2\pi\sqrt{2}F\omega\sin(\omega t_0)\mathrm{sech}\left(\frac{\omega\pi}{2}\right)\,.$$

Therefore if

$$F > \frac{4\cosh(\omega\pi/2)}{3\pi\sqrt{2}\omega}\mu$$

$M(t_0)$ has simple zeros implying that transverse homoclinic crossings occur while, in the opposite condition, there is no crossing. In the equality situation $M(t_0)$ has a double zero corresponding to a tangential contact between $W^s(P_\epsilon)$ and $W^u(P_\epsilon)$. Note that in the case of non dissipative perturbation $\mu = 0$, Melnikov's method predicts chaos for any value of the parameter F.

7.6 Exercises

Exercise 7.1: Consider the standard map

$$\begin{aligned} \mathrm{I}(t+1) &= \mathrm{I}(t) + K\sin(\theta(t)) \\ \theta(t+1) &= \theta(t) + \mathrm{I}(t+1) \qquad \mathrm{mod}\ 2\pi\,, \end{aligned}$$

write a numerical code to compute the action diffusion coefficient $D = \lim_{t\to\infty}\frac{1}{2t}\langle(\mathrm{I}(t) - \mathrm{I}_0)^2\rangle$ where the average is over a set of initial values $\mathrm{I}(0) = \mathrm{I}_0$. Produce a plot of D versus the map parameter K and compare the result with Random Phase Approximation, consisting in assuming $\theta(t)$ as independent random variables, which gives $D_{RPA} = K^2/4$ [Lichtenberg and Lieberman (1992)]. Note that for some specific values of K (e.g $K = 6.9115$) the diffusion is anomalous, since the mean square displacement scales with time as $\langle(\mathrm{I}(t) - \mathrm{I}_0)^2\rangle \sim t^{2\nu}$, where $\nu > 1/2$ (see Castiglione *et al.* (1999)).

Exercise 7.2: Using some numerical algorithm for ODE to integrate the Duffing oscillator Eq. (7.15). Check that for small ϵ
(1) trajectories starting from initial conditions close to the separatrix have $\lambda_1 > 0$;
(2) trajectories with initial conditions far enough from the separatrix exhibit regular motion ($\lambda_1 = 0$).

Exercise 7.3: Consider the time-dependent Hamiltonian

$$H(q,p,t) = -V_2\cos(2\pi p) - V_1\cos(2\pi q)K(t) \quad \text{with} \quad K(t) = \tau\sum_{n=-\infty}^{\infty}\delta(t - n\tau)$$

called the kicked Harper model. Show that integrating over the time of a kick (as for the standard map in Sec. 2.2.1.2) it reduces to the Harper map

$$p(n+1) = p(n) - \gamma_1 \sin(2\pi q(n))$$
$$q(n+1) = q(n) + \gamma_2 \sin(2\pi p(n+1)),$$

with $\gamma_i = 2\pi V_i \tau$, which is symplectic. For $\tau \to 0$ this is an exact integration of the original Hamiltonian system. Fix $\gamma_{1,2} = \gamma$ and study the qualitative changes of the dynamics as γ becomes larger than 0. Find the analogies with the standard map, if any.

Exercise 7.4: Consider the ODE

$$\frac{dx}{dt} = -a(t)\frac{\partial \psi}{\partial y}, \qquad \frac{dy}{dt} = a(t)\frac{\partial \psi}{\partial x}$$

where $\psi = \psi(x, y)$ is a smooth function periodic on the square $[0:L] \times [0:L]$ and $a(t)$ an arbitrary bounded function. Show that the system is not chaotic.
Hint: *Show that the system is integrable, thus non chaotic.*

Exercise 7.5: Consider the system defined by the Hamiltonian

$$H(x, y) = U \sin x \sin y$$

which is integrable and draw some trajectories, you will see counter-rotating square vortices. Then consider a time-dependent perturbation of the following form

$$H(x, y, t) = U \sin(x + B\sin(\omega t)) \sin y$$

study the qualitative changes of the dynamics at varying B and ω. You will recognize that now trajectories can travel in the x-direction, then fix $B = 1/3$ and study the behavior of the diffusion coefficient $D = \lim_{t\to\infty} \frac{1}{2t}\langle (x(t) - x(0))^2 \rangle$ as a function of ω. This system can be seen as a two-dimensional model for the motion of particles in a convective flow [Solomon and Gollub (1988)]. Compare your findings with those reported in Sec. 11.2.2.2. See also Castiglione *et al.* (1999).

Exercise 7.6: Consider a variant of the Hénon-Heiles system defined by the potential energy

$$V(q_1, q_2) = \frac{q_1^2}{2} + \frac{q_2^2}{2} + q_1^4 q_2 - \frac{q_2^2}{4}.$$

Identify the stationary points of $V(q_1, q_2)$ and their nature. Write the Hamilton equations and integrate numerically the trajectory for $E = 0.06$, $q_1(0) = -0.1$, $q_2(0) = -0.2$, $p_1(0) = -0.05$. Construct and interpret the Poincaré section on the plane $q_1 = 0$, by plotting q_2, p_2 when $p_1 > 0$.

PART 2

Advanced Topics and Applications: From Information Theory to Turbulence

Chapter 8

Chaos and Information Theory

You should call it entropy, for two reasons. In the first place your uncertainty function has been used in statistical mechanics under that name, so it already has a name. In the second place, and more important, no one really knows what entropy really is, so in a debate you will always have the advantage.

John von Neumann (1903-1957)

In the first part of the book, it has been stated many times that chaotic trajectories are aperiodic and akin to random behaviors. This Chapter opens the second part of the book attempting to give a quantitative meaning to the notion of *deterministic randomness* through the framework of information theory.

8.1 Chaos, randomness and information

The basic ideas and tools of this Chapter can be illustrated by considering the Bernoulli shift map (Fig. 8.1a)

$$x(t+1) = f(x(t)) = 2x(t) \mod 1 \,. \tag{8.1}$$

This map generates chaotic orbits for generic initial conditions and is ergodic with uniform invariant distribution $\rho^{inv}(x) = 1$ (Sec. 4.2). The Lyapunov exponent λ can be computed as in Eq. (5.24) (see Sec. 5.3.1)

$$\lambda = \int \mathrm{d}x \, \rho^{inv}(x) \, \ln |f'(x)| = \ln 2 \,. \tag{8.2}$$

Looking at a typical trajectory (Fig. 8.1b), the absence of any apparent regularity suggests to call it random, but how is *randomness* defined and quantified? Let's simplify the description of the trajectory to something closer to our intuitive notion of random process. To this aim we introduce a coarse-grained description $s(t)$ of the trajectory by recording whether $x(t)$ is larger or smaller than $1/2$

$$s(t) = \begin{cases} 0 & \text{if} \;\; 0 \leq x(t) < 1/2 \\ 1 & \text{if} \;\; 1/2 \leq x(t) \leq 1 \,, \end{cases} \tag{8.3}$$

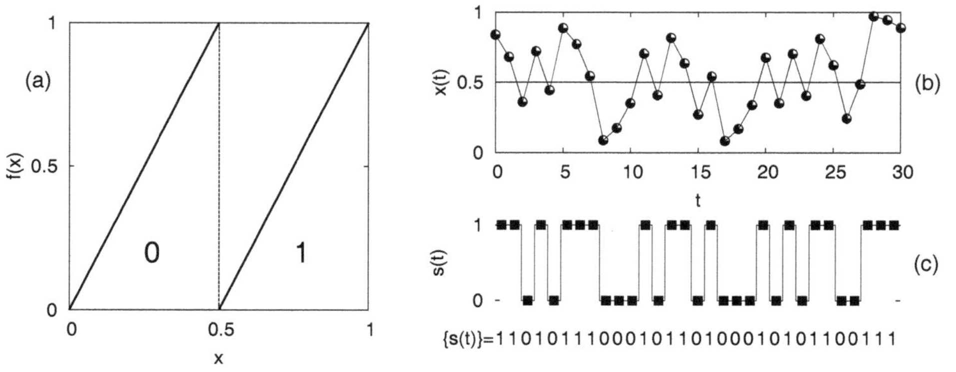

Fig. 8.1 (a) Bernoulli shift map (8.1), the vertical tick line at 1/2 defines a partition of the unit interval to which we can associate two symbols $s(t) = 0$ if $0 \leq x(t) < 1/2$ and $s(t) = 1$ if $1/2 \leq x(t) \leq 1$. (b) A typical trajectory of the map with (c) the associated symbolic sequence.

a typical symbolic sequence obtained with this procedure is shown in Fig. 8.1c. From Section 4.5 we realize that (8.3) defines a Markov partition for the Bernoulli map, characterized by transition matrix $\mathbb{W}_{ij} = 1/2$ for all i and j, which is actually a (memory-less) Bernoulli process akin to a fair coin flipping: with probability 1/2 showing heads (0) or tails (1).[1] This analogy seems to go in the desired direction, the coin tossing being much closer to our intuitive idea of random process. We can say that trajectories of the Bernoulli map are random because akin, once a proper coarse-grained description is adopted, to coin tossing.

However, an operative definition of randomness is still missing. In the following, we attempt a first formalization of randomness by focusing on the coin tossing.

Let's consider an ensemble of sequences of length N resulting from a fair coin tossing game. Each string of symbols will typically looks like

$$110100001001001010101001101010100001111001\ldots.$$

Intuitively, we shall call such a sequence random because given the nth symbol, $s(n)$, we are *uncertain* about the $n+1$ outcome, $s(n+1)$. Therefore, quantifying randomness amounts to quantify such an *uncertainty*. Slightly changing the point of view, assume that two players play the coin tossing game in Rome and the result of each flipping is transmitted to a friend in Tokyo, e.g. by a teletype. After receiving the symbol $s(n) = 1$, the friend in Tokyo will be in suspense waiting for the next uncertain result. When receiving $s(n{+}1) = 0$, she/he will gain *information* by removing the uncertainty. If an unfair coin, displaying 1 and 0 with probability $p_0 = p \neq 1/2$ and $p_1 = 1 - p$, is thrown and, moreover, if $p \ll 1/2$, the sequence of heads and tails will be akin to

$$00000000001000001000000000000000000001000000001\ldots.$$

[1] This is not a mere analogy, the Bernoulli shift map is indeed equivalent, in the probabilistic world, to a Bernoulli process, hence its name.

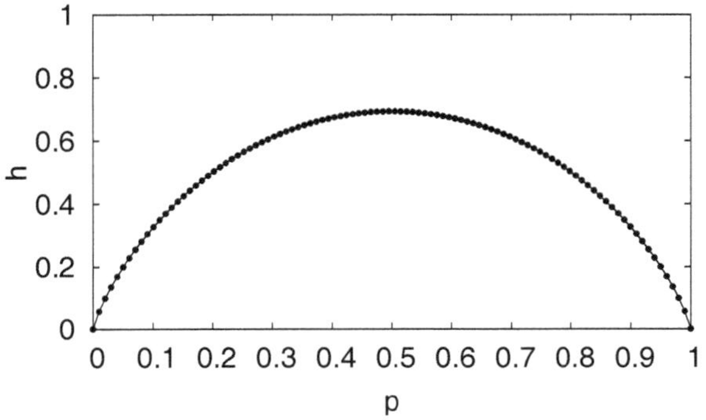

Fig. 8.2 Shannon entropy h versus p for the Bernoulli process.

This time, the friend in Tokyo will be less surprised to see that the nth symbol $s(n)=0$ and, bored, would expect that also $s(n+1) = 0$, while she/he will be more surprised when $s(n+1) = 1$, as it appears more rarely. In summary, on average, she/he will gain less information, being less uncertain about the outcome.

The above example teaches us two important aspects of the problem:
I) randomness is connected to the amount uncertainty we have prior the symbol is received or, equivalently, to the amount of information we gain once we received it;
II) our *surprise* in receiving a symbol is the larger the less probable is to observe it.

Let's make more precise these intuitive observations. We start quantifying the surprise u_i to observe a symbol α_i. For a fair coin, the symbols $\{0, 1\}$ appear with the same probability and, naively, we can say that the uncertainty (or surprise) is 2 — i.e. the number of possible symbols. However, this answer is unsatisfactory: the coin can be unfair ($p \neq 1/2$), still two symbols would appear, but we consider more surprising that appearing with lower probability. A possible definition overcoming this problem is $u_i = -\ln p_i$, where p_i is the probability to observe $\alpha_i \in \{0, 1\}$ [Shannon (1948)]. This way, the uncertainty is the average surprise associated to a long sequence of N outcomes extracted from an alphabet of M symbols ($M = 2$ in our case). Denoting with n_i the number of times the i-th symbol appears (note that $\sum_{i=0}^{M-1} n_i = N$), the average surprise per symbol will be

$$h = \frac{\sum_{i=0}^{M-1} n_i u_i}{N} = \sum_{i=0}^{M-1} \frac{n_i}{N} u_i \underset{N\to\infty}{\longrightarrow} - \sum_{i=0}^{M-1} p_i \ln p_i \,,$$

where the last step uses the law of large numbers ($n_i/N \to p_i$ for $N \to \infty$), and the convention $0 \ln 0 = 0$. For an unfair coin tossing with $M = 2$ and $p_0 = p$, $p_1 = 1-p$, we have $h(p) = -p \ln p - (1-p) \ln(1-p)$ (Fig. 8.2). The uncertainty per symbol h is known as the entropy of the Bernoulli process [Shannon (1948)]. If the outcome is certain $p = 0$ or $p = 1$, the entropy vanishes $h = 0$, while it is positive for a random processes $p \neq 0$, attaining its maximum $h = \ln 2$ for a fair coin $p = 1/2$ (Fig. 8.2). The Bernoulli map (8.1), once coarse-grained, gives rise to sequences of 0's and 1's characterized by an entropy, $h = \ln 2$, equal to the Lyapunov exponent λ (8.2).

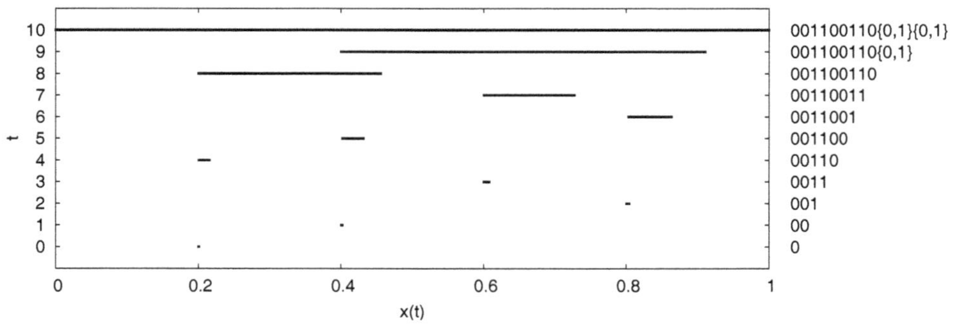

Fig. 8.3 Spreading of initially localized trajectories in the Bernoulli map, with the associated symbolic sequences (right). Until the 8th iteration a unique symbolic sequence describes all trajectories starting from $\mathcal{I}_0 = [0.2 : 0.201]$. Later, different symbols $\{0, 1\}$ appear for different trajectories.

It thus seems that we now possess an operative definition of randomness in terms of the entropy h which, if positive, well quantifies how random the process is. Furthermore, entropy seems to be related to the Lyapunov exponent; a pleasant fact as LEs quantify the most connotative property of chaotic systems, namely the sensitive dependence on initial conditions.

A simple, sketchy, way to understand the connection between entropy per symbol and Lyapunov exponent in the Bernoulli shift map is as follows (see also Fig. 8.3). Consider an ensemble of trajectories with initial conditions such that $x(0) \in \mathcal{I}_0 \subset [0:1]$, e.g., $\mathcal{I}_0 = [0.2:0.201]$. In the course of time, trajectories exponentially spread with a rate $\lambda = \ln 2$, so that the interval $\mathcal{I}_t$ containing the iterates $\{x(t)\}$ doubles its length $|\mathcal{I}_t|$ at each iteration, $|\mathcal{I}_{t+1}| = 2|\mathcal{I}_t|$. Being $|\mathcal{I}_0| = 10^{-3}$, in only ten iterations, a trajectory that started in $\mathcal{I}_0$ can be anywhere in the interval $[0:1]$, see Fig. 8.3. Now let's switch the description from actual (real valued) trajectories to symbolic strings. The whole ensemble of initial conditions $x(0) \in \mathcal{I}_0$ is uniquely coded by the symbol 0, after a step $\mathcal{I}_1 = [0.4:0.402]$ so that again 0 codes all $x(1) \in \mathcal{I}_1$. As shown on the right of Fig. 8.3, till the 8th iterate all trajectories are coded by a single string of nine symbols 001100110. At the next step most of the trajectories are coded by adding 1 to the symbolic string and the rest by adding 0. After the 10th iterate symbols $\{0, 1\}$ appear with equal probability. Thus the sensitive dependence on initial conditions makes us unable to predict the next outcome (symbol).[2] Chaos is then a source of uncertainty/information and, for the shift map, the rate at which information is produced — the entropy rate — equals the Lyapunov exponent.

It seems we found a satisfactory, mathematically well grounded, definition of randomness that links to the Lyapunov exponents. However, there is still a vague

[2]From Sec. 3.1, it should be clear that the symbols obtained from the Bernoulli map with the chosen partition correspond to the binary digit expansion of $x(0)$. Longer we wait more binary digits we know, gaining information on the initial condition $x(0)$. Such a correspondence between initial value and the symbolic sequences only exists for special partitions called "generating" (see below).

sense of incomplete contentment. Consider again the fair coin tossing, two possible realizations of N matches of the game are

$$0010011101100010101001110010011100 10\ldots \quad (8.4)$$

$$001100110011001100110011001100110011\ldots \quad (8.5)$$

The source of information — here, the fair coin tossing — is characterized by an entropy $h = \ln 2$ and generates these strings with the same probability, suggesting that entropy characterizes the source in a *statistical* sense, but does not say much on specific sequences emitted by the source. In fact, while we find natural to call sequence (8.4) random and highly informative, our intuition cannot qualify in the same way sequence (8.5). The latter is indeed "simple" and can be transmitted to Tokyo easily and efficiently by simply saying to a friend of us

$$\text{PRINT ``0011} \quad \text{for} \quad N/4 \quad \text{times''}\,, \quad (8.6)$$

thus we can *compress* sequence (8.5) providing a *shorter* (with respect to N) description. This contrasts with sequence (8.4) for which we can only say

$$\text{PRINT ``001001110110001010100111001001110010\ldots''}\,, \quad (8.7)$$

which amounts to use roughly the same number of symbols of the sequence.

The two descriptions (8.6) and (8.7) may be regarded as two programs that, running on a computer, produce on output the sequences (8.5) and (8.4), respectively. For $N \gg 1$, the former program is much shorter ($\mathcal{O}(\log_2 N)$ symbols) than the output sequence, while the latter has a length comparable to that of the output. This observation constitutes the basis of *Algorithmic Complexity* [Solomonoff (1964); Kolmogorov (1965); Chaitin (1966)], a notion that allows us to define randomness for a given sequence $\mathcal{W}_N$ of N symbols without any reference to the (statistical properties of the) source which emitted it. Randomness is indeed quantified in terms of the binary length $\mathcal{K}_{\mathcal{M}}(\mathcal{W}(N))$ of the shortest algorithm which, implemented on a machine $\mathcal{M}$, is able to reproduce the entire sequence $\mathcal{W}(N)$, which is called random when the algorithmic complexity per symbol $\kappa_{\mathcal{M}}(\mathcal{S}) = \lim_{N\to\infty} \mathcal{K}_{\mathcal{M}}(\mathcal{W}(N))/N$ is positive. Although the above definition needs some specifications and contains several pitfalls, for instance, $\mathcal{K}_{\mathcal{M}}$ could at first glance be machine dependent, we can anticipate that algorithmic complexity is a very useful concept able to overcome the notion of statistical ensemble needed to the entropic characterization.

This brief excursion put forward a few new concepts as information, entropy, algorithmic complexity and their connection with Lyapunov exponents and chaos. The rest of the Chapter will deepen these aspects and discuss connected ideas.

8.2 Information theory, coding and compression

Information has found a proper characterization in the framework of Communication Theory, pioneered by Shannon (1948) (see also Shannon and Weaver (1949)).

The fundamental problem of communication is the faithful reproduction at a place of messages emitted elsewhere. The typical process of communication involves several components as illustrated in Fig. 8.4.

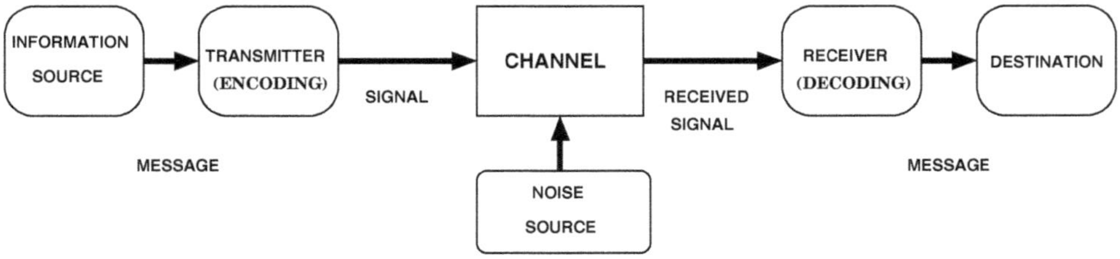

Fig. 8.4 Sketch of the processes involved in communication theory. [After Shannon (1948)]

In particular, we have:
An *information source* emitting messages to be communicated to the receiving terminal. The source may be discrete, emitting messages that consist of a sequence of "letters" as in teletypes, or continuous, emitting one (or more) function of time, of space or both, as in radio or television.
A *transmitter* which acts on the signal, for example digitalizing and/or *encoding* it, in order to make it suitable for cheap and efficient transmissions.
The transmission *channel* is the medium used to transmit the message, typically a channel is influenced by environmental or other kinds of noise (which can be modeled as a *noise source*) degrading the message.
Then a *receiver* is needed to recover the original message. It operates in the inverse mode of the transmitter by *decoding* the received message, which can eventually be delivered to its *destination*.

Here we are mostly concerned with the problem of characterizing the information source in terms of Shannon entropy, and with some aspects of coding and compression of messages. For the sake of simplicity, we consider discrete information sources emitting symbols from a finite alphabet. We shall largely follow Shannon's original works and Khinchin (1957), where a rigorous mathematical treatment can be found.

8.2.1 *Information sources*

Typically, interesting messages carry a meaning that refers to certain physical or abstract entities, e.g. a book. This requires the devices and involved processes of Fig. 8.4 to be adapted to the specific category of messages to be transmitted. However, in a mathematical approach to the problem of communication the semantic aspect is ignored in favor of a the generality of transmission protocol. In this respect we can, without loss of generality, limit our attention to discrete sources emitting sequences of random objects α_i out of a finite set — the alphabet — $\mathcal{A} = \{\alpha_0, \alpha_2, \ldots, \alpha_{M-1}\}$, which can be constituted, for instance, of letters as in

English language or numbers, and which we generically call letters or symbols. In this framework defining a source means to provide its complete probabilistic characterization. Let $\mathcal{S} = \ldots s(-1)s(0)s(1)\ldots$ be an infinite (on both sides) sequence of symbols ($s(t) = \alpha_k$ for some $k = 0, \ldots, M-1$) emitted by the source and thus representing one of its possible "life history". The sequence $\mathcal{S}$ corresponds to an elementary event of the (infinite) probability space Ω. The source $\{\mathcal{A}, \mu, \Omega\}$ is then defined in terms of the alphabet $\mathcal{A}$ and the probability measure μ assigned on Ω.

Specifically, we are interested in *stationary* and *ergodic* sources. The former property means that if σ is the *shift operator*, defined by

$$\sigma\mathcal{S} = \ldots s'(-1)s'(0)s'(1)\ldots \quad \text{with} \quad s'(n) = s(n+1)\,,$$

then the source is stationary if $\mu(\sigma\Xi) = \mu(\Xi)$ for any $\Xi \subset \Omega$: the sequences obtained translating by an arbitrary number of steps the symbols are statistically equivalent to the original ones. A set $\Xi \in \Omega$ is called *invariant* when $\sigma\Xi = \Xi$ and the source is ergodic if for any invariant set $\Xi \in \Omega$ we have $\mu(\Xi) = 0$ or $\mu(\Xi) = 1$.[3] Similarly to what we have seen in Chapter 4, ergodic sources are particularly useful as they allow the exchange of averages over the probability space with averages performed over a long typical sequence (i.e. the equivalent of time averages):

$$\int_\Omega \mathrm{d}\mu\, F(\mathcal{S}) = \lim_{n\to\infty} \frac{1}{n} \sum_{k=1}^{n} F(\sigma^k \mathcal{S})\,,$$

where F is a generic function defined in the space of sequences.

A string of N consecutive letters emitted by the source $\mathcal{W}_N = s(1), s(2), \ldots, s(N)$ is called a N-string or N-word. Therefore, at a practical level, the source is known once we know the (joint) probabilities $P(s(1), s(2), \ldots, s(N)) = P(\mathcal{W}_N)$ of all the set of the N-words it is able to emit, i.e., $P(\mathcal{W}_N)$ for each $N = 1, \ldots, \infty$, these are called N-block probabilities. For memory-less processes, as Bernoulli, the knowledge of $P(\mathcal{W}_1)$ fully characterizes the source, i.e. to know the probabilities of each letter α_i which is indicated by p_i with $i = 0, \ldots, M-1$ (with $p_i \geq 0$ for each i and $\sum_{i=0}^{M-1} p_i = 1$). In general, we need all the joint probabilities $P(\mathcal{W}_N)$ or the conditional probabilities $p(s(N)|s(N-1), \ldots, s(N-k), \ldots)$. For Markovian sources (Box B.6), a complete characterization is achieved through the conditional probabilities $p(s(N)|s(N-1), \ldots, s(N-k))$, if k is the order of the Markov process.

8.2.2 *Properties and uniqueness of entropy*

Although the concept of entropy appeared in information theory with Shannon (1948) work, it was long known in thermodynamics and statistical mechanics. The statistical mechanics formulation of entropy is essentially equivalent to that used in information theory, and conversely the information theoretical approach enlightens

[3]The reader may easily recognize that these notions coincide with those of Chap. 4, provided the translation from sequences to trajectories.

many aspects of statistical mechanics [Jaynes (1957a,b)]. At the beginning of this Chapter, we provided some heuristic arguments to show that entropy can properly measure the information content of messages, here we summarize its properties.

Given a finite probabilistic scheme A characterized by an alphabet $\mathcal{A} = \{\alpha_0, \ldots, \alpha_{M-1}\}$ of M letters and the probabilities $p_0, \ldots, p_{M-1}$ of occurrence for each symbol, the entropy of A is given by:

$$H(A) = H(p_0, \ldots, p_{M-1}) = -\sum_{i=0}^{M-1} p_i \ln p_i \tag{8.8}$$

with $\sum_{i=0}^{M-1} p_i = 1$ and the convention $0 \ln 0 = 0$.

Two properties can be easily recognized. First, $H(A) = 0$ if and only if for some k, $p_k = 1$ while $p_i = 0$ for $i \neq k$. Second, as $x \ln x$ $(x > 0)$ is convex

$$\max_{p_0,\ldots,p_{M-1}} \{H(p_0, \ldots, p_{M-1})\} = \ln M \quad \text{for} \quad p_k = 1/M \quad \text{for all} \quad k\,, \tag{8.9}$$

i.e. entropy is maximal for equiprobable events.[4]

Now consider the composite events $\alpha_i \beta_j$ obtained from two probabilistic schemes: A with alphabet $\mathcal{A} = \{\alpha_0, \ldots, \alpha_{M-1}\}$ and probabilities $p_0, \ldots, p_{M-1}$, and B with alphabet $\mathcal{B} = \{\beta_0, \ldots, \beta_{K-1}\}$ and probabilities $q_0, \ldots, q_{K-1}$; the alphabet sizes M and K being arbitrary but finite.[5] If the schemes are mutually independent, the composite event $\alpha_i \beta_j$ has probability $p(i,j) = p_i q_j$ and, applying the definition (8.8), the entropy of the scheme AB is just the sum of the entropies of the two schemes

$$H(A;B) = H(A) + H(B)\,. \tag{8.10}$$

If they are not independent, the joint probability $p(i,j)$ can be expressed in terms of the conditional probability $p(\beta_j|\alpha_i) = p(j|i)$ (with $\sum_k p(k|i) = 1$) through $p(i,j) = p_i p(j|i)$. In this case, for any outcome α_i of scheme A, we have a new probabilistic scheme, and we can introduce the *conditional entropy*

$$H_i(B|A) = -\sum_{k=0}^{K-1} p(k|i) \ln p(k|i)\,,$$

and Eq. (8.10) generalizes to[6]

$$H(A;B) = H(A) + \sum_{i=0}^{M-1} p_i\, H_i(B|A) = H(A) + H(B|A)\,. \tag{8.11}$$

The meaning of the above quantity is straightforward: the information content of the composite event $\alpha\beta$ is equal to that of the scheme A plus the average information

[4]Hint for the demonstration: notice that if $g(x)$ is convex then $g(\sum_{k=0}^{n-1} a_k/n) \leq (1/n)\sum_{k=0}^{n-1} g(a_k)$, then put $a_i = p_i$, $n = M$ and $g(x) = x \ln x$.

[5]The scheme B may also coincide with A meaning that the composite event $\alpha_i \beta_j = \alpha_i \alpha_j$ should be interpreted as two consecutive outcomes of the same random process or measurement.

[6]Hint: use the definition of entropy with $p(i,j) = p_i p(j|i)$.

needed to specify β once α is known. Furthermore, still thanks to the convexity of $x \ln x$, it is easy to prove the inequality

$$H(B|A) \leq H(B) \tag{8.12}$$

whose interpretation is: the knowledge of the outcome of A cannot increase our uncertainty on that of B.

Properties (8.9) and (8.11) constitute two natural requests for any quantity aiming to characterize the uncertainty (information content) of a probabilistic scheme: maximal uncertainty should be always obtained for equiprobable events, and the information content of the combination of two schemes should be additive, or better, the generalization (8.11) for correlated events which implies through (8.12) the sub-additive property

$$H(A; B) \leq H(A) + H(B).$$

As shown by Shannon (1948), see also Khinchin (1957), these two requirements plus the obvious condition that $H(p_0, \ldots, p_{M-1}, 0) = H(p_0, \ldots, p_{M-1})$ imply that H has to be of the form $H = -\kappa \sum p_i \ln p_i$, where κ is a positive factor fixing the units in which we measure information. This result, known as *uniqueness theorem*, is of great aid as it tells us that, once the desired (natural) properties of entropy as a measure of information are fixed, the choice (8.8) is unique but for a multiplicative factor.

A complementary concept is that of *mutual information* (sometimes called *redundancy*) defined by

$$I(A; B) = H(A) + H(B) - H(A; B) = H(B) - H(B|A), \tag{8.13}$$

where the last equality derives from Eq. (8.11). The symmetry of $I(A; B)$ in A and B implies also that $I(A; B) = H(A) - H(A|B)$. First we notice that inequality (8.12) implies $I(A; B) \geq 0$ and, moreover, $I(A; B) = 0$ if and only if A and B are mutually independent. The meaning of $I(A; B)$ is rather transparent: $H(B)$ measures the uncertainty of scheme B, $H(B|A)$ measures what the knowledge of A does not say about B, while $I(A; B)$ is the amount of uncertainty removed from B by knowing A. Clearly, $I(A; B) = 0$ if A says nothing about B (mutually independent events) and is maximal and equal to $H(B) = H(A)$ if knowing the outcome of A completely determines that of B.

8.2.3 *Shannon entropy rate and its meaning*

Consider an ergodic and stationary source emitting symbols from a finite alphabet of M letters, denote with $s(t)$ the symbol emitted at time t and with $P(\mathcal{W}_N) = P(s(1), s(2), \ldots, s(N))$ the probability of finding the N consecutive symbols (N-word) $\mathcal{W}_N = s(1)s(2)\ldots s(N)$. We can extend the definition (8.8) to N-tuples of random variables, and introduce the N-block entropies:

$$H_N = -\sum_{\mathcal{W}_N} P(\mathcal{W}_N) \ln P(\mathcal{W}_N) = -\sum_{s(1)=\alpha_0}^{\alpha_{M-1}} \cdots$$
$$\cdots \sum_{s(N)=\alpha_0}^{\alpha_{M-1}} P(s(1), s(2), \ldots, s(N)) \ln P(s(1), s(2), \ldots, s(N)) \,, \tag{8.14}$$

with $H_{N+1} \geq H_N$ as from Eqs. (8.11) and (8.12). We then define the differences

$$h_N = H_N - H_{N-1} \qquad \text{with} \quad H_0 = 0 \,,$$

measuring the average information supplied by (or needed to specify) the N-th symbol when the $(N-1)$ previous ones are known. One can directly verify that $h_N \leq h_{N-1}$, as also their meaning suggests: more knowledge on past history cannot increase the uncertainty about the future.

For stationary and ergodic sources the limit

$$h_{Sh} = \lim_{N\to\infty} h_N = \lim_{N\to\infty} \frac{H_N}{N} \tag{8.15}$$

exists and defines the Shannon entropy, i.e. the average information amount per symbol emitted by (or rate of information production of) the source.

To better understand the meaning to this quantity, it is worth analyzing some examples. Back to the Bernoulli process (the coin flipping model of Sec. 8.1) it is easy to verify that $H_N = Nh$ with $h = -p \ln p - (1-p)\ln(1-p)$, therefore the limit (8.15) is attained already for $N \geq 1$ and thus the Shannon entropy is $h_{Sh} = h = H_1$. Intuitively, this is due to the absence of memory in the process, in contrast to the presence of correlations in generic sources. This can be illustrated considering as an information source a Markov Chain (Box B.6) where the random emission of the letters $\alpha_0, \ldots, \alpha_{M-1}$ is determined by the $(M \times M)$ transition matrix $\mathbb{W}_{ij} = p(i|j)$. By using repeatedly Eq. (8.11), it is not difficult to see that $H_N = H_1 + (N-1)h_{Sh}$ with $H_1 = -\sum_{i=0}^{M-1} p_i \ln p_i$ and $h_{Sh} = -\sum_{i=0}^{M-1} p_i \sum_{j=0}^{M-1} p(j|i) \ln p(j|i)$, $(p_0, \ldots, p_{M-1}) = \boldsymbol{p}$ being the invariant probabilities, i.e. $\mathbb{W}\boldsymbol{p} = \boldsymbol{p}$. It is straightforward to generalize the above reasoning to show that a generic k-th order Markov Chain, which is determined by the transition probabilities $P(s(t)|s(t-1), s(t-2), \ldots, s(t-k))$, is characterized by block entropies behaving as:

$$H_{k+n} = H_k + n h_{Sh}$$

meaning that h_N equals the Shannon entropy for $N > k$.

From the above examples, we learn two important lessons: first, the convergence of h_N to h_{Sh} is determined by the degree of memory/correlation in the symbol emission, second using h_N instead of H_N/N ensures a faster convergence to h_{Sh}.[7]

[7] It should however be noticed that the difference entropies h_N may be affected by larger statistical errors than H_N/N. This is important for correctly estimating the Shannon entropies from finite strings. We refer to Schürmann and Grassberger (1996) and references therein for a throughout discussion on the best strategies for unbiased estimations of Shannon entropy.

Actually the convergence behavior of h_N may highlight important features of the source (see Box B.15 and Grassberger (1986, 1991)).

Shannon entropy quantifies the richness (or "complexity") of the source emitting the sequences, providing a measure of the "surprise" the source reserves to us. This can be better expressed in terms of a fundamental theorem, first demonstrated by Shannon (1948) for Markov sources and then generalized by McMillan (1953) to generic ergodic stationary sources (see also Khinchin (1957)):

If N is large enough, the set of all possible N-words, $\Omega(N) \equiv \{\mathcal{W}_N\}$ can be partitioned into two classes $\Omega_1(N)$ and $\Omega_0(N)$ such that if $\mathcal{W}_N \in \Omega_1(N)$ then $P(\mathcal{W}_N) \sim \exp(-N h_{Sh})$ and

$$\sum_{\mathcal{W}_N \in \Omega_1(N)} P(\mathcal{W}_N) \underset{N\to\infty}{\longrightarrow} 1$$

while

$$\sum_{\mathcal{W}_N \in \Omega_0(N)} P(\mathcal{W}_N) \underset{N\to\infty}{\longrightarrow} 0\,.$$

In principle, for an alphabet composed by M letters there are M^N different N-words, although some them can be forbidden (see the example below), so that, in general, the number of possible N-words is $\mathcal{N}(N) \sim \exp(N\, h_T)$ where

$$h_T = \lim_{N\to\infty} \frac{1}{N} \ln \mathcal{N}(N)$$

is named *topological entropy* and has as the upper bound $h_T \leq \ln M$ (the equality being realized if all words are allowed).[8]

The meaning of Shannon-McMillan theorem is that among all the permitted N-words, $\mathcal{N}(N)$, the number of typical ones ($\mathcal{W}_N \in \Omega_1(N)$), that are effectively observed, is

$$\mathcal{N}_{\text{eff}}(N) \sim e^{N\, h_{Sh}}\,.$$

As $\mathcal{N}_{\text{eff}}(N) \leq \mathcal{N}(N)$ it follows

$$h_{Sh} \leq h_T \leq \ln M\,.$$

The fair coin tossing, examined in the previous section, corresponds to $h_{Sh} = h_T = \ln 2$, the unfair coin to $h_{Sh} = -p\ln p - (1-p)\ln(1-p) < h_T = \ln 2$ (where $p \neq 1/2$). A slightly more complex and instructive example is obtained by considering a random source constituted by the two states (say 0 and 1) Markov Chain with transition matrix

$$\mathbb{W} = \begin{pmatrix} p & 1 \\ 1-p & 0 \end{pmatrix}. \tag{8.16}$$

Being $\mathbb{W}_{11} = 0$ when 1 is emitted with probability one the next emitted symbol is 0, meaning that words with two or more consecutive 1 are forbidden (Fig. 8.5). It is

[8]Notice that, in the case of memory-less processes, Shannon-McMillan theorem is nothing but the law of large numbers.

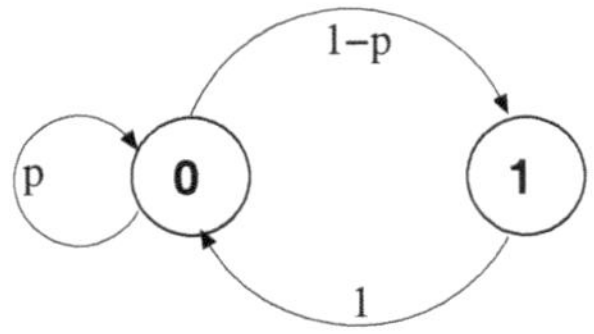

Fig. 8.5 Graph representing the coin-tossing process described by the matrix (8.16).

easy to show (see Ex.8.2) that the number of allowed N-words, $\mathcal{N}(N)$, is given by the recursion $\mathcal{N}(N)=\mathcal{N}(N-1)+\mathcal{N}(N-2)$ for $N\geq 2$ with $\mathcal{N}(0)=1$, $\mathcal{N}(1)=2$, which is nothing but the famous Fibonacci sequence.[9] The ratios of Fibonacci numbers are known, since Kepler, to have as a limit the *golden ratio*

$$\frac{\mathcal{N}(N)}{\mathcal{N}(N-1)} \underset{N\to\infty}{\longrightarrow} \mathcal{G} = \frac{1+\sqrt{5}}{2}\,,$$

so that the topological entropy of the above Markov chain is simply $h_T = \ln \mathcal{G} = 0.48121\ldots$. From Eq. (8.11), we have $h_{Sh} = -[p\ln p + (1-p)\ln(1-p)]/(2-p) \leq h_T = \ln\phi$ with the equality realized for $p = \mathcal{G} - 1$.

We conclude by stressing that h_{Sh} is a property inherent to the source and that, thanks to ergodicity, it can be derived analyzing just one single, long enough sequence in the ensemble of the typical ones. Therefore, h_{Sh} can also be viewed as a property of typical sequences, allowing us to, with a slight abuse of language, speak about Shannon entropy of a sequence.

Box B.15: Transient behavior of block-entropies

As underlined by Grassberger (1986, 1991) the transient behavior of N-block entropies H_N reveals important features of the complexity of a sequence. The N-block entropy H_N is a non-decreasing concave function of N, so that the difference

$$h_N = H_N - H_{N-1} \qquad (\text{with} \quad H_0 = 0)$$

is a decreasing function of N representing the average amount of information needed to predict $s(N)$ given $s(1),\ldots,s(N-1)$. Now we can introduce the quantity

$$\delta h_N = h_{N-1} - h_N = 2H_{N-1} - H_N - H_{N-1} \qquad (\text{with} \quad H_{-1} = H_0 = 0)\,,$$

which, due to the concavity of H_N, is a positive non-increasing function of N, vanishing for $N \to \infty$ as $h_N \to h_{Sh}$. Grassberger (1986) gave an interesting interpretation of δh_N as the amount by which the uncertainty on $s(N)$ decreases when one more symbol of the past is known, so that $N\delta h_N$ measures the difficulty in forecasting an N-word, and

$$C_{EMC} = \sum_{k=1}^{\infty} k\delta h_k$$

[9] Actually it is a shift by 2 of the Fibonacci sequence.

is called the *effective measure of complexity* [Grassberger (1986, 1991)]: the average usable part of information on the past which has to be remembered to reconstruct the sequence. In this respect, it measures the difficulty of forecasting. Noticing that $\sum_{k=1}^{N} k\delta h_k = \sum_{k=1}^{N} h_k - (N+1)h_N = H_N - (N+1)(H_N - H_{N-1})$ we can rewrite C_{EMC} as

$$C_{EMC} = \lim_{N\to\infty} H_N - (N+1)(H_N - H_{N-1}) = C + h_{Sh}$$

where C is nothing but the intercept of the tangent to H_N as $N \to \infty$. In other words this shows that, for large N, the block-entropies grow as:

$$H_N \simeq C + N\, h_{Sh}\,, \tag{B.15.1}$$

therefore C_{EMC} is essentially a measure of C.[10] In processes without or with limited memory such as, e.g., for Bernoulli schemes or Markov chain of order 1, $C = 0$ and $h_{Sh} > 0$, while in a periodic sequence of period $\mathcal{T}$, $h_{Sh} = 0$ and $C \sim \ln(\mathcal{T})$. The quantity C has a number of interesting properties. First of all within all stochastic processes with the same H_k, for $k \leq N$, C is minimal for the Markov processes of order $N-1$ compatible with the block entropies of order $k \leq N$. It is remarkable that even systems with $h_{Sh} = 0$ can have a nontrivial behavior if C is large. Actually, C or C_{EMC} are minimal for memory-less stochastic processes, and a high value of C can be seen as an indication of a certain level of organizational complexity [Grassberger (1986, 1991)].

As an interesting application of systems with a large C, we mention the use of chaotic maps as pseudo-random numbers generators (PRNGs) [Falcioni *et al.* (2005)]. Roughly speaking, a sequence produced by a PRNG is considered good if it is practically indistinguishable from a sequence of independent "true" random variables, uniformly distributed in the interval $[0\!:\!1]$. From an entropic point of view this means that if we make a partition, similarly to what has been done for the Bernoulli map in Sec. 8.1, of $[0:1]$ in intervals of length ε and we compute the Shannon entropy $h(\varepsilon)$ at varying ε (this quantity, called ε-entropy, is studied in details in the next Chapter), then $h(\varepsilon) \simeq \ln(1/\varepsilon)$.[11]

Consider the lagged Fibonacci map [Green Jr. *et al.* (1959)]

$$x(t) = ax(t-\tau_1) + bx(t-\tau_2) \qquad \mathrm{mod}\ 1\,, \tag{B.15.2}$$

with a and b $\mathcal{O}(1)$ constants and $\tau_1 < \tau_2$. Such a map, can be written in the form

$$\boldsymbol{y}(t) = \mathbb{F}\boldsymbol{y}(t-1) \qquad \mathrm{mod}\ 1 \tag{B.15.3}$$

$\mathbb{F}$ being the $\tau_2 \times \tau_2$ matrix

$$\mathbb{F} = \begin{pmatrix} 0 & \dots & a & \dots & b \\ 1 & 0 & 0 & \dots & 0 \\ 0 & 1 & 0 & \dots & 0 \\ \dots & \dots & \dots & \dots & \dots \\ 0 & \dots & \dots & 1 & 0 \end{pmatrix}$$

---

[10] We remark that this is true only if h_N converges fast enough to h_{Sh}, otherwise C_{EMC} may also be infinite, see [Badii and Politi (1997)]. We also note that the faster convergence of h_N with respect to H_N/N is precisely due to the cancellation of the constant C.

[11] For any ε the number of symbols in the partition is $M = (1/\varepsilon)$. Therefore, the request $h(\varepsilon) \simeq \ln(1/\varepsilon)$ amounts to require that for any ε-partition the Shannon entropy is maximal.

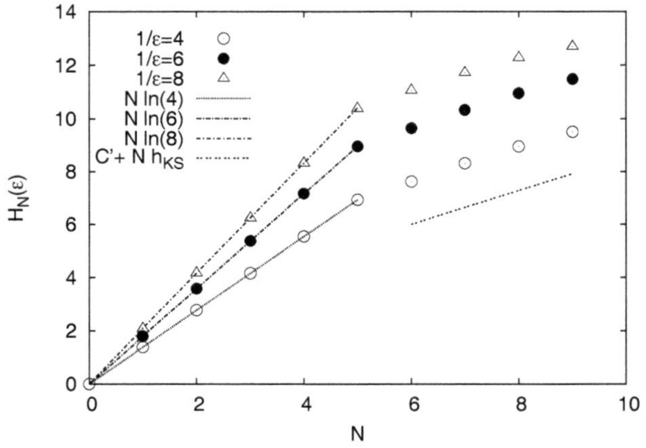

Fig. B15.1 N-block entropies for the Fibonacci map (B.15.2) with $\tau_1 = 2, \tau_2 = 5, a = b = 1$ for different values of ε as in label. The change of the slope from $-\ln(\varepsilon)$ to h_{KS} is clearly visible for $N \sim \tau_2 = 5$. For large τ_2 ($\sim \mathcal{O}(10^2)$) C becomes so huge that only an extremely long sequence of $\mathcal{O}(e^{\tau_2})$ (likely outside the capabilities of modern computers) may reveal that h_{Sh} is indeed small.

which explicitly shows that the map (B.15.2) has dimension τ_2. It is easily proved that this system is chaotic when a and b are positive integers and that the Shannon entropy does not depend on τ_1 and τ_2; this means that to obtain high values of h_{Sh} we are forced to use large values of a, b. The lagged Fibonacci generators are typically used with $a = b = 1$. In spite of the small value of the resulting h_{Sh} is a reasonable PRNG. The reason is that the N-words, built up by a single variable (y_1) of the τ_2-dimensional system (B.15.3), have the maximal allowed block-entropy, $H_N(\varepsilon) = N \ln(1/\varepsilon)$, for $N < \tau_2$, so that:

$$H_N(\varepsilon) \simeq \begin{cases} -N \ln \varepsilon & \text{for } N < \tau_2 \\ -\tau_2 \ln \varepsilon + h_{Sh}(N - \tau_2) & \text{for } N \geq \tau_2 \,. \end{cases}$$

For large N one can write the previous equation in the form (B.15.1) with

$$C = \tau_2 \left[\ln \left(\frac{1}{\varepsilon} \right) - h_{Sh} \right] \approx \tau_2 \ln \left(\frac{1}{\varepsilon} \right) .$$

Basically, a long transient is observed in N-block ε-entropies, characterized by a maximal (or almost maximal) value of the slope, and then a crossover to a regime with the slope of h_{Sh} of the system. Notice that, although the h_{Sh} is small, it can be computed only using large $N > \tau_2$, see Fig. B15.1.

8.2.4 *Coding and compression*

In order to optimize communications, by making them cheaper and faster, it is desirable to have encoding of messages which shorten their length. Clearly, this is possible when the source emits messages with some extent of the redundancy (8.13), whose reduction allows the message to *compressed* while preserving its integrity. In this case we speak of *lossless* encoding or compression.[12]

Shannon demonstrated that there are intrinsic limits in compressing sequences emitted by a given source, and these are connected with the entropy of the source. Consider a long sequence of symbols $\mathcal{S}(T) = s(1)s(2)\dots s(n)\dots s(T)$ having length $\mathfrak{L}(\mathcal{S}) = T$, and suppose that the symbols are emitted by a source with an alphabet of M letters and Shannon entropy h_{Sh}. Compressing the sequence means generating another one $\mathcal{S}'(T') = s'(1)s'(2)\dots s'(T')$ of length $\mathfrak{L}(\mathcal{S}') = T'$ with $\mathcal{C} = \mathfrak{L}(\mathcal{S}')/\mathfrak{L}(\mathcal{S}) < 1$, $\mathcal{C}$ being the *compression coefficient*, such that the original sequence can be recovered exactly. Shannon's compression theorem states that, if the sequence is generic and T large enough if, in the coding, we use an alphabet with the same number of letters M, then $\mathcal{C} \geq h_{Sh}/\ln M$, that is the compression coefficient has a lower bound given by the ratio between the actual and the maximal allowed value $\ln M$ of Shannon entropy of the source .

The relationship between Shannon entropy and the compression problem is well illustrated by the Shannon-Fano code [Welsh (1989)], which maps $\mathcal{N}$ objects into sequences of binary digits $\{0, 1\}$ as follows. For example, given a number $\mathcal{N}$ of N-words $\mathcal{W}_N$, first determine their probabilities of occurrence. Second, sort the N-words in a descending order according to the probability value, $P(\mathcal{W}_N^1) \geq P(\mathcal{W}_N^2) \geq \dots \geq P(\mathcal{W}_N^{\mathcal{N}})$. Then, the most compressed description corresponds to the faithful code $E(\mathcal{W}_N^k)$, which codifies each $\mathcal{W}_N^k$ in terms of a string of zeros and ones, producing a compressed message with minimal expected length $\langle \mathfrak{L}_N \rangle = \sum_{k=1}^{\mathcal{N}} \mathfrak{L}(E(\mathcal{W}_N^k))P(\mathcal{W}_N^k)$. The minimal expected length is clearly realized with the choice

$$- \log_2 P(\mathcal{W}_N^k) \leq \mathfrak{L}(E(\mathcal{W}_N^k)) \leq - \log_2 P(\mathcal{W}_N^k) + 1\,,$$

where [...] denotes the integer part and $\log_2$ the base-2 logarithm, the natural choice for binary strings. In this way, highly probable objects are mapped into short code words whereas low probability ones into longer code words. Averaging over the probabilities $P(\mathcal{W}_N^k)$, we thus obtain:

$$\frac{H_N}{\ln 2} \leq \sum_{k=1}^{\mathcal{N}} \mathfrak{L}(E(\mathcal{W}_N^k))P(\mathcal{W}_N^k) \leq \frac{H_N}{\ln 2} + 1\,.$$

which in the limit $N \to \infty$ prescribes

$$\lim_{N\to\infty} \frac{\langle \mathfrak{L}_N \rangle}{N} = \frac{h_{Sh}}{\ln 2}\,,$$

N-words are thus mapped into binary sequences of length $N h_{Sh}/\ln 2$. Although the Shannon-Fano algorithm was rather simple and powerful, it is of little practical use

[12]In certain circumstances, we may relax the requirement of *fidelity* of the code, that is to content ourselves with a compressed message which is fairly close the original one but with less information, this is what we commonly do using, e.g., the jpeg format in digital images. We shall postpone this problem to the next Chapter.

when N-word probabilities are not known a priori. Powerful compression schemes, not needing prior knowledge on the source, can however be devised. We will see an example of them later in Box B.16.

We end by remarking that compression theorem has to be understood within the ergodic theory framework. For a given source, there will exist specific sequences which might be compressed more efficiently than expected from the theorem, as, for instance, the sequence (8.5) with respect to (8.4). However, the probability to actually observe such sequences is zero. In other words, these atypical sequences are the N-words belonging to set $\Omega_0(N)$ of the Shannon-McMillan theorem.

8.3 Algorithmic complexity

The Shannon entropy sets the limits of how efficiently an ensemble of messages emitted by an ergodic and stationary source can be compressed, but says nothing about single sequences. Sometimes we might be interested in a specific sequence and not in an ensemble of them. Moreover, not all interesting sequences belong to a stationary ensemble think of, for example, the case of the DNA of a given individual. As anticipated in Sec. 8.1, the single-sequence point of view can be approached in terms of the algorithmic complexity, which precisely quantifies the difficulty to reproduce a given string of symbols on a computer. This notion was independently introduced by Kolmogorov (1965), Chaitin (1966) and Solomonoff (1964), and can be formalized as follows.

Consider a binary digit (this does not constitute a limitation) sequence of length N, $\mathcal{W}_N = s(1), s(2), \ldots, s(N)$, its algorithmic complexity, or algorithmic information content, $\mathcal{K}_{\mathcal{M}}(\mathcal{W}_N)$ is the bit length $\mathfrak{L}(\wp)$ of the shortest computer program $\wp$ that running on a machine $\mathcal{M}$ is able to re-produce that N-sequence and stop afterward,[13] in formulae

$$\mathcal{K}_{\mathcal{M}}(\mathcal{W}_N) = \min_{\wp}\{\mathfrak{L}(\wp) \ : \ \mathcal{M}(\wp) = \mathcal{W}_N\} . \tag{8.17}$$

In principle, the program length depends not only on the sequence but also on the machine $\mathcal{M}$. However, as shown by Kolmogorov (1965), thanks to the conceptual framework developed by Turing (1936), we can always use a universal computer $\mathcal{U}$ that is able to perform the same computation program $\wp$ makes on $\mathcal{M}$, with a modification of $\wp$ that depends on $\mathcal{M}$ only. This implies that for all finite strings:

$$\mathcal{K}_{\mathcal{U}}(\mathcal{W}_N) \leq \mathcal{K}_{\mathcal{M}}(\mathcal{W}_N) + c_{\mathcal{M}} , \tag{8.18}$$

where $\mathcal{K}_{\mathcal{U}}(\mathcal{W}_N)$ is the complexity with respect to the universal computer $\mathcal{U}$ and $c_{\mathcal{M}}$ is a constant only depending on the machine $\mathcal{M}$. Hence, from now on, we consider the algorithmic complexity with respect to $\mathcal{U}$, neglecting the machine dependence.

[13]The halting constraint is not requested by all authors, and entails many subtleties related to computability theory, here we refrain from entering this discussion and refer to Li and Vitányi (1997) for further details.

Typically, we are interested in the algorithmic complexity per unit symbol

$$\kappa(\mathcal{S}) = \lim_{N\to\infty} \frac{\mathcal{K}(\mathcal{W}_N)}{N}$$

for very long sequences $\mathcal{S}$ which, thanks to Eq. (8.18), is an intrinsic quantity independent of the computer. For instance, non-random sequences as (8.5) admit very short descriptions (programs) like (8.6), so that $\kappa(\mathcal{S}) = 0$; while random ones as (8.4) cannot be compressed in a description shorter than they are, like (8.6), so that $\kappa(\mathcal{S}) > 0$. In general, we call algorithmically complex or random all those sequences $\mathcal{S}$ for which $\kappa(\mathcal{S}) > 0$.

Although information and algorithmic approaches originate from two rather different points of view, Shannon entropy h_{Sh} and algorithmic complexity κ are not unrelated. In fact, it is possible to show that given an ensemble of N-words $\mathcal{W}_N$ occurring with probabilities $P(\mathcal{W}_N)$, we have [Chaitin (1990)]

$$\lim_{N\to\infty} \frac{\sum \mathcal{K}(\mathcal{W}_N)P(\mathcal{W}_N)}{H_N} \equiv \lim_{N\to\infty} \frac{\langle \mathcal{K}(\mathcal{W}_N)\rangle}{H_N} = \frac{1}{\ln 2}\,. \tag{8.19}$$

In other words, the algorithmic complexity averaged over the ensemble of sequences $\langle\kappa\rangle$ is equal to h_{KS}, but for a $\ln 2$ factor, only due to the different units used to measure the two quantities. The result (8.19) stems from Shannon-McMillan theorem about the two classes $\Omega_1(N)$ and $\Omega_0(N)$ of N-words: in the limit of very large N, the probability to observe a sequence in $\Omega_1(N)$ goes to 1, and the algorithmic complexity per symbol κ of such a sequence equals the Shannon entropy.

Despite the numerical coincidence of κ and $h_{Sh}/\ln 2$, information and algorithmic complexity theory are conceptually very different. This difference is well illustrated considering the sequence of the digits of $\pi = \{314159265358\ldots\}$. On the one hand, any statistical criterion would say that these digits look completely random [Wagon (1985)]: all digits are equiprobable as also digit pairs, triplets etc., meaning that the Shannon entropy is close to the maximum allowed value for an alphabet of $M = 10$ letters. On the other hand, very efficient programs $\wp$ are known for computing an arbitrary number N of digits of π and $\mathfrak{L}(\wp) = \mathcal{O}(\log_2 N)$, from which we would conclude that $\kappa(\pi) = 0$. Thus the question "is π random or not?" remains open. The solution to this paradox is in the true meaning of entropy and algorithmic complexity. Technically speaking $\mathcal{K}(\pi[N])$ (where $\pi[N]$ denotes the first N digits of π) measures the amount of information needed to specify the first N digits of π, while h_{Sh} refers to the average information necessary for designating *any* consecutive N digits: it is easier to determine the first 100 digits than the 100 digits, between, e.g., 40896 and 40996 [Grassberger (1986, 1989)].

From a physical perspective, statistical quantities are usually preferable with respect to non-statistical ones, due to their greater robustness. Therefore, in spite of the theoretical and conceptual interest of algorithmic complexity, in the following we will mostly discuss the information theory approach. Readers interested in a systematic treatment of algorithmic complexity, information theory and data compression may refer to the exhaustive monograph by Li and Vitányi (1997).

It is worth concluding this brief overview pointing out that the algorithmic complexity concept is very rich and links to deep pieces of mathematics and logic as Gödel's incompleteness theorem [Chaitin (1974)] and Turing's 1936 theorem of uncomputability [Chaitin (1982, 1990)]. As a result the true value of the algorithmic complexity of a N-sequence is uncomputable. This problem is hidden in the very definition of algorithmic complexity (8.17), as illustrated by the famous Berry's paradox: "*Let N be the smallest positive integer that cannot be defined in fewer than twenty English words*" which *de facto* defines N by using 17 English words only! Contradictory statements similar to Berry's paradox stand at the basis of the proof uncomputability of the algorithmic complexity by Chaitin. Although theoretically uncomputable, in practice, a fair upper bound to the true (uncomputable) algorithmic complexity of a sequence can be estimated in terms of the length of a compressed version of it produced by the powerful Ziv and Lempel (1977, 1978) compression algorithms (Box B.16), on which commonly employed digital compression tools are based.

Box B.16: Ziv-Lempel compression algorithm

A way to circumvent the problem of the uncomputability of the algorithmic complexity of a sequence is to relax the requirement of finding the shortest description, and to content us with a "reasonably" short one. Probably the best known and elegant encoding procedure, adapt to any kind of alpha-numeric sequence, is due to Ziv and Lempel (1977, 1978), as sketched in the following.

Consider a string $s(1)s(2)\dots s(L)$ of L characters with $L \gg 1$ and unknown statistics. To illustrate how the encoding of such a sequence can be implemented we can proceed as follows. Assume to have already encoded it up to $s(m)$ with $1 < m < L$, how to proceed with the encoding of $s(m+1)\dots s(L)$. The best way to provide a concise description is to search for the longest sub-string (i.e. consecutive sequence of symbols) in $s(1)\dots s(m)$ matching a sub-string starting at $s(m+1)$. Let k be the length of such sub-sequence for some $j < m-k+1$, we thus have $s(j)s(j+1)\dots s(j+k-1) = s(m+1)s(m+2)\dots s(m+k)$ and we can code the string $s(m+1)s(m+2)\dots s(m+k)$ with a pointer to the previous one, i.e. the pair $(m-j, k)$ which identifies the distance between the starting point of the previous strings and its length. In the absence of matching the character is not encoded, so that a typical coded string would read

input sequence: ABRACADABRA *output sequence*: ABR(3,1)C(2,1)D(7,4)

In such a way, the original sequence of length L is converted into a new sequence of length L_{ZL}, and the Ziv-Lempel algorithmic complexity of the sequence is defined as

$$\mathcal{K}_{ZL} = \lim_{L\to\infty} \frac{L_{ZL}}{L} \, .$$

Intuitively, low (resp. high) entropy sources will emit sequences with many (resp. few) repetitions of long sub-sequences producing low (resp. high) values for $\mathcal{K}_{ZL}$. Once the sequence has been compressed, it can be readily decompressed (decoded) just by replacing sub-string occurrences following the pointer *(position,length)*.

A better understanding of the link between $\mathcal{K}_{ZL}$ and the Shannon entropy can be obtained thanks to the Shannon-McMillan theorem (Sec. 8.2.3). If we encoded the sequence up to $s(m)$, as the probability of typical sequences of length n is $p \approx \exp(-nh_{Sh})$ (where h_{Sh} is the Shannon entropy of the source that emitted the string of characters) we can estimate to be able to encode a string starting in $s(m+1)$ of typical length $n = \log_2(m)/h_{Sh}$. Thus the Ziv and Lempel algorithm, on average, encodes the $n = \log_2(m)/h_{Sh}$ characters of the string using the pair $(m-j, n)$ using $\log_2(m-j) \approx \log_2 m$ characters[14] plus $\log_2 n = \log_2(\log_2 m/h_{Sh})$ characters needed to code the string length, so that

$$\mathcal{K}_{ZL} \approx \frac{\log_2 m + \log_2(\log_2 m/h_{Sh})}{\log_2 m/h_{Sh}} = h_{Sh} + \mathcal{O}\left(\frac{\log_2(\log_2 m)}{\log_2 m}\right),$$

which is the analogous of Eq. (8.19) and conveys two important messages. First, in the limit of infinitely long sequences $\mathcal{K}_{ZL} = h_{Sh}$, providing another method to estimate the entropy, see e.g. Puglisi *et al.* (2003). Second, the convergence to h_{Sh} is very slow, e.g. for $m = 2^{20}$ we have a correction order $\log_2(\log_2 m)/\log_2 m \approx 0.15$, independently of the value of h_{Sh}.

Although very efficient, the above described algorithm presents some difficulties of implementation and can be very slow. To overcome such difficulties Ziv and Lempel (1978) proposed another version of the algorithm. In a nutshell the idea is to break a sequence into words $w_1, w_2 \ldots$ such that $w_1 = s(1)$ and w_{k+1} is the shortest new word immediately following w_k, e.g. 110101001111010 ... is broken in (1)(10)(101)(0)(01)(11)(1010) .... Clearly in this way each word w_k is an extension of some previous word w_j $(j < k)$ plus a new symbol s' and can be coded by using a pointer to the previous word j plus the new symbol, i.e. by the pair (j, s'). This version of the algorithm is typically faster but presents similar problems of convergence to the Shannon entropy [Schürmann and Grassberger (1996)].

8.4 Entropy and complexity in chaotic systems

We now exploit the technical and conceptual framework of information theory to characterize chaotic dynamical systems, as heuristically anticipated in Sec. 8.1.

8.4.1 *Partitions and symbolic dynamics*

Most of the introduced tools are based on symbolic sequences, we have thus to understand how chaotic trajectories, living in the world of real numbers, can be properly encoded into (discrete) symbolic sequences. As for the Bernoulli map (Fig. 8.1), the encoding is based on the introduction of a partition of phase space Ω, but not all partitions are good, and we need to choose the appropriate one.

From the outset, notice that it is not important whether the system under consideration is time-discrete or continuous. In the latter case, a time discretization

[14] For m sufficiently large it will be rather probable to find the same character in a not too far past, so that $m - j \approx m$.

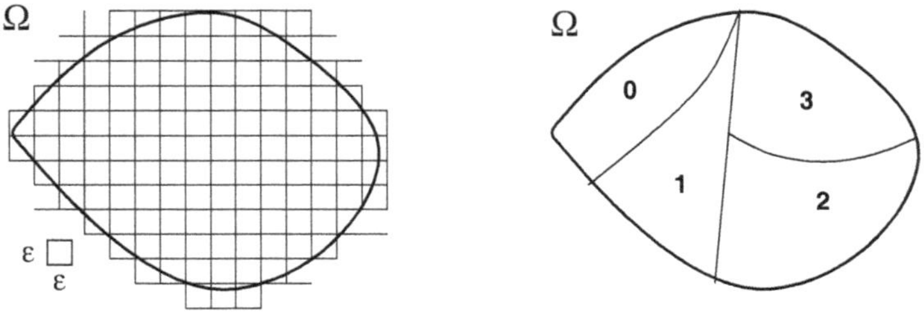

Fig. 8.6 Generic partitions with same-size elements (here square elements of side ε) (left) or with elements having arbitrary size and/or shape (right).

can be introduced either by means of a Poincaré map (Sec. 2.1.2) or by fixing a sampling time τ and recording a trajectory at times $t_j = j\tau$. Therefore, without loss of generality, in the following, we can limit the analysis to maps $\boldsymbol{x}(t+1) = \boldsymbol{F}(\boldsymbol{x}(t))$.

We consider partitions $A = \{A_0, \dots, A_{M-1}\}$ of Ω made of disjoint elements, $A_j \cap A_k = \emptyset$ if $j \neq k$, such that $\cup_{k=0}^{M-1} A_k = \Omega$. The set $\mathcal{A} = \{0, 1, \dots, M-1\}$ of $M < \infty$ symbols constitutes the alphabet induced by the partition. Then any trajectory $\boldsymbol{X} = \{\boldsymbol{x}(0)\boldsymbol{x}(1)\dots\boldsymbol{x}(n),\dots\}$ can be encoded in the symbolic sequence $\mathcal{S} = \{s(1)s(2)\dots s(n)\dots\}$ with $s(j) = k$ if $\boldsymbol{x}(j) \in A_k$.

In principle, the number, size and shape of the partition elements can be chosen arbitrarily (Fig. 8.6), provided the encoding does not lose relevant information on the original trajectory. In particular, given the knowledge of the symbolic sequence, we would like to reconstruct the trajectory itself. This is possible when the infinite symbolic sequence $\mathcal{S}$ unambiguously identifies a single trajectory, in this case we speak about a *generating partition*.

To better understand the meaning of a generating partition, it is useful to introduce the notion of *dynamical refinement*. Given two partitions $A = \{A_0, \dots, A_{M-1}\}$ and $B = \{B_0, \dots, B_{M'-1}\}$ with $M' > M$, we say that B is a *refinement* of A if each element of A is a union of elements of B. As shown in Fig. 8.7 for the case of the Bernoulli and tent map, the partition can be suitably chosen in such a way that the first N symbols of $\mathcal{S}$ identify the subset where the initial condition $\boldsymbol{x}(0)$ of the original trajectory $\boldsymbol{X}$ is contained, this is indeed obtained by the intersection:

$$A_{s_0} \cap \boldsymbol{F}^{-1}(A_{s(1)}) \cap \dots \cap \boldsymbol{F}^{-(N-1)}(A_{s(N-1)})\,.$$

It should be noticed that the above subset becomes smaller and smaller as N increases, making a refinement of the original partition that allows for a better and better determination of the initial condition. For instance, from the first two symbols of a trajectory of the Bernoulli or tent map 01, we can say that $x(0) \in [1/4 : 1/2]$ for both maps; knowing the first three 011, we recognize that $x(0) \in [3/8 : 1/2]$ and $x(0) \in [1/4 : 3/8]$ for the Bernoulli and tent map, respectively (see Fig. 8.7). As time proceeds, the successive divisions in sub-intervals shown in Fig. 8.7 constitute a refinement of the previous step. With reference to the figure as representative of a generic binary partition of a set, if we call $A^{(0)} = \{A_0^{(0)}, A_1^{(0)}\}$ the original partition,

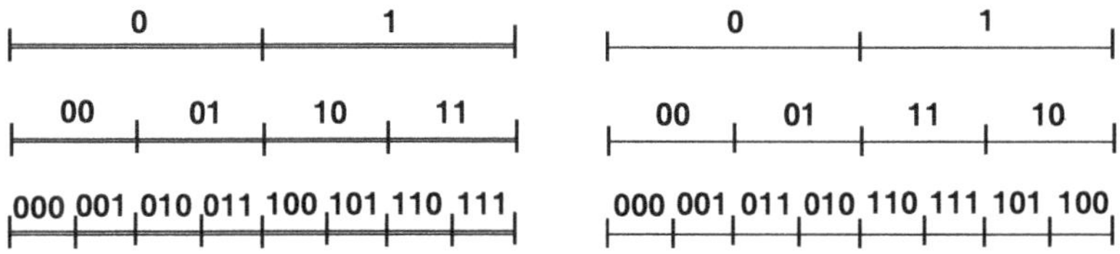

Fig. 8.7 From top to bottom, refinement of the partition $\{[0:1/2],[1/2:1]\}$ induced by the Bernoulli (left) and tent (right) map, only the first two refinements are shown.

in one step the dynamics generates the refinement $A^{(1)} = \{A^{(1)}_{00}, A^{(1)}_{01}, A^{(1)}_{10}, A^{(1)}_{11}\}$ where $A^{(1)}_{ij} = A^{(0)}_i \cap \boldsymbol{F}^{-1}(A^{(0)}_j)$. So the first refinement is indicated by two symbols, and the n-th one by $n+1$ symbols. The successive refinements of a partition A induced by the dynamics $\boldsymbol{F}$ are indicated by

$$A^{(n)} = \bigvee_{k=0}^{n} \boldsymbol{F}^{-k} A = A \vee \boldsymbol{F}^{-1} A \vee \ldots \vee \boldsymbol{F}^{-n} A \tag{8.20}$$

where $\boldsymbol{F}^{-k}A = \{\boldsymbol{F}^{-k}A_0, \ldots, \boldsymbol{F}^{-k}A_{M-1}\}$ and $A \vee B$ denotes the join of two partitions, i.e. $A \vee B = \{A_i \cap B_j$ for all $i = 0, \ldots, M-1$ and $j = 0, \ldots, M'-1\}$. If a partition G, under the effect of the dynamics, indefinitely refines itself according to Eq. (8.20) in such a way that the partition

$$\bigvee_{k=0}^{\infty} \boldsymbol{F}^{-k} G$$

is constituted by points, then an infinite symbolic string unequivocally identifies the initial condition of the original trajectory and the partition is said to be *generating*. As any refinement of a generating partition is also generating, there are an infinite number of generating partitions, the optimal one being constituted by the minimal number of elements, or generating a simpler dynamics (see Ex. 8.3).

Thanks to the link of the Bernoulli shift and tent map to the binary decomposition of numbers (see Sec. 3.1) it is readily seen that the partition $G = \{[0:1/2],[1/2:1]\}$ (Fig. 8.7) is a generating partition. However, for generic dynamical systems, it is not easy to find a generating partition. This task is particularly difficult in the (generic) case of non-hyperbolic systems as the Hénon map, although good candidates have been proposed [Grassberger and Kantz (1985); Giovannini and Politi (1992)].

Typically, the generating partition is not known, and a natural choice amounts to consider partitions in hypercubes of side ε (Fig. 8.6 left). When $\varepsilon \ll 1$, the partition is expected to be a good approximation of the generating one. We call these ε-partitions and indicate them with A_ε. As a matter of fact, a generating partition is usually recovered by the limit $\lim_{\varepsilon\to 0} A_\varepsilon$ (see Exs. 8.4, 8.6 and 8.7).

When a generating partition is known, the resulting symbolic sequences faithfully encode the system trajectories, and we can thus focus on the *Symbolic Dynamics* in order to extract information on the system [Alekseev and Yakobson (1981)].

One should be however aware that the symbolic dynamics resulting from a dynamical system is always due to the combined effect of the evolution rule and the chosen partition. For example, the dynamics of a map can produce rather simple sequences with Markov partitions (Sec. 4.5), in these cases we can achieve a complete characterization of the system in terms of the transition matrix, though the characterization is faithful only if the partition, besides being Markov, is generating [Bollt *et al.* (2001)] (see Exs. 8.3 and 8.5).

We conclude mentioning that symbolic dynamics can be also interpreted in the framework of language theory, allowing for the use of powerful methods to characterize the dynamical complexity of the system (see, e.g., Badii and Politi (1997)).

8.4.2 *Kolmogorov-Sinai entropy*

Consider the symbolic dynamics resulting from a partition A of the phase space Ω of a discrete time ergodic dynamical systems $\boldsymbol{x}(t+1) = \boldsymbol{F}(\boldsymbol{x}(t))$ with invariant measure μ^{inv}. We can associate a probability $P(A_k) = \mu^{inv}(A_k)$ to each element A_k of the partition. Taking the $(N-1)$-refinement $A^{(N-1)} = \vee_{k=0}^{N-1} \boldsymbol{F}^{-1} A$, $P(A_k^{(N-1)}) = \mu^{inv}(A_k^{(N-1)})$ defines the probability of N-words $P(\mathcal{W}_N(A))$ of the symbolic dynamics induced by A, from which we have the N-block entropies

$$H_N(A) = H(\vee_{k=0}^{N-1} A) = - \sum_{\{\mathcal{W}_N(A)\}} P(\mathcal{W}_N(A)) \ln P(\mathcal{W}_N(A))$$

and the difference entropies

$$h_N(A) = H_N(A) - H_{N-1}(A) .$$

The Shannon entropy characterizing the system with respect to the partition A,

$$h(A) = \lim_{N\to\infty} \frac{H(\vee_{k=0}^{N-1} A)}{N} = \lim_{N\to\infty} \frac{H_N(A)}{N} = \lim_{N\to\infty} h_N(A) ,$$

exists and depends on both the partition A and the invariant measure [Billingsley (1965); Petersen (1990)]. It quantifies the average uncertainty per time step on the partition element visited by the trajectories of the system. As the purpose is to characterize the source and not a specific partition A, it is desirable to eliminate the dependence of the entropy on A, this can be done by considering the supremum over all possible partitions:

$$h_{KS} = \sup_A \{h(A)\} , \tag{8.21}$$

which defines the *Kolmogorov-Sinai* (KS) entropy [Kolmogorov (1958); Sinai (1959)] (see also Billingsley, 1965; Eckmann and Ruelle, 1985; Petersen, 1990) of the dynamical system under consideration, that only depends on the invariant measure, hence the other name *metric entropy.* The supremum in the definition (8.21) is necessary because misplaced partitions can eliminate uncertainty even if the system is chaotic (Ex. 8.5). Furthermore, the supremum property makes the quantity invariant with respect to isomorphisms between dynamical systems. Remarkably, if the partition G is generating the supremum is automatically attained and $h(G) = h_{KS}$ [Kolmogorov (1958); Sinai (1959)]. Actually for invertible maps Krieger (1970) theorem ensures that a generating partition with $e^{h_{KS}} < k \leq e^{h_{KS}} + 1$ elements always exists, although the theorem does not specify how to build it. When the generating partition is not known, due to the impossibility to practically compute the supremum (8.21), KS-entropy can be defined as

$$h_{KS} = \lim_{\varepsilon \to 0} h(A_\varepsilon) \tag{8.22}$$

where A_ε is an ε-partition. It is expected that $h(A_\varepsilon)$ becomes independent of ε when the partition is so fine ($\varepsilon \ll 1$) to be contained in a generating one (see Ex. 8.7).

For time continuous systems, we introduce a time discretization in terms either of a fixed time lag τ or by means of a Poincaré map, which defines an average return time $\langle\tau\rangle$. Then $h_{KS} = \sup_A\{h(A)\}/\tau$ or $h_{KS} = \sup_A\{h(A)\}/\langle\tau\rangle$, respectively. Note that, at a theoretical level, the rate $h(A)/\tau$ does not depend on τ [Billingsley (1965); Eckmann and Ruelle (1985)], however the optimal value of τ may be important in practice (Chap. 10).

We can define the notion of algorithmic complexity $\kappa(\boldsymbol{X})$ of a trajectory $\boldsymbol{X}(t)$ of a dynamical system. Analogously to the KS-entropy, this requires to introduce a finite covering $\mathfrak{C}$[15] of the phase space. Then the algorithmic complexity per symbol $\kappa_{\mathfrak{C}}(\boldsymbol{X})$ has to be computed for the resulting symbolic sequences on each $\mathfrak{C}$. Finally $\kappa(\boldsymbol{X})$ corresponds to the supremum over the coverings [Alekseev and Yakobson (1981)]. Then it can be shown — Brudno (1983) and White (1993) theorems — that for almost all (with respect to the natural measure) initial conditions

$$\kappa(\boldsymbol{X}) = \frac{h_{KS}}{\ln 2},$$

which is equivalent to Eq. (8.19). Therefore, KS-entropy quantifies not only the richness of the system dynamics but also the difficulty of describing (almost) everyone of the resulting symbolic sequences. Some of these aspects can be illustrated with the Bernoulli map, discussed in Sec. 8.1. In particular, as the symbolic dynamics resulting from the partition of the unit interval in two halves is nothing but the binary expansion of the initial condition, it is possible to show that $\mathcal{K}(\mathcal{W}_N) \simeq N$ for almost all trajectories [Ford (1983, 1986)]. Let us consider $x(t)$ with accuracy 2^{-k} and $x(0)$ with accuracy 2^{-l}, of course $l = t + k$. This means that, in order to obtain the k binary digits of the output solution of the shift map, we must use a program of length no less than $l = t + k$. Martin-Löf (1966) proved a remarkable theorem stating that, with respect to the Lebesgue measure, almost all the binary sequences representing a real number in $[0:1]$ have maximum complexity, i.e. $\mathcal{K}(\mathcal{W}_N) \simeq N$.

We stress that, analogously to information dimension and Lyapunov exponents, the Kolmogorov-Sinai entropy provides a characterization of typical trajectories, and does not take into account fluctuations, which can be accounted by introducing

[15] A covering is like a partition with cells that may have a non-zero intersection.

the Rényi (1960, 1970) entropies (Box B.17). Moreover, metric entropy, as the Lyapunov exponents (Sec. 5.3.2.1), is an invariant characteristic quantity of a dynamical system, meaning that isomorphisms leave the KS-entropy unchanged [Kolmogorov (1958); Sinai (1959); Billingsley (1965)].

We conclude examining the connection between the KS-entropy and LEs, which was anticipated in the discussion of Fig. 8.3. Lyapunov exponents measure the rate at which infinitesimal errors, corresponding to maximal observation resolution, grow with time. Assuming the same resolution ε for each degree of freedom of a d-dimensional system amounts to consider an ε-partition of the phase space with cubic cells of volume ε^d, so that the state of the system at $t = 0$ belongs to a region of volume $V_0 = \varepsilon^d$ around the initial condition $\boldsymbol{x}(0)$. Trajectories starting from V_0 and sampled at discrete times, $t_j = j\tau$ ($\tau = 1$ for maps), generate a symbolic dynamics over the ε-partition. What is the number of sequences $N(\varepsilon, t)$ originating from trajectories which start in V_0?

From information theory (Sec. 8.2.3) we expect:

$$h_T = \lim_{\varepsilon\to 0}\lim_{t\to\infty}\frac{1}{t}\ln N(\varepsilon) \quad \text{and} \quad h_{KS} = \lim_{\varepsilon\to 0}\lim_{t\to\infty}\frac{1}{t}\ln N_{\text{eff}}(\varepsilon)$$

to be the topological and KS-entropies,[16] $N_{\text{eff}}(\varepsilon)$ ($\leq N(\varepsilon)$) being the effective (in the measure sense) number of sequences, which should be proportional to the coarse-grained volume $V(\varepsilon, t)$ occupied by the trajectories at time t. From Equation (5.19), we expect $V(t) \sim V_0 \exp(t\sum_{i=1}^d \lambda_i)$, but this holds true only in the limit $\varepsilon \to 0$.[17] In this limit, $V(t) = V_0$ for a conservative system ($\sum_{i=1}^d \lambda_i = 0$) and $V(t) < V_0$ for a dissipative system ($\sum_{i=1}^d \lambda_i < 0$). On the contrary, for any finite ε, the effect of contracting directions, associated with negative LEs, is completely wiped out. Thus only expanding directions, associated with positive LEs, matter in estimating the coarse-grained volume that behaves as

$$V(\varepsilon, t) \sim V_0\, e^{(\sum_{\lambda_i>0}\lambda_i)\, t},$$

when V_0 is small enough. Since $N_{\text{eff}}(\varepsilon, t) \propto V(\varepsilon, t)/V_0$, one has

$$h_{KS} = \sum_{\lambda_i>0}\lambda_i\,. \tag{8.23}$$

The above equality does not hold in general, actually it can be proved only for systems with SRB measure (Box B.10), see e.g. Eckmann and Ruelle (1985). However, for generic systems it can be rigorously proved the the Pesin (1976) relation [Ruelle (1978a)]

$$h_{KS} \leq \sum_{\lambda_i>0}\lambda_i\,.$$

We note that only in low dimensional systems a direct numerical computation of h_{KS} is feasible. Therefore, the knowledge of the Lyapunov spectrum provides, through Pesin relation, the only estimate of h_{KS} for high dimensional systems.

[16]Note that the order of the limits, first $t \to \infty$ and then $\varepsilon \to 0$, cannot be exchanged, and that they are in the opposite order with respect to Eq. (5.17), which defines LEs.

[17]I.e. if the limit $\varepsilon \to 0$ is taken first than that $t \to \infty$

Box B.17: Rényi entropies

The Kolmogorov-Sinai entropy characterizes the rate of information generation for typical sequences. Analogously to the generalized LE (Sec. 5.3.3), it is possible to introduce a generalization of the KS-entropy to account for (finite-time) fluctuations of the entropy rate. This can be done in terms of the Rényi (1960, 1970) entropies which generalize Shannon entropy. However it is should be remarked that these quantities do not possess the (sub)-additivity property (8.11) and thus are not unique (Sec. 8.2.2).

In the context of dynamical systems, the generalized Rényi entropies [Paladin and Vulpiani (1987); Badii and Politi (1997)], $h^{(q)}$, can be introduced by observing that KS-entropy is nothing but the average of $-\ln P(\mathcal{W}_N)$ and thus, as done with the generalized dimensions $D(q)$ for multifractals (Sec. 5.2.3), we can look at the moments:

$$h^{(q)} = - \lim_{\varepsilon \to 0} \lim_{N \to \infty} \frac{1}{N(q-1)} \ln \left(\sum_{\{\mathcal{W}_N(A_\varepsilon)\}} P(\mathcal{W}_N(A_\varepsilon))^q \right) .$$

We do not repeat here all the considerations we did for generalized dimensions, but it is easy to derive that $h_{KS} = \lim_{q \to 1} h^{(q)} = h^{(1)}$ and that the topological entropy corresponds to $q = 0$, i.e. $h_T = h^{(0)}$; in addition from general results of probability theory, one can show that $h^{(q)}$ is monotonically decreasing with q. Essentially $h^{(q)}$ plays the same role of $D(q)$.

Finally, it will not come as a surprise that the generalized *Rényi entropies* can be related to the generalized Lyapunov exponents $L(q)$. Denoting with $n^\star$ the number of non-negative Lyapunov exponents (i.e. $\lambda_{n^\star} \geq 0$, $\lambda_{n^\star+1} < 0$), the Pesin relation (8.23) can be written as

$$h_{KS} = \sum_{i=1}^{n^\star} \lambda_i = \left. \frac{\mathrm{d}L_{n^\star}(q)}{\mathrm{d}q} \right|_{q=0}$$

where $\{L_i(q)\}_{i=1}^d$ generalize the Lyapunov spectrum $\{\lambda_i\}_{i=1}^d$ [Paladin and Vulpiani (1986, 1987)]. Moreover, under some restrictions [Paladin and Vaienti (1988)]:

$$h^{(q+1)} = \frac{L_{n^\star}(-q)}{-q} .$$

We conclude this Box noticing that the generalized dimensions, Lyapunov exponents and Rényi entropies can be combined in an elegant common framework: the *Thermodynamic Formalism* of chaotic systems. The interested reader may refer to the two monographs Ruelle (1978b); Beck and Schlögl (1997).

8.4.3 *Chaos, unpredictability and uncompressibility*

In summary, Pesin relation together with Brudno and White theorems show that unpredictability of chaotic dynamical systems, quantified by the Lyapunov exponents, has a counterpart in information theory. Deterministic chaos generates messages that cannot be coded in a concise way, due to the positiveness of the Kolmogorov-Sinai entropy, thus chaos can be interpreted as a source of information and chaotic trajectories are algorithmically complex. This connection is further illustrated by the following example inspired by Ford (1983, 1986).

Let us consider a one-dimensional chaotic map

$$x(t+1) = f(x(t))\,. \tag{8.24}$$

Suppose that we want to transmit a portion of one of its trajectories $X(T) = \{x(t),\ t = 1, 2, \ldots, T\}$ to a remote friend (say on Mars) with an error tolerance Δ. Among the possible strategies, we can use the following one [Boffetta *et al.* (2002)]:

(1) Transmit the rule (8.24), which requires a number of bits independent of the length T of the sequence.
(2) Transmit the initial condition $x(0)$ with a precision δ_0, this means using a finite number of bits independent of T.

Steps (1) and (2) allows our friend to evolve the initial condition and start reproducing the trajectory. However, in a short time, $\mathcal{O}(\ln(\Delta/\delta_0)/\lambda)$, her/his trajectory will differ from our by an amount larger than the acceptable tolerance Δ. We can overcome this trouble by adding two further steps in the transmission protocol.

(3) Besides the trajectory to be transmitted, we evolve another one to check whether the error exceeds Δ. At the first time τ_1 the error equals Δ, we transmit the new initial condition $x(\tau_1)$ with precision δ_0.
(4) Let the system evolve and repeat the procedure (2)-(3), i.e. each time the error acceptance tolerance is reached we transmit the new initial condition, $x(\tau_1 + \tau_2)$, $x(\tau_1 + \tau_2 + \tau_3)$ ..., with precision δ_0.

By following the steps (1)-(4) the fellow on Mars can reconstruct within a precision Δ the sequence $X(T)$ simply iterating on a computer the system (8.24) between 0 and $\tau_1 - 1$, τ_1 and $\tau_1 + \tau_2 - 1$, and so on.

Let us now compute the amount of bits necessary to implement the above procedure (1)-(4). For the sake of notation simplicity, we introduce the quantities

$$\gamma_i = \frac{1}{\tau_i} \ln\left(\frac{\Delta}{\delta_0}\right)$$

equivalent to the *effective* Lyapunov exponents (Sec. 5.3.3). The Lyapunov Exponent λ is given by

$$\lambda = \langle \gamma_i \rangle = \frac{\sum_i \tau_i \gamma_i}{\sum_i \tau_i} = \frac{1}{\overline{\tau}} \ln\left(\frac{\Delta}{\delta_0}\right) \qquad \text{with} \quad \overline{\tau} = \frac{1}{N}\sum_{i=1}^{N} \tau_i\,, \tag{8.25}$$

where $\overline{\tau}$ is the average time after which we have to transmit the new initial condition and $N = T/\overline{\tau}$ is the total number of such transmissions. Let us observe that since the τ_i's are not constant, λ can be obtained from γ_i's by performing the average (8.25). If T is large enough, the number of transmissions is $N = T/\overline{\tau} \simeq$

$\lambda T/\ln(\Delta/\delta_0)$. Each transmission requires $\ln_2(\Delta/\delta_0)$ bits to reduce the error from Δ to δ_0, hence the amount of bits used in the whole transmission is

$$\frac{T}{\overline{\tau}}\ln_2\left(\frac{\Delta}{\delta_0}\right) = \frac{\lambda}{\ln 2}T\,. \tag{8.26}$$

In other words the number of bits for unit time is proportional to λ.[18]

In more than one dimension, we have simply to replace λ with h_{KS} in (8.26). Intuitively, this point can be understood by repeating the above transmission procedure in each of the expanding directions.

8.5 Concluding remarks

In conclusions, the Kolmogorov-Sinai entropy of chaotic systems is strictly positive and finite, in particular $0 < h_{KS} \leq \sum_{\lambda_i>0}\lambda_i < \infty$, while for truly (non-deterministic) random processes with continuous valued random variables $h_{KS} = +\infty$ (see next Chapter). We thus have another definition of chaos as positiveness of the KS-entropy, i.e. chaotic systems, viewed as sources of information, generate algorithmically complex sequences, that cannot be compressed. Thanks to the Pesin relation, we know that this is equivalent to require that at least one Lyapunov exponent is positive and thus that the system is unpredictable. These different points of view with which we can approach the definition of chaos suggest the following chain of equivalences.

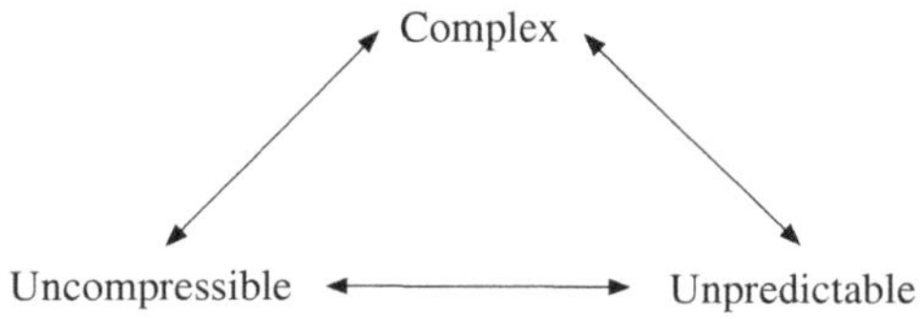

This view based on dynamical systems and information theory characterizes the complexity of a sequence considering each symbol relevant, but does not capture the structural level, for instance: on the one hand, a binary sequence obtained with a coin tossing is, from the information and algorithmic complexity points of view, complex since it cannot be compressed (i.e. it is unpredictable); on the other hand, the sequence is somehow trivial, i.e. with low "organizational" complexity. According to this example, we should define complex something "less random than a random object but more random than a regular one". Several attempts to introduce quantitative measures of this intuitive idea have been tried and it is difficult to say that a unifying point of view has been reached so far. For instance, the effective measure of complexity discussed in Box B.15 represents one possible approach towards such a definition, indeed C_{EMC} is minimal for memory-less (structureless) random processes, while it can be high for nontrivial zero-entropy sequences. We

[18] Of course, the costs of specifying the times τ_i should be added but this is negligible as we just need $\log_2 \tau_i$ bits each time.

just mention some of the most promising proposals as the logical depth [Bennet (1990)] and the sophistication [Koppel and Atlan (1991)], for throughout surveys on this subject we refer to Grassberger (1986, 1989); Badii and Politi (1997).

Some deterministic system gives rise to complex, seemingly random, dynamical behavior but without sensitivity to initial conditions ($\lambda_i \leq 0$). This happens, e.g., in quantum systems [Gutzwiller (1990)], cellular automata [Wolfram (1986)] and also some high-dimensional dynamical systems [Politi *et al.* (1993); Cecconi *et al.* (1998)] (Box B.29). In all these cases, although Pesin's relation cannot be invoked, at least in some limits (typically when the number of degrees of freedom goes to infinity), the system is effectively a source of information with a positive entropy. For this reason, there have been proposals to define "chaos" or "deterministic randomness" in terms of the positiveness of the KS-entropy which should be considered the "fundamental" quantity. This is, for instance, the perspective adopted in a quantum mechanical context by Gaspard (1994). In classical systems with a finite number of degrees of freedom, as consequence of Pesin's formula, the definition in terms of positiveness of KS-entropy coincides with that provided by Lyapunov exponents. The proposal of Gaspard (1994) is an interesting open possibility for quantum and classical systems in the limit of infinite number of degrees of freedom.

As a final remark, we notice that both KS-entropy and LEs involve both the limit of infinite time and infinite "precision"[19] meaning that these are asymptotic quantities which, thanks to ergodicity, globally characterize a dynamical system. From an information theory point of view this corresponds to the request of lossless recovery of information produced by a chaotic source.

8.6 Exercises

Exercise 8.1: Compute the topological and the Kolmogorov-Sinai entropy of the map defined in Ex.5.12 using as a partition the intervals of definition of the map;

Exercise 8.2:

Consider the one-dimensional map defined by the equation:

$$x(t+1) = \begin{cases} 2x(t) & x(t) \in [0:1/2) \\ x(t) - 1/2 & x(t) \in [1/2:1] \,. \end{cases}$$

and the partition $A_0 = [0:1/2]$, $A_1 = [1/2:1]$, which is a Markov and generating partition. Compute:

(1) the topological entropy;

(2) the KS entropy.

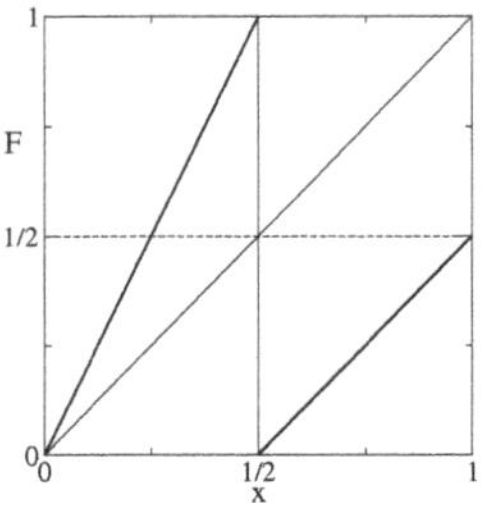

Hint: *Use the Markov property of the partition.*

[19]Though the order of the limits is inverted.

Exercise 8.3: Compute the topological and the Kolmogorov-Sinai entropy of the roof map defined in Ex.4.10 using the partitions: (1) $[0\!:\!1/2[$, $[1/2\!:\!1[$ and (2) $[0\!:\!x_1[$, $[x_1\!:\!1/2[$, $[1/2\!:\!x_2[$, $[x_2\!:\!1]$. Is the result the same? If yes or not explain why.
Hint: *Remember the definition of refinement of a partition and that of generating partition.*

Exercise 8.4: Consider the one-dimensional map

$$x(t+1) = \begin{cases} 8x(t) & 0 \le x < 1/8 \\ 1 - 8/7(x(t) - 1/8) & 1/8 \le x \le 1 \end{cases}$$

Compute the Shannon entropy of the symbolic sequences obtained using the family of partitions $A_i^{(k)} = \{x_i^{(k)} \le x < x_{i+1}^{(k)}\}$, with $x_{i+1}^{(k)} = x_i^{(k)} + 2^{-k}$, use $k = 1, 2, 3, 4, \ldots$. How does the entropy depend on k? Explain what does happen for $k \ge 3$. Compare the result with the Lyapunov exponent of the map and determine for which partitions the Shannon entropy equals the Kolmogorov-Sinai entropy of the map.
Hint: *Note that $A^{(k+1)}$ is a refinement of $A^{(k)}$.*

Exercise 8.5: Numerically compute the Shannon and topological entropy of the symbolic sequences obtained from the tent map using the partition $[0 : z[$ and $[z : 1]$ varying $z \in]0 : 1[$. Plot the results as a function of z. For which value of z does the Shannon entropy coincide the KS-entropy of the tent map? and why?

Exercise 8.6: Numerically compute the Shannon entropy for the logistic map at $r = 4$ using a ε-partition obtained dividing the unit interval in equal intervals of size $\varepsilon = 1/N$. Check the convergence of the entropy changing N, compare the results when N is odd or even, and explain the difference if any. Finally compare with the Lyapunov exponent.

Exercise 8.7: Numerically estimate the Kolmogorov-Sinai entropy h_{KS} of the Hénon map, for $b = 0.3$ and a varying in the range $[1.2, 1.4]$, as a partition divide the portion of x-axis spanned by the attractor in sets $A_i = \{(x, y) : x_i < x < x_{i+1}\}$, $i = 1, \ldots, N$. Choose, $x_1 = -1.34$, $x_{i+1} = x_i + \Delta$, with $\Delta = 2.68/N$. Observe above which values of N the entropy approach the correct value, i.e. that given by the Lyapunov exponent.

Chapter 9

Coarse-Grained Information and Large Scale Predictability

It is far better to foresee even without certainty than not to foresee at all.
Jules Henri Poincaré (1854–1912)

In the previous Chapter, we saw that the transmission rate (compression efficiency) for *lossless* transmission (compression) of messages is constrained by the Shannon entropy of the source emitting the messages. The Kolmogorov-Sinai entropy characterizes the rate of information production of chaotic sources and coincides with the sum of positive Lyapunov exponents, which determines the predictability of *infinitesimal* perturbations. If the initial state is known with accuracy δ ($\to 0$) and we ask for how long the state of the system can be predicted within a tolerance Δ, exponential amplification of the initial error implies that

$$T_p = \frac{1}{\lambda_1} \ln\left(\frac{\Delta}{\delta}\right) \sim \frac{1}{\lambda_1}, \tag{9.1}$$

i.e. the predictability time T_p is given by the inverse of maximal LE but for a weak logarithmic dependence on the ratio between threshold tolerance and initial error. Therefore, a precise link exists between predictability skill against infinitesimal uncertainties and possibility to compress/transmit "chaotic" messages.

In this Chapter we discuss what happens when we relax the constraints and are content with some (controlled) loss in the message and with finite[1] perturbations.

9.1 Finite-resolution versus infinite-resolution descriptions

Often, lossless transmission or compression of a message is impossible. This is the case of continuous random sources, where entropy is infinite as illustrated in the following. For simplicity, consider discrete time and focus on a source X emitting continuous valued random variables x characterized by a probability distribution

[1]Technically speaking the Lyapunov analysis deals with infinitesimal perturbations, i.e. both δ and Δ are infinitesimally small, in the sense of errors so small that can be approximated as evolving in the tangent space. Therefore, here and in the following finite should always be interpreted as outside the tangent space dynamics.

function $p(x)$. A natural candidate for the entropy of continuous sources is the naive generalization of the definition (8.8)

$$h(X) = -\int \mathrm{d}x\, p(x) \ln p(x)\,, \tag{9.2}$$

called *differential entropy*. However, despite $h(X)$ shares many of the properties of discrete entropy, several caveats make its use problematic. In particular, the differential entropy is not an intrinsic quantity and may be unbounded or negative.[2]

Another possibility is to discretize the source by introducing a set of discrete variables $x_k(\varepsilon) = k\varepsilon$, meaning that $x \in [k\varepsilon : (k+1)\varepsilon]$, having probability $p_k(\varepsilon) \approx p(x_k(\varepsilon))\varepsilon$. We can then use the mathematically well founded definition (8.8) obtaining

$$h(X_\varepsilon) = -\sum_k p_k(\varepsilon) \ln[p_k(\varepsilon)] = -\varepsilon \sum_k p(x_k(\varepsilon)) \ln p(x_k(\varepsilon)) - \ln \varepsilon\,.$$

However, problems arise when performing the limit $\varepsilon \to 0$: while the first term approximates the differential entropy $h(X)$, the second one diverges to $+\infty$. Therefore, lossy representation is unavoidable whenever we work with continuous sources.[3] Then, as it will be discussed in the next section, the problem turns into the request of providing a controlled *lossy* description of messages [Shannon (1948, 1959); Kolmogorov (1956)], see also Cover and Thomas (1991); Berger and Gibson (1998).

In practical situations lossy compression are useful to decrease the rate at which information needs to be transmitted, provided we can control the error and we do not need a faithful representation of the message. This can be illustrated with the following example. Consider a Bernoulli binary source which emits 1 and 0 with probabilities p and $1-p$, respectively. A typical message is a N-word which will, on average, be composed by Np ones and $N(1-p)$ zeros with an information content per symbol equal to $h_B(p) = -p \ln p - (1-p)\ln(1-p)$ (B stays for Bernoulli). Assume $p < 1/2$ for simplicity, and consider the case where a certain amount of error can be tolerated. For instance, 1's in the original message will be mis-coded/transmitted as 0's, with probability α. This means that typically a N-word contains $N(p-\alpha)$ ones, becoming equivalent to a Bernoulli binary source with $p \to p - \alpha$, which can be compressed more efficiently than the original one, as $h_B(p-\alpha) < h(p)$.

The fact that we may renounce to an infinitely accurate description of a message is often due, ironically, to our intrinsic limitations. This is the case of digital images with jpeg or other (lossy) compressed formats. For example, in Fig. 9.1 we show two pictures of the Roman forum with different levels of compression. Clearly, the image on the right is less accurate than that on the left, but we can still recognize

[2]For example, choosing $p(x) = \nu \exp(-\nu x)$ with $x \geq 0$, i.e. the exponential distribution, it is easily checked that $h(X) = -\ln \nu + 1$, becoming negative for $\nu > \mathrm{e}$. Moreover, the differential entropy is not invariant under a change of variable. For instance, consider the source Y linked to X by $y = ax$ with a constant, we have $h(Y) = h(X) - \ln |a|$.

[3]This problem is absent if we consider the mutual information between two continuous signals which remains well defined as discussed in the next section.

Fig. 9.1 (left) High resolution image (1424Kb) of the Roman Forum, seen from Capitoline Hill; (right) lossy compressed version (128Kb) of the same image.

several details. Therefore, unless we are interested in studying the effigies on the architrave (epistyle), the two photos are essentially equivalent. In this example, we exploited our limitation in detecting image details, on a first glance. To identify an image we just need a rough understanding of the main patterns.

Summarizing, in many practical cases, we do not need an arbitrarily high-resolution description of an object (message, image etc.) to grasp relevant information about it. Further, in some physical situations, considering a system at a too accurate observation scale may be not only unnecessary but also misleading as illustrated by the following example.

Consider the coupled map model [Boffetta *et al.* (1996)]

$$\begin{cases} \boldsymbol{x}(t+1) = \mathbb{R}[\theta]\,\boldsymbol{x}(t) + c\boldsymbol{f}(y(t)) \\ y(t+1) \;= g(y(t))\,, \end{cases} \tag{9.3}$$

where $\boldsymbol{x} \in \mathbb{R}^2$, $y \in \mathbb{R}$, $\mathbb{R}[\theta]$ is the rotation matrix of an arbitrary angle θ, $\boldsymbol{f}$ is a vector function and g is a chaotic map. For simplicity we consider a linear coupling $\boldsymbol{f}(y) = (y, y)$ and the logistic map at the Ulam point $g(y) = 4y(1-y)$.

For $c = 0$, Eq. (9.3) describes two independent systems: the predictable and regular $\boldsymbol{x}$-subsystem with $\lambda_x(c=0) = 0$ and the chaotic y-subsystem with $\lambda_y = \lambda_1 = \ln 2$. Switching on a small coupling, $0 < c \ll 1$, we have a single three-dimensional chaotic system with a positive "global" LE

$$\lambda_1 = \lambda_y + \mathcal{O}(c)\,.$$

A direct application of Eq. (9.1) would imply that the predictability time of the $\boldsymbol{x}$-subsystem is

$$T_p^{(x)} \sim T_p \sim \frac{1}{\lambda_y}\,,$$

contradicting our intuition as the predictability time for $\boldsymbol{x}$ would be basically independent of the coupling strength c. Notice that this paradoxical circumstance is not an artifact of the chosen example. For instance, the same happens considering the

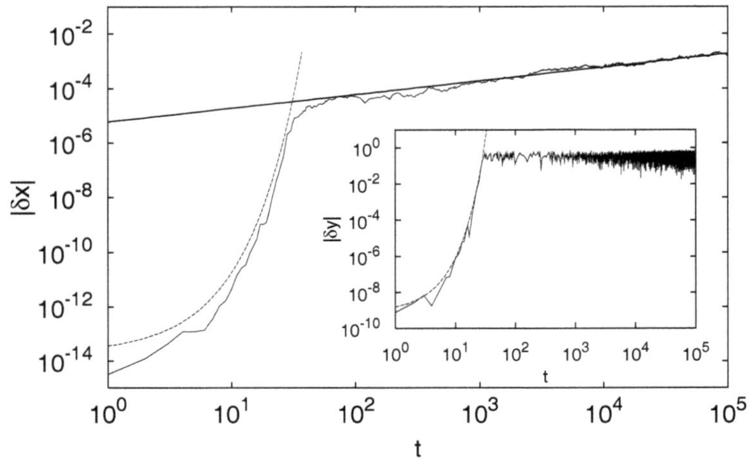

Fig. 9.2 Error growth $|\delta \boldsymbol{x}(t)|$ for the map (9.3) with parameters $\theta = 0.82099$ and $c = 10^{-5}$. Dashed line $|\delta \boldsymbol{x}(t)| \sim e^{\lambda_1 t}$ where $\lambda_1 = \ln 2$, solid line $|\delta \boldsymbol{x}(t)| \sim t^{1/2}$. Inset: evolution of $|\delta y(t)|$, dashed line as in the main figure. Note error saturation at the same time at which the diffusive regime establishes for the error on $\boldsymbol{x}$. The initial error only on the y variable is $\delta y = \delta_0 = 10^{-10}$.

gravitational three-body problem with one body (asteroid) of mass m much smaller than the other two (planets). If the gravitational feedback of the asteroid on the two planets is neglected (restricted problem), it results a chaotic asteroid with fully predictable planets. Whilst if the feedback is taken into account ($m > 0$ in the example) the system becomes the fully chaotic non-separable three-body problem (Sec. 11.1). Intuition correctly suggests that it should be possible to forecast planet evolutions for very long times if the asteroid has a negligible mass ($m \to 0$).

The paradox arises from the misuse of formula (9.1), which is valid only for the tangent-vector dynamics, i.e. with both δ and Δ infinitesimal. In other words, it stems from the application of the correct formula (Eq. (9.1)) to a wrong regime, because as soon as the errors become large, the full nonlinear error evolution has to be taken into account (Fig. 9.2). The evolution of $\delta \boldsymbol{x}$ is given by

$$\delta \boldsymbol{x}(t+1) = \mathbb{R}[\theta]\delta \boldsymbol{x}(t) + c\,\delta \boldsymbol{f}(y)\,, \tag{9.4}$$

where, with our choice, $\delta \boldsymbol{f} = (\delta y, \delta y)$. At the beginning, both $|\delta \boldsymbol{x}|$ and $|\delta y|$ grow exponentially. However, the available phase space for y is bounded leading to a saturation of the uncertainty $|\delta y| \sim \mathcal{O}(1)$ in a time $t^\star = \mathcal{O}(1/\lambda_1)$. Therefore, for $t > t^\star$, the two realizations of the y-subsystem are completely uncorrelated and their difference δy acts as noise in Eq. (9.4), which becomes a sort of discrete time Langevin equation driven by chaos, instead of noise. As a consequence, the growth of the uncertainty on $\boldsymbol{x}$-subsystem becomes diffusive with a diffusion coefficient proportional to c^2, i.e. $|\delta \boldsymbol{x}(t)| \sim c\,t^{1/2}$ implying [Boffetta *et al.* (1996)]

$$T_p^{(x)} \sim \left(\frac{\Delta}{c}\right)^2 , \tag{9.5}$$

which is much longer than the time expected on the basis of tangent-space error growth (now Δ is not constrained to be infinitesimal). The above example shows that, in some circumstances, the Lyapunov exponent is of little relevance for the predictability. This is expected to happen when different characteristic times are present (Sec. 9.4.2), as in atmospheric predictability (see Chap. 13), where additionally our knowledge on the current meteorological state is very inaccurate due to our inability to measure at each point the relevant variables (temperature, wind velocity, humidity etc.) and moreover, the models we use are both imperfect and at very low resolution [Kalnay (2002)].

The rest of the Chapter will introduce the proper tools to develop a finite-resolution description of dynamical processes from both the information theory and dynamical systems point of view.

9.2 ε-entropy in information theory: lossless versus lossy coding

This section focuses on the problem of an imperfect representation in the information-theory framework. We first briefly discuss how a communication channel (Cfr. Fig. 8.4) can be characterized and then examine lossy compression/transmission in terms of the *rate distortion theory* (RDT) originally introduced by Shannon (1948, 1959), see also Cover *et al.* (1989); Berger and Gibson (1998).

As the matter is rather technical, the reader mostly interested in dynamical systems may skip this section and go directly to the next section when RDT is studied in terms of the equivalent concept of ε-entropy, due to Kolmogorov (1956), in the dynamical-system context.

9.2.1 *Channel capacity*

Entropy also characterizes the communication channel. With reference to Fig. 8.4 we denote with S the source emitting the input sequences $s(1)s(2)\ldots s(k)\ldots$ which enter the channel (i.e. the transmitter) and with $\widehat{S}$ the source (represented by the receiver) generating the output messages $\hat{s}(1)\hat{s}(2)\ldots\hat{s}(k)\ldots$. The channel associates an output symbol $\hat{s}$ to each input s symbol. We thus have the entropies characterizing the input/output sources. $h(S) = \lim_{N\to\infty} H_N(\mathcal{W}_N)/N$ and $h(\widehat{S}) = \lim_{N\to\infty} H_N(\widehat{\mathcal{W}_N})/N$ (the subscript $_{Sh}$ has been removed for the sake of notation simplicity). From Eq. (8.11), for the channel we have

$$h(S;\widehat{S}) = h(S) + h(\widehat{S}|S) = h(\widehat{S}) + h(S|\widehat{S}) \,,$$

then, the conditional entropies can be obtained as

$$\begin{aligned} h(\widehat{S}|S) &= h(S;\widehat{S}) - h(S) \\ h(S|\widehat{S}) &= h(S;\widehat{S}) - h(\widehat{S}) \,, \end{aligned}$$

where $h(S)$ provides a measure of the uncertainty per symbol associated to the input sequence $s(1)s(2)\ldots$, $h(S|\widehat{S})$ quantifies the conditional uncertainty per symbol on the same sequence given that it entered the channel giving as an output the sequence $\hat{s}(1)\hat{s}(2)\ldots$. In other terms $h(S|\widehat{S})$ indicates how uncertain is the symbol s when we receive $\hat{s}$, often the term *equivocation* is used for this quantity. For *noiseless* channels there is no equivocation and $h(S|\widehat{S}) = 0$ while, in general, $h(S|\widehat{S}) > 0$ due to the presence of noise in the transmission channel.

In the presence of errors the input signal cannot be known with certainty from the knowledge of the output solely, and a correction protocol should be added. Although the correction protocol is out of the scope of this book, it is however interesting to wonder about the rate the channel can transmit information in such a way that the message-recovery strategy can be implemented.

Shannon (1948) considered a *gedanken* experiment consisting in sending an error-correcting message parallel to the transmission of the input, and showed that the amount of information needed to transmit the original message without errors is precisely given by $h(S|\widehat{S})$. Therefore, for corrections to be possible, the channel has to transmit at a rate, i.e. with a capacity, equal to the mutual information between input and output sources

$$I(S;\widehat{S}) = h(S) - h(S|\widehat{S})\,.$$

If the noise is such that the input and output signals are completely uncorrelated $I(S;\widehat{S}) = 0$ no reliable transmission is possible. On the other extreme, if the channel is noiseless, $h(S|\widehat{S}) = 0$ and thus $I(S;\widehat{S}) = h(\widehat{S})$, and we can transmit at the same rate at which information is produced.

Specifically, as the communication apparatus should be suited for transmitting any kind of message, the channel capacity $\mathcal{C}$ is defined by taking the supremum over all possible input sources [Cover and Thomas (1991)]

$$\mathcal{C} = \sup_{S}\{I(S;\widehat{S})\}\,.$$

Messages can be sent through a channel with capacity $\mathcal{C}$ and recovered without errors only if the source entropy is smaller than the capacity of the channel, i.e. if information is produced at a rate less than the maximal rate sustained by the channel. When the source entropy becomes larger than the channel capacity unavoidable errors will be present in the received signal, and the question becomes to estimate the errors for a given capacity (i.e. available rate of information transmission), this naturally lead to the concept of rate distortion theory.

Before discussing RDT, it is worth remarking that the notion of channel capacity can be extended to continuous sources, indeed, despite the entropy Eq. (9.2) is an ill-defined quantity, the mutual information

$$I(X;\widehat{X}) = h(X) - h(X|\widehat{X}) = \int \mathrm{d}x\,\mathrm{d}\hat{x}\,p(x,\hat{x})\ln\left(\frac{p(x,\hat{x})}{p_x(x)p_{\hat{x}}(\hat{x})}\right),$$

remains well defined (see Kolmogorov (1956)) as verified by discretizing the integral ($p(x,\hat{x})$ is the joint probability density to observe x and $\hat{x}$, and $p_x(x) = \int \mathrm{d}\hat{x}\,p(x,\hat{x})$ while $p_{\hat{x}}(\hat{x}) = \int \mathrm{d}x\,p(x,\hat{x})$).

9.2.2 *Rate distortion theory*

Rate distortion theory was originally formulated by Shannon (1948) and can be stated in two equivalent ways.

Consider a (continuous or discrete[4]) random source X emitting messages $x(1), x(2), \ldots$ which are then codified into the messages $\hat{x}(1), \hat{x}(2), \ldots$ that can be seen as emitted by the output source $\widehat{X}$. Now assume that due to unrecoverable errors, the output message is not a faithful representation of the original one. The error can be measured in terms of a distortion/distance function, $d(x, \hat{x})$, depending on the context, e.g.

Squared error distortion	$d(x, \hat{x}) = (x - \hat{x})^2$;
Absolute error	$d(x, \hat{x}) = \lvert x - \hat{x} \rvert$;
Hamming distance	$d(x, \hat{x}) = 0$ if $\hat{x} = x$ and 1 otherwise;

where the last one is more appropriate in the case of discrete sources. For sequences $\mathcal{W}_N = x(1), x(2), \ldots, x(N)$, $\widehat{\mathcal{W}}_N = \hat{x}(1), \hat{x}(2), \ldots, \hat{x}(N)$ we define the distortion per symbol as

$$d(\mathcal{W}_N, \widehat{\mathcal{W}}_N) = \frac{1}{N} \sum_{i=1}^{N} d(x(i), \hat{x}(i)) \overset{N \to \infty}{=} \langle d(x, \hat{x}) \rangle = \iint \mathrm{d}x \, \mathrm{d}\hat{x} \, p(x, \hat{x}) \, d(x, \hat{x})$$

where ergodicity is assumed to hold in the last two equalities. Message transmission may fall into one of the the following two cases:

(1) We may want to fix the rate R for transmitting a message from a given source and being interested in the maximal average error/distortion $\langle d(x, \hat{x}) \rangle$ in the received message. This is, for example, a relevant situation when we have a source with entropy larger than the channel capacity $\mathcal{C}$ and so we want to fix the transmission rate to a value $R \leq \mathcal{C}$ which can be sustained by the channel.

(2) We may decide to accept an average error below a given threshold $\langle d(x, \hat{x}) \rangle \leq \varepsilon$ and being interested in the minimal rate R at which the messages can be transmitted ensuring that constraint. This is nothing but an optimal coding request: given the error tolerance ε find the best compression, i.e. the way to encode messages with the lowest entropy rate per symbol R. Said differently, given the accepted distortion, what is the channel with minimal capacity to convey the information.

We shall briefly discuss only the second approach which is better suited to applications of RDT to dynamical systems. The interested reader can find exhaustive discussions about the whole conceptual and technical apparatus of RDT in, e.g., Cover and Thomas (1991); Berger and Gibson (1998).

In the most general formulation, the problem of computing the rate $R(\varepsilon)$ associated to an error tolerance $\langle d(x, \hat{x}) \rangle \leq \varepsilon$ — fidelity criterion in Shannon's words —

[4]In the following we shall use the notation for continuous variables, where obvious modifications (such as integrals into sums, probability densities into probabilities, etc.) are left to the reader.

can be cast as a constrained optimization problem, as sketched in the following. Denote with x and $\hat{x}$ the random variable associated to the source X and its representation $\widehat{X}$, we know the probability density $p_x(x)$ of the random variables emitted by X and we want to find the representation (coding) of x, i.e. the conditional density $p(\hat{x}|x)$. Equivalently we can use either $p(x|\hat{x})$ or the joint distribution $p(x,\hat{x})$, which minimizes the transmission rate that is, from the previous subsection, the mutual information $I(X;\widehat{X})$. This is mathematically expressed by

$$\begin{aligned} R(\varepsilon) &= \min_{p(x,\hat{x}):\langle d(x,\hat{x})\rangle \le \varepsilon} I(X;\widehat{X}) = \min_{p(x,\hat{x}):\langle d(x,\hat{x})\rangle \le \varepsilon} \{h(X) - h(X|\widehat{X})\} \\ &= \min_{p(x,\hat{x}):\langle d(x,\hat{x})\rangle \le \varepsilon} \left\{ \iint \mathrm{d}x\, \mathrm{d}\hat{x}\, p(x,\hat{x}) \ln \left(\frac{p(x,\hat{x})}{p_x(x) p_{\hat{x}}(\hat{x})} \right) \right\} , \end{aligned} \tag{9.6}$$

where $p(x,\hat{x}) = p_x(x)p(\hat{x}|x) = p_{\hat{x}}(\hat{x})p(x|\hat{x})$ and $\langle d(x,\hat{x})\rangle = \iint \mathrm{d}x\mathrm{d}\hat{x}\, p(x,\hat{x})\, d(x,\hat{x})$. Additional constraints to Eq. (9.6) are imposed by the requests $p(x,\hat{x}) \ge 0$ and $\iint \mathrm{d}x\mathrm{d}\hat{x}\, p(x,\hat{x}) = 1$.

The definition (9.6) applies to both continuous and (with the proper modification) discrete sources. However, as noticed by Kolmogorov (1956), it is particularly useful when considering continuous sources as it allows to overcome the problem of the inconsistency of the differential entropy (9.2) (see also Gelfand *et al.* (1958); Kolmogorov and Tikhomirov (1959)). For this reason he proposed the term ε-entropy for the entropy of signals emitted by a source that are observed with ε-accuracy. While in this section we shall continue to use the information theory notation, $R(\varepsilon)$, in the next section we introduce the symbol $h(\varepsilon)$ to stress the interpretation put forward by Kolmogorov, which is better suited to a dynamical system context.

The minimization problem (9.6) is, in general, very difficult, so that we shall discuss only a lower bound to $R(\varepsilon)$, due to Shannon (1959). Shannon's idea is illustrated by the following chain of relations:

$$\begin{aligned} R(\varepsilon) &= \min_{p(x,\hat{x}):\langle d(x,\hat{x})\rangle \le \varepsilon} \{h(X) - h(X|\widehat{X})\} = h(X) - \max_{\langle d(x,\hat{x})\rangle \le \varepsilon} h(X|\widehat{X}) \\ &= h(X) - \max_{\langle d(x,\hat{x})\rangle \le \varepsilon} h(X - \widehat{X}|\widehat{X}) \ge h(X) - \max_{\langle d(x,\hat{x})\rangle \le \varepsilon} h(X - \widehat{X}) , \end{aligned} \tag{9.7}$$

where the second equality is trivial, the third comes from the fact $h(X - \widehat{X}|\widehat{X}) = h(X|\widehat{X})$ (here $X - \widehat{X}$ represents a suitable difference between the messages originating from the sources X and $\widehat{X}$). The last step is a consequence of the fact that the conditional entropy is always lower than the unconstrained one, although we stress that assuming the error independent of the output is generally wrong.

The lower bound (9.7) to can be used to derive $R(\varepsilon)$ in some special cases. In the following we discuss two examples to illustrate the basic properties of the ε-entropy for discrete and continuous sources, the derivation details, summarized in Box B.18, can be found in Cover and Thomas (1991).

We start from a memory-less binary source X emitting a Bernoulli signal $x = 1, 0$ with probability p and $1 - p$, in which we tolerate errors $\le \varepsilon$ as measured by the

Hamming distance. In this case one can prove that the ε-entropy $R(\varepsilon)$ is given by

$$R(\varepsilon) = \begin{cases} h_B(p) - h_B(\varepsilon) & 0 \le \varepsilon \le \min\{p, 1-p\} \\ 0 & \varepsilon > \min\{p, 1-p\}\,, \end{cases} \tag{9.8}$$

with $h_B(x) = -x \ln x - (1-x)\ln(1-x)$.

Another instructive example is the case of a (continuous) memory-less Gaussian source X emitting random variables x having zero mean and variance σ^2 with the square distance function $d(x, \hat{x}) = (x - \hat{x})^2$. As we cannot transmit the exact value, because it would require an infinite amount of information and thus infinite rate, we are forced to accept a tolerance ε allowing us to decrease the transmission rate to [Kolmogorov (1956); Shannon (1959)]

$$R(\varepsilon) = \begin{cases} \frac{1}{2}\ln\left(\frac{\sigma^2}{\varepsilon}\right) & \varepsilon \le \sigma^2 \\ 0 & \varepsilon > \sigma^2\,. \end{cases} \tag{9.9}$$

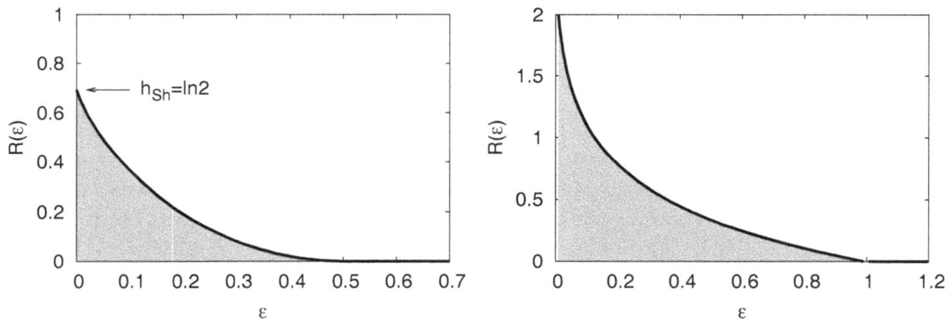

Fig. 9.3 $R(\varepsilon)$ vs ε for the Bernoulli source with $p = 1/2$ (a) and the Gaussian source with $\sigma = 1$ (b). The shaded area is the unreachable region, meaning that fixing e.g. a tolerance ε we cannot transmit with a rate in the gray region. In the discrete case the limit $R(\varepsilon) \to 0$ recovers the Shannon entropy of the source here $h_{Sh} = \ln 2$, while in the continuous case $R(\varepsilon) \to \infty$ for $\varepsilon \to 0$.

In Fig. 9.3 we show the behavior $R(\varepsilon)$ in these two cases. We can extract the following general properties:

- $R(\varepsilon) \ge 0$ for any $\varepsilon \ge 0$;
- $R(\varepsilon)$ is a non-increasing convex function of ε;
- $R(\varepsilon) < \infty$ for any finite ε, making it a well defined quantity also for continuous processes, so in contrast to the Shannon entropy it can be defined also for continuous stochastic processes;
- in the limit of lossless description, $\varepsilon \to 0$, $R(\varepsilon) \to h_{Sh}$, which is finite for discrete sources and infinite for continuous ones.

Next section will reexamine the same object from a slightly different point of view, specializing the discussion to dynamical systems and stochastic processes.

Box B.18: ε-entropy for the Bernoulli and Gaussian source

We sketch the steps necessary to derive results (9.8) and (9.9) following [Cover and Thomas (1991)] with some slight changes.

Bernoulli source

Be X a binary source emitting $x = 1, 0$ with probability p and $1 - p$, respectively. For instance, take $p < 1/2$ and assume that, while coding or transmitting the emitted messages, errors are present. We want to determine the minimal rate R such that the average Hamming distortion is bounded by $\langle d(x, \hat{x})\rangle \leq \varepsilon$, meaning that we accept a probability of error $\text{Prob}(x \neq \hat{x}) \leq \varepsilon$. To simplify the notation, it is useful to introduce the modulo 2 addition denoted by $\oplus$, which is equivalent to the XOR binary operand, i.e. $x \oplus \hat{x} = 1$ if $x \neq \hat{x}$. From Eq. (9.7), we can easily find a lower bound to the mutual information, i.e. $I(X; \widehat{X}) = h(X) - h(X|\widehat{X}) = h_B(p) - h(X \oplus \widehat{X}|\widehat{X}) \geq h_B(p) - h(X \oplus \widehat{X}) \geq h_B(p) - h_B(\varepsilon)$ where $h_B(x) = -x \ln x - (1-x)\ln(1-x)$. The last step stems from the accepted probability of error. The above inequality translates into an inequality for the rate function

$$R(\varepsilon) \geq h_B(p) - h_B(\varepsilon) \tag{B.18.1}$$

which, of course, makes sense only for $0 \leq \varepsilon \leq p$. The idea is to find a coding from x to $\hat{x}$ such that this rate is actually achieved, i.e. we have to prescribe a conditional probability $p(x|\hat{x})$ or equivalently $p(\hat{x}|x)$ for which the rate (B.18.1) is achieved. An easy computation shows that choosing the transition probabilities as in Fig. B18.1, i.e. replacing p with $(p - \varepsilon)/(1 - 2\varepsilon)$, the bound (B.18.1) is actually reached. If $\varepsilon > p$ we can fix $\text{Prob}(\hat{x} = 0) = 1$ obtaining $R(\varepsilon) = 0$, meaning that messages can be transmitted at any rate with this tolerance (as the message will anyway be unrecoverable). If $p > 1/2$ we can repeat the same reasoning for $p \to (1 - p)$ ending with the result (9.8). Notice that the so obtained rate is lower than $h_B(p - \varepsilon)$, suggested by the naive coding discussed on Sect. 8.1.

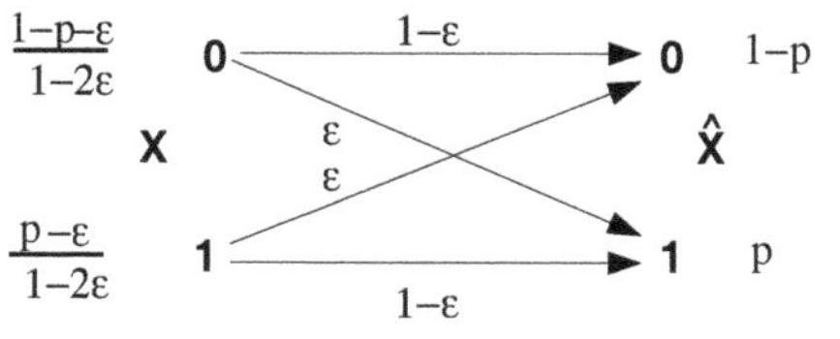

Fig. B18.1 Schematic representation of the probabilities involved in the coding scheme which realizes the lower bound for the Bernoulli source. [After Cover and Thomas (1991)]

Gaussian source

Be X a Gaussian source emitting random variables with zero mean and variance σ^2, i.e. $p_x(x) = \mathcal{G}(x, \sigma) = \exp[-x^2/(2\sigma^2)]/\sqrt{2\pi\sigma^2}$ for which an easy computation shows that the differential entropy (9.2) is equal to $h(X) = h(\mathcal{G}(x,\sigma)) = (1/2)\ln(2\pi e\sigma^2)$. Further let's assume that we can tolerate errors, measured by the square function, less than ε, i.e.$\langle(x-\hat{x})^2\rangle \leq \varepsilon$. Simple dimensional argument [Aurell *et al.* (1997)] suggests that

$$R(\varepsilon) = A\ln\left(\frac{\sigma}{\sqrt{\varepsilon}}\right) + B\,.$$

Indeed typical fluctuations of x will be of order σ and we need about $\ln(\sigma/\sqrt{\varepsilon})$ bits for coding them within an accuracy ε. However, this dimensional argument cannot determine the constants A and B. To obtain the correct result (9.9) we can proceed in a way very similar to the Bernoulli case. Consider the inequality $I(X;\widehat{X}) = h(X) - h(X|\widehat{X}) = h(\mathcal{G}(x,\sigma)) - h(X-\widehat{X}|\widehat{X}) \geq h(\mathcal{G}(x,\sigma)) - h(X-\widehat{X}) \geq h(\mathcal{G}(x,\sigma)) - h(\mathcal{G}(x,\sqrt{\varepsilon}))$, where the last one stems from the fact that if we fix the variance of the distribution $\langle(x-\hat{x})^2\rangle$ entropy is maximal for a Gaussian source, and then using that $\langle(x-\hat{x})^2\rangle \leq \varepsilon$ as required by the admitted error. Therefore, we can immediately derive

$$R(\varepsilon) \geq h(\mathcal{G}(x,\sigma)) - h(\mathcal{G}(x,\sqrt{\varepsilon})) = \frac{1}{2}\ln\left(\frac{\sigma^2}{\varepsilon}\right)\,.$$

Now, again, to prove Eq. (9.9) we simply need to find the appropriate coding from X to $\widehat{X}$ that makes the lower bound achievable. An easy computation shows that this is possible by choosing $p(x|\hat{x}) = \mathcal{G}(x-\hat{x},\sqrt{\varepsilon})$ and so $p_{\hat{x}}(\hat{x}) = \mathcal{G}(x,\sqrt{\sigma^2-\varepsilon})$ when $\varepsilon < \sigma^2$, while for $\varepsilon > \sigma^2$ we can choose $\text{Prob}(\hat{x}=0) = 1$, which gives $R = 0$.

9.3 ε-entropy in dynamical systems and stochastic processes

The Kolmogorov-Sinai entropy h_{KS}, Eq. (8.21) or equivalently Eq. (8.22), measures the amount of information per unit time necessary to record without ambiguity a generic trajectory of a chaotic system. Since the computation of h_{KS} involves the limit of arbitrary fine resolution and infinite times (8.22), in practice, for most systems it cannot be computed. However, as seen in the previous section, the ε-entropy, measuring the amount of information to reproduce a trajectory with ε-accuracy, is a measurable and valuable indicator, at the price of renouncing to arbitrary accuracy in monitoring the evolution of trajectories. This is the approach put forward by Kolmogorov (1956) see also [Kolmogorov and Tikhomirov (1959)].

Consider a continuous (in time) variable $\boldsymbol{x}(t) \in \mathbb{R}^d$, which represents the state of a d-dimensional system which can be either deterministic or stochastic.[5] Discretize the time by introducing an interval τ and consider, in complete analogy with the procedure of Sec. 8.4.1, a partition A_ε of the phase-space in cells with edges (diameter) $\leq \varepsilon$. The partition may be composed of unequal cells or, as typically done in

[5]In experimental studies, typically, the dimension d of the phase-space is not known. Moreover, usually only a scalar variable $u(t)$ can be measured. In such a case, for deterministic systems, a reconstruction of the original phase space can be done with the *embedding* technique which is discussed in the next Chapter.

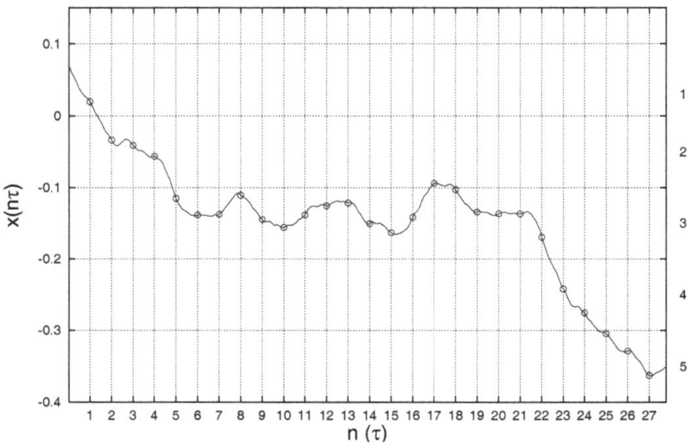

Fig. 9.4 Symbolic encoding of a one-dimensional signal obtained starting from an equal cell ε-partition (here $\varepsilon = 0.1$) and time discretization $\tau = 1$. In the considered example we have $\mathcal{W}_{27}(\varepsilon, \tau) = (1, 2, 2, 2, 3, 3, 3, 3, 3, 3, 3, 3, 3, 3, 3, 3, 2, 3, 3, 3, 3, 3, 4, 4, 5, 5, 5)$.

practical computations, of identical cells, e.g. hypercubes of side ε (see Fig. 9.4 for an illustration for a one-dimensional trajectory). The partition induces a symbolic dynamics (Sec. 8.4.1), for which a portion of trajectory, i.e. the vector

$$\boldsymbol{X}^{(N)}(t) \equiv \{\boldsymbol{x}(t), \boldsymbol{x}(t+\tau), \ldots \boldsymbol{x}(t+(N-1)\tau)\} \in \mathbb{R}^{Nd}, \tag{9.10}$$

can be coded into a word of length N, from a finite alphabet:

$$\boldsymbol{X}^{(N)}(t) \longrightarrow \mathcal{W}_N(\varepsilon, t) = (s(\varepsilon, t), s(\varepsilon, t+\tau), \ldots, s(\varepsilon, t+(N-1)\tau)),$$

where $s(\varepsilon, t+j\tau)$ labels the cell in $\mathbb{R}^d$ containing $\boldsymbol{x}(t+j\tau)$. The alphabet is finite for bounded motions that can be covered by a finite number of cells.

Assuming ergodicity, we can estimate he probabilities $P(\mathcal{W}_N(\varepsilon))$ of the admissible words $\{\mathcal{W}_N(\varepsilon)\}$ from a long time record of $\boldsymbol{X}^{(N)}(t)$. Following Shannon (1948), we can thus introduce the (ε, τ)-entropy per unit time,[6] $h(A_\varepsilon, \tau)$ associated to the partition A_ε

$$h_N(A_\varepsilon, \tau) = \frac{1}{\tau}[H_N(A_\varepsilon, \tau) - H_{N-1}(A_\varepsilon, \tau)] \tag{9.11}$$

$$h(A_\varepsilon, \tau) = \lim_{N\to\infty} h_N(A_\varepsilon, \tau) = \frac{1}{\tau} \lim_{N\to\infty} \frac{H_N(A_\varepsilon, \tau)}{N}, \tag{9.12}$$

where H_N is the N-block entropy (8.14). Similarly to the KS-entropy, we would like to obtain a partition-independent quantity, and this can be realized by defining the (ε, τ)-entropy as the infimum over all partitions with cells of diameter smaller

[6]The dependence on τ is retained as in some stochastic systems the ε-entropy may also depend on it [Gaspard and Wang (1993)]. Moreover, τ may be important in practical implementations.

than ε [Gaspard and Wang (1993)]:[7]

$$h(\varepsilon, \tau) = \inf_{A:\mathrm{diam}(A)\leq\varepsilon} \{h(A_\varepsilon, \tau)\} \,. \tag{9.13}$$

It should be remarked that, for $\varepsilon \neq 0$, $h(\varepsilon, \tau)$ depends on the actual definition of diameter which is, in the language of previous section, the distance function used in computing the rate distortion function.

For deterministic systems, Eq. (9.13) can be shown to be independent of τ [Billingsley (1965); Eckmann and Ruelle (1985)] and, in the limit $\varepsilon \to 0$, the KS-entropy is recovered

$$h_{KS} = \lim_{\varepsilon\to 0} h(\varepsilon, \tau) \,,$$

in this respect a deterministic chaotic system behaves similarly to a discrete random processes such as the Bernoulli source the ε-entropy of which is shown in Fig. 9.3a.

Differently from the KS-entropy, which is a number, the ε-entropy is a function of the observation scale and its behavior as a function of ε provides information on the dynamical properties of the underlying system [Gaspard and Wang (1993); Abel *et al.* (2000b)]. Before discussing the behavior of $h(\varepsilon)$ in specific examples, it is useful to briefly recall some of the most used methods for its evaluation.

A first possibility is, for any fixed ε, to compute the Shannon entropy by using the symbolic dynamics which results from an equal cells partition. Of course, taking the infimum over all partitions is impossible and thus some of the nice properties of the "mathematically well defined" ε-entropy will be lost, but this is often the best it can be done in practice. However, implementing directly the Shannon definition is sometimes rather time consuming, and faster estimators are necessary.

Two of the most widely employed estimators are the correlation entropy $h^{(2)}(\varepsilon, \tau)$ (i.e. the Rényi entropy of order 2, see Box B.17), which can be obtained by a slight modification of the Grassberger and Procaccia (1983a) algorithm (Sec. 5.2.4) and the Cohen and Procaccia (1985) entropy estimator (see next Chapter for a discussion of the estimation of entropy and other quantities from experimental data).

The former estimate is based on the correlation integral (5.14) which is now applied to the N-vectors (9.10). Assuming to have M points of the trajectory $\boldsymbol{x}(t_i)$ with $i = 1, \ldots, M$ at times $t_i = i\tau$, we have $(M - N + 1)$ N-vectors $\boldsymbol{X}^{(N)}(t_j)$ for which the correlation integral (5.14) can be written

$$C_N(\varepsilon) = \frac{1}{M - N + 1} \sum_{i,j>i} \Theta(\varepsilon - ||\boldsymbol{X}^{(N)}(t_i) - \boldsymbol{X}^{(N)}(t_j)||) \tag{9.14}$$

where we dropped the dependence on M, assumed to be large enough, and used ε in place of ℓ to adhere to the current notation. The correlation, ε-entropy can be computed from the $N \to \infty$ behavior of (9.14). In fact, it can be proved that [Grassberger and Procaccia (1983a)]

$$C_N(\varepsilon) \sim \varepsilon^{D_2(\varepsilon,\tau)} \exp[-N\tau h^{(2)}(\varepsilon, \tau)] \tag{9.15}$$

[7] For continuous stochastic processes, for any ε, $\sup_{A:\mathrm{diam}(A)\leq\varepsilon} \{h(A_\varepsilon, \tau)\} = \infty$ as it recovers the Shannon entropy of an infinitely refined partition, which is infinite. This explains the rationale of the infimum in the definition (9.13).

so that we can estimate the entropy as

$$h^{(2)}(\varepsilon,\tau) = \frac{1}{\tau}\lim_{N\to\infty} h_N^{(2)}(\varepsilon,\tau) = \frac{1}{\tau}\lim_{N\to\infty}\frac{C_N(\varepsilon)}{C_{N+1}(\varepsilon)}\,. \tag{9.16}$$

In the limit $\varepsilon \to 0$, $h^{(2)}(\varepsilon) \to h^{(2)}$, which for a chaotic system is independent of τ and provides a lower bound to the Kolmogorov-Sinai entropy. We notice that Eq. (9.15) can also be used to define a correlation dimension which depends on the observation scale, whose behavior as a function of ε can also be rather informative [Olbrich and Kantz (1997); Olbrich *et al.* (1998)] (see also Sec. 12.5.1). In practice, as the limit $N \to \infty$ cannot be performed, one has to use different values of N and search for a collapse of $h_N^{(2)}$ as N increases (see Chap. 10).

Cohen and Procaccia (1985) proposal to estimate the ε-entropy is based on the observation that

$$n_j^{(N)}(\varepsilon) = \frac{1}{M-N}\sum_{i\neq j}\Theta(\varepsilon - ||\boldsymbol{X}^{(N)}(t_i) - \boldsymbol{X}^{(N)}(t_j)||)$$

estimates the probability of N-words $P(\mathcal{W}^N(\varepsilon,\tau))$ obtained from an ε-partition of the original trajectory, so that, the N-block entropy $H_N(\varepsilon,\tau)$ is given by

$$H_N(\varepsilon,\tau) = -\frac{1}{(M-N+1)}\sum_j \ln n_j^{(N)}(\varepsilon)\,.$$

The ε-entropy can thus be estimated as in Eq. (9.11) and Eq. (9.12). From a numerical point of view, the correlation ε-entropies are sometimes easier to compute.

Another method to estimate the ε-entropy, particularly useful in the case of intermittent systems or in the presence of many characteristic time-scales, is based on exit times statistics [Abel *et al.* (2000a,b)] and it is discussed, together with some examples in Box B.19.

9.3.1 *Systems classification according to ε-entropy behavior*

The dependence of $h(\varepsilon,\tau)$ on ε and in certain cases from τ, as for white-noise where $h(\varepsilon,\tau) \propto (1/\tau)\ln(1/\varepsilon)$ [Gaspard and Wang (1993)], can give some insights into the underlying stochastic process. For instance, in the previous section we found that a memory-less Gaussian process is characterized by $h(\varepsilon) \sim \ln(1/\varepsilon)$. Gelfand *et al.* (1958) (see also Kolmogorov (1956)) showed that for stationary Gaussian processes with spectrum $S(\omega) \propto \omega^{-2}$

$$h(\varepsilon) \propto \frac{1}{\varepsilon^2}\,, \tag{9.17}$$

which is also expected in the case of Brownian motion [Gaspard and Wang (1993)], though it is often difficult to detect mainly due to problems related to the choice of τ (see Box B.19). Equation (9.17) can be generalized to stationary Gaussian process with spectrum $S(\omega) \propto \omega^{-(2\alpha+1)}$ and fractional Brownian motions with

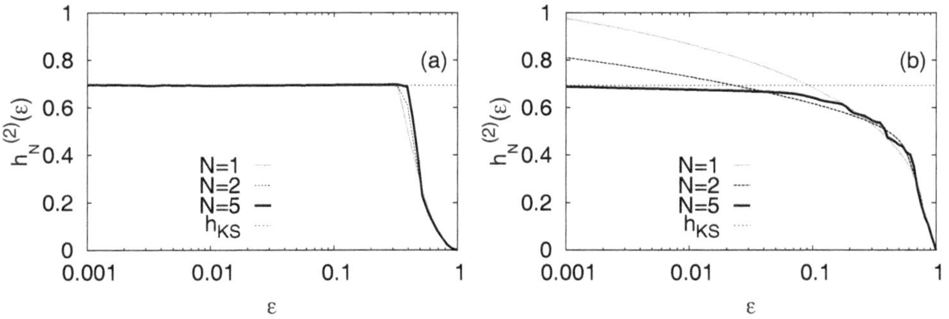

Fig. 9.5 Correlation ε-entropy $h_N^{(2)}(\varepsilon)$ vs ε for different block lengths N for the Bernoulli map (a) and logistic map with $r = 4$ (b).

Hurst exponent $0 < \alpha < 1$, meaning that $|x(t + \Delta t) - x(t)| \sim \Delta t^\alpha$, α is also called Hölder exponent [Metzler and Klafter (2000)], and reads

$$h(\varepsilon) \sim \frac{1}{\varepsilon^{1/\alpha}} .$$

As far as chaotic deterministic systems are concerned, in the limit $\varepsilon \to 0$, $h(\varepsilon) \to h_{KS}$ (see Fig. 9.5) while the large-ε behavior is system dependent. Having access to the ε-dependence of $h(\varepsilon)$, in general, provides information on the macroscale behavior of the system. For instance, it may happens that at large scales the system displays a diffusive behavior recovering the scaling (9.17) (see the first example in Box B.19). In Fig. 9.5, we show the behavior of $h_N^{(2)}(\varepsilon)$ for a few values of N as obtained from the Grassberger-Procaccia method (9.16) in the case of the Bernoulli and logistic maps.

Table 9.1 Classification of systems according to the ε-entropy behavior [After Gaspard and Wang (1993)]

Deterministic Processes	$h(\varepsilon)$
Regular	0
Chaotic	$h(\varepsilon) \leq h_{KS}$ and $0 < h_{KS} < \infty$
Stochastic Processes	$h(\varepsilon, \tau)$
Time discrete bounded Gaussian process	$\sim \ln(1/\varepsilon)$
White Noise	$\sim (1/\tau)\ln(1/\varepsilon)$
Brownian Motion	$\sim (1/\varepsilon)^2$
Fractional Brownian motion	$\sim (1/\varepsilon)^{1/\alpha}$

As clear from the picture, the correct value of Kolmogorov-Sinai entropy is attained for enough large block lengths, N, and sufficiently small ε. Moreover, for the Bernoulli map, which is memory-less (Sec. 8.1) the correct value is obtained already for $N = 1$, while in the logistic map it is necessary $N \gtrsim 5$ before approaching

h_{KS}. In general, only the lower bound $h^{(2)} \le h_{KS}$ is approached: for instance for the Hénon map with parameters $a = 1.4$ and $b = 0.3$, we find $h^{(2)}(\varepsilon) \approx 0.35$ while $h_{KS} \approx 0.42$ (see, e.g. Grassberger and Procaccia, 1983a). A common feature of this kind of computation is the appearance of a plateau for ε small enough which is usually recognized as the signature of deterministic chaos in the dynamics (see Sec. 10.3). However, the quality and extension of the plateau usually depends on many factors such as the number of points, the value of N, the presence of noise, the value of τ etc. Some of these aspects will be discussed in the next Chapter.

We conclude by stressing that the detailed dependence of the (ε, τ)-entropy on both ε and τ can be used to classify the character of the stochastic or dynamical process as, e.g., in Table 9.1 (see also Gaspard and Wang (1993)).

Box B.19: ε-entropy from exit-times statistics

This Box presents an alternative method for computing the ε-entropy, which is particularly useful and efficient when the system of interest is characterized by several scales of motion as in turbulent fluids or diffusive stochastic processes [Abel *et al.* (2000a,b)]. The idea is that in these cases an efficient coding procedure reduces the redundancy improving the quality of the results. This method is based on the exit times coding as shown below for a one-dimensional signal $x(t)$ (Fig. B19.1).

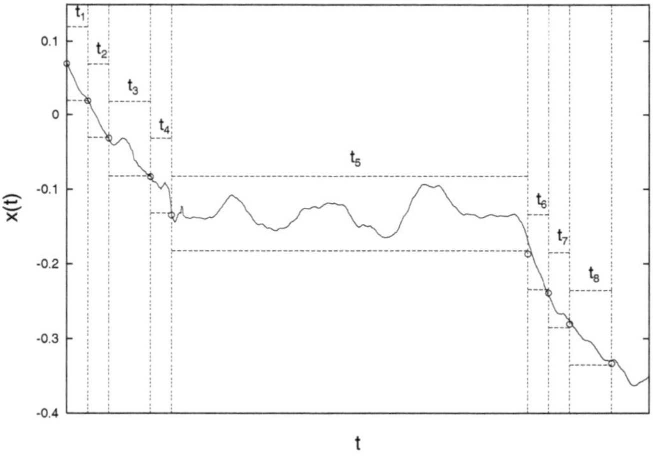

Fig. B19.1 Symbolic encoding of the signal shown in Fig. 9.4 based on the exit-time described in the text. For the specific signal here analyzed the symbolic sequence obtained with the exit time method is $\Omega_0^{27} = [(t_1, -1); (t_2, -1); (t_3, -1); (t_4, -1); (t_5, -1); (t_6, -1); (t_7, -1); (t_8, -1)]$.

Given a reference starting time $t = t_0$, measure the first exit-time from a cell of size ε, i.e. the first time t_1 such that $|x(t_0 + t_1) - x(t_0)| \ge \varepsilon/2$. Then from $t = t_0 + t_1$, look for the next exit-time t_2 such that $|x(t_0 + t_1 + t_2) - x(t_0 + t_1)| \ge \varepsilon/2$ and so on. This way from the signal a sequence of exit-times, $\{t_i(\varepsilon)\}$ is obtained together with the labels $k_i = \pm 1$, distinguishing the upward or downward exit direction from the cell. Therefore, as illustrated in Fig. B19.1, the trajectory is coded without ambiguity, with the required accuracy ε, by the sequence $\{(t_i, k_i),\ i = 1, \dots, M\}$, where M is the total number of exit-time events observed during time T. Finally, performing a coarse-graining of the values assumed by $t(\varepsilon)$ with a resolution time τ_r, we accomplish the goal of obtaining a symbolic sequence. We can now study the "exit-time N-words" $\Omega_i^N(\varepsilon, \tau_r) = ((\eta_i, k_i), (\eta_{i+1}, k_{i+1}), \dots, (\eta_{i+N-1}, k_{i+N-1}))$, where η_j labels the time-window (of width τ_r) containing the exit-time t_j. Estimating the probabilities of these words, we can compute the block entropies at the given time resolution, $H_N^\Omega(\varepsilon, \tau_r)$, and from them the exit-time (ε, τ_r)-entropy is given by:

$$h^\Omega(\varepsilon, \tau_r) = \lim_{N\to\infty} H_{N+1}^\Omega(\varepsilon, \tau_r) - H_N^\Omega(\varepsilon, \tau_r)\,.$$

The limit of infinite time-resolution gives us the ε-entropy *per exit*, i.e.:

$$h^\Omega(\varepsilon) = \lim_{\tau_r\to 0} h^\Omega(\varepsilon, \tau_r)\,.$$

The link between $h^\Omega(\varepsilon)$ and the ε-entropy (9.13) is established by noticing that there is a one-to-one correspondence between the exit-time histories and the (ε, τ)-histories (in the limit $\tau \to 0$) originating from a given ε-cell. Shannon-McMillan theorem (Sec. 8.2.3) grants that the number of the typical (ε, τ)-histories of length N, $\mathcal{N}(\varepsilon, N)$, is such that: $\ln \mathcal{N}(\varepsilon, N) \simeq h(\varepsilon) N\tau = h(\varepsilon)T$. For the number of typical (exit-time)-histories of length M, $\mathcal{M}(\varepsilon, M)$, we have: $\ln \mathcal{M}(\varepsilon, M) \simeq h^\Omega(\varepsilon)M$. If we consider $T = M\langle t(\varepsilon)\rangle$, where $\langle t(\varepsilon)\rangle = 1/M \sum_{i=1}^M t_i = T/M$, we must obtain the same number of (very long) histories. Therefore, from the relation $M = T/\langle t(\varepsilon)\rangle$ we finally obtain

$$h(\varepsilon) = \frac{M h^\Omega(\varepsilon)}{T} = \frac{h^\Omega(\varepsilon)}{\langle t(\varepsilon)\rangle} \simeq \frac{h^\Omega(\varepsilon, \tau_r)}{\langle t(\varepsilon)\rangle}\,. \tag{B.19.1}$$

The last equality is valid at least for small enough τ_r [Abel *et al.* (2000a)]. Usually, the leading ε-contribution to $h(\varepsilon)$ in (B.19.1) is given by the mean exit-time $\langle t(\varepsilon)\rangle$, though computing $h^\Omega(\varepsilon, \tau_r)$ is needed to recover zero entropy for regular signals.

It is worth noticing that an upper and a lower bound for $h(\varepsilon)$ can be easily obtained from the exit time scheme [Abel *et al.* (2000a)]. We use the following notation: for given ε and τ_r, $h^\Omega(\varepsilon, \tau_r) \equiv h^\Omega(\{\eta_i, k_i\})$, and we indicate with $h^\Omega(\{k_i\})$ and $h^\Omega(\{\eta_i\})$, respectively the Shannon entropy of the sequence $\{k_i\}$ and $\{\eta_i\}$. From standard information theory results, we have the inequalities [Abel *et al.* (2000a,b)]:

$$h^\Omega(\{k_i\}) \le h^\Omega(\{\eta_i, k_i\}) \le h^\Omega(\{\eta_i\}) + h^\Omega(\{k_i\}).$$

Moreover, $h^\Omega(\{\eta_i\}) \le H_1^\Omega(\{\eta_i\})$, where $H_1^\Omega(\{\eta_i\})$ is the entropy of the probability distribution of the exit-times measured on the scale τ_r) which reads

$$H_1^\Omega(\{\eta_i\}) = c(\varepsilon) + \ln\left(\frac{\langle t(\varepsilon)\rangle}{\tau_r}\right)\,,$$

where $c(\varepsilon) = -\int p(z)\ln p(z)\mathrm{d}z$, and $p(z)$ is the probability distribution function of the rescaled exit-time $z(\varepsilon) = t(\varepsilon)/\langle t(\varepsilon)\rangle$. Using the previous relations, the following bounds for the ε-entropy hold

$$\frac{h^{\Omega}(\{k_i\})}{\langle t(\varepsilon)\rangle} \leq h(\varepsilon) \leq \frac{h^{\Omega}(\{k_i\}) + c(\varepsilon) + \ln(\langle t(\varepsilon)\rangle/\tau_r)}{\langle t(\varepsilon)\rangle} . \tag{B.19.2}$$

These bounds are easy to compute and provide good estimate of $h(\varepsilon)$.
We consider below two examples in which the ε-entropy can be efficiently computed via the exit times strategy.

Diffusive maps

Consider the one-dimensional chaotic map:

$$x(t+1) = x(t) + p\sin[2\pi x(t)] , \tag{B.19.3}$$

which, for $p > 0.7326\ldots$, produces a large scale diffusive behavior [Schell *et al.* (1982)]

$$\langle (x(t) - x(0))^2 \rangle \simeq 2D\,t \qquad \text{for} \qquad t \to \infty , \tag{B.19.4}$$

where D is the diffusion coefficient. In the limit $\varepsilon \to 0$, we expect $h(\varepsilon) \to h_{KS} = \lambda$ (λ being the Lyapunov exponent) while for large ε, being the motion diffusive, a simple dimensional argument suggests that the typical exit time over a threshold of scale ε should scale as ε^2/D as obtained by using (B.19.4), so that

$$h(\varepsilon) \simeq \lambda \text{ for } \varepsilon \lesssim 1 \qquad \text{and} \qquad h(\varepsilon) \propto \frac{D}{\varepsilon^2} \text{ for } \varepsilon \gtrsim 1,$$

in agreement with (9.17).

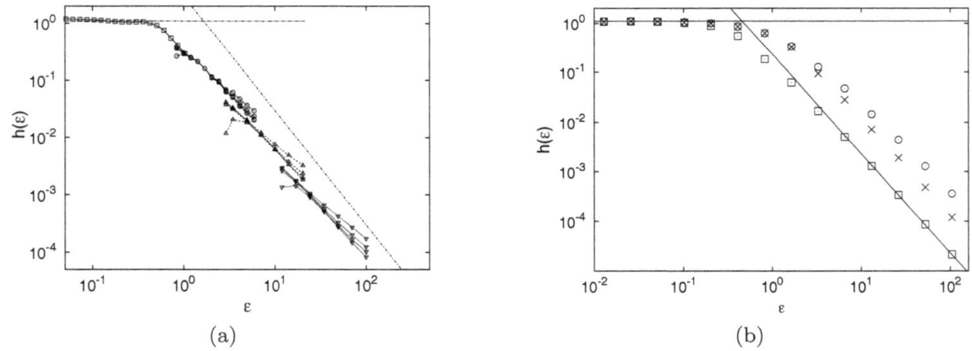

Fig. B19.2 (a) ε-entropy for the map (B.19.3) with $p = 0.8$ computed with GP algorithm and sampling time $\tau = 1$ ($\circ$), 10 ($\triangle$) and 100 ($\triangledown$) for different block lengths ($N = 4, 8, 12, 20$). The computation assumes periodic boundary conditions over a large interval $[0 : L]$ with L an integer. This is necessary to have a bounded phase space. Boxes refer to entropy computed with $\tau = 1$ and periodic boundary conditions on $[0 : 1]$. The straight lines correspond to the asymptotic behaviors, $h(\varepsilon) = h_{KS}$ and $h(\varepsilon) \sim \varepsilon^{-2}$, respectively. (b) Lower bound ($\square$) and upper bound ($\circ$) for the ε-entropy as obtained from Eq. (B.19.2), for the sine map with parameters as in (a). The straight (solid) lines correspond to the asymptotic behaviors $h(\varepsilon) = h_{KS}$ and $h(\varepsilon) \sim \varepsilon^{-2}$. The ε-entropy $h^{\Omega}(\varepsilon, \tau_e)/\langle t(\varepsilon)\rangle$ with $\tau_e = 0.1\langle t(\varepsilon)\rangle$ correspond to the $\times$ symbols.

Computing $h(\varepsilon)$ with standard techniques based on the Grassberger-Procaccia or Cohen-Procaccia methods requires to consider several measurements in which the sampling time τ is varied and the correct behavior is recovered only through the envelope of all these curves (Fig. B19.2a) [Gaspard and Wang (1993); Abel *et al.* (2000a)]. In fact, by looking at any single (small) value of τ (e.g. $\tau = 1$) one obtains a rather inconclusive result. This is due to the fact that one has to consider very large block lengths, N, in order to obtain a good convergence for $H_N - H_{N-1}$. In the diffusive regime, a dimensional argument shows that the characteristic time of the system at scale ε is $T_\varepsilon \approx \varepsilon^2/D$. If we consider for example, $\varepsilon = 10$ and $D \simeq 10^{-1}$, the characteristic time, T_ε, is much larger than the elementary sampling time $\tau = 1$. On the contrary, the exit time strategy does not require any fine tuning of the sampling time and provides the clean result shown in Fig. B19.2b. The main reason for which the exit time approach is more efficient than the usual one is that at fixed ε, $\langle t(\varepsilon)\rangle$ automatically gives the typical time at that scale. As a consequence, it is not necessary to reach very large block sizes — at least if ε is not too small.

Intermittent maps

Several systems display intermittency characterized by very long laminar intervals separating short intervals of bursting activity, as in Fig. B19.3a. It is easily realized that coding the trajectory of Fig. B19.3a at fixed sampling times is not very efficient compared with the exit times method, which codifies a very long quiescent period with a single symbol. As a specific example, consider the one-dimensional intermittent map [Bergé *et al.* (1987)]:

$$x(t+1) = x(t) + ax^z(t) \mod 1\,, \tag{B.19.5}$$

with $z > 1$ and $a > 0$, which is characterized by an invariant density with power law singularity near the marginally stable fixed point $x = 0$, i.e. $\rho(x) \propto x^{1-z}$. For $z \geq 2$, the density is not normalizable and the so-called *sporadic chaos* appears [Gaspard and Wang (1988); Wang (1989)], where the separation between two close trajectories diverge as a stretched exponential. For $z < 2$, the usual exponential divergence is observed. Sporadic chaos is thus intermediate between chaotic and regular motion, as obtained from the algorithmic complexity computation [Gaspard and Wang (1988); Wang (1989)] or by studying the mean exit time, as shown in the sequel.

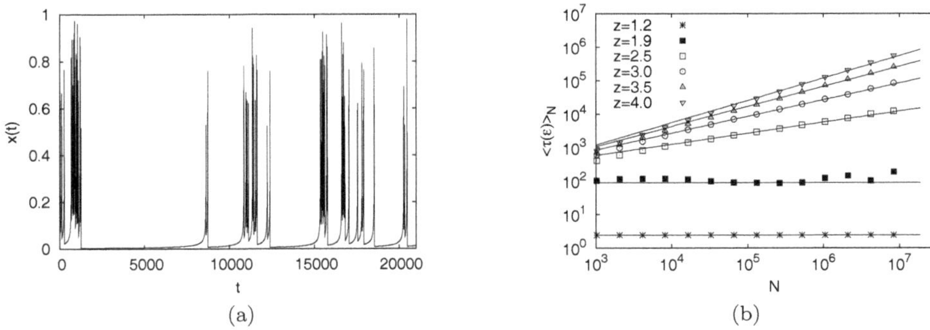

Fig. B19.3 (a) Typical evolution of the intermittent map Eq. (B.19.5) for $z = 2.5$ and $a = 0.5$. (b) $\langle t(\varepsilon)\rangle_N$ versus N for the map (B.19.5) at $\varepsilon = 0.243$, $a = 0.5$ and different z. The straight lines indicate the power law (B.19.6). $\langle t(\varepsilon)\rangle_N$ is computed by averaging over 10^4 different trajectories of length N. For $z < 2$, $\langle t(\varepsilon)\rangle_N$ does not depend on N, the invariant measure $\rho(x)$ is normalizable, the motion is chaotic and H_N/N is constant. Different values of ε provide equivalent results.

Neglecting the contribution of $h^{\Omega}(\varepsilon)$, and considering only the mean exit time, the total entropy H_N of a trajectory of length N can be estimated as

$$H_N \propto \frac{N}{\langle t(\varepsilon)\rangle_N} \qquad \text{for large } N\,,$$

where $\langle[...]\rangle_N$ indicates the mean exit time computed on a sequence of length N. The dependence of H_N on ε can be neglected as exit times at scale ε are dominated by the first exit from a region of size ε around the origin, so that, $\langle t(\varepsilon)\rangle_N$ approximately gives the duration of the laminar period and does not depend on ε (this is exact for ε large enough). Further, the power law singularity at the origin implies $\langle t(\varepsilon)\rangle_N$ to diverge with N.

In Fig. B19.3b, $\langle t(\varepsilon)\rangle_N$ is shown as a function of N and z. For large enough N the behavior is almost independent of ε, and for $z \geq 2$ one has

$$\langle t(\varepsilon)\rangle_N \propto N^{\alpha}\,, \qquad \text{where} \qquad \alpha = \frac{z-2}{z-1}\,. \tag{B.19.6}$$

For $z < 2$, as expected for usual chaotic motion, $\langle t(\varepsilon)\rangle \approx \text{const}$ at large N.

Exponent α can be estimated via the following argument: the power law singularity entails $x(t) \approx 0$ most of the time. Moreover, near the origin the map (B.19.5) is well approximated by the differential equation $dx/dt = ax^z$ [Bergé *et al.* (1987)]. Therefore, denoting with x_0 the initial condition, we obtain $(x_0+\varepsilon)^{1-z} - x_0^{1-z} = a(1-z)t(\varepsilon)$, where the first term can be neglected as, due to the singularity, x_0 is typically much smaller than $x_0+\varepsilon$, so that the exit time is $t(\varepsilon) \propto x_0^{1-z}$. From the probability density of x_0, $\rho(x_0) \propto x_0^{1-z}$, one obtains the probability distribution of the exit times $\rho(t) \sim t^{1/(1-z)-1}$, the factor t^{-1} takes into account the non-uniform sampling of the exit time statistics. The average exit time on a trajectory of length N is thus given by

$$\langle t(\varepsilon)\rangle_N \sim \int_0^N t\,\rho(t)\,dt \sim N^{\frac{z-2}{z-1}}\,,$$

and for block-entropies we have $H_N \sim N^{\frac{1}{z-1}}$, that behaves as the algorithmic complexity [Gaspard and Wang (1988)]. Note that though the entropy per symbol is zero, it converges very slowly with N, $H_N/N \sim 1/\langle t(\varepsilon)\rangle_N \sim N^{\frac{2-z}{z-1}}$, due to sporadicity.

9.4 The finite size lyapunov exponent (FSLE)

We learned from the example (9.3) that the Lyapunov exponent is often inadequate to quantify our ability to predict the evolution of a system, indeed the predictability time (9.1) derived from the LE

$$T_p(\delta,\Delta) = \frac{1}{\lambda_1}\ln\left(\frac{\Delta}{\delta}\right)$$

requires both δ and Δ to be infinitesimal, moreover it excludes the presence of fluctuations (Sec. 5.3.3) as the LE is defined in the limit of infinite time. As argued by Keynes "*In the long run everybody will be dead*" so that we actually need to quantify predictability relying on finite-time and finite-resolution quantities.

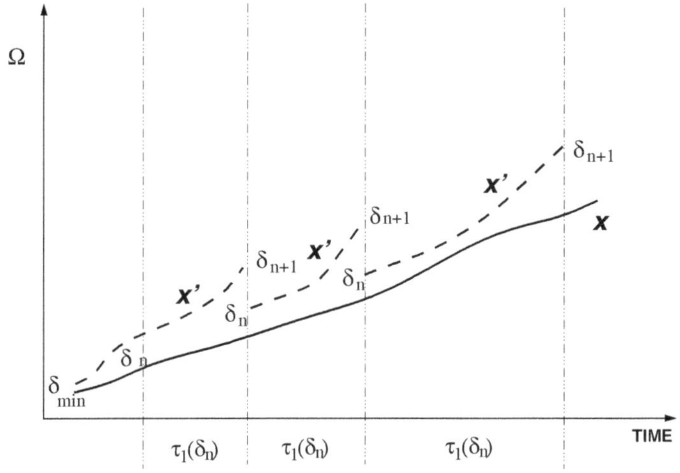

Fig. 9.6 Sketch of the first algorithm for computing the FSLE.

At some level of description such a quantity may be identified in the ε-entropy which, though requiring the infinite time limit, is able to quantify the rate of information creation (and thus the loss of predictability) also at non-infinitesimal scales. However, it is usually quite difficult to estimate the ε-entropy especially when the dimensionality of the state space increases, as it happens for system of interest like atmospheric weather. Finally, we have seen that a relationship (8.23) can be established between KS-entropy and positive LEs. This may suggest that something equivalent could be hold in the case of the ε-entropy for finite ε.

In this direction, it is useful here to discuss an indicator — the Finite Size Lyapunov Exponent (FSLE) — which fulfills some of the above requirements. The FSLE has been originally introduced by Aurell *et al.* (1996) (see also for a similar approach Torcini *et al.* (1995)) to quantify the predictability in turbulence and has then been successfully applied in many different contexts [Aurell *et al.* (1997); Artale *et al.* (1997); Boffetta *et al.* (2000b, 2002); Cencini and Torcini (2001); Basu *et al.* (2002); d'Ovidio *et al.* (2004, 2009)].

The main idea is to quantify the average growth rate of error at different scales of observations, i.e. associated to non-infinitesimal perturbations. Since, unlike the usual LE and the ε-entropy, such a quantity has a less firm mathematical ground, we will introduce it in an operative way through the algorithm used to compute it. Assume that a system has been evolved for long enough that the transient dynamics has lapsed, e.g., for dissipative systems the motion has settled onto the attractor. Consider at $t = 0$ a "reference" trajectory $\boldsymbol{x}(0)$ supposed to be on the attractor, and generate a "perturbed" trajectory $\boldsymbol{x}'(0) = \boldsymbol{x}(0) + \delta\boldsymbol{x}(0)$. We need the perturbation to be initially very small (essentially infinitesimal) in some chosen norm $\delta(t = 0) = ||\delta\boldsymbol{x}(t = 0)|| = \delta_{min} \ll 1$ (typically $\delta_{min} = \mathcal{O}(10^{-6} - 10^{-8})$).

Then, in order to study the perturbation growth through different scales, we can define a set of thresholds δ_n, e.g., $\delta_n = \delta_0 \varrho^n$ with $\delta_{min} \ll \delta_0 \ll 1$, where δ_0 can still be considered infinitesimal and $n = 0, \dots, N_s$. To avoid saturation on the maximum allowed separation (i.e. the attractor size) attention should be payed to have $\delta_{N_s} < \langle ||\boldsymbol{x} - \boldsymbol{y}|| \rangle_\mu$ with $\boldsymbol{x}, \boldsymbol{y}$ generic points on the attractor. The factor ϱ should be larger than 1 but not too large in order to avoid interferences of different length scales: typically, $\varrho = 2$ or $\varrho = \sqrt{2}$.

The purpose is now to measure the perturbation growth rate at scale δ_n. After a time t_0 the perturbation has grown from δ_{min} up to δ_n ensuring that the perturbed trajectory relaxes on the attractor and aligns along the maximally expanding direction. Then, we measure the time $\tau_1(\delta_n)$ needed to the error to grow up to δ_{n+1}, i.e. the first time such that $\delta(t_0) = ||\delta \boldsymbol{x}(t_0)|| = \delta_n$ and $\delta(t_0 + \tau_1(\delta_n)) = \delta_{n+1}$. After, the perturbation is rescaled to δ_n, keeping the direction $\boldsymbol{x}' - \boldsymbol{x}$ constant. This procedure is repeated $\mathcal{N}_d$ times for each thresholds obtaining the set of the "doubling"[8] times $\{\tau_i(\delta_n)\}$ for $i = 1, \dots, \mathcal{N}_d$ error-doubling experiments. Note that $\tau(\delta_n)$ generally may also depend on ϱ. The doubling rate

$$\gamma_i(\delta_n) = \frac{1}{\tau_i(\delta_n)} \ln \varrho \,,$$

when averaged defines the FSLE $\lambda(\delta_n)$ through the relation

$$\lambda(\delta_n) = \langle \gamma(\delta_n) \rangle_t = \frac{1}{T} \int_0^T \mathrm{d}t \; \gamma = \frac{\sum_i \gamma_i \tau_i}{\sum_i \tau_i} = \frac{\ln \varrho}{\langle \tau(\delta_n) \rangle_d} \,, \tag{9.18}$$

where $\langle \tau(\delta_n) \rangle_d = \sum \tau_i / \mathcal{N}_d$ is the average over the doubling experiments and the total duration of the trajectory is $T = \sum_i \tau_i$.

Equation (9.18) assumes the distance between the two trajectories to be continuous in time. This is not true for maps or time-continuous systems sampled at discrete times, for which the method has to be slightly modified defining $\tau(\delta_n)$ as the minimum time such that $\delta(\tau) \geq \varrho \delta_n$. Now $\delta(\tau)$ is a fluctuating quantity, and from (9.18) we have

$$\lambda(\delta_n) = \frac{1}{\langle \tau(\delta_n) \rangle_d} \left\langle \ln \left(\frac{\delta(\tau(\delta_n))}{\delta_n} \right) \right\rangle_d \,. \tag{9.19}$$

When δ_n is infinitesimal $\lambda(\delta_n)$ recovers the maximal LE

$$\lim_{\delta \to 0} \lambda(\delta) = \lambda_1 \tag{9.20}$$

indeed the algorithm is equivalent to the procedure adopted in Sec. 8.4.3.

However, it is worth discussing some points.

At difference with the standard LE, $\lambda(\delta)$, for finite δ, depends on the chosen norm, as it happens also for the ε-entropy which depends on the distortion function. This apparently ill-definition tells us that in the nonlinear regime the predictability time depends on the chosen observable, which is somehow reasonable (the same happens for the ε-entropy and in infinite dimensional systems [Kolmogorov and Fomin (1999)]).

[8] Strictly speaking the name applies for $\varrho = 2$ only.

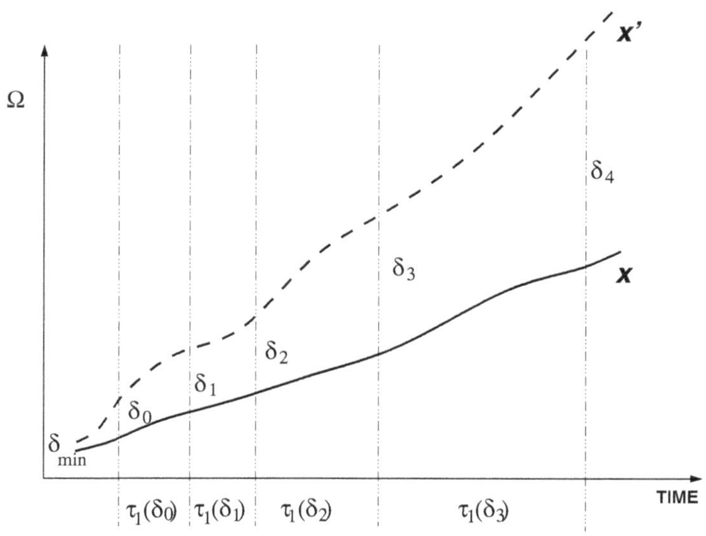

Fig. 9.7 Sketch of the second algorithm cor computing the FSLE.

A possible problem with the above described method is that we have implicitly assumed that the statistically stationary state of the system is homogeneous with respect to finite perturbations. Typically the attractor is fractal and not equally dense at all distances, this may cause an incorrect sampling of the doubling times at large δ_n. To cure such a problem the algorithm can be modified to avoid the rescaling of the perturbation at finite δ_n. This can be accomplished by the following modification of the previous method (Fig. 9.7). The thresholds $\{\delta_n\}$ and the initial perturbation ($\delta_{min} \ll \delta_0$) are chosen as before, but now the perturbation growth is followed from δ_0 to δ_{N_s} without rescaling back the perturbation once the threshold is reached (see Fig. 9.7). In practice, after the system reaches the first threshold δ_0, we measure the time $\tau_1(\delta_0)$ to reach δ_1, then following the same perturbed trajectory we measure the time $\tau_1(\delta_1)$ to reach δ_2, and so on up to δ_{N_s}, so to register the time $\tau(\delta_n)$ for going from δ_n to δ_{n+1} for each value of n. The evolution of the error from the initial value δ_{min} to the largest threshold δ_N carries out a single error-doubling experiment, and the FSLE is finally obtained by using Eq. (9.18) or Eq. (9.19), which are accurate also in this case, according to the continuous-time or discrete-time nature of the system, respectively. As finite perturbations are realized by the dynamics (i.e. the perturbed trajectory is on the attractor), the problems related to the attractor inhomogeneity are not present anymore. Even though some differences between the two methods are possible for large δ they should coincide for $\delta \to 0$ and, in any case, in most numerical experiments they give the same result.[9]

[9] Another possibility for computing the FSLE is to remove the threshold condition and simply compute the average error growth rate at every time step. Thus, at every integration time step Δt, the perturbed trajectory $\boldsymbol{x}'(t)$ is rescaled to the original distance δ, keeping the direction $\boldsymbol{x} - \boldsymbol{x}'$

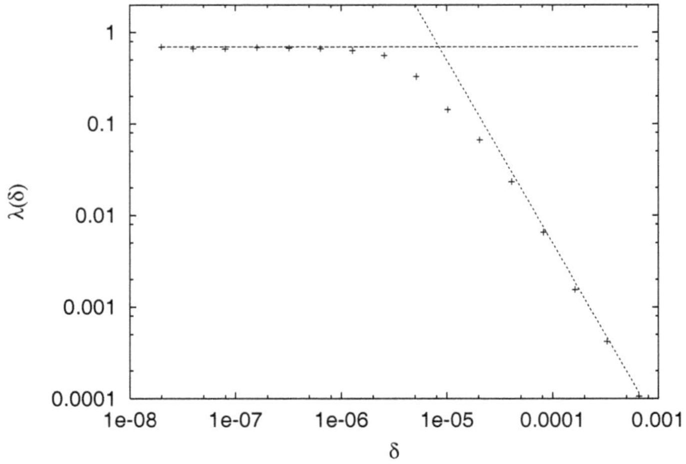

Fig. 9.8 $\lambda(\delta)$ vs δ for the coupled map (9.3) with the same parameters of Fig. 9.2. For $\delta \to 0$, $\lambda(\delta) \simeq \lambda_1$ (solid line). The dashed line displays the behavior $\lambda(\delta) \sim \delta^{-2}$.

With reference to example (9.3), we show in Fig. 9.8 the result of the computation of the FSLE with the above algorithm. For $\delta \ll 1$ a plateau at the value of maximal Lyapunov exponent λ_1 is recovered as from the limit (9.20), while for finite δ the behavior of $\lambda(\delta)$ depends on the details of the nonlinear dynamics which is diffusive (see Fig. 9.2 and Eq. (9.5)) and leads to

$$\lambda(\delta) \sim \delta^{-2}\,, \tag{9.21}$$

as suggested by dimensional analysis. Notice that (9.21) corresponds to the scaling behavior (9.17) expected for the ε-entropy.

We mention that other approaches to finite perturbations have been proposed by Dressler and Farmer (1992); Kantz and Letz (2000), and conclude this section with a final remark on the FSLE. Be $\boldsymbol{x}(t)$ and $\boldsymbol{x}'(t)$ a reference and a perturbed trajectory of a given dynamical system with $R(t) = |\boldsymbol{x}(t) - \boldsymbol{x}'(t)|$, naively one could be tempted to define a scale dependent growth rate also using

$$\tilde{\lambda}(\delta) = \frac{1}{2\,\langle R^2(t)\rangle}\,\frac{\mathrm{d}\,\langle R^2(t)\rangle}{\mathrm{d}t}\bigg|_{\langle R^2\rangle=\delta^2} \qquad \text{or} \qquad \tilde{\lambda}(\delta) = \frac{\mathrm{d}\,\langle \ln R(t)\rangle}{\mathrm{d}t}\bigg|_{\langle \ln R(t)\rangle = \ln\delta}\,.$$

---

constant. The FSLE is then obtained by averaging at each time step the growth rate, i.e.

$$\lambda(\delta) = \frac{1}{\Delta t}\left\langle \ln\left(\frac{||\delta\boldsymbol{x}(t+\Delta t)||}{||\delta\boldsymbol{x}(t)||}\right)\right\rangle_t\,,$$

which, if non negative, is equivalent to the definition (9.18). Such a procedure is nothing but the finite scale version of the usual algorithm of [Benettin *et al.* (1978b, 1980)] for the LE. The one-step method can be, in principle, generalized to compute the sub-leading finite-size Lyapunov exponent following the standard ortho-normalization method. However, the problem of homogeneity of the attractor and, perhaps more severely, that of isotropy may invalidate the procedure.

However, $\tilde{\lambda}(\delta)$ should not be confused with the FSLE $\lambda(\delta)$, as $\langle R^2(t)\rangle$ usually depends on $\langle R^2(0)\rangle$ while $\lambda(\delta)$ depends only on δ. This difference has an important conceptual and practical consequence, for instance, when considering the relative dispersion of two tracer particles in turbulence or geophysical flows [Boffetta *et al.* (2000a); Lacorata *et al.* (2004)].

9.4.1 *Linear vs nonlinear instabilities*

In Chapter 5, when introducing the Lyapunov exponents we quoted that they generalize the linear stability analysis (Sec. 2.4) to aperiodic motions. The FSLE can thus be seen as an extension of the stability analysis to nonlinear regimes. Passing from the linear to the nonlinear realm interesting phenomena may happen. In the following we consider two simple one-dimensional maps for which the computation of the FSLE can be analytically performed [Torcini *et al.* (1995)]. These examples, even if extremely simple, highlight some peculiarities of the nonlinear regime of perturbation growth.

Let us start with the tent map $f(x) = 1 - 2|x - 1/2|$, which is piecewise linear with uniform invariant density in the unit interval, i.e. $\rho(x) = 1$, (see Chap. 4). By using the tools of Sec. 5.3, the Lyapunov exponent can be easily computed as

$$\lambda = \lim_{\delta\to 0}\left\langle \ln\left|\frac{f(x+\delta/2) - f(x-\delta/2)}{\delta}\right|\right\rangle = \int_0^1 dx\, \rho(x) \ln|f'(x)| = \ln 2\,.$$

Relaxing the request $\delta \to 0$, we can compute the FSLE as:

$$\lambda(\delta) = \left\langle \ln\left|\frac{f(x+\delta/2) - f(x-\delta/2)}{\delta}\right|\right\rangle = \langle I(x,\delta)\rangle\,, \tag{9.22}$$

where (for $\delta < 1/2$) $I(x,\delta)$ is given by:

$$I(x,\delta) = \begin{cases} \ln 2 & x \in [0:1/2-\delta/2[\ \ \cup\ \ ]1/2+\delta/:1] \\ \ln\frac{|2(2x-1)|}{\delta} & \text{otherwise}\,. \end{cases}$$

The average (9.22) yields, for $\delta < 1/2$,

$$\lambda(\delta) = \ln 2 - \delta\,,$$

in very good agreement with the numerically computed[10] $\lambda(\delta)$ (Fig. 9.9 left). In this case, the error growth rate decreases for finite perturbations.

However, under certain circumstances the finite size corrections due to the higher order terms may lead to an enhancement of the separation rate for large perturbations [Torcini *et al.* (1995)]. This effect can be dramatic in marginally stable systems ($\lambda = 0$) and even in stable systems ($\lambda < 0$) [Cencini and Torcini (2001)]. An example of the latter situation is given by the Bernoulli shift map $f(x) = 2x$

[10] No matter of the used algorithm.

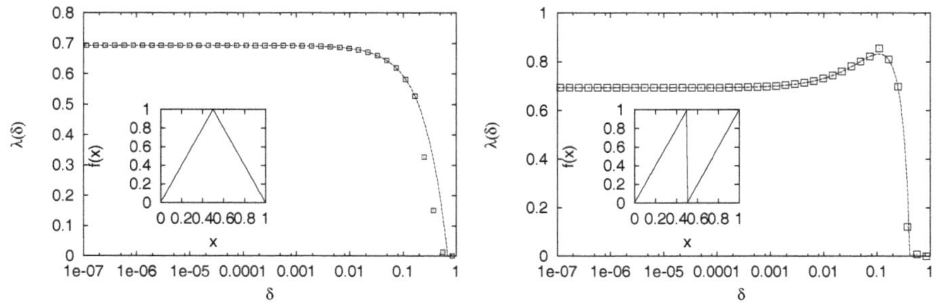

Fig. 9.9 $\lambda(\delta)$ versus δ for the tent map (left) and the Bernoulli shift map (right). The continuous lines are the analytical estimation of the FSLE. The maps are shown in the insets.

mod 1. By using the same procedure as before, we easily find that $\lambda = \ln 2$, and for δ not too large

$$I(x,\delta) = \begin{cases} \ln\left[\frac{(1-2\delta)}{\delta}\right] & x \in [1/2-\delta/2, 1/2+\delta/2] \\ \ln 2 & \text{otherwise}. \end{cases}$$

As the invariant density is uniform, the average of $I(x,\delta)$ gives

$$\lambda(\delta) = (1-\delta)\ln 2 + \delta \ln\left(\frac{1-2\delta}{\delta}\right).$$

In Fig. 9.9 right we show the analytic FSLE compared with the numerically evaluated $\lambda(\delta)$. In this case, we have an anomalous situation that $\lambda(\delta) \geq \lambda$ for some $\delta > 0$.[11] The origin of this behavior is the presence of the discontinuity at $x = 1/2$ which causes trajectories residing on the left (resp.) right of it to experience very different histories no matter of the original distance between them. Similar effects can be very important when many of such maps are coupled together [Cencini and Torcini (2001)]. Moreover, this behavior may lead to seemingly chaotic motions even in the absence of chaos (i.e. with $\lambda \leq 0$) due to such finite size instabilities [Politi *et al.* (1993); Cecconi *et al.* (1998); Cencini and Torcini (2001); Boffetta *et al.* (2002); Cecconi *et al.* (2003)].

9.4.2 *Predictability in systems with different characteristic times*

The FSLE is particularly suited to quantify the predictability of systems with different characteristic times as illustrated from the following example with two characteristic time scales, taken by [Boffetta *et al.* (1998)] (see also Boffetta *et al.* (2000b) and Peña and Kalnay (2004)).

Consider a dynamical system in which we can identify two different classes of degrees of freedom according to their characteristic time. The interest for this class of models is not merely academic, for instance, in climate studies a major relevance

[11] This is not possible for the ε-entropy as $h(\varepsilon)$ is a non-increasing function of ε.

is played by models of the interaction between Ocean and Atmosphere where the former is known to be much slower than the latter. Assume the system to be of the form

$$\frac{d\boldsymbol{x}^{(s)}}{dt} = \boldsymbol{f}(\boldsymbol{x}^{(s)}, \boldsymbol{x}^{(f)}) \qquad \frac{d\boldsymbol{x}^{(f)}}{dt} = \boldsymbol{g}(\boldsymbol{x}^{(s)}, \boldsymbol{x}^{(f)}),$$

where $\boldsymbol{f}, \boldsymbol{x}^{(s)} \in \mathbb{R}^{d_1}$ and $\boldsymbol{g}, \boldsymbol{x}^{(f)} \in \mathbb{R}^{d_2}$, in general $d_1 \neq d_2$. The label (s, f) identifies the slow/fast degrees of freedom.

For the sake of concreteness we can, e.g., consider the following two coupled Lorenz models

$$\begin{cases} \frac{dx_1^{(s)}}{dt} = \sigma(x_2^{(s)} - x_1^{(s)}) \\ \frac{dx_2^{(s)}}{dt} = (-x_1^{(s)}x_3^{(s)} + r_s x_1^{(s)} - x_2^{(s)}) - \epsilon_s x_1^{(f)} x_2^{(f)} \\ \frac{dx_3^{(s)}}{dt} = x_1^{(s)}x_2^{(s)} - bx_3^{(s)} \\ \\ \frac{dx_1^{(f)}}{dt} = c\sigma(x_2^{(f)} - x_1^{(f)}) \\ \frac{dx_2^{(f)}}{dt} = c(-x_1^{(f)}x_3^{(f)} + r_f x_1^{(f)} - x_2^{(f)}) + \epsilon_f x_1^{(f)} x_2^{(s)} \\ \frac{dx_3^{(f)}}{dt} = c(x_1^{(f)}x_2^{(f)} - bx_3^{(f)}), \end{cases} \tag{9.23}$$

where the constant $c > 1$ sets the time scale of the fast degrees of freedom, here we choose $c = 10$. The parameters have the values $\sigma = 10$, $b = 8/3$, the customary choice for the Lorenz model (Sec. 3.2),[12] while the Rayleigh numbers are taken different, $r_s = 28$ and $r_f = 45$, in order to avoid synchronization effects (Sec. 11.4). With the present choice, the two uncoupled systems ($\epsilon_s = \epsilon_f = 0$) display chaotic dynamics with Lyapunov exponent $\lambda^{(f)} \simeq 12.17$ and $\lambda^{(s)} \simeq 0.905$ respectively and thus a relative intrinsic time scale of order 10.

By switching the couplings on, e.g. $\epsilon_s = 10^{-2}$ and $\epsilon_f = 10$, the resulting dynamical system has maximal LE λ_{max} close (for small couplings) to the Lyapunov exponent of the fastest decoupled system ($\lambda^{(f)}$), indeed $\lambda_{max} \simeq 11.5$ and $\lambda^{(f)} \approx 12.17$.

A natural question is how to quantify the predictability of the slowest system. Using the maximal LE of the complete system leads to $T_p \approx 1/\lambda_{max} \approx 1/\lambda^{(f)}$, which seems rather inappropriate because, for small coupling ϵ_s, the slow component of the system $\boldsymbol{x}^{(s)}$ should remain predictable up to its own characteristic time $1/\lambda^{(s)}$. This apparent difficulty stems from the fact that we did not specified neither the

[12]The form of the coupling is constrained by the physical request that the solution remains in a bounded region of the phase space. Since

$$\frac{d}{dt}\left\{\epsilon_f\left(\frac{x_1^{(f)\,2}}{2\sigma}+\frac{x_2^{(f)\,2}}{2}+\frac{x_3^{(f)\,2}}{2}-(r_f+1)x_3^{(f)}\right)+\epsilon_s\left(\frac{x_1^{(s)\,2}}{2\sigma}+\frac{x_2^{(s)\,2}}{2}+\frac{x_3^{(s)\,2}}{2}-(r_s+1)x_3^{(s)}\right)\right\}<0,$$

if the trajectory is far enough from the origin, it evolves in a bounded region of phase space.

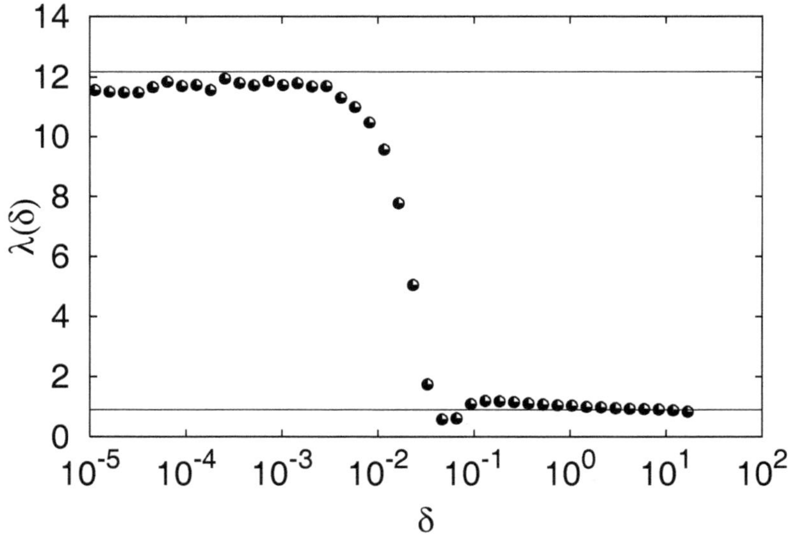

Fig. 9.10 $\lambda(\delta)$ vs δ for the two coupled Lorenz systems (9.23) with parameters as in the text. The error is computed only on the slow degrees of freedom (9.24), while the initial perturbation is set only on the fast degrees of freedom $|\delta x^{(f)}| = 10^{-7}$. As for the FLSE, the second algorithm has been used with $\varrho = \sqrt{2}$ and $N_s = 49$, the first threshold is at $\delta_0 = 10^{-6}$ and $\delta_{min} = 0$ as at the beginning the slow degrees of freedom are error-free. The straight lines indicate the value of the Lyapunov exponents of the uncoupled models $\lambda^{(f,s)}$. The average is over $\mathcal{O}(10^4)$ doubling experiments.

size of the initial perturbation nor the error we are going to accept. This point is well illustrated by the behavior of the Finite Size Lyapunov exponent $\lambda(\delta)$ which is computed from two trajectories of the system (9.23) — the reference $\boldsymbol{x}$ and the forecast or perturbed trajectory $\boldsymbol{x}'$ — subjected to an initial (very tiny) error $\delta(0)$ in the fast degrees of freedom, i.e. $||\delta \boldsymbol{x}^{(f)}|| = \delta(0)$.[13] Then the evolution of the error is monitored looking only at the slow degrees of freedom using the norm

$$||\delta \mathbf{x}^{(s)}(t)|| = \left[\sum_{i=1}^{3} \left(x_i'^{(s)} - x_i^{(s)} \right)^2 \right]^{1/2} \tag{9.24}$$

In Figure 9.10, we show $\lambda(\delta)$ obtained by averaging over many error-doubling experiments performed with the second algorithm (Fig. 9.7). For very small δ, the FSLE recovers the maximal LE λ_{max}, indicating that in small scale predictability, the fast component plays indeed the dominant role. As soon as the error grows above the coupling ϵ_s, $\lambda(\delta)$ drops to a value close to $\lambda^{(s)}$ and the characteristic time of small scale dynamics is no more relevant.

[13]Adding an initial error also in the slow degrees of freedom causes no basic difference to the presented behavior of the FSLE, and also using the norm in the full phase-space it is not so relevant due to the fast saturation of the fast degrees of freedom.

9.5 Exercises

Exercise 9.1: Consider the one-dimensional map $x(t+1) = [x(t)] + F(x(t) - [x(t)])$ with

$$F(z) = \begin{cases} az & \text{if} \quad 0 \le z \le 1/2 \\ 1 + a(z-1) & \text{if} \quad 1/2 < z \le 0\,, \end{cases}$$

where $a > 2$ and $[\ldots]$ denotes the integer part of a real number. This map produces a dynamics similar to a one-dimensional Random Walk. Following the method used to obtain Fig. B19.2, choose a value of a, compute the ε-entropy using the Grassberger-Procaccia and compare the result with a computation performed with the exit-times. Then, being the motion diffusive, compute the the diffusion coefficient as a function of a and plot $D(a)$ as a function of a (see Klages and Dorfman (1995)). Is it a smooth curve?

Exercise 9.2: Consider the one-dimensional intermittent map

$$x(t+1) = x(t) + ax^z(t) \mod 1$$

fix $a = 1/2$ and $z = 2.5$. Look at the symbolic dynamics obtained by using the partition identified by the two branches of the map. Compute the N-block entropies as introduced in Chap. 8 and compare the result with that obtained using the exit-time ϵ-entropy (Fig. B19.3b). Is there a way to implement the exit time idea with the symbolic dynamics obtained with this partition?

Exercise 9.3: Compute the FSLE using both algorithms described in Fig. 9.7 and Fig. 9.6 for both the logistic maps ($r = 4$) and the tent map. Is there any appreciable difference?

Hint: *Be sure to use double precision computation. Use* $\delta_{min} = 10^{-9}$ *and define the thresholds as* $\delta_n = \delta_0 \varrho^n$ *with* $\varrho = 2^{1/4}$ *and* $\delta_0 = 10^{-7}$.

Exercise 9.4: Compute the FSLE for the generalized Bernoulli shift map $F(x) = \beta x \mod 1$ at $\beta = 1.01,\ 1.1,\ 1.5,\ 2$. What does changes with β?

Hint: *Follow the hint of Ex.9.3*

Exercise 9.5: Consider the two coupled Lorenz models as in Eq. (9.23) with the parameters as described in the text, compute the full Lyapunov spectrum $\{\lambda_i\}_{i=1,6}$ and reproduce Fig. 9.10.

Chapter 10

Chaos in Numerical and Laboratory Experiments

> Science is built up with facts, as a house is with stones. But a collection of facts is no more a science than a heap of stones is a house.
>
> *Jules Henri Poincaré (1854–1912)*

In the previous Chapters, we illustrated the main techniques for computing Lyapunov exponents, fractal dimensions of strange attractors, Kolmogorov-Sinai and ε-entropy in dynamical systems whose evolution laws are known in the form of either ordinary differential equations or maps. However, we did not touch any practical aspects, unavoidable in numerical and experimental studies, such as:

- Any numerical study is affected by "errors" due to discretization of number representation and of the algorithmic procedures. We may thus wonder in which sense numerical trajectories represent "true" ones;
- In typical experiments, the variables $(x_1, \ldots, x_d)$ describing the system state are unknown and, very often, the phase-space dimension d is unknown too;
- Usually, experimental measurements provide just a time series $u_1, u_2, \cdots, u_M$ (depending on the state vector $\boldsymbol{x}$ of the underlying system) sampled at discrete times $t_1 = \tau, t_2 = 2\tau, \cdots, t_M = M\tau$. How to compute from this series quantities such as Lyapunov exponents or attractor dimensions? Or, more generally, to assess the deterministic or stochastic nature of the system, or to build up from the time series a mathematical model enabling predictions.

Perhaps, to someone the above issues may appear relevant just to practitioners, working in applied sciences. We do not share such an opinion. Rather, we believe that mastering the outcomes of experiments and numerical computations is as important as understanding chaos foundations.

10.1 Chaos in silico

A part from rather special classes of systems amenable to analytical treatments, when studying nonlinear systems, numerical computations are mandatory. It is thus natural to wonder to what extent *in silico* experiments, unavoidably affected by round-off errors due to the finite precision of real number representation on computers (Box B.20), reflect the "true" dynamics of the actual system, expressed in terms of ODEs or maps whose solution is carried out by the computer algorithm.

Without loss of generality, consider a map

$$\boldsymbol{x}(t+1) = \boldsymbol{g}(\boldsymbol{x}(t)) \tag{10.1}$$

representing the "true" evolution law of the system, $\boldsymbol{x}(t) = S^t\boldsymbol{x}(0)$. Any computer implementation of Eq. (10.1) is affected by round-off errors, meaning that the computer is actually implementing a slightly modified evolution law

$$\boldsymbol{y}(t+1) = \tilde{\boldsymbol{g}}_\epsilon(\boldsymbol{y}(t)) = \boldsymbol{g}(\boldsymbol{y}(t)) + \epsilon\boldsymbol{h}(\boldsymbol{y}(t)), \tag{10.2}$$

where ϵ is a small number, say $\mathcal{O}(10^{-a})$ with a being the number of digits in the floating-point representation (Box B.20). The $\mathcal{O}(1)$ function $\boldsymbol{h}(\boldsymbol{y})$ is typically unknown and depends on computer hardware and software, algorithmic implementation and other technical details. However, for our purposes, the exact knowledge of $\boldsymbol{h}$ is not crucial.[1] In the following, Eq. (10.1) will be dubbed the "true" dynamics and Eq. (10.2) the "false" one, $\boldsymbol{y}(t) = \widetilde{S}^t\boldsymbol{y}(0)$.

It is worth remarking that understanding the relationship between the "true" dynamics of a system and that obtained with a small change of the evolution law is a general problem, not restricted to computer simulations. For instance, in weather forecasting, this problem is known as *predictability of the second kind* [Lorenz (1996)], where first kind is referred to the predictability limitations due to an imperfect knowledge on initial conditions. In general, the problem is present whenever the evolution laws of a system are not known with arbitrary precision, e.g. the determination of the parameters of the equations of motion is usually affected by measurement errors. We also mention that, at a conceptual level, this problem is related to the *structural stability* problem (see Sec. 6.1.2). Indeed, if we cannot determine with arbitrary precision the evolution laws, it is highly desirable that, at least, a few properties were not too sensitive to details of the equations [Berkooz (1994)]. For example, in a system with a strange attractor, small generic changes of the evolution laws should not drastically modify the the dynamics.

When $\epsilon \ll 1$, from Eqs. (10.1)-(10.2), it is easy to derive the evolution law for the difference between true and false trajectories

$$\boldsymbol{y}(t) - \boldsymbol{x}(t) = \boldsymbol{\Delta}(t) \simeq \mathbb{L}[\boldsymbol{x}(t-1)]\boldsymbol{\Delta}(t-1) + \epsilon\,\boldsymbol{h}[\boldsymbol{x}(t-1)] \tag{10.3}$$

where we neglected terms $\mathcal{O}(|\boldsymbol{\Delta}|^2)$ and $\mathcal{O}(\epsilon|\boldsymbol{\Delta}|)$, and $\mathbb{L}_{ij}[\boldsymbol{x}(t)] = \partial g_i/\partial x_j|_{\boldsymbol{x}(t)}$ is the usual stability matrix computed in $\boldsymbol{x}(t)$. Iterating Eq. (10.3) from $\boldsymbol{\Delta}(0) = 0$, for $t \geq 2$, we have

$$\begin{aligned}\boldsymbol{\Delta}(t) &= \epsilon\,\mathbb{L}[t-1]\mathbb{L}[t-2]\cdots\mathbb{L}[2]\,\boldsymbol{h}(\boldsymbol{x}(1)) + \epsilon\,\mathbb{L}[t-1]\mathbb{L}[t-2]\cdots\mathbb{L}[3]\,\boldsymbol{h}(\boldsymbol{x}(2)) + \cdots\\ &\quad\cdots + \epsilon\,\mathbb{L}[t-1]\mathbb{L}[t-2]\,\boldsymbol{h}(\boldsymbol{x}(t-2)) + \epsilon\,\mathbb{L}[t-1]\,\boldsymbol{h}(x(t-1)) + \epsilon\,\boldsymbol{h}(\boldsymbol{x}(t)),\end{aligned}$$

where $\mathbb{L}[j]$ is a shorthand for $\mathbb{L}[\boldsymbol{x}(j)]$.

[1] Notice that ODEs are practically equivalent to discrete time maps: the rule (10.1) can be seen as the exact evolution law between t and $t + dt$, while (10.2) is actually determined by the used algorithm (e.g. the Runge-Kutta), the round-off truncation, etc.

The above equation is similar in structure to that ruling the tangent vector dynamics (5.18), where ϵ plays the role of the uncertainty on the initial condition. As the "forcing term" $\epsilon \boldsymbol{h}[\boldsymbol{x}(t-1)]$ does not change the asymptotic behavior, for large times, the difference between "true" and "false" trajectories $|\boldsymbol{\Delta}(t)|$ will grow as [Crisanti *et al.* (1989)]

$$|\boldsymbol{\Delta}(t)| \sim \epsilon \, \mathrm{e}^{\lambda_1 t} \,.$$

Summarizing, an uncertainty on the evolution law has essentially the same effect of an uncertainty on the initial condition when the dynamical law is perfectly known. This does not sound very surprising but may call into question the effectiveness of computer simulations of chaotic systems: as a small uncertainty on the evolution law leads to an exponential separation between "true" and "false" trajectories, does a numerical ("false") trajectory reproduce the correct features of the "true" one?

Box B.20: Round-off errors and floating-point representation

Modern computers deal with real numbers using the *floating-point representation.* A floating-point number consists of two sequences of bits

(1) one representing the digits in the number, including its sign;

(2) the other characterizes the magnitude of the number and amounts to a signed exponent determining the position of the radix point.

For example, by using base-10, i.e. the familiar decimal notation, the number 289658.0169 is represented as $+2.896580169 \times 10^{+05}$.

The main advantage of the floating-point representation is to permit calculations over a wide range of magnitudes via a fixed number of digits. The drawback is, however, represented by the unavoidable errors inherent to the use of a limited amount of digits, as illustrated by the following example. Assume to use a decimal floating-point representation with 3 digits only, then the product $P = 0.13 \times 0.13$ which is equal to 0.0169 will be represented as $\tilde{P} = 1.6 \times 10^{-2} = 0.016$ or, alternatively, as $\tilde{P} = 1.7 \times 10^{-2}$.[2] The difference between the calculated approximation $\tilde{P}$ and its exact value P is known as *round-off error*. Obviously, increasing the number of digits reduces the magnitude of round-off errors, but any finite-digit representation will necessarily entails an error.

The main problem in floating-point arithmetic is that small errors can grow when the number of consecutive operations increases. In order to avoid miscomputations, it is thus crucial, when possible, to rearrange the sequence of operations to get a mathematically equivalent result but with the smallest round-off error. As an example, we can mention Archimedes' evaluation of π through the successive approximation of a circle by inscribed or circumscribed regular polygons with an increasing number of sides. Starting from a hexagon circumscribing a unit-radius circle and, then, doubling the number of sides, we

[2]There are, at least, two ways of approximating a number with a limited amount of digits: *truncation* corresponding to drop off the digits from a position on, i.e. 1.6×10^{-2} in the example, and *rounding*, i.e. 1.7×10^{-2}, that is to truncate the digits to the nearest floating number.

have a sequence of regular polygons with 6×2^n sides each of length t_n from which

$$\pi = 6 \lim_{n\to\infty} 2^n t_n, \quad \text{with} \quad t_{n+1} = \frac{\sqrt{t_n^2+1}-1}{t_n},$$

where $t_0 = 1/\sqrt{3}$. The above sequence $\{t_n\}$ can be also evaluated via the equivalent recursion:

$$t_{n+1} = \frac{t_n}{\sqrt{t_n^2+1}+1},$$

which is more convenient for floating-point computations as the propagation of round-off error is limited. Indeed it allows a 16-digit precision for π, by using 53 bits of significance. The former sequence, on the contrary, is affected by cancellation errors in the numerator, thus when the recurrence is applied, first accuracy improves, but then it deteriorates spoiling the result.

10.1.1 *Shadowing lemma*

A first mathematical answer to the above question, satisfactory at least for a certain class of systems, is given by the *shadowing lemma* [Katok and Hasselblatt (1995)] stating that, for hyperbolic systems (Box B.10), a computer may not calculate the true trajectory generated by $\boldsymbol{x}(0)$, but it nevertheless finds an approximation of a true trajectory starting from an initial state close to $\boldsymbol{x}(0)$.

Before enunciating the shadowing lemma, it is useful to introduce two definitions:
a) the orbit $\boldsymbol{y}(t)$ with $t = 0, 1, 2, \dots, T$ is an ϵ*−pseudo orbit* for the map (10.1) if $|\boldsymbol{g}(\boldsymbol{y}(t)) - \boldsymbol{y}(t+1)| < \epsilon$ for any t.
b) the "true" orbit $\boldsymbol{x}(t)$ with $t = 0, 1, 2, \dots, T$ is a δ−shadowing orbit for $\boldsymbol{y}(t)$ if $|\boldsymbol{x}(t) - \boldsymbol{y}(t)| < \delta$ for all t.

Shadowing lemma: If the invariant set of the map (10.1) is compact, invariant and hyperbolic, for all sufficiently small $\delta > 0$ there exists $\epsilon > 0$ such that each ϵ-pseudo orbit is δ-shadowed by a unique true orbit.

In other words, even if the trajectory of the perturbed map $\boldsymbol{y}(t)$ which starts in $\boldsymbol{x}(0)$, i.e. $\boldsymbol{y}(t) = \tilde{S}^t\boldsymbol{x}(0)$, does not reproduce the true trajectory $S^t\boldsymbol{x}(0)$, there exists a true trajectory with initial condition $\boldsymbol{z}(0)$ close to $\boldsymbol{x}(0)$ that remains close to (shadows) the false trajectory, i.e. $|S^t\boldsymbol{z}(0) - \tilde{S}^t\boldsymbol{x}(0)| < \delta$ for any t, as illustrated in Fig. 10.1.

The importance of the previous result for numerical computations is rather transparent, when applied to an ergodic system. Although the true trajectory obtained from $\boldsymbol{x}(0)$ and the false one from the same initial condition become very different after a time $\mathcal{O}(1/\lambda_1 \ln(1/\epsilon))$, the existence of a shadowing trajectory along with ergodicity imply that time averages computed on the two trajectories will be equivalent. Thus shadowing lemma and ergodicity imply "statistical reproducibility" of the true dynamics by the perturbed one [Benettin *et al.* (1978a)].

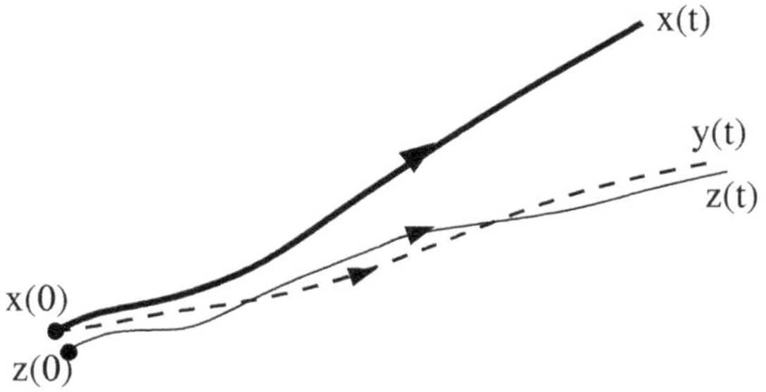

Fig. 10.1 Sketch of the shadowing mechanism: the tick line indicates the "true" trajectory from $\boldsymbol{x}(0)$ (i.e. $\boldsymbol{x}(t) = S^t\boldsymbol{x}(0)$), the dashed line the "false" one from $\boldsymbol{x}(0)$ (i.e. $\boldsymbol{y}(t) = \tilde{S}^t\boldsymbol{x}(0)$), while the solid line is the "true" trajectory from $\boldsymbol{z}(0)$ (i.e. $\boldsymbol{z}(t) = S^t\boldsymbol{z}(0)$) shadowing the "false" one.

We now discuss an example that, although specific, well illustrates the main aspects of the shadowing lemma. Consider as "true" dynamics the shift map

$$x(t+1) = 2x(t) \mod 1\,, \tag{10.4}$$

and the perturbed dynamics

$$y(t+1) = 2y(t) + \epsilon(t+1) \mod 1$$

where $\epsilon(t)$ represents a small perturbation, meaning that $|\epsilon(t)| \leq \epsilon$ for each t.

The trajectory $y(t)$ from $t = 0$ to $t = T$ can be expressed in terms of the initial condition $x(0)$ noticing that

$$\begin{aligned}
y(0) &= x(0) + \epsilon(0) \\
y(1) &= 2x(0) + 2\epsilon(0) + \epsilon(1) \mod 1 \\
&\vdots \\
y(T) &= 2^T x(0) + \sum_{j=0}^{T} 2^{T-j}\epsilon(j) \mod 1\,.
\end{aligned}$$

Now we must determine a $z(0)$ which, evolved according to the map (10.4), generates a trajectory that δ-shadows the perturbed one $(y(0), y(1), \ldots, y(T))$. Clearly, this require that $S^k z(0) = (\,2^k z(0) \mod 1\,)$ is close to $\tilde{S}^k x(0) = 2^k x(0) + \sum_{j=0}^{k} 2^{k-j}\epsilon(j) \mod 1$, for $k \leq T$. An appropriate choice is

$$z(0) = x(0) + \sum_{j=0}^{T} 2^{-j}\epsilon(j) \mod 1\,.$$

In fact, the "true" evolution from $z(0)$ is given by

$$z(k) = 2^k x(0) + \sum_{j=0}^{T} 2^{k-j}\epsilon(j) \mod 1\,,$$

and computing the difference $\Delta(k) = y(k) - z(k) = \sum_{j=k+1}^{T} 2^{k-j}\epsilon(j)$, for each $k \leq T$, we have

$$|\Delta(k)| \leq \sum_{j=k+1}^{T} 2^{k-j}|\epsilon(j)| \leq \sum_{j=k+1}^{T} 2^{k-j}\epsilon \leq \epsilon\,,$$

which confirms that the difference between the true trajectory starting from $z(0)$ and that one obtained by the perturbed dynamics remains small at any time. However, it is should be clear that determining the proper $z(0)$ for δ-shadowing the perturbed trajectory up to a given time T requires the knowledge of the perturbed trajectory in the whole interval $[0\!:\!T]$.

The shadowing lemma holds in hyperbolic chaotic systems, but generic chaotic systems are not hyperbolic, so that the existence of a $\delta-$shadowing trajectory is not granted, in general. There are some interesting results which show, with the help of computers and interval arithmetic,[3] the existence of $\epsilon-$pseudo orbit which is $\delta-$shadowed by a true orbit up to a large time T. For instance Hammel *et al.* (1987) have shown that for the logistic map with $r = 3.8$ and $x(0) = 0.4$ for $\delta = 10^{-8}$ it results $\epsilon = 3 \times 10^{-14}$ and $T = 10^7$, while for the Hénon map with $a = 1.4\,,\ b = 0.3\,,\ \boldsymbol{x}(0) = (0,0)$ for $\delta = 10^{-8}$ one has $\epsilon = 10^{-13}$ and $T = 10^6$.

10.1.2 *The effects of state discretization*

The above results should have convinced the reader that round-off errors do not represent a severe limitation to computer simulations of chaotic systems. There is, however, an apparently more serious problem inherent to floating-point computations (Box B.20). Because of the finite number of digits, when iterating dynamical systems, one basically deals with discrete systems having a finite number $\mathcal{N}$ of states. In this respect, simulating a chaotic system on a computer is not so different from investigating a deterministic cellular automaton [Wolfram (1986)].

A direct consequence of phase-space discreteness and finiteness is that any numerical trajectory must become periodic, questioning the very existence of chaotic trajectories in computer experiments.

To understand why finiteness and discreteness imply periodicity, consider a system of N elements, each assuming an integer number k of distinct values. Clearly, the total number of possible states is $\mathcal{N} = k^N$. A deterministic rule to pass from one state to another can be depicted in terms of oriented graphs: a set of points, representing the states, are connected by arrows, indicating the time evolution (Fig. 10.2). Determinism implies that each point has one, and only one, outgoing arrow, while

[3] An interval is the set of all real numbers between and including the interval's lower and upper bounds. Interval arithmetic is used to evaluate arithmetic expressions over sets of numbers contained in intervals. Any interval arithmetic result is a new interval that is guaranteed to contain the set of all possible resulting values. Interval arithmetic allow the uncertainty in input data to be dealt with and round-off errors to be rigorously taken into account, for some examples see Lanford (1998).

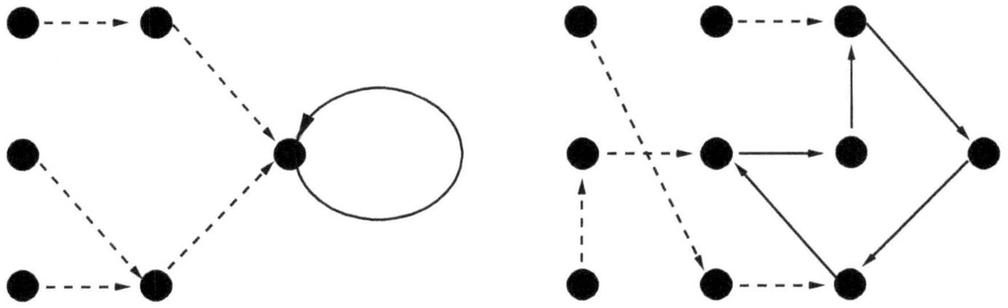

Fig. 10.2 Schematic representation of the evolution of a deterministic rule with a finite number of states: (a) with a fixed point, (b) with a periodic cycle.

different arrows can end at the same point. It is then clear that, for any system with a finite number of states, each initial condition evolves to a definite attractor, which can be either a fixed point, or a periodic orbit, see Fig. 10.2.

Having understood that discrete state systems are necessarily asymptotically trivial, in the sense of being characterized by a periodic orbit, a rather natural question concerns how the period $\mathcal{T}$ of such orbit depends on the number of states $\mathcal{N}$ and eventually on the initial state [Grebogi *et al.* (1988)]. For deterministic discrete state systems, such a dependence is a delicate issue. A possible approach is in terms of random maps [Coste and Hénon (1986)]. As described in Box B.21, if the number of states of the system is very large, $\mathcal{N} \gg 1$, the basic result for the average period is

$$\mathcal{T}(\mathcal{N}) \sim \sqrt{\mathcal{N}}. \tag{10.5}$$

We have now all the instruments to understand whether discrete state computers can simulate continuous-state chaotic trajectories. Actually the proper question can be formulated as follows. How long should we wait before recognizing that a numerical trajectory is periodic?

To answer, assume that n is the number of digits used in the floating-point representation, and $D(2)$ the correlation dimension of the attractor of the chaotic system under investigation, then the number of states $\mathcal{N}$ can reasonably be expected to scale as $\mathcal{N} \sim 10^{nD(2)}$ [Grebogi *et al.* (1988)], and thus from Eq. (10.5) we get

$$\mathcal{T} \sim 10^{\frac{nD(2)}{2}} .$$

For instance, for $n = 16$ and $D(2) \approx 1.4$ as in the Hénon map we should typically wait more than 10^{10} iterations before recognizing the periodicity. The larger $D(2)$ or the number of digits, the longer numerical trajectories can be considered chaotic.

To better illustrate the effect of discretization, we conclude this section discussing the generalized Arnold map

$$\begin{pmatrix} \boldsymbol{x}(t+1) \\ \boldsymbol{y}(t+1) \end{pmatrix} = \begin{pmatrix} \mathbb{I} & \mathbb{A} \\ \mathbb{B} & \mathbb{I}+\mathbb{B}\mathbb{A} \end{pmatrix} \begin{pmatrix} \boldsymbol{x}(t) \\ \boldsymbol{y}(t) \end{pmatrix} \mod 1 , \tag{10.6}$$

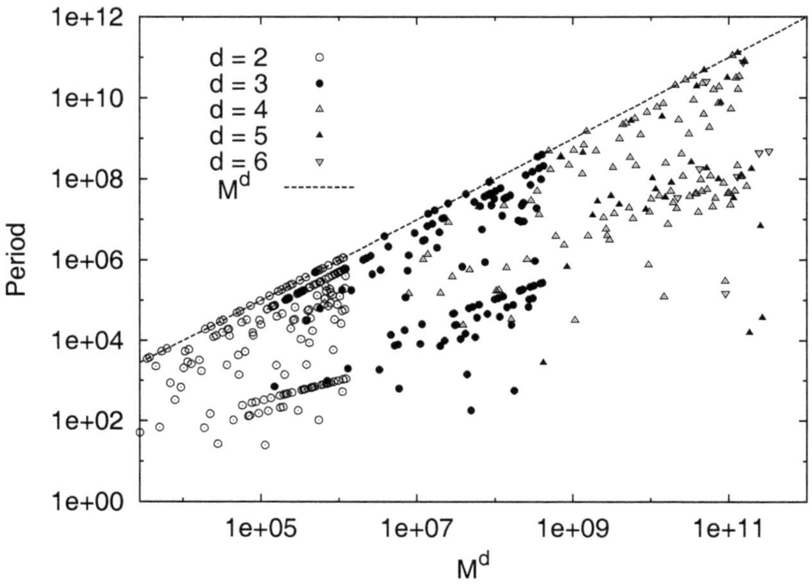

Fig. 10.3 Period $\mathcal{T}$ as a function of the dimensionality d of the system (10.7) and different initial conditions. The dashed line corresponds to the prediction (10.5).

where $\mathbb{I}$ denotes the $(d \times d)$-identity matrix, and $\mathbb{A}$, $\mathbb{B}$ are two $(d \times d)$−symmetric matrices whose entries are integers. The discretized version of map (10.6) is

$$\begin{bmatrix} \boldsymbol{z}(t+1) \\ \boldsymbol{w}(t+1) \end{bmatrix} = \begin{bmatrix} \mathbb{I} & \mathbb{A} \\ \mathbb{B} & \mathbb{I}+\mathbb{B}\mathbb{A} \end{bmatrix} \begin{bmatrix} \boldsymbol{z}(t) \\ \boldsymbol{w}(t) \end{bmatrix} \mod M \tag{10.7}$$

where each component z_i and $w_i \in \{0, 1, \ldots, M-1\}$. The number of possible states is thus $\mathcal{N} = M^{2d}$ and the probabilistic argument (10.5) gives $\mathcal{T} \sim M^d$. Figure 10.3 shows the period $\mathcal{T}$ for different values of M and d and various initial conditions. Large fluctuations and strong sensitivity of $\mathcal{T}$ on initial conditions are well evident.

These features are generic both in symplectic and dissipative systems [Grebogi *et al.* (1988)], and the estimation Eq. (10.5) gives just an upper bound to the typical number of meaningful iterations of a map on a computer. On the other hand, the period $\mathcal{T}$ is very large for almost all practical purposes, but for one or two dimensional maps with few digits in the floating-point representation.

It should be remarked that entropic measurements (of e.g. the N-block ε-entropies) of the sequences obtained by the discretized map have shown that the asymptotic regularity can be accessed only for large N and small ε, meaning that for large times ($< \mathcal{T}$) the trajectories of the discretized map can be considered chaotic. This kind of discretized map can be used to build up very efficient pseudo-random number generators [Falcioni *et al.* (2005)].

Box B.21: Effect of discretization: a probabilistic argument

Chaotic indicators, such as LEs and KS-entropy, cannot be used in deterministic discrete-state systems because their definitions rely on the continuous character of the system states. Moreover, the asymptotic periodic behavior seems to force the conclusion that discrete states systems are trivial, from an entropic or algorithmic complexity point of view.

The above mathematically correct conclusions are rather unsatisfactory from a physical point of view, indeed from this side the following questions are worth of investigations:

(1) What is the "typical" period $\mathcal{T}$ for systems with N elements, each assuming k distinct values?

(2) When $\mathcal{T}$ is very large, how can we characterize the (possible) irregular behavior of the trajectories, on times that are large enough but still much smaller than $\mathcal{T}$?

(3) What does it happen in the limit $k^N \to \infty$?

Point (1) will be treated in a statistical context, using random maps [Coste and Hénon (1986)], while for a discussion of (2) and (3) we refer to Boffetta *et al.* (2002) and Wolfram (1986).

It is easy to realize that the number of possible deterministic evolutions for system composed by N elements each assuming k distinct values is finite. Let us now assume that all the possible rules are equiprobable. Denoting with $I(t)$ the state of the system, for a certain map we have a periodic attractor of period m if $I(p+m) = I(p)$ and $I(p+j) \neq I(p)$, for $j < m$. The probability, $\omega(m)$, of this periodic orbit is obtained by specifying that the first $(p+m-1)$ consecutive iterates of the map are distinct from all the previous ones, and the $(p+m)$-th iterate coincides with the p-th one. Since one has $I(p+1) \neq I(p)$, with probability $(1-1/\mathcal{N})$; $I(p+2) \neq I(p)$, with probability $(1-2/\mathcal{N})$; ......; $I(p+m-1) \neq I(p)$, with probability $(1-(m-1)/\mathcal{N})$; and, finally, $I(p+m) = I(p)$ with probability $(1/\mathcal{N})$, one obtains

$$\omega(m) = \left(1 - \frac{1}{\mathcal{N}}\right)\left(1 - \frac{2}{\mathcal{N}}\right)\cdots\left(1 - \frac{m-1}{\mathcal{N}}\right)\frac{1}{\mathcal{N}}.$$

The average number, $M(m)$, of cycles of period m is

$$M(m) = \frac{\mathcal{N}}{m}\,\omega(m) \overset{(\mathcal{N}\gg 1)}{\approx} \frac{e^{-m^2/2\mathcal{N}}}{m},$$

from which we obtain $\mathcal{T} \sim \sqrt{\mathcal{N}}$ for the average period.

10.2 Chaos detection in experiments

The practical contribution of chaos theory to "real world" interpretation stems also from the possibility to detect and characterize chaotic behaviors in experiments and observations of naturally occurring phenomena. This and the next section will focus on the main ideas and methods able to detect chaos and quantify chaos indicators from experimental signals.

Typically, experimental measurements have access only to scalar observables $u(t)$ depending on the state $(x_1(t), x_2(t), \dots, x_d(t))$ of the system, whose dimensionality d is unknown. For instance, $u(t)$ can be the function $u = x_1^2 + x_2^2 + x_3^2$ of the coordinates (x_1, x_2, x_3) of Lorenz's system. Assuming that the dynamics of the system underlying the experimental investigation is ruled by ODEs, we expect that the observable u obeys a differential equation as well

$$\frac{\mathrm{d}^d u}{\mathrm{d}t^d} = G\left[u, \frac{\mathrm{d}u}{\mathrm{d}t}, \frac{\mathrm{d}^2 u}{\mathrm{d}t^2}, \dots, \frac{\mathrm{d}^{d-1} u}{\mathrm{d}t^{d-1}}\right]$$

where the phase space is determined by the d–dimensional vector

$$\left(u, \frac{\mathrm{d}u}{\mathrm{d}t}, \frac{\mathrm{d}^2 u}{\mathrm{d}t^2}, \dots, \frac{\mathrm{d}^{d-1} u}{\mathrm{d}t^{d-1}}\right).$$

Therefore, in principle, if we were able to compute from the signal $u(t)$ a sufficient number of derivatives, we might reconstruct the underlying dynamics. As the signal is typically known only in the form of discrete-time sequence $u_1, u_2, \dots, u_M$ (with $u_i = u(i\tau)$ and $i = 1, \dots M$) its derivatives can be determined in terms of finite differences, such as

$$\left.\frac{\mathrm{d}u}{\mathrm{d}t}\right|_{t=k\tau} \simeq \frac{u_{k+1} - u_k}{\tau}, \qquad \left.\frac{\mathrm{d}^2 u}{\mathrm{d}t^2}\right|_{t=k\tau} \simeq \frac{u_{k+1} - 2u_k + u_{k-1}}{\tau^2}.$$

As a consequence, the knowledge of $(u, \mathrm{d}u/\mathrm{d}t)$ is equivalent to (u_j, u_{j-1}); while $(u, \mathrm{d}u/\mathrm{d}t, \mathrm{d}^2u/\mathrm{d}t^2)$ corresponds to (u_j, u_{j-1}, u_{j-2}), and so on. This suggests that information on the underlying dynamics can be extracted in terms of the *delay-coordinate vector* of dimension m

$$\boldsymbol{Y}_k^m = (u_k, u_{k-1}, u_{k-2}, \dots, u_{k-(m-1)}),$$

which stands at the basis of the so-called *embedding technique* [Takens (1981); Sauer *et al.* (1991)]. Of course, if m is too small,[4] the delay-coordinate vector cannot catch all the features of the system. While, we can fairly expect that when m is large enough, the vector $\boldsymbol{Y}_k^m$ can faithfully reconstruct the properties of the underlying dynamics. Actually, a powerful mathematical result from Takens (1981) ensures that an attractor with box counting dimension D_F can always be reconstructed if the embedding dimension m is larger than $2[D_F] + 1$,[5] see also Sauer *et al.* (1991); Ott *et al.* (1994); Kantz and Schreiber (1997). This result lies at the basis of the embedding technique, and, at least in principle, gives an answer to the problem of experimental signals treatment.

[4]In particular, if $m < [D_F] + 1$ where D_F is the box counting dimension of the attractor and the $[s]$ indicate the integer part of the real number s.

[5]Notice that this does not mean that with a lower m it is not possible to obtain a faithful reconstruction.

If m is large enough to ensure phase-space reconstruction then the embedding vectors sequence $(\boldsymbol{Y}_1^m, \boldsymbol{Y}_2^m, \ldots, \boldsymbol{Y}_M^m)$ bears the same information of the sequence $(\boldsymbol{x}_1, \boldsymbol{x}_2, \ldots, \boldsymbol{x}_M)$ obtained with the state variables, sampled at discrete time interval $\boldsymbol{x}_j = \boldsymbol{x}(j\tau)$. In particular, this means that we can achieve a quantitative characterization of the dynamics by using essentially the same methods discussed in Chap. 5 and Chap. 8 applied to the embedded dynamics.

Momentarily disregarding the unavoidable practical limitations, to be discussed later, once the embedding vectors have been derived from the experimental time series, we can proceed as follows. For each value of m, we have the proxy vectors $\boldsymbol{Y}_1^m, \boldsymbol{Y}_2^m, \ldots, \boldsymbol{Y}_M^m$ for the system states, from which we can evaluate the generalized dimensions $D_m(q)$ and entropies $h_m^{(q)}$, and study their dependence on m.

The procedure to compute the generalized dimensions is rather simple and essentially coincides with the Grassberger-Procaccia method (Sec. 5.2.4). For each m, we compute the number of points in a sphere of radius ε around the point $\boldsymbol{Y}_k^m$:

$$n_k^{(m)}(\varepsilon) = \frac{1}{M-m} \sum_{j \neq k} \Theta(\varepsilon - |\boldsymbol{Y}_k^m - \boldsymbol{Y}_j^m|)$$

from which we estimate the generalized correlation integrals

$$C_m^{(q)}(\varepsilon) = \frac{1}{M-m+1} \sum_{k=1}^{M-m+1} \left[n_k^{(m)}(\varepsilon) \right]^q , \tag{10.8}$$

and hence the generalized dimensions

$$D_m(q) = \lim_{\varepsilon \to 0} \frac{1}{q-1} \frac{\ln C_m^{(q-1)}(\varepsilon)}{\ln \varepsilon} . \tag{10.9}$$

The correlation integral also allows the generalized or Renyi's entropies $h_m^{(q)}$ to be determined as (see Eq. (9.15)) [Grassberger and Procaccia (1983a)]

$$h_m^{(q)} = \lim_{\varepsilon \to 0} \frac{1}{(q-1)\tau} \ln \left[\frac{C_m^{(q-1)}(\varepsilon)}{C_{m+1}^{(q-1)}(\varepsilon)} \right] , \tag{10.10}$$

or alternatively we can use the method proposed by Cohen and Procaccia (1985) (Sec. 9.3). Of course, for finite ε, we have an estimator for the generalized (ε, τ)-entropies. For instance, Fig. 10.4 shows the correlation dimension extracted from a Rayleigh-Bénard experiment: as m increases and the phase-space reconstruction becomes effective, $D_m(2)$ converges to a finite value corresponding to the correlation dimension of the attractor of the underlying dynamics. In the same figure it is also displayed the behavior of $D_m(2)$ for a simple stochastic (non-deterministic) signal, showing that no saturation to any finite value is obtained in that case. This difference between deterministic and stochastic signals seems to suggest that it is possible to discern the character of the dynamics from quantities like $D_m(q)$ and $h_m^{(q)}$. This is indeed a crucial aspect, as the most interesting application of the embedding method is the study of systems whose dynamics is not known *a priori*.

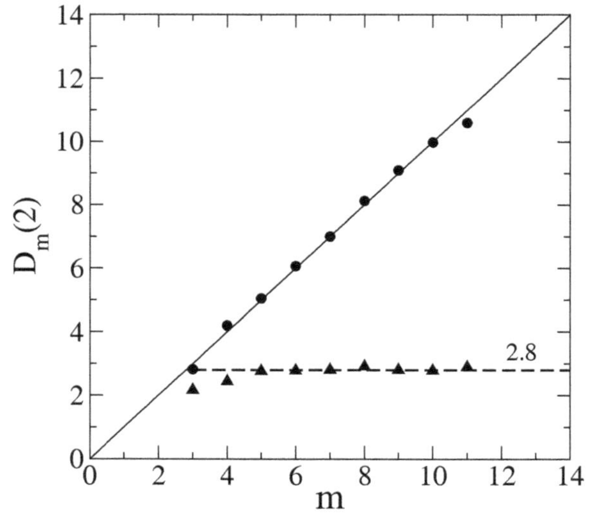

Fig. 10.4 $D_m(2)$ vs. m for Rayleigh-Bénard convection experiment (triangles), and for numerical white noise (dots). [After Malraison *et al.* (1983)]

Unfortunately, however, the detection of saturation to a finite value for $D_m(2)$ from a signal is generically not enough to infer the presence of deterministic chaos. For instance, Osborne and Provenzale (1989) provided examples of stochastic processes showing a spurious saturation of $D_m(2)$ for increasing m. We shall come back to the problem of distinguishing deterministic chaos from noise in experimental signals in the next section.[6]

Before examining the practical limitations, always present in experimental or numerical data analysis, we mention that embedding approach can be useful also for computing the Lyapunov exponents [Wolf *et al.* (1985); Eckmann *et al.* (1986)] (as briefly discussed in Box B.22).

Box B.22: Lyapunov exponents from experimental data

In numerical experiments we know the dynamics of the system and thus also the stability matrix along a given trajectory necessary to evaluate the tangent dynamics and the Lyapunov exponents of the system (Sec. 5.3). These are, of course, unknown in typical experiments, so that we need to proceed differently. In principle to compute the maximal LE would be enough to follow two trajectories which start very close to each other. Since, a part from a few exception [Espa *et al.* (1999); Boffetta *et al.* (2000d)], it is not easy to have two close states $\boldsymbol{x}(0)$ and $\boldsymbol{x}'(0)$ in a laboratory experiment, even the evaluation of

[6] We remark however that Theiler (1991) demonstrated that such a behavior should be ascribed to the non-stationarity and correlations of the analyzed time series, which make critically important the number of data points. The artifact indeed disappears when a sufficient number of data points is considered.

the first LE λ_1 from the growth of the distance $|\boldsymbol{x}(t) - \boldsymbol{x}'(t)|$ does not appear to be so simple. However, once identified the proper embedding dimension, it is possible to compute, at least in principle, λ_1 from the data. There are several methods [Kantz and Schreiber (1997)], here we briefly sketch that proposed by Wolf *et al.* (1985).

Assume that a point $\boldsymbol{Y}_j^m$ is observed close enough to another point $\boldsymbol{Y}_i^m$, i.e. if they are two "analogues" we can say that the two trajectories $\boldsymbol{Y}_{i+1}^m, \boldsymbol{Y}_{i+2}^m, \ldots$ and $\boldsymbol{Y}_{j+1}^m, \boldsymbol{Y}_{j+2}^m, \ldots$ evolve from two close initial conditions. Then one can consider $\delta(k) = |\boldsymbol{Y}_{i+k}^m - \boldsymbol{Y}_{j+k}^m|$ as a small quantity, so that monitoring the time evolution of $\delta(k)$, which is expected to grow as $\exp(\lambda_1 \tau k)$, the first Lyapunov exponent can be determined. In practice, one computes

$$\Lambda_m(k) = \left\langle \frac{1}{N_{ij}(\varepsilon)} \sum_{j:|\boldsymbol{Y}_i^m - \boldsymbol{Y}_j^m| < \varepsilon} \ln\left(\frac{|\boldsymbol{Y}_{i+k}^m - \boldsymbol{Y}_{j+k}^m|}{|\boldsymbol{Y}_i^m - \boldsymbol{Y}_j^m|}\right)\right\rangle_i ,$$

where $N_{ij}(\varepsilon)$ is the number of $\boldsymbol{Y}_j^m$ such that $|\boldsymbol{Y}_i^m - \boldsymbol{Y}_j^m| < \varepsilon$, and the average $\langle\,\rangle_i$ is over the points $\boldsymbol{Y}_i^m$ corresponding to an ergodic average. For k not too large, the nonlinear terms are expected to be negligible and we have

$$\frac{1}{k\tau}\Lambda_m(k) \simeq \lambda_1 .$$

The computation of the other Lyapunov exponents requires considerable more effort than just the first one. We do not enter the details, however the basic idea due to Eckmann *et al.* (1986) is to estimate the local Jacobian matrix around a point $\boldsymbol{Y}_i^m$, looking at the closest points (at least m), and then using the Benettin *et al.* (1978b, 1980) method (see Box B.9). The reader can find a detailed discussion about the methods to extract the Lyapunov exponents, and other indicators, from time series analysis in the book by Kantz and Schreiber (1997).

10.2.1 *Practical difficulties*

When applying the above mentioned ideas and methods to true experimental time series a number of limitations and delicate issues should be considered, as usual when passing from theory to practice. In this respect, *time series analysis* requires a long training to master the field. Several research papers, essays and books have been written on this subject so that, here, we will limit the discussion to some specific aspects, referring to the main literature in the field for more detailed discussions [Abarbanel (1996); Kantz and Schreiber (1997); Hegger *et al.* (1999)].

10.2.1.1 *Choice of delay time*

In principle, the sampling times τ is an irrelevant free parameter of the embedding reconstruction technique [Takens (1981)]. For instance, if τ is the minimum sampling time of the experimental apparatus, we can use any multiple of $n\tau$, and reconstruct the phase space in terms of another delay vector:

$$\boldsymbol{Y}_k^{m,n} = (u_k, u_{k-n}, u_{k-2n}, \ldots, u_{k-(m-1)n}) .$$

However, Takens' mathematical result refers to arbitrarily long, noise-free signals, while in practice this is not the case and careful values of n have to be chosen. If $n\tau$ is too small, the variables $\{u_k\}$ might be too correlated (redundant), which implies the need of very large embedding dimensions m for properly sampling the dynamics. Similarly, if $n\tau$ is too large, the variables $\{u_k\}'s$ are almost independent, and again a huge M is necessary to observe the dynamical dependencies among them. These intuitive ideas suggest the existence of an optimal delay time, to be determined.

A first natural attempt to determine the optimal n is from the correlation function

$$C_{uu}(k) = \frac{\langle u_{j+k} u_j \rangle - \langle u \rangle^2}{\langle u^2 \rangle - \langle u \rangle^2} \, .$$

For instance, n can be determined as the value k^* at which $C_{uu}(k^*)$ first passes through zero or goes below a certain threshold. In this way, we use neither too correlated nor completely independent variables. While this prescription is typically reasonably good [Abarbanel (1996); Kantz and Schreiber (1997)], it is unsatisfactory as it is based on a linear approach.

Another, usually well performing, proposal [Fraser and Swinney (1986)] is based on information theory indicators. In practice, one looks for the first minimum of the average mutual information (8.13) between the measurements at time t and those at time $t + n\tau$:

$$I(n\tau) = \int \mathrm{d}u(t)\, \mathrm{d}u(t+n\tau)\; P(u(t), u(t+n\tau)) \ln \left[\frac{P(u(t), u(t+n\tau))}{P((u(t))P(u(t+n\tau))} \right] \, ,$$

where $P(u(t))$ is the pdf of the variable u and $P(u(t), u(t+n\tau))$ the joint probability density of u at time t and $t+n\tau$. Note that $I(n\tau) \geq 0$ and $I(n\tau) = 0$ if $P(u(t), u(t+n\tau)) = P(u(t))P(u(t+n\tau))$. Typically, the choice based on the first minimum of the average mutual information is a good compromise between values that are not too small and those which are not too large [Kantz and Schreiber (1997)]. Its main advantage is that, unlike the autocorrelation function, the mutual information takes into account also nonlinear correlations.

10.2.1.2 *Choice of the embedding dimension*

As intuition may suggest, properly choosing the embedding dimension, together with the aforementioned delay time, is not only a crucial aspect of the embedding technique but also one of the most discussed in the literature. From a mathematical point of view, the embedding theorem [Takens (1981); Sauer *et al.* (1991)] states that $m \geq 2[D_F] + 1$ should ensure a perfect phase-space reconstruction. However such a bound is by no means very strict and, as discussed before, does not account for the presence of noise or finiteness of the data set.

Here, there is not enough space for a throughout review of all the proposals for determining the optimal m, their advantages and shortcomings, so that we will limit

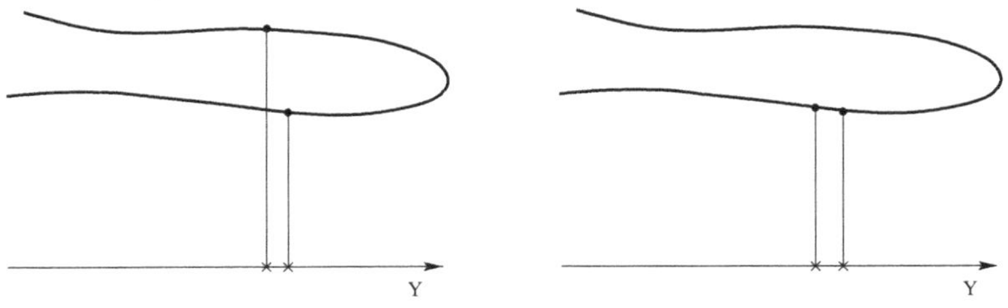

Fig. 10.5 False neighbors (left), true neighbors (right), for an attractor embedded in a plane with $m = 1$.

the discussion to one of the most used, which is the *false nearest neighbors search* proposed by Kennel *et al.* (1992). We should warn the reader that, most likely, an optimal choice of the delay time and embedding dimension can — if at all — only be defined relative to the specific purpose for which embedding is used [Kantz and Schreiber (1997); Hegger *et al.* (1999)].

The basic idea of the false nearest neighbors search method is the following. Suppose that $\widetilde{m}$ is the minimal embedding dimension required for faithfully reconstructing the system phase space. Then in a $(m > \widetilde{m})$-dimensional delay space, the reconstructed attractor is a perfect one-to-one version of the original phase space. In particular, neighbors of a given point are mapped onto neighbors in the embedded space. On the contrary, if $m < \widetilde{m}$, the attractor of the m-dimensional delay space is a projection of the "true" attractor. Therefore, points which are close in the embedding space may correspond to points which are not close on the true attractor, as illustrated in Fig. 10.5. When this happens, we are in the presence of false neighbors (FN). The fraction $F(m)$ of FN decreases with m and vanishes for $m \geq \widetilde{m}$. Of course, the presence of some noise may prevent $F(m)$ from vanishing. Therefore, in practice, $\widetilde{m}$ is determined by requiring $F(\widetilde{m})$ to be below a certain threshold, say 1%.

To complete the description we should now explain how to determine if two close points $\boldsymbol{Y}_i^m$ and $\boldsymbol{Y}_j^m$ in embedding space are actually distant in the true phase space, and thus false neighbor. Suppose that the distance $|\boldsymbol{Y}_i^m - \boldsymbol{Y}_j^m|$ is very small with respect to the linear size of the attractor. Then we can look at the two points after one step, and compute

$$R_{ij} = \frac{|\boldsymbol{Y}_{i+1}^m - \boldsymbol{Y}_{j+1}^m|}{|\boldsymbol{Y}_i^m - \boldsymbol{Y}_j^m|} \,.$$

If $\boldsymbol{Y}_i^m$ and $\boldsymbol{Y}_j^m$ correspond to states which are close on the true attractor R_{ij} will be close to 1, on the contrary R_{ij} will be a "large" number for FN. Indeed we expect that close points remain close when seen at successive times. Typically a threshold condition should be used to decide if R_{ij} is close to or far from 1.

10.2.1.3 *The necessary amount of data*

In principle, the embedding method can work for deterministic system having an attractor with an arbitrarily large but finite dimension. However, as a matter of facts, the use the method is beyond any practical possibility already for $D_F \gtrsim 5-6$.

The origin of such restriction can be traced back to the difficulties encountered in computing $D(2)$ from the correlation integral $C_m^{(2)}(\varepsilon)$ (10.8). For each m, $D_m(2)$ is determined as the slope of the plot $\ln C_m^{(2)}(\varepsilon)$ vs $\ln \varepsilon$. In practice, this procedure is meaningful if $\mathrm{d} \ln C_m^{(2)}(\varepsilon)/\mathrm{d} \ln \varepsilon$ is approximately constant on a certain range $\varepsilon_1 < \varepsilon < \varepsilon_2$, with $\varepsilon_2/\varepsilon_1$ large enough. Convincing estimates require, at least, $\varepsilon_2/\varepsilon_1 = \mathcal{O}(10)$. We should now wonder about the minimum amount of data M_{min} necessary to estimate $D_m(2)$ in such a range. A minimal requirement is $M_{min} \sim (\varepsilon_2/\varepsilon_1)^{D_m(2)}$, therefore M_{min} to detect an attractor with correlation dimension $D(2)$ increases exponentially with $D(2)$. As a rule of thumb, Smith (1988) proposed that $M_{min} \approx 42^{D(2)}$, which corresponds roughly to one decade and half of scaling. For $D(2) = 5$ or 6, the above rule imposes to use from hundreds of millions to billions of measurement data, too large for typical experiments.[7]

The previous argument can be repeated for the computation of the Kolmogorov-Sinai entropy: the Shannon-McMillan theorem states that the number of different trajectories giving contribution to h_{KS} increases as $\exp(m\tau h_{KS})$, therefore one needs $M_{min} \gg \exp(m\tau h_{KS})$. On the other hand, m must be at least $D(2)$, giving another severe limitations for the practical use of the embedding methods in high dimensional systems. Refined arguments show that, in general, if $D(1)$ is the information dimension (Sec. 5.2.3) of the attractor we want to reconstruct, M the number of data, m the embedding dimension and τ the time delay, the following inequality holds [Olbrich and Kantz (1997)]:

$$\frac{\varepsilon_2}{\varepsilon_1} \leq \left(M e^{-m\tau h_{KS}}\right)^{1/D(1)} . \tag{10.11}$$

The above arguments strictly limits the applicability of the phase-space reconstruction method to low dimensional systems, i.e. to systems with attractor's dimension $\leq 4-5$. However, when nonlinear time series analysis started to be massively employed in experimental data analysis, perhaps as a consequence of the enthusiasm for the availability of new tools, these limitations were overlooked by many researchers and a number of misleading papers appeared (see Ruelle (1990) for a discussion of some of these works).[8]

[7]The choice 42 has not a particular meaning. Other authors proposed slightly different recipes, for instance [Essex and Nerenberg (1991)] gave $10^{D(2)/2}$. However, replacing 42 with $\sqrt{10}$ does not change much the conclusion.

[8]Tsonis *et al.* (1993) noted that *Smith's result effectively "killed" all hopes for estimating the dimension of low-dimensional attractors irrespective of the availability of data*, and tried to give a less severe bound: $M_{min} \sim 10^{[2+0.4D(2)]}$. However, even this more optimistic ansatz does not change too much the negative conclusions on the plethora of papers on this issue at the end of '80s/beginning of '90s.

10.2.1.4 *Role of noise*

The unavoidable presence of noise in experiments can spoil, at least partially, the results of nonlinear analysis. There are two main sources of noise:

(a) interactions of the system under investigation, and/or of the experimental set up, with the external environment;
(b) uncertainties on the measurement procedure, so that the measured signal $u_j = u_j^T + \eta_j$ $(j = 1, \dots, M)$ differs from the true one u_j^T by an amount η_j which we denote as "noise".

In the case a), we speak of dynamical noise, meaning that we have a random dynamical system [Arnold (1998)], inherently stochastic in character.[9] In such a case, for small noise and in the presence of a low dimensional attractor the scenario is basically clear. We discuss, for instance, what happens for the correlation integral $C_m^{(2)}(\varepsilon)$. Let us, for example, consider the van der Pol equation (Box B.12) subjected to a small random forcing. Instead of a pure limit cycle, we will have a smooth distribution of points around the limit cycle of the noiseless system, having a thickness ε_c increasing with the strength of the random noise. In generic chaotic systems, the presence of noise induces a smoothing of the fractal structure of the attractor at scale smaller that ε_c. The typical scenario is the following: for $\varepsilon > \varepsilon_c$ the presence of the noise does not affect too much the fractal structure. On the contrary for $\varepsilon < \varepsilon_c$ one sees the noisy nature of the system and the logarithmic slope $D_m(2)$ of $C_m^{(2)}(\varepsilon)$ increases linearly with m.

In the case b) we speaks of measurement noise because it is not part of the dynamics but it affects the estimation of chaos indicators and masks the nonlinear deterministic dynamics underlying the system. In such cases, the main aim of nonlinear time series analysis is to extract the deterministic character of the noisy signal. There are several ways to achieve this purpose by different methods of filtering and noise reduction strategies, the demanding reader may consult, e.g., Kantz and Schreiber (1997); Hegger *et al.* (1999) and references therein.

10.3 Can chaos be distinguished from noise?

Possibly the most important, at least conceptually, goal of nonlinear data analysis is to determine whether the system under investigation is deterministic and chaotic or stochastic. More precisely, we would like to understand whether a given experimental signal (a time series of a certain observable) originates from a chaotic deterministic or stochastic dynamics, i.e. we would like to have a method for accomplish such a distinction without any *a priori* knowledge on the system which generated the signal. Despite this longstanding problem has been subject of many

[9] At least if we do not include in the "deterministic" description also the environment or the details of the experimental apparatus.

investigations it is still largely unsolved [Nicolis and Nicolis (1984); Osborne and Provenzale (1989); Sugihara and May (1990); Casdagli and Roy (1991); Kaplan and Glass (1992); Kubin (1995); Cencini *et al.* (2000)](see also Abarbanel (1996); Kantz and Schreiber (1997)).

In the following, we discuss how the analysis of signals and observables at various resolutions may be used to answer, at least partially, to the question posed in the title of this section. We also discuss some examples able to highlight the difficulties inherent to such a distinction.

10.3.1 *The finite resolution analysis*

If we were able to measure the maximum Lyapunov exponent (λ) and/or the Kolmogorov-Sinai entropy (h_{KS}) from a given experimental signal, we could, in principle, ascertain whether the time series has been generated by a deterministic law ($\lambda, h_{KS} < \infty$) or a stochastic process ($\lambda, h_{KS} \to \infty$). However, as previously discussed (see Sec. 10.2.1.3), many practical limitations make problematic the correct determination of chaos indicators, especially of h_{KS} and λ, due to the infinite time averages and the limit of arbitrary fine resolution required for their evaluation. Furthermore, besides being unreachable in experiments, the infinite time and arbitrary resolution limits may also result uninteresting in many physical contexts, e.g. in the presence of intermittent behaviors [Benzi *et al.* (1985)] or many degrees of freedom [Grassberger (1991); Aurell *et al.* (1996)].

Part of these restrictions can be, to some extent, circumvented by using quantities such as the (ε, τ)-entropy per unit time, $h(\varepsilon, \tau)$, (see Sec. 9.3) or the finite size Lyapunov exponent, $\lambda(\varepsilon)$,[10] (see Sec. 9.4) which allow for a scale dependent description of a given signal. When these quantities are properly defined, we have $\lambda = \lim_{\varepsilon \to 0} \lambda(\varepsilon)$ and $h_{KS} = \lim_{\varepsilon \to 0} h(\varepsilon)$, so that they can, in principle, be used to answer the question about the deterministic or stochastic character of the dynamical law that generated the signal. In addition, being defined at each observation scale ε, they give us the opportunity to recast the question about the noisy or chaotic character of a signal at each observation scale [Cencini *et al.* (2000)], as discussed in the following.

10.3.2 *Scale-dependent signal classification*

For classifying signals in terms of resolution dependent quantities it is convenient to introduce an indicator complementary to the ε-entropy which is the ε-redundancy:

$$r_m(\varepsilon, \tau) = \frac{1}{\tau} \left[H_1(\varepsilon, \tau) - (H_{m+1}(\varepsilon, \tau) - H_m(\varepsilon, \tau)) \right] = \frac{1}{\tau} H_1(\varepsilon, \tau) - h_m(\varepsilon, \tau)$$

where m is the embedding dimension, H_m the block entropies, in particular, $H_1(\varepsilon, \tau))$ quantifies the uncertainty of the single outcome of the measurement,

[10]For uniformity of notation, here the argument of the FSLE has been denoted ε instead of δ.

Table 10.1 Signal classification in terms of ε-entropy and ε-redundancy.

Deterministic ($m > D$)		Stochastic	
$r_m(\varepsilon) \to \infty$		$h_m(\varepsilon) \to \infty$	
chaotic	non-chaotic	white noise	colored noise
$\lim_{m\to\infty} h_m(\varepsilon) > 0$	$\lim_{m\to\infty} h_m(\varepsilon) = 0$	$r_m(\varepsilon) = 0$	$r_m(\varepsilon) > 0$

disregarding any correlation. The redundancy, which is nothing but the mutual information (8.13), measures the amount of uncertainty that can be removed from future observations by taking into account the information accumulated in the past. The redundancy $r_m(\varepsilon, \tau)$ can be easily computed from $h_m(\varepsilon, \tau)$ noticing that $H_1(\varepsilon) \sim -\ln \varepsilon$ for bounded continuous valued non-periodic signals.

The redundancy $r_m(\varepsilon, \tau)$ vanishes for a time uncorrelated stochastic process and tends to infinity for a deterministic one, while the entropy $h_m(\varepsilon, \tau)$ vanishes for a regular deterministic signal and is infinite for a stochastic one. Moreover $r_m(\varepsilon, \tau)$ and $h_m(\varepsilon, \tau)$ are finite and positive for stochastic signals with correlation or deterministic chaotic signals, respectively. Generic signals can thus be classified, at any given scale of observation ε, according to behavior of the entropy and the redundancy, as shown in Table 10.1 (see Kubin (1995); Cencini *et al.* (2000) for further details).

Of course, in order to ascertain the "nature" of the signal we should analyze the behavior of the entropy $h_m(\varepsilon)$, or equivalently of the FSLE $\lambda(\varepsilon)$, and of the redundancy $r_m(\varepsilon)$ for $\varepsilon \to 0$. However, in practical situations, we have access only to a finite amount of data (finite time series) and we cannot take the limit $\varepsilon \to 0$. Indeed, as discussed in Sec. 10.2.1.3, in general, we have a lower resolution cutoff $\varepsilon_1 > 0$ below which we are blind on the behavior of these quantities. Of course, on any finite scale, and hence-force at ε_1, both entropy and redundancy are always finite, so that we are unable to decide which one, for $\varepsilon \to 0$, will extrapolate to infinity.

Figure 10.6 shows the typical behavior of the entropy $h_m(\varepsilon)$ and the redundancy $r_m(\varepsilon)$ in case of a chaotic deterministic model and a stochastic process obtained from long enough a time series. As shown in the figure, although constrained by inequality (10.11), a saturation range can be detected for the entropy or the redundancy as summarized in Table 10.2.[11]

According to Tables 10.1 and 10.2, we can classify the character of a signal as deterministic or stochastic according to the following criterion: when on some

[11]It is however worth recalling that Table 10.2 does not exhaust all the possible behaviors: the ε-entropy can indeed exhibit power law behaviors, e.g. in the diffusive processes Eq. (9.17), or other behaviors when correlations are present, see Gaspard and Wang (1993) and Abel *et al.* (2000b) for further details.

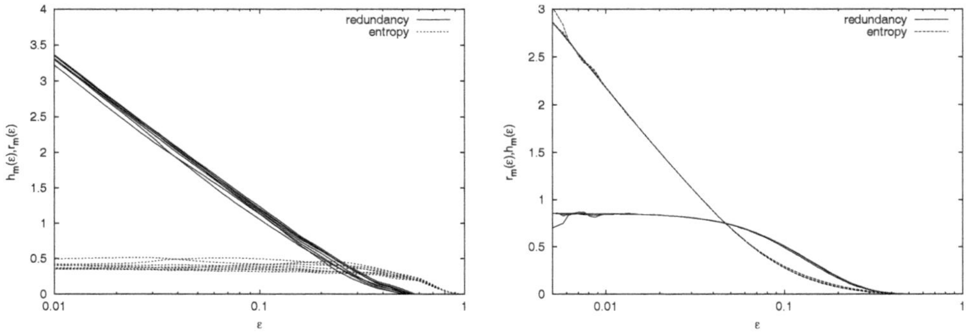

Fig. 10.6 (a) ε-entropy $h_m(\varepsilon)$ (dashed lines) and ε-redundancy $r_m(\varepsilon)$ (solid lines) for the Hénon map with $a = 1.5$ and $b = 0.3$, at various embedding dimensions $m = 2, \ldots, 9$. (b) Same as (a) for a first order auto-regressive stochastic process AR(1) (see Sec. 10.4.2 for details), with $m = 1, \ldots, 5$ and fixed τ. The behaviors of the two quantities are summarized in Table 10.2.

range of length scales, either the entropy $h_m(\varepsilon)$ or the redundancy $r_m(\varepsilon)$ displays a plateau to a constant value, we call the signal deterministic or stochastic on those scales, respectively. Such a definition is free from the necessity to specify a model for the system which generated the signal, so that we are no longer obliged to answer the "metaphysical" question on whether the system which produced the data was deterministic or a stochastic [Cencini *et al.* (2000)].

Table 10.2 Complementary behavior of entropy and redundancy for stochastic and chaotic signals.

Deterministic	Stochastic
$r_m(\varepsilon) \propto -\ln \varepsilon$	$h_m(\varepsilon) \propto -\ln \varepsilon$
$h_m(\varepsilon) \approx \text{const}$	$r_m(\varepsilon) \approx \text{const}$

The distinction between chaos and noise based on (ε, τ)-entropy (or the FSLE) complements previous attempts based on correlation dimension estimation, where a finite value of that dimension was regarded as a mark for the deterministic nature of the signal [Grassberger and Procaccia (1983b)].

Before examining some specific examples, let us mention other attempts to distinguish chaos from noise based on prediction algorithms [Sugihara and May (1990); Casdagli and Roy (1991)] or on the smoothness of the signal [Kaplan and Glass (1992, 1993)]. Finally, we stress that, despite their differences, all approaches for distinguishing chaos from noise share the necessity to specify a particular length scale and embedding dimension m.

10.3.3 *Chaos or noise? A puzzling dilemma*

Having a practical signal classification method, we find now instructive to analyze some specific examples highlighting the extent up to which the chaos-noise distinction is far from being sharp even in simple models, when only finite resolution or finite amount of data are available. These simple examples should be seen as proxies of the typical difficulties encountered in real systems, and as illustrations of the classification scheme above discussed. We shall briefly reconsider the scale dependent description signals in the context of high dimensional systems (see Sec. 12.5.1).

10.3.3.1 *Indeterminacy due to finite resolution*

We now illustrate the difficulties due to finite resolution effects by discussing the behavior of two systems that display large scale diffusion [Cencini *et al.* (2000)].

As first, consider the map (Fig. 10.7)

$$x(t+1) = [x(t)] + F\left(x(t) - [x(t)]\right), \tag{10.12}$$

where $[u]$ denotes the integer part of u and $F(y)$ is given by:

$$F(y) = \begin{cases} (2+\Delta)y & \text{if} \quad y \in [0:1/2[\\ (2+\Delta)y - (1+\Delta) & \text{if} \quad y \in]1/2:1]\,. \end{cases}$$

The above system is chaotic, with maximum Lyapunov exponent $\lambda = \ln|F'| = \ln(2+\Delta)$, and gives rise to a diffusive behavior on the large scales [Schell *et al.* (1982)]. As a consequence, the ε-entropy $h(\varepsilon)$ (or equivalently the FSLE $\lambda(\varepsilon)$) behaves as (Fig. 10.8):

$$h(\varepsilon) \approx \begin{cases} \lambda & \text{for} \quad \varepsilon < 1 \\ \dfrac{D}{\varepsilon^2} & \text{for} \quad \varepsilon > 1 \end{cases}$$

where $D = \lim_{t\to\infty}\langle [x(t)-x(0)]^2\rangle/(2t)$ is the diffusion coefficient.

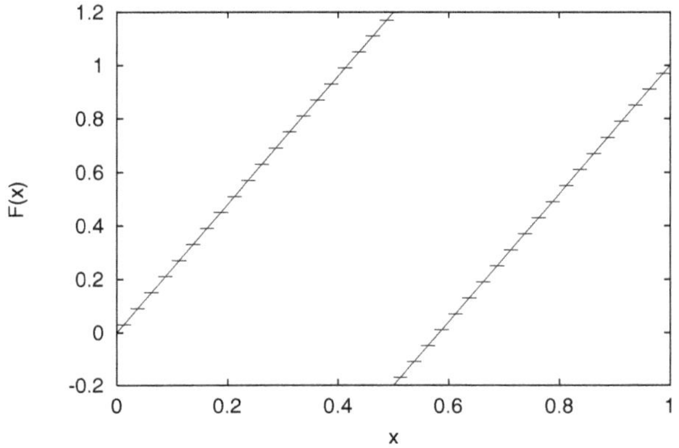

Fig. 10.7 The map $F(x)$ used in (10.13) for $\Delta = 0.4$ is shown with superimposed the approximating (regular) map $G(x)$ used in (10.14), here obtained by using 40 intervals of slope 0.

While the numerical computation of $\lambda(\varepsilon)$ is rather straightforward, that of $h(\varepsilon)$ is more delicate but can be efficiently handled by means of the exit times encoding, as discussed in Box B.19 (see also Abel *et al.* (2000a,b)).

As a second system, consider the noisy map

$$x(t+1) = [x(t)] + G\left(x(t) - [x(t)]\right) + \sigma\eta_t, \tag{10.14}$$

where η_t is a time uncorrelated noise with uniform distribution in the interval $[-1, 1]$, and σ a free parameter controlling its intensity. As shown in Fig. 10.7, now the deterministic component of the dynamics $G(y)$ is chosen to be a piecewise linear map approximating $F(y)$ in Eq. (10.13). In particular, we can choose $|\mathrm{d}G/\mathrm{d}y| \leq 1$ so that the map (10.14) without noise, gives a non-chaotic time evolution.

Now one can compare the chaotic dynamics (10.12) with the non-chaotic plus noise dynamics (10.14). For example, let us start with the computation of the finite size Lyapunov exponent for the two cases.

From a data analysis point of view, one should compute the FSLE by reconstructing the dynamics by embedding. However, if one is interested only in discussing the resolution effects, the FSLE can be directly computed by integrating the evolution equations for two (initially) very close trajectories, in the case of noisy maps using two different realizations of the noise [Cencini *et al.* (2000)]. Figure 10.8 shows the behavior of $\lambda(\varepsilon)$ (left) and $h(\varepsilon)$ (right) versus ε for both systems (10.12) and (10.14). The two observables essentially convey the same message, we thus limit ourselves to the discussion of the FSLE, where we can distinguish three different regimes. On the large length scales, $\varepsilon \gg 1$, we observe diffusive behavior in both models. On intermediate (small) length scales $\sigma < \varepsilon < 1$ both models show chaotic deterministic behavior, because the entropy and the FSLE are independent of ε and larger than zero. Finally we see the stochastic behavior for the system (10.14) on the smallest length scales $\varepsilon < \sigma$, while the system (10.12) still displays chaotic behavior.

Clearly, extrapolating character of the signal generated by these two systems would change a lot depending on the smaller cutoff ε_1 being smaller or larger than σ or of 1. However, the above described scale-dependent classification scheme gives us the freedom to call deterministic the signal produced by Eq. (10.14) when observed in $\sigma < \varepsilon < 1$, refraining from accounting its "true" nature, i.e. its $\varepsilon \to 0$ behavior. Practically, this means that, on these scales, Eq. (10.12) can be considered as an appropriate model for Eq. (10.14).

10.3.3.2 *Indeterminacy due to finite block length effects*

While the previous example has clearly shown the difficulties in achieving a unambiguous distinction between chaos and noise due to finite resolution, here we examine an example where the finite amount of data generates an even more striking situation, in which a non-chaotic deterministic system may produce a signal practically indistinguishable from a stochastic one.

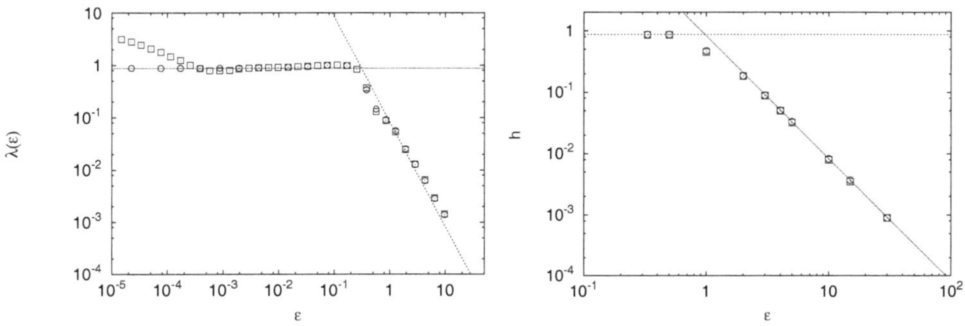

Fig. 10.8 Left: $\lambda(\varepsilon)$ versus ε for the (10.13) with $\Delta = 0.4$ (∘) and with the noisy (regular) map (□) (10.14), with 10^4 intervals of slope $dG/dy = 0.9$ and noise intensity $\sigma = 10^{-4}$. The straight lines indicates the Lyapunov exponent $\lambda = \ln(2.4)$ and the diffusive behavior $\lambda(\varepsilon) \sim \varepsilon^{-2}$. Right: (ε, τ)-entropy for the noisy (□) and the chaotic maps (∘). The straight lines indicates the KS-entropy $h_{KS} = \lambda = \ln(2.4)$ and the diffusive behavior $h(\varepsilon) \sim \varepsilon^{-2}$. The region $\varepsilon < \sigma$ has not be explored for the high computational costs.

A simple way to generate a non-chaotic (regular) signal having statistical properties similar to a stochastic one is by considering the Fourier expansion of a random signal

$$x(t) = \sum_{i=1}^{M} A_i \sin(\Omega_i t + \phi_i) \tag{10.15}$$

where the frequencies are such that $\Omega_i = \Omega_0 + i\Delta\Omega$, the phases ϕ_i are random variables uniformly distributed in $[0 : 2\pi]$ and the amplitudes A_i are chosen to produce a definite power spectrum. The expression (10.15) represents the Fourier expansion of a stochastic signal only if one considers a set of $2M$ points such that $M\Delta\Omega = \pi/\Delta t$, where Δt is the sampling time [Osborne and Provenzale (1989)]. In a more physical context, the signal (10.15) can also be interpreted as the displacement of an harmonic oscillator linearly coupled to a bath of harmonic oscillators [Mazur and Montroll (1960)].[12] In Fig 10.9a, we show an output of the signal (10.15) and, for a qualitative comparison, in Fig 10.9b, we also plot an artificial continuous time Brownian motion obtained integrating the stochastic equation

$$\frac{dx}{dt} = \xi(t) \tag{10.16}$$

[12] In particular, the signal (10.15) represents the displacement of an oscillator coupled to other oscillators provided the frequencies Ω_i are derived in the limit of small mass [Mazur and Montroll (1960)] and phases ϕ_i are uniformly distributed random variables in $[0 : 2\pi]$ and the amplitudes A_i are such that

$$A_i = C\Omega_i^{-1}$$

where the C is an arbitrary constant and the Ω dependence is just to obtain a diffusive-like behavior. Notice that the proposal by Mazur and Montroll (1960) to mimic Brownian motion with a superposition of trigonometric functions (10.15) is somehow similar to Landau's suggestion to explain the "complex behavior" of turbulent fluids as a combination of many simple elements (Sec. 6.1.1).

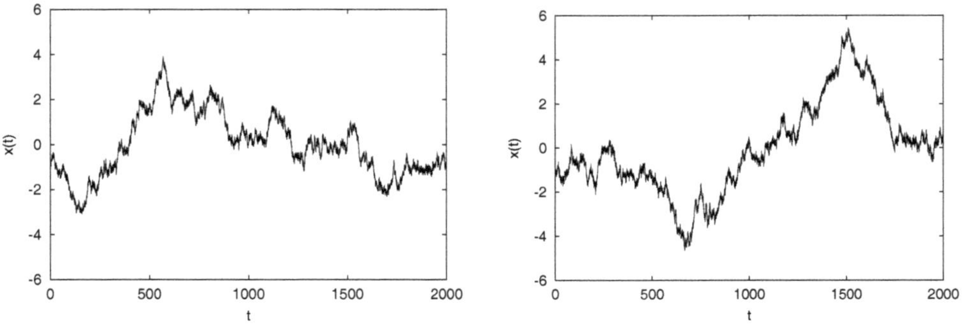

(a) (b)

Fig. 10.9 (a) Time record obtained from (Eq. (10.15)) with the frequencies chosen as discussed in Cencini *et al.* (2000), the numerically computed diffusion constant is $D \approx 0.007$. Data are sampled with $\Delta t = 0.02$ for a total of 10^5 points. (b) Time record obtained from an artificial Brownian motion (10.16) tuned to have the diffusion constant as in (a).

where $\xi(t)$ is a Gaussian white noise whose variance is tuned as to mimic the signal obtained by Eq. (10.15).[13]

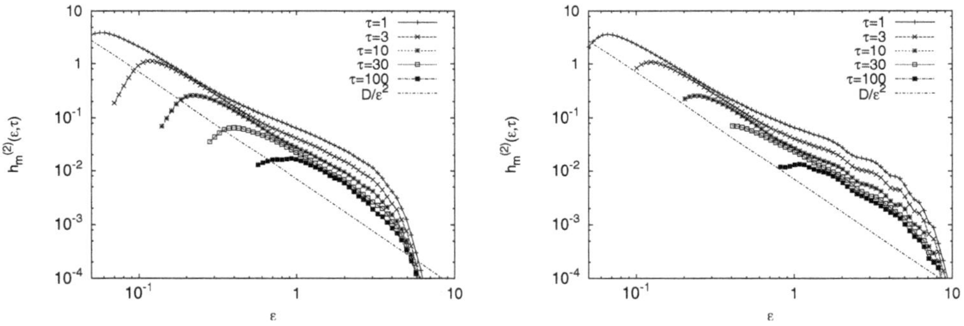

(a) (b)

Fig. 10.10 ε-entropy calculated with the Grassberger-Procaccia algorithm using using 10^5 points from the time series shown in Fig. 10.9. We show the results for embedding dimension $m = 50$. The two straight-lines show the D/ε^2 behavior. Note that $h_m^{(2)}(\varepsilon, \tau)$ is preferred to $h_m^{(1)}(\varepsilon, \tau)$ because it guarantees a better statistics and convergence.

As it is possible to see, the two signals appears to be very similar already at a first sight. The observed similarity is confirmed by Fig. 10.10 which shows the ε-entropy computed for the signals in Fig. 10.9, indeed both develop the ε^{-2} behavior typical of diffusive processes.[14] One may question that if $M < \infty$ the signals obtained

[13]To be precise, in a computer ξ is obtained through a pseudo-random number generator, i.e. a high entropic one-dimensional deterministic map. Thus, in principle, we should consider this an example of a high entropic low dimensional system, which produces stochastic behavior. However, in the text we will ignore this subtleties and consider the signal as a genuinely stochastic.

[14]Notice that the power law only emerges as the envelope of different computations with different delay times for the reasons discussed in Box B.19.

form Eq. (10.15) cannot develop a true Brownian motion because regularities must manifest in the trajectory of $x(t)$ for a long enough record. However, even increasing the length of the time record the result would not change too much, because a very large embedding dimension would be needed to discern the character of the two signals, as the deterministic behavior could manifest only if m is larger than the dimension of the manifold where the motion takes place, which is M for M harmonic oscillators.

This simple example proves that the impossibility of reaching high enough embedding dimensions severely limits our ability to make definite statements about the "true" character of the system which generated a given time series as well as the already analyzed problem of the lack of resolution.

10.4 Prediction and modeling from data

Predicting future evolution of a system and modeling complex phenomena had been natural desires in the development of science. In this Section we briefly discuss these problems in the general framework of time series analysis. Of course, prediction and modeling are closely related: being able to build a good model usually lead to the possibility to predict.

10.4.1 *Data prediction*

As far as we know, at least in modern times, one of the first methods proposed to forecast future evolution of a system from the knowledge of its past is due to Lorenz (1969), who put forward the use of "analogues" for weather forecasting. The idea is rather simple. Given a known sequence of "states" $\boldsymbol{x}_1, \boldsymbol{x}_2, \ldots, \boldsymbol{x}_M$,[15] the "analogues" provide a proxy for the next state $\boldsymbol{x}_{M+1}$. By analogous we designate two states, say $\boldsymbol{x}_i$ and $\boldsymbol{x}_j$, which (in Lorenz words) *resemble each other closely*, meaning that $|\boldsymbol{x}_i - \boldsymbol{x}_j| \leq \varepsilon$, with ε reasonably small. If $\boldsymbol{x}_k$ is an analogous of $\boldsymbol{x}_M$, the forecasting rule is rather obvious:

$$\boldsymbol{x}_{M+1} = \boldsymbol{x}_{k+1} \; . \tag{10.17}$$

In the presence of $l > 1$ analogues: $\boldsymbol{x}_{k_1}, \ldots, \boldsymbol{x}_{k_l}$, Eq. (10.17) can be generalized to

$$\boldsymbol{x}_{M+1} = \sum_{n=1}^{l} a_n \boldsymbol{x}_{k_n+1} \; , \tag{10.18}$$

where the coefficients $\{a_n\}$ are computed with suitable interpolations.

Unfortunately, as noticed by Lorenz himself, at least for atmospheric prediction, the method does not seem really useful as *there are numerous mediocre analogous but not truly good ones.* However, atmosphere evolution is rather complex and,

[15] In the work of Lorenz the "states" are height values of the 200 mb, 500 mb and 850 mb surfaces at a grid of 1003 point over the Northern Hemisphere.

moreover, it is unclear which would be the best choice for the "states" to be used. Therefore, the failure of the method in such a case is not an obvious mark that the proposal cannot be used in other, simpler, contexts.

It is quite natural to combine the method of the embedding and the idea of the analogues to build a method for the prediction of u_{M+1} from a sequence $u_1, u_2, \ldots, u_M$ [Kostelich and Lathrop (1994)], essentially this is the same idea exploited by Wolf *et al.* (1985) to compute the Lyapunov exponents from data (see Box B.22). Once estimated the value m of the embedding dimension, and therefore computed the series of the delay-vectors $\boldsymbol{Y}_j^m$ with $j = 1, \ldots, M$; the prediction of the state at time $M+1$ is obtained by using Eq. (10.17) or Eq. (10.18), replacing $\boldsymbol{x}$ with $\boldsymbol{Y}^m$. If m is large enough, the use of embedding vectors should circumvent the problem of choosing the proper states. Of course, the method can properly work only if analogues are found. We should then wonder which is the probability to find such analogues. For instance, in a system characterized by a strange attractor with correlation dimension $D(2)$, the probability to find analogous within a tolerance ε is $\mathcal{O}(\varepsilon^{D(2)})$. Therefore, it is rather clear that the possibility to predict the future from the past using the analogues has its practical validity only for low dimensional systems. More than one century after, scientists working on prediction problems basically re-discovered the conclusion of Maxwell *the same antecedents never again concur, and nothing ever happens twice*, discussed in the Introduction.

10.4.2 *Data modeling*

The ambitious aim of modeling is to find an algorithmic procedure which allows the determination of a suitable model (i.e. a deterministic or stochastic equation) from a long time series of an observable, $\{u_k\}_{k=0}^M$, extracted by a system whose evolution rule is not known. We stress that here we are not concerned with model building based on prior knowledge of the system, physical intuition or from first principles understanding of the phenomenon under consideration. We only have access to the time series of an unknown system.

Given the time series $\{u_k\}_{k=0}^M$, assume that the true dynamics that produced the sequence is a map in $\mathbb{R}^d$, i.e. the "true" state $\boldsymbol{x}_k \in \mathbb{R}^d$ evolves as $\boldsymbol{x}_k = \boldsymbol{g}(\boldsymbol{x}_k)$. Then the states can be linked to the observable $\{u_k\}$ by a smooth map from $\mathbb{R}^d$ to $\mathbb{R}$: $u_k = f[\boldsymbol{x}_k]$. Notice that even if the true state variables are unknown, thanks to Takens (1981) theorem (see alsoOtt *et al.* (1994); Sauer *et al.* (1991)), we can always reconstruct the state from the vector obtained with the time-embedding delay with m large enough.

In principle, a simple algorithmic procedure to determine $\boldsymbol{g}$ is represented by the previously discussed analogues method, i.e. for any test point $\boldsymbol{x}$ in phase space, find the closest data point $\boldsymbol{x}_k$ and then $\boldsymbol{g}[\boldsymbol{x}] = \boldsymbol{x}_{k+1}$. Besides the above mentioned difficulties, the main disadvantage of the analogues method is that it is local, while a "global" approach which use all data, would be surely preferable.

In the modeling process basically we have to face two aspects:

(a) model selection problem, i.e. to choose the "best" model (in a certain class) that it should capture the essential dynamics of the time series, of course, without "over-fitting";
(b) fit of the parameters of the model in a).

Usually step a) is accomplished by choosing simple classes of models such as global polynomials, linear combination of other basis functions containing some parameters. The simplest cases are, of course, the linear modeling procedures: the so called auto-regressive (AR) and the auto-regressive moving average (ARMA) methods [Gershenfeld and Weigend (1994)].

In the AR method one has $m + 1$ parameters

$$u_t = \sum_{j=1}^{m} a_j u_{t-j} + b_0 e_t \ ,$$

where $\{e_t\}$ are standard, independent, Gaussian noises, the parameters $\{a_j\}$ and b_0 are obtained with a best fit. Clearly, m is not completely arbitrary as it has the same status of the embedding dimension. In the ARMA, which is a rather natural generalization of AR, one has $m + n$ parameters

$$u_t = \sum_{j=1}^{m} a_j u_{t-j} + \sum_{k=0}^{n-1} b_k e_{t-k} \ .$$

Also for ARMA the parameters $\{a_j\}_{j=1}^{m}$ and $\{b_j\}_{j=1}^{n}$ are obtained via a best fit procedure, the choice of m and n depends on the available data and the system under investigation [Gershenfeld and Weigend (1994)].

The work of Rissanen (1989) on the minimum description length is one of the few attempts which provides, at least, a partial answer for a systematic approach to data modeling. The basic idea, which is a mathematical version of the Occam's Razor principle, is that the best model is that one, among those able to compress the known data, with the minimum description length of the parameters and rules. Of course, in practice, such idea works only by selecting (with intuition and previous knowledge of the problem) the proper class of models. For the use of the minimum description length approach in specific cases see Judd and Mees (1995).

We conclude mentioning another possibility, which can be considered the most direct approach to reconstruct the dynamics from data. The idea is to determine a map $\boldsymbol{F}$ for the delay embedding space:

$$\boldsymbol{Y}_{k+1} = \boldsymbol{F}(\boldsymbol{Y}_k) \ ,$$

where $\boldsymbol{Y}_j$ is the usual delay-vector, and, for sake of notation simplicity, we again considered the discrete time case and did not explicitly indicate the embedding dimension. The first step is to have an ansatz for the map $\boldsymbol{F}$ which depends on a set of tunable parameters $\{\boldsymbol{p}\}$. Then, such parameters are determined by minimizing the prediction error

$$\text{err} = \sqrt{\frac{1}{M}\sum_{k=1}^{M}\left|\boldsymbol{Y}_{k+1} - \boldsymbol{F}_{\boldsymbol{p}}(\boldsymbol{Y}_k)\right|^2}$$

with respect to $\{\boldsymbol{p}\}$ and where M denotes the number of data in the time series. Of course, unlike AR and ARMA which only rely on the data sequence, here the choice of the ansatz for $\boldsymbol{F}$ requires some prior knowledge on the physics of the problem under investigations. This method is rather powerful and can also be applied to high dimensional systems. For instance, it has been used to reconstruct the PDE of reaction diffusion and other high-dimensional systems, whose functional structure were known [Voss *et al.* (1998, 1999); Bär *et al.* (1999)]. We mention also the work by Hegger *et al.* (1998) who inferred an ODE able to model the dynamics of ferroelectric capacitors.

Chapter 11

Chaos in Low Dimensional Systems

We can only see a short distance ahead, but we can see plenty there that needs to be done.

Alan Turing (1912–1954)

This Chapter encompasses several phenomena and illustrates some basic issues of diverse disciplines where chaos is at work: celestial mechanics, transport in fluid flows, chemical reactions and population dynamics, to finish with chaotic synchronization. Each section of this Chapter could be a book for itself, in the impossibility of any exhaustive treatment, we will follow two main guidelines. On the one side, we illustrate the basic methodology of several research subjects in which chaos controls the main phenomena. On the other side, we will exploit the opportunity of new examples to deepen some aspects already introduced in the first part of the book.

11.1 Celestial mechanics

A typical problem in celestial mechanics is the computation of the *ephemeris* which consists in building a table of the positions and velocities of all celestial bodies (Sun, planets, asteroids, comets etc.) as function of time. In principle, to obtain an ephemeris of the Solar System it is required to solve the equations of motion for the full many-body problem of N celestial bodies involved, given their masses $(m_1, m_2, ..., m_N)$, initial values of positions $(\boldsymbol{q}_1(0), \boldsymbol{q}_2(0), ..., \boldsymbol{q}_N(0))$ and velocities $(\boldsymbol{p}_1(0)/m_1, \mathbf{p}_2(0)/m_2, ..., \boldsymbol{p}_N(0)/m_N)$. As they mutually interact by means of the gravitational force the ODE to be solved is the second Newton's law of dynamics[1]

$$\frac{\mathrm{d}^2 \boldsymbol{q}_j}{\mathrm{d}t^2} = -G \sum_{k \neq j} m_k \frac{\boldsymbol{q}_j - \boldsymbol{q}_k}{|\boldsymbol{q}_j - \boldsymbol{q}_k|^3} \quad j = 1, 2, ..., N \; , \tag{11.1}$$

[1] In the following we consider almost always the celestial bodies as points. In some circumstances, e.g. when considering the motion of spacecrafts or small satellites, it is necessary to be more accurate. For instance, later we will see that to properly describe the motion of Hyperion (a small moon of Saturn) we need to account for its non-spherical shape.

where G is the universal gravitational constant. Equation (11.1) defines a Hamiltonian system whose solution, depending on the number of bodies N involved the problem, may constitute a formidable task.

The two-body problem ($N = 2$) is completely solvable as the system is integrable (Box B.1). Working in the reference frame of the center of mass at rest, it can be easily derived that each body moves along a conic section with focus at the center of mass, and the two orbits are coplanar. The type of conic (ellipse, parabola or hyperbola) is determined by the energy[2] value E so that: $E < 0$ corresponds to two ellipses; $E = 0$ to two parabolas; $E > 0$ to two hyperbolas.

The first is the most interesting case and it applies, e.g., to the simplest Solar system with the Sun (of mass M_S) and a unique planet (of mass $m_p \ll M_S$) which follows the well known Kepler's laws:

Law 1: The planet moves, relatively to the Sun, in an elliptical orbit with major and minor semi-axes a and b, respectively (the eccentricity $e = \sqrt{a^2 - b^2}/a$, which vanishes for a circular orbit, measures the deviation from the circle), with the Sun in one of the two foci of the ellipse;[3]

Law 2: The motion in the elliptical orbit is such that the vector from the Sun to the planet spans equal areas in equal times;

Law 3: The orbital period of the planet is such that $T \propto a^{3/2}$.

As soon as $N \geq 3$,[4] the system is no more integrable and despite more than three centuries of investigations, there is still an intense research activity. Of special interest, both from a theoretical and historical point of view, is the three-body problem ($N = 3$), which was the most studied since the 18-th century. We mention two classical results which are, still nowadays, among the few explicit solutions valid for arbitrary masses: Euler found a periodic motion in which the bodies are collinear and move in ellipses (Fig. 11.1a); Lagrange found periodic solutions in which the bodies lie at the vertexes of an equilateral triangle that rotates, changing size periodically (Fig. 11.1b).

The origin of the difficulties in solving the problem can be appreciated considering an interesting limiting case of the three-body problem. Assume that the third body has a very small mass compared with the other two ($m_3 \ll m_2 < m_1$). Such a situation is rather common in astronomy, for instance the system Sun, Jupiter and asteroid (or Earth, Moon and an artificial satellite). Neglecting the interaction with Jupiter, the asteroid and the Sun are nothing but a two body problem (H_0), which is integrable. Thus the three-body problem can be represented as an almost integrable Hamiltonian system, that in action-angle variables would read

$$H(\mathbf{I}, \boldsymbol{\phi}) = H_0(\mathbf{I}) + \epsilon H_1(\mathbf{I}, \boldsymbol{\phi}) ,$$

[2]With the usual convention that the potential energy at infinite distance is zero.

[3]As $M_S \gg m_p$ the barycenter basically coincides with the Sun position

[4]Consider that in the Solar system, besides the Sun and the 8 major planets with their (more than sixty) moons, there are thousands of asteroids and comets.

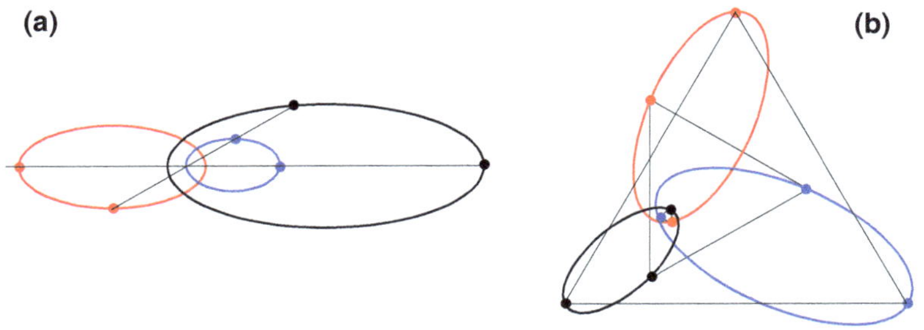

Fig. 11.1 Sketch of the Euler collinear (a) and the Lagrange equilateral triangle (b) solutions to the three-body problem.

where the $H_0(\mathbf{I})$ is the integrable part and the strength of the perturbation is given by $\epsilon = m_2/m_1$.[5] Unfortunately, the *problem of small denominators* [Poincaré (1892, 1893, 1899)] (see Sec. 7.1.1) frustrates any naive perturbative attempt to approach the above problem. However, although non-integrability leaves room for chaotic orbits to exist, thanks to KAM theorem (Sec. 7.2) we know that the non-existence of (global) integral of motion does not imply the complete absence of regular motions.

11.1.1 *The restricted three-body problem*

Some insights into the three-body problem can be obtained considering a simplification in which the third body (the asteroid) does not induce any feedback on the two principal ones (Sun and Jupiter). Due to the small mass of asteroids (with respect to the Sun and Jupiter) this approximation — called the *restricted three-body problem* — is reasonable and can be used to understand some observations made in the Solar system. Here, for the sake of simplicity, we further assume a circular orbit for the principal bodies (for instance, Jupiter's eccentricity is $e \approx 0.049$ and thus the circular approximation is reasonable). Finally, we restrict the analysis to an asteroid moving on the plane determined by the Sun and the planet orbits, ending in the *circular, planar, restricted three-body problem* (CPR3BP).

Working in the rotating frame with the center of mass at the origin, (x, y) denotes the position of the asteroid while the Sun (of mass $M_S = 1 - \mu$) and Jupiter (of mass $m_J = \mu$)[6] are in the fixed positions $(-\mu, 0)$ and $(1 - \mu, 0)$, respectively. In this frame of reference, taking into account gravitational, Coriolis and centrifugal forces, the evolution equations read[7] [Szebehely (1967)]

$$\begin{aligned} \frac{\mathrm{d}^2 x}{\mathrm{d}t^2} - 2\frac{\mathrm{d}y}{\mathrm{d}t} &= -\frac{\partial V}{\partial x} \\ \frac{\mathrm{d}^2 y}{\mathrm{d}t^2} + 2\frac{\mathrm{d}x}{\mathrm{d}t} &= -\frac{\partial V}{\partial y}\,, \end{aligned} \tag{11.2}$$

[5] I.e. by the ratio of the mass of Jupiter and the Sun.
[6] The total mass of the Sun plus the planet has been normalized to 1.
[7] These equations can be put in Hamiltonian form by a change of variables [Koon *et al.* (2000)].

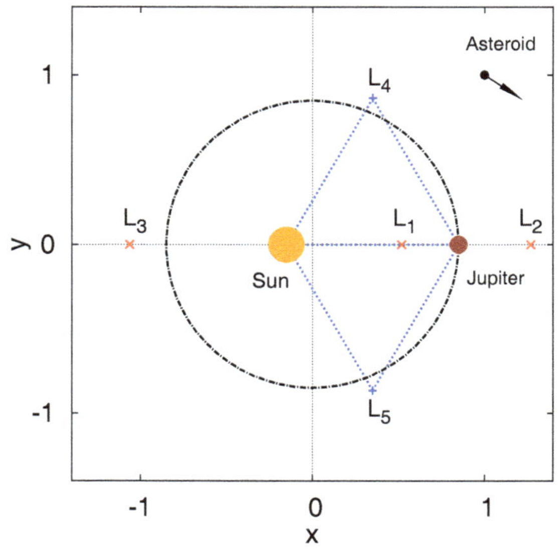

Fig. 11.2 Equilibrium points of the CPR3BP in the rotating frame (for $\mu = 0.15$).

with the effective potential (including centrifugal and gravitational forces)

$$V(x,y) = -\frac{x^2+y^2}{2} - \left[\frac{1-\mu}{r_1} + \frac{\mu}{r_2}\right] , \tag{11.3}$$

where $r_1 = \sqrt{(x+\mu)^2+y^2}$ and $r_2 = \sqrt{(x-1+\mu)^2+y^2}$ are the distances of the third body from the Sun and Jupiter, respectively. In Eqs. (11.2) and (11.3) suitably rescaled time and length units have been used. It is easily checked that the system admits the conservation law (Jacobi integral):[8]

$$J = \frac{1}{2}\left[\left(\frac{\mathrm{d}x}{\mathrm{d}t}\right)^2 + \left(\frac{\mathrm{d}y}{\mathrm{d}t}\right)^2\right] + V(x,y) = \text{const} .$$

The two equations (11.2) have five fixed points (shown in Fig. 11.2) corresponding to the solutions of $\partial V/\partial x = \partial V/\partial y = 0$, in particular:
L_1, L_2, and L_3: are collinear and lie on the Sun-Jupiter (x-)axis: L_1 is between the two principal bodies but closer to Jupiter, L_2 is on Jupiter side (close to it) while L_3 is on the far side of the Sun;
L_4 and L_5: are at the same distance from the Sun and Jupiter forming two equilateral triangles; in the limit $\mu \ll 1$ they lie on the circle of radius ~ 1.
These fixed points correspond, in the CPR3BP limit, to the solutions discovered by Euler and Lagrange (Fig. 11.1), and are usually termed Lagrangian points.

Due to the positivity of the kinetic energy, $(\mathrm{d}x/\mathrm{d}t)^2 + (\mathrm{d}y/\mathrm{d}t)^2) \geq 0$, the third body can only move in the region $J - V \geq 0$, which is called Hill's region and is

[8] A part from a proportionality factor, given by the third body mass, J is nothing but the total energy of the asteroid in the rotating frame, i.e. kinetic plus centrifugal and gravitational energies.

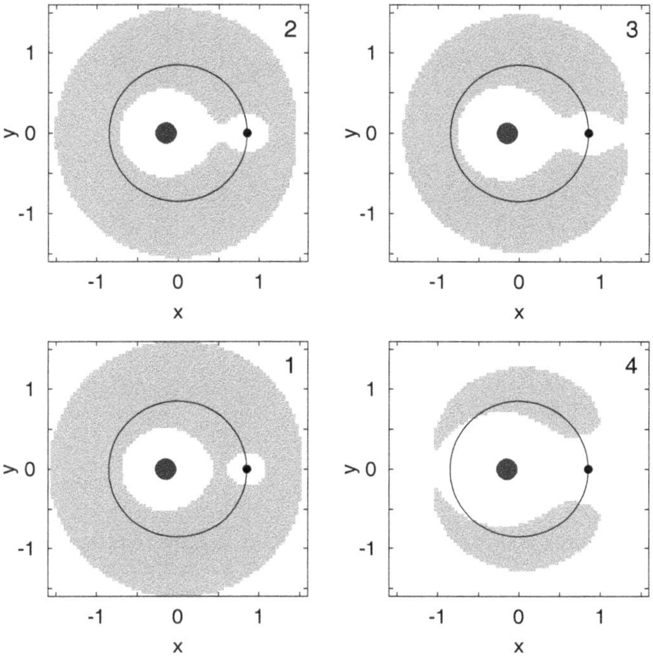

Fig. 11.3 The four basic configurations of the Hill's regions, shaded areas are the forbidden regions. The plot is done for $\mu = 0.15$ and (1) $J < J_1$ (2) $J_1 < J < J_2$ (3) $J_2 < J < J_3$ (4) $J_3 < J < J_4$, see text for further explanations.

determined by $\mathcal{H}(J) = \{x, y| \, V(x, y) \leq J\}$, where the equality is realized at the points of zero velocity. Different cases can be realized depending on the value of J with respect to four critical values of the Jacobi constant, which correspond to the equilibrium points $J_i = V(L_i)$ (with $J_4 = J_5$). As depicted in Fig. 11.3, the third body can move:
(1) for $J < J_1$, either close to the Sun realm, the Jupiter realm or the exterior realm, which are disconnected;
(2) for $J_1 < J < J_2$, in the Sun and Jupiter realms, which are connected at the neck close to L_1, or in the (disconnected) exterior realm;
(3) for $J_2 < J < J_3$, in the three realms, in particular the third body can pass from the interior to the exterior, and viceversa, through the neck around L_1 and L_2;
(4) for $J_3 < J < J_4$, in the whole plane a part from two disconnected forbidden regions around L_4 and L_5;
(5) for $J > J_4$, in the whole plane.

An example of the case 1) is the motion of the Jovian moons. More interesting is the case 3), for which a representative orbit is shown in Fig. 11.4. As shown in the figure, in the rotating frame, the trajectory of the third body behaves qualitatively as a ball in a billiard where the walls are replaced by the complement of the Hill's

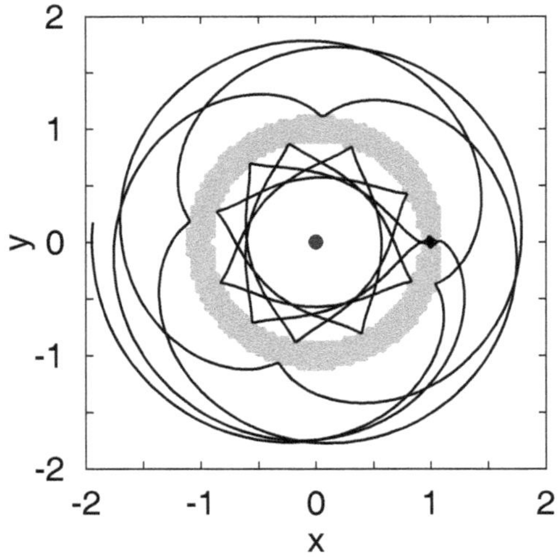

Fig. 11.4 Example of orbit which executes revolutions around the Sun passing both in the interior and exterior of Jupiter's orbit. This example has been generated integrating Eq. (11.2) with $\mu = 0.0009537$ which is the ratio between Jupiter and Sun masses. The gray region as in Fig. 11.3 displays the forbidden region according to the Jacobian value.

region, this schematic idea was actually used by Hénon (1988) to develop a simplified model for the motion of a satellite. Due to the small channel close to L_2 the body can eventually exit Sun realm and bounce on the external side of Hill's region, till it re-enters and so hence so forth. It should be emphasized that a number of Jupiter comets, such as Oterma, make rapid transitions from heliocentric orbits outside the orbit of Jupiter to heliocentric orbits inside the orbit of Jupiter (similarly to the orbit shown in Fig. 11.4). In the rotation reference frame, this transition happens trough the bottleneck containing L_1 and L_2. The interior orbit of Oterma is typically close to a 3:2 resonance (3 revolutions around the Sun in 2 Jupiter periods) while the exterior orbit is nearly a 2:3 resonance.

In spite of the severe approximations, the CPR3BR is able to predict very accurately the motion of Oterma [Koon *et al.* (2000)]. Yet another example of the success of this simplified model is related to the presence of two groups of asteroids, called Trojans, orbiting around Jupiter which have been found to reside around the points L_4 and L_5 of the system Sun-Jupiter, which are marginally stable for $\mu < \mu_c = \frac{1}{2} - \sqrt{\frac{23}{108}} \simeq 0.0385$. These asteroids follow about Jupiter orbit but 60° ahead of or behind Jupiter.[9] Also other planets may have their own Trojans, for instance, Mars has 4 known Trojan satellites, among which Eureka was the first to be discovered.

[9] The asteroids in L_4 are named Greek heroes (or "Greek node"), and those in L_5 are the Trojan node. However there is some confusion with "misplaced" asteroids, e.g. Hector is among the Greeks while Patroclus is in the Trojan node.

In general the CPR3BR system generates regular and chaotic motion at varying the initial condition and the value of J, giving rise to Poincaré maps typical of Hamiltonian system as, e.g., the Hénon-Heiles system (Sec. 3.3).

It is worth stressing that the CPR3BP is not a mere academic problem as it may look at first glance. For instance, an interesting example of its use in practical problem has been the *Genesis Discovery Mission* (2001-2004) to collect ions of Solar origin in a region sufficiently far from Earth's geomagnetic field. The existence of a heteroclinic connection between pairs of periodic orbits having the same energy: one around L_1 and the other around L_2 (of the system Sun-Earth), allowed for a consistent reduction of the necessary fuel [Koon *et al.* (2000)]. In a more futuristic context, the Lagrangian points L_4 and L_5 of the system Earth-Moon are, in a future space colonization project, the natural candidates for a colony or a manufacturing facility. We conclude by noticing that there is a perfect parallel between the governing equations of atomic physics (for the hydrogen ionization in crossed electric and magnetic field) and celestial mechanics; this has induced an interesting cross fertilization of methods and ideas among mathematicians, chemists and physicists [Porter and Cvitanovic (2005)].

11.1.2 *Chaos in the Solar system*

The Solar system consists of the Sun, the 8 main planets (Mercury, Venus, Earth, Mars, Jupiter, Saturn, Uranus, Neptune[10]) and a very large number of minor bodies (satellites, asteroids, comets, etc.), for instance, the number of asteroids of linear size larger than 1Km is estimated to be $\mathcal{O}(10^6)$.[11]

11.1.2.1 *The chaotic motion of Hyperion*

The first striking example (both theoretical and observational) of chaotic motion in our Solar system is represented by the rotational motion of Hyperion. This small moon of Saturn, with a very irregular shape (a sort of deformed hamburger), was detected by Voyager spacecraft in 1981. It was found that Hyperion is spinning along neither its largest axis nor the shortest one, suggesting an unstable motion. Wisdom *et al.* (1984, 1987) proposed the following Hamiltonian, which is good model, under suitable conditions, for any satellite of irregular shape:[12]

$$H = \frac{p^2}{2} - \frac{3}{4}\frac{I_B - I_A}{I_C}\left[\frac{a}{r(t)}\right]^3 \cos(2q - 2v(t)) \,, \tag{11.4}$$

[10] The dwarf planet Pluto is now considered an asteroid, member of the so-called Kuiper belt.

[11] However the total mass of all the minor bodies is rather small compared with that one of Jupiter, therefore is is rather natural to study separately the dynamics of the small bodies and the motion of the Sun and the planets. This is the typical approach used in celestial mechanics as described in the following.

[12] As, for instance, for Deimos and Phobos which are two small satellites of Mars.

where the generalized coordinate q represents the orientation of the satellite's longest axis with respect to a fixed direction and $p = dq/dt$ the associated velocity; $I_C > I_B > I_A$ are the principal moments of inertia so that $(I_B - I_A)/I_C$ measures the deviation from a sphere; $r(t)$ gives the distance of the moon from Saturn and $q - v(t)$ measures the orientation of Hyperion's longest axis with respect to the line Saturn-to-Hyperion; finally Hyperion's orbit is assumed to be a fixed ellipse with semi-major axis of length a. The idea behind the derivation of such a Hamiltonian is that due to the non-spherical mass distribution of Hyperion the gravitational field of Saturn can produce a net torque which can be modeled, at the lowest order, by considering a quadrupole expansion of the mass distribution.

It can be easily recognized that the Hamiltonian (11.4) describes a nonlinear oscillator subject to a periodic forcing, namely the periodic variation of $r(t)$ and $v(t)$ along the orbit of the satellite around Saturn. In analogy with the vertically forced pendulum of Chapter 1, chaos may not be unexpected in such a system. It should be, however, remarked that crucial for the appearance of chaos in Hyperion is the fact that its orbit around Saturn deviates from a circle, the eccentricity being $e \approx 0.1$. Indeed, for $e = 0$ one has $r(t) = a$ and, eliminating the time dependence in $v(t)$ by a change of variable, the Hamiltonian can be reduced to that of a simple nonlinear pendulum which always gives rise to periodic motion. To better appreciate this point, we can expand H with respect to the eccentricity e, retaining only the terms of first order in e [Wisdom *et al.* (1984)], obtaining

$$H = \frac{p^2}{2} - \frac{\alpha}{2}\cos(2x - 2t) + \frac{\alpha e}{2}\left[\cos(2x - t) - 7\cos(2x - 3t)\right] ,$$

where we used suitable time units and $\alpha = 3(I_B - I_A)/(2I_C)$. Now it is clear that, for circular orbits, $e = 0$, the system is integrable, being basically a pendulum with possibility of libration and circulation motion. For $\alpha e \neq 0$, the Hamiltonian is not integrable and, because of the perturbation terms, irregular transitions occur between librational and rotational motion. For large value of αe the overlap of the resonances (14) gives rise to large scale chaotic motion; for Hyperion this appears for $\alpha e \geq 0.039...$ [Wisdom *et al.* (1987)].

11.1.2.2 *Asteroids*

Between the orbits of Mars and Jupiter there is the so-called asteroid belt[13] containing thousands of small celestial objects, the largest asteroid Ceres (which was the first to be discovered)[14] has a diameter $\sim 10^3$ km.

[13]Another belt of small objects — the Kuiper belt — is located beyond Neptune orbit.

[14]The first sighting of an asteroid occurred on Jan. 1, 1801, when the Italian astronomer Piazzi noticed a faint, star-like object not included in a star catalog that he was checking. Assuming that Piazzi's object circumnavigated the Sun on an elliptical course and using only three observations of its place in the sky to compute its preliminary orbit, Gauss calculated what its position would be when the time came to resume observations. Gauss spent years refining his techniques for handling planetary and cometary orbits. Published in 1809 in a long paper *Theoria motus corporum coelestium in sectionibus conicis solem ambientium* (Theory of the motion of the heavenly bodies

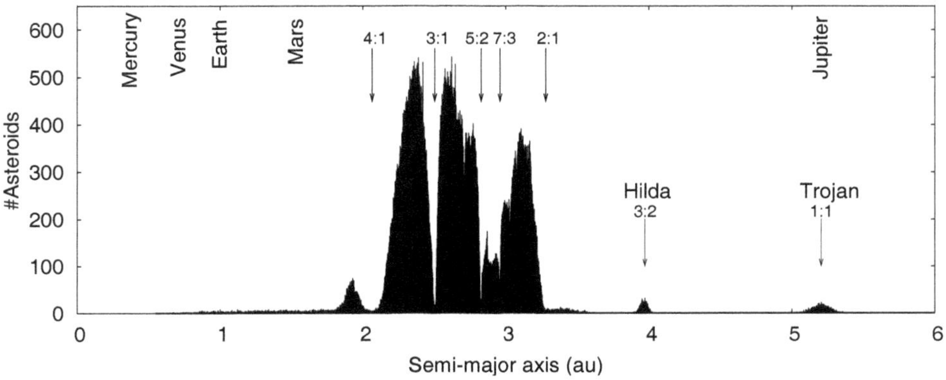

Fig. 11.5 Number of asteroids as a function of the distance from the Sun, measured in au. Note the gaps at the resonances with Jupiter orbital period (top arrow) and the "anomaly" represented by Hilda group.

Since the early work of Kirkwood (1888), the distribution of asteroids has been known to be not uniform. As shown in Fig. 11.5, clear gaps appear in the histogram of the number of asteroids as function of the major semi-axis expressed in astronomical units (au),[15] the clearest ones being $4:1$, $3:1$, $5:2$, $7:3$ and $2:1$ (where $n:m$ means that the asteroid performs n revolutions around the Sun in m Jupiter periods). The presence of these gaps cannot be caught using the crudest approximation — the CPR3BP — as it describes an almost integrable $2d$ Hamiltonian system where the KAM tori should prevent the spreading of asteroid orbits. On the other hand, using the full three-body problem, since the gaps are in correspondence to precise resonances with Jupiter orbital period, it seems natural to interpret their presence in terms of a rather generic mechanism in Hamiltonian system: the destruction of the resonant tori due to the perturbation of Jupiter (see Chap. 7). However, this simple interpretation, although not completely wrong, does not explain all the observations. For instance, we already know the Trojans are in the stable Lagrangian points of the Sun-Jupiter problem, which correspond to the $1:1$ resonance. Therefore, being in resonance is not equivalent to the presence of a gap in the asteroid distribution. As a further confirmation, notice the presence of asteroids (the Hilda group) in correspondence of the resonances $3:2$ (Fig. 11.5). One is thus forced to increase the complexity of the description including the effects of other planets. For instance, detailed numerical and analytical computations show that sometimes, as for the resonance $3:1$, it is necessary to account for the perturbation due to Saturn (or Mars) [Morbidelli (2002)].

---

moving about the sun in conic sections), this collection of methods still plays an important role in modern astronomical computation and celestial mechanics.

[15]the Astronomical unit (au) is the mean Sun-Earth distance, the currently accepted value of the is $1\text{au} = 149.6 \times 10^6$ km.

Assuming that at the beginning of the asteroid belt the distribution of the bodies was more uniform than now, it is interesting to understand the dynamical evolution which lead to the formation of the gaps. In this framework, numerical simulations, in different models, show that the Lyapunov time $1/\lambda_1$ and the escape time t_e, i.e. the time necessary to cross the orbit of Mars, computed as function of the initial major semi-axis, have minima in correspondence of the observed Kirkwood gaps. For instance, test particles initially located near the 3:1 resonance with low eccentricity orbits, after a transient of about 2×10^5 years increase the eccentricity, setting their motions on Mars crossing orbits which produce an escape from the asteroid belt [Wisdom (1982); Morbidelli (2002)].

The above discussion should have convinced the reader that the rich features of the asteroid belt (Fig. 11.5) are a vivid illustration of the importance of chaos in the Solar system. An uptodate review of current understanding, in terms of dynamical systems, of Kirkwood's gaps and other aspects of small bodies motion can be found in the monograph by Morbidelli (2002). We conclude mentioning that chaos also characterizes the motion of other small bodies such as comets (see Box B.23 where we briefly describe an application of symplectic maps to the motion of Halley comet).

Box B.23: A symplectic map for Halley comet

The major difficulties in the statistical study of long time dynamics of comets is due to the necessity of accounting for a large number ($\mathcal{O}(10^6)$) of orbits over the life time of the Solar system ($\mathcal{O}(10^{10})$ys), a task at the limit of the capacity of existing computers. Nowadays the common belief is that certain kind of comets (like those with long periods and others, such as Halley's comet) originate from the hypothetical Oort cloud, which surrounds our Sun at a distance of $10^4 - 10^5$ au. Occasionally, when the Oort cloud is perturbed by passing stars, some comets can enter the Solar system with very eccentric orbits. The minimal model for this process amounts to consider a test particle (the comet) moving on a circular orbit under the combined effect of the gravitational field of the Sun and Jupiter, i.e. the CPR3BP (Sec. 11.1.1). Since most of the discovered comets have perihelion distance smaller than few au, typically the perihelion is inside the Jupiter orbit (5.2au), the comet is significantly perturbed by Jupiter only in a small fraction of time. Therefore, it sounds reasonable to approximate the perturbations by Jupiter as impulsive, and thus model the comet dynamics in terms of discrete time maps. Of course, such a map, as consequence of the Hamiltonian character of the original problem, must be symplectic. In the sequel we illustrate how such a kind of model can be build up.

Define the running "period" of the comet as $P_n = t_{n+1} - t_n$, t_n being the perihelion passage time, and introduce the quantities

$$x(n) = \frac{t_n}{T_J}\,, \qquad w(n) = \left(\frac{P_n}{T_J}\right)^{-2/3}\,, \tag{B.23.1}$$

where T_J is Jupiter orbital period. The quantity $x(n)$ can be interpreted as Jupiter's phase when the comet is at its perihelion. From the third Kepler's law, the energy E_n of the comet considering only the interaction with the Sun, which is reasonable far from Jupiter, is proportional to $-w(n)$, within the interval (t_{n-1}, t_n). Thus, in order to have an elliptic orbit, $w(n)$ must be positive.

The changes of $w(n)$ are induced by the perturbation by Jupiter and thus depends on the phase $x(n)$, so that we can write the equations for $x(n)$ and $w(n)$ as

$$\begin{aligned} x(n+1) &= x(n) + w(n)^{-3/2} \mod 1 \\ w(n+1) &= w(n) + F(x(n+1)) \quad , \end{aligned}$$

where the first amounts to a simple rewriting of (B.23.1), while the second contains the nontrivial contribution of Jupiter, $F(x)$, for which some models have been proposed in specific limits [Petrosky (1986)].

In the following we summarize the results of an interesting study which combines astronomical observations and theoretical ideas. This choice represents a tribute to Boris V. Chirikov (1928–2008) who passed away during the writing of this book and has a pedagogical intent in showing how dynamical systems can be used in modeling and applications. In this perspective we shall avoid to enter the details of the delicate issues of the origins and dynamics of comets.

Halley's comet is perhaps the most famous minor celestial body, whose observation dates back to the year 12 BC till its last passage close to Earth in 1986. From the available observations, Chirikov and Vecheslavov (1989) build up a simple model describing the chaotic evolution of Halley comet. They fitted the unknown function $F(x)$ using the known 46 values of t_n: since 12 BC there are historical data, mainly from Chinese astronomers; while for the previous passages, they used the prediction from numerical orbit simulations of the comet [Yeomans and Kiang (1981)]. Then studied the map evolution by means of numerical simulations which, as typical in two-dimensional symplectic map, show a coexistence of ordered and chaotic motion. In the time unit of the model, the Lyapunov exponent (in the chaotic region) was estimated as $\lambda_1 \sim 0.2$ corresponding to a physical Lyapunov time of about 400ys.

However, from an astronomical point of view, it is more interesting the value of the diffusion coefficient $D = \lim_{n\to\infty} \langle (w(n) - w(0))^2 \rangle / (2n)$ which allows the sojourn time N_s of the comet in the Solar system to be estimated. When the comet enters the Solar system it usually has a negative energy corresponding to a positive w (the typical value is estimated to be $w_c \approx 0.3$). At each passage t_n, the perturbation induced by Jupiter changes the value of w, which performs a sort of random walk. When $w(n)$ becomes negative, energy becomes positive converting the orbit from elliptic to hyperbolic and thus leading to the expulsion of the comet from the Solar system. Estimating $w(n) - w_c \sim \sqrt{Dn}$ the typical time to escape, and thus the sojourn time, will be $N_S \sim w_c^2/D$. Numerical computations give $D = \mathcal{O}(10^{-5})$, in the units of the map, i.e. $N_s = \mathcal{O}(10^5)$ corresponding to a sojourn time of $\mathcal{O}(10^7)$ys. Such time seems to be of the same order of magnitude of the hypothetical comet showers in Oort cloud as conjectured by Hut *et al.* (1987).

11.1.2.3 *Long time behavior of the Solar system*

The "dynamical stability" of the Solar system has been a central issue of astronomy for centuries. The problem has been debated since Newton's age and had attracted the interest of many famous astronomers and mathematicians over the years, from

Lagrange and Laplace to Arnold. In Newton's opinion the interactions among the planets were enough to destroy the stability, and a divine intervention was required, from time to time, to tune the planets on the Keplerian orbits. Laplace and Lagrange tried to show that Newton's laws and the gravitational force were sufficient to explain the movement of the planets throughout the known history. Their computations, based on a perturbation theory, have been able to explained the observed motion of the planets over a range of some thousand years.

Now, as illustrated in the previous examples, we know that in the Solar system chaos is at play, a fact in apparent in contradiction with the very idea of "stability".[16] Therefore, before continuing the discussion, it is worth discussing a bit more about the concept of chaos and "stability". On the one hand, sometimes the presence of chaos is associated with very large excursion of the variables of the system which can induce "catastrophic" events as, for instance, the expulsion of asteroids from the Solar system or their fall on the Sun or, this is very scary, on a planet. On the other hand, as we know from Chap. 7, chaos may also be bounded in small regions of the phase space, giving rise to much less "catastrophic" outcomes. Therefore, in principle, the Solar system can be chaotic, i.e. with positive Lyapunov exponents, but not necessarily this implies events such as collisions or escaping of planets. In addition, from an astronomical point of view, it is important the value of the maximal Lyapunov exponent.

In the following, for Solar system we mean Sun and planets, neglecting all the satellites, the asteroids and the comets. A first, trivial (but reassuring) observation is that the Solar system is "macroscopically" stable, at least for as few as 10^9 years, this just because it is still there! But, of course, we cannot be satisfied with this "empirical" observation.

Because of the weak coupling between the four outer planets (Jupiter, Saturn, Uranus and Neptun) with the four inner ones (Mercury, Venus, Earth and Mars), and their rather different time scales, it is reasonable to study separately the internal Solar system and the external one. Computations had been performed both with the integration of the equations from first principles (using special purpose computers) [Sussman and Wisdom (1992)] and the numerical solution of averaged equations [Laskar *et al.* (1993)], a method which allows to reduce the number of degrees of freedom. Interestingly, the two approaches give results in good agreement.[17]

As a result of these studies, the outer planets system is chaotic with a Lyapunov time $1/\lambda \sim 2 \times 10^7$ys[18] while the inner planets system is also chaotic but with a Lyapunov time $\sim 5 \times 10^6$ys [Sussman and Wisdom (1992); Laskar *et al.*

[16]Indeed, in a strict mathematical sense, the presence of chaos is inconsistent with the stability of given trajectories.

[17]As a technical details, we note that the masses of the planets are not known with very high accuracy. This is not a too serious problem, as it gives rise to effects rather similar to those due to an uncertainty on the initial conditions (see Sec. 10.1).

[18]A numerical study of Pluto, assumed as a zero-mass test particles, under the action of the Sun and the outer planets, shows a chaotic behavior with a Lyapunov time of about 2×10^7ys.

(1993)].[19] However, there is evidence that the Solar system is "astronomically" stable, in the sense that the 8 largest planets seem to remain bound to the Sun in low eccentricity and low inclination orbits for time $\mathcal{O}(10^9)$ys. In this respect, chaos mostly manifest in the irregular behavior of the eccentricity and inclination of the less massive planets, Mercury and Mars. Such variations are not large enough to provoke catastrophic events before extremely very large time. For instance, recent numerical investigations show that for catastrophic events, such as "collisions" between Mercury and Venus or Mercury failure onto the Sun, we should wait at least $\mathcal{O}(10^9)$ys [Batygin and Laughlin (2008)]. We finally observe that the results of detailed numerical studies of the whole Solar system (i.e. Sun and the 8 largest planets) are basically in agreement with those obtained considering as decoupled the internal and external Solar system, confirming the basic correctness of the approach [Sussman and Wisdom (1992); Laskar *et al.* (1993); Batygin and Laughlin (2008)].

11.2 Chaos and transport phenomena in fluids

In this section, we discuss some aspects of the transport properties in fluid flows, which are of great importance in many engineering and natural occurring settings, we just mention pollutants and aerosols dispersion in the atmosphere and oceans [Arya (1998)], the transport of magnetic field in plasma physics [Biskamp (1993)], the optimization of mixing efficiency in several contexts [Ottino (1990)].

Transport phenomena can be approached, depending on the application of interest, in two complementary formulations.

The *Eulerian approach* concerns with the advection of fields such as a scalar $\theta(\boldsymbol{x}, t)$ like the temperature field whose dynamics, when the feedback on the fluid can be disregarded, is described by the equation[20]

$$\partial_t \theta + \boldsymbol{u} \cdot \boldsymbol{\nabla} \cdot \theta = D\, \nabla^2 \theta + \Phi \tag{11.5}$$

where D is the molecular diffusion coefficient, and $\boldsymbol{v}$ the velocity field which may be given or dynamically determined by the Navier-Stokes equations. The source term Φ may or may not be present as it relates to the presence of an external mechanism responsible of, e.g., warming the fluid when θ is the temperature field.

The *Lagrangian approach* instead focuses on the motion of particles released in the fluid. As for the particles, we must distinguish *tracers* from *inertial particles.* The former class is represented by point-like particles, with density equal to the fluid one, that, akin to fluid elements, move with the fluid velocity. The latter kind of particles is characterized by a finite-size and/or density contrast with the

[19]We recall that because of the Hamiltonian character of the system under investigation, the Lyapunov exponent can, and usually does, depend on the initial condition (Sec. 7). The above estimates indicate the maximal values of λ, in some phase-space regions the Lyapunov exponent is close to zero.

[20]When the scalar field is conserved as, e.g., the particle density field the l.h.s. of the equation reads $\partial_t \theta + \boldsymbol{\nabla} \cdot (\theta \boldsymbol{u})$. However for incompressible flows, $\boldsymbol{\nabla} \cdot \boldsymbol{u} = 0$, the two formulations coincide.

fluid, which due to inertia have their own velocity dynamics. Here, we mostly concentrate on the former case, leaving the latter to a short subsection below. The tracer position $\boldsymbol{x}(t)$ evolves according to the Langevin equation

$$\frac{\mathrm{d}\boldsymbol{x}}{\mathrm{d}t} = \boldsymbol{u}(\boldsymbol{x}(t), t) + \sqrt{2D}\boldsymbol{\eta}(t) \tag{11.6}$$

where $\boldsymbol{\eta}$ is a Gaussian process with zero mean and time uncorrelated accounting for the, unavoidable, presence of thermal fluctuations.

In spite of the apparent differences, the two approaches are tightly related as Eq. (11.5) (with $\Phi = 0$) is nothing but the Fokker-Planck equation associated to the Langevin one (11.6) [Gardiner (1982)]. The relationship between these two formulations will be briefly illustrated in a specific example (see Box B.24), while in the rest of the section we shall focus on the Lagrangian approach, which well illustrates the importance of dynamical system theory in the context of transport.

Clearly, Eq. (11.6) defines a dynamical systems with an external randomness. In many realistic situations, however, D is so small (as, e.g., for a powder particle[21] embedded in a fluid, provided that its density equals the fluid one and its size is small not to perturb the velocity field, but large enough not to perform a Brownian motion) that it is enough to consider the limit $D = 0$

$$\frac{\mathrm{d}\boldsymbol{x}}{\mathrm{d}t} = \boldsymbol{u}(\boldsymbol{x}(t), t)\,, \tag{11.7}$$

which defines a standard ODE.

The properties of the dynamical system (11.7) are related to those of $\boldsymbol{u}$. If the flow is incompressible $\boldsymbol{\nabla} \cdot \boldsymbol{u} = 0$ (as typical in laboratory and geophysical flows, where the velocity is usually much smaller than the sound velocity) particle dynamics is conservative; while for compressible flows $\boldsymbol{\nabla} \cdot \boldsymbol{u} < 0$ (as in, e.g. supersonic motions) it is dissipative and particle motions asymptotically evolve onto an attractor. As in most applications we are confronted with incompressible flows, in the following we focus on the former case and, as an example of the latter, we just mention the case of neutrally buoyant particles moving on the surface of a three-dimensional incompressible flow. In such a case the particles move on an effectively compressible two-dimensional flow (see, e.g., Cressman *et al.*, 2004), offering the possibility to visualize a strange attractor in real experiments [Sommerer and Ott (1993)].

Box B.24: Chaos and passive scalar transport

Tracer dynamics in a given velocity field bears information on the statistical features of advected scalar fields, as we now illustrate in the case of passive fields, e.g. a colorant dye, which do not modify the advecting velocity field[Falkovich *et al.* (2001)]. In particular, we focus on the small scale features of a passive field (as, e.g., in Fig. B24.1a) evolving in a

[21]This kind of particles are commonly employed in, e.g. flow visualization [Tritton (1988)].

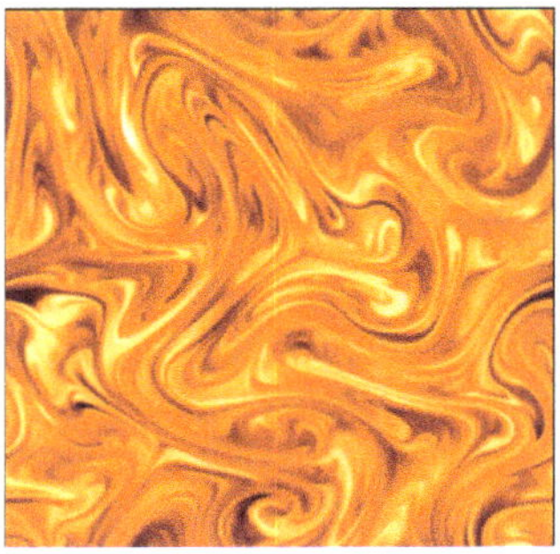

(a)

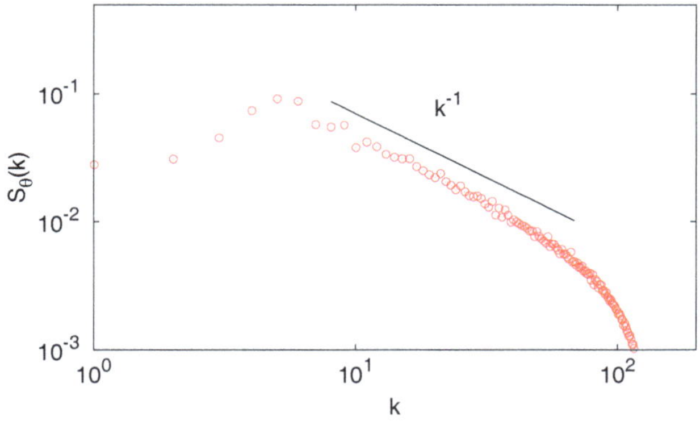

(b)

Fig. B24.1 (a) Snapshot of a passive scalar evolving in a smooth flow obtained by a direct numerical simulation of the two-dimensional Navier-Stokes equation in the regime of enstrophy cascade [Kraichnan (1967)] (see Sec. 13.2.4). Scalar input Φ is obtained by means of a Gaussian uncorrelated in time processes of zero mean concentrated in a small shell of Fourier modes $\sim 2\pi/L_\Phi$. (b) Scalar energy spectrum $S_\theta(k)$. The k^{-1} behavior is shown by the straight line.

laminar flow and, specifically, on the two-point correlation function or, equivalently, the Fourier spectrum of the scalar field.

The equation for a passive field $\theta(\boldsymbol{x})$ can be written as

$$\partial_t \theta(\boldsymbol{x},t) + \boldsymbol{u}(\boldsymbol{x},t)\cdot\boldsymbol{\nabla}\theta(\boldsymbol{x},t) = D\Delta\theta(\boldsymbol{x},t) + \Phi(\boldsymbol{x},t)\,, \qquad \text{(B.24.1)}$$

where molecular diffusivity D is assumed to be small and the velocity $\boldsymbol{u}(\boldsymbol{x},t)$ to be differentiable over a range of scales, i.e. $\delta_{\boldsymbol{R}}\boldsymbol{u} = \boldsymbol{u}(\boldsymbol{x}+\boldsymbol{R},t) - \boldsymbol{u}(\boldsymbol{x},t) \sim R$ for $0 < R < L$, where L is the flow correlation length. The velocity $\boldsymbol{u}$ can be either prescribed or dynamically obtained, e.g., by stirring (not too violently) a fluid. In the absence of a scalar input θ decays in time so that, to reach stationary properties, we need to add a source of tracer fluctuations, Φ, acting at a given length scale $L_\Phi \lesssim L$.

The crucial step is now to recognize that Eq. (B.24.1) can be solved in terms of particles evolving in the flow,[22] [Celani *et al.* (2004)], i.e.

$$\begin{aligned}
\vartheta(\boldsymbol{x},t) &= \int_{-\infty}^{t} \mathrm{d}s\, \Phi(\boldsymbol{x}(s;t),s) \\
\tfrac{\mathrm{d}\boldsymbol{x}}{\mathrm{d}s}(s;t) &= \boldsymbol{u}(\boldsymbol{x}(s;t),s) + \sqrt{2D}\,\boldsymbol{\eta}(s)\,, \qquad \boldsymbol{x}(t;t) = \boldsymbol{x}\,;
\end{aligned}$$

we remark that in the Langevin equation the final position is assigned to be $\boldsymbol{x}$. The noise term $\boldsymbol{\eta}(t)$ is the Lagrangian counterpart of the diffusive term, and is taken as a Gaussian, zero mean, random field with correlation $\langle \eta_i(t)\eta_j(s)\rangle = \delta_{ij}\delta(t-s)$.

Essentially to determine the field $\theta(\boldsymbol{x},t)$ we need to look at all trajectories $\boldsymbol{x}(s;t)$ which land in $\boldsymbol{x}$ at time t and to accumulate the contribution of the forcing along each path. The field $\theta(\boldsymbol{x},t)$ is then obtained by averaging over all these paths, i.e. $\theta(\boldsymbol{x},t) = \langle \vartheta(\boldsymbol{x},t)\rangle_\eta$, where the subscript η indicates that the average is over noise realizations.

[22] I.e. solving (B.24.1) via the method of characteristics [Courant and Hilbert (1989)].

A straightforward computation allows us to connect the dynamical features of particle trajectories to the correlation functions of the scalar field. For instance, the simultaneous two-point correlations can be written as

$$\langle \theta(\boldsymbol{x}_1,t)\theta(\boldsymbol{x}_2,t)\rangle = \int_{-\infty}^{t} \mathrm{d}s_1 \int_{-\infty}^{t} \mathrm{d}s_2 \, \langle \Phi(\boldsymbol{x}_1(s_1;t),s_1)\Phi(\boldsymbol{x}_2(s_2;t),s_2)\rangle_{u,\eta,\Phi} \,, \tag{B.24.2}$$

with $\boldsymbol{x}_1(t;t) = \boldsymbol{x}_1$ and $\boldsymbol{x}_2(t;t) = \boldsymbol{x}_2$. The symbol $\langle[\ldots]\rangle_{u,\eta,\Phi}$ denotes the average over the noise and the realizations of both the velocity and the scalar input term. To ease the computation we assume the forcing to be a random, Gaussian process with zero mean and correlation function $\langle\Phi(\boldsymbol{x}_1,t_1)\Phi(\boldsymbol{x}_2,t_2)\rangle = \chi(|\boldsymbol{x}_1-\boldsymbol{x}_2|)\delta(t_1-t_2)$.

Exploiting space homogeneity, Eq. (B.24.2) can be further simplified in[23]

$$C_2(R) = \langle \theta(\boldsymbol{x},t)\theta(\boldsymbol{x}+\boldsymbol{R},t)\rangle = \int_{-\infty}^{t} \mathrm{d}s \int \mathrm{d}\boldsymbol{r}\, \chi(\boldsymbol{r})\, p(\boldsymbol{r},s|\boldsymbol{R},t)\,. \tag{B.24.3}$$

where $p(\boldsymbol{r},s|\boldsymbol{R},t)$ is the probability density function for a particle pair to be at separation $\boldsymbol{r}$ at time s, under the condition to have separation $\boldsymbol{R}$ at time t. Note that $p(\boldsymbol{r},s|\boldsymbol{R},t)$ only depends on the velocity field demonstrating, at least for the passive problem, the fundamental role of the Lagrangian dynamics in determining the scalar field statistics.

Finally, to grasp the physical meaning of (B.24.3) it is convenient to choose a simplified forcing correlation, $\chi(r)$, which vanishes for $r > L_\Phi$ and stays constant to $\chi(0) = \chi_0$ for $r < L_\Phi$. It is then possible to recognize that Eq. (B.24.3) can be written as

$$C_2(R) \approx \chi_0 T(R; L_\Phi)\,, \tag{B.24.4}$$

where $T(R; L_\Phi)$ is the average time the particle pair employs (backward evolving in time) to reach a separation $\mathcal{O}(L_\Phi)$ starting from a separation R. In typical laminar flows, due to Lagrangian chaos[24] (Sec. 11.2.1) we have an exponentially growth of the separation, $R(t) \approx R(0)\exp(\lambda t)$. As a consequence, $T(R;L_\Phi) \propto (1/\lambda)\ln(L_\Phi/R)$ meaning a logarithmic dependence on R for the correlation function, which translates in a passive scalar spectrum $S_\theta(k) \propto k^{-1}$ as exemplified in Fig. B24.1b. Chaos is thus responsible for the k^{-1} behavior of the spectrum [Monin and Yaglom (1975); Yuan *et al.* (2000)]. This is contrasted by diffusion which causes an exponential decreasing of the spectrum at high wave numbers (very small scales).

We emphasize that the above idealized description is not far from reality and is able to catch the relevant aspects of experimentally observations pioneered by Batchelor (1959) (see also, e.g, Jullien *et al.*, 2000).

We conclude mentioning the result (B.24.4) does not rely on the smoothness of the velocity field, and can thus be extended to generic flows and that the above treatment can be extended to correlation functions involving more than two points which may be highly non trivial [Falkovich *et al.* (2001)]. More delicate is the extension of this approach to active, i.e. having a feedback on the fluid velocity, fields [Celani *et al.* (2004)].

---

[23] The passivity of the field allows us to separate the average over velocity from that over the scalar input [Celani *et al.* (2004)].

[24] This is true regardless we consider the forward or backward time evolution. For instance, in two dimensions $\boldsymbol{\nabla}\cdot\boldsymbol{u} = 0$ implies $\lambda_1 + \lambda_2 = 0$, meaning that forward and backward separation take place with the same rate $\lambda = \lambda_1 = |\lambda_2|$. In three dimensions, the rate may be different.

11.2.1 *Lagrangian chaos*

Everyday experience, when preparing a cocktail or a coffee with milk, teaches us that fluid motion is crucial for mixing substances. The enhanced mixing efficiency is clearly linked to the presence of the stretching and folding mechanism typical of chaos (Sec. 5.2.2). Being acquainted with the basics of dynamical systems theory, it is not unexpected that in laminar velocity field the motion of fluid particles may be very irregular, even in the absence of *Eulerian Chaos*, i.e. even in regular velocity field.[25] However, in spite of several early studies by Arnold (1965) and Hénon (1966) already containing the basic ideas, the importance of chaos in the transport of substances was not widely appreciated before Aref's contribution [Aref (1983, 1984)], when terms as *Lagrangian chaos* or *chaotic advection* have been coined.

The possibility of an irregular behavior of test particles even in regular velocity fields had an important technological impact, as it means that we can produce a well controlled velocity field (as necessary for the safe maintenance of many devices) but still able to efficiently mix transported substances. This has been somehow a small revolution in the geophysical and engineering community. In this respect, it is worth mentioning that chaotic advection is now experiencing a renewed attention due to development of microfluidic devices [Tabeling and Cheng (2005)]. At micrometer scale, the velocity fields are extremely laminar, so that it is becoming more and more important to devise systems able to increase the mixing efficiency for building, e.g., microreactor chambers. In this framework, several research groups are proposing to exploit chaotic advection to increase the mixing efficiency (see, e.g., Stroock *et al.*, 2002). Another recent application of Lagrangian Chaos is in biology, where the technology of DNA microarrays is flourishing [Schena *et al.* (1995)]. An important step accomplished in such devices is the hybridization that allows single-stranded nucleic acids to find their targets. If the single-stranded nuclei acids have to explore, by simple diffusion, the whole microarray in order to find their target, hybridization last for about a day and often is so inefficient to severely diminish the signal to noise ratio. Chaotic advection can thus be used to speed up the process and increase the signal to noise ratio (see, e.g., McQuain *et al.*, 2004).

11.2.1.1 *Eulerian vs Lagrangian chaos*

To exemplify the difference between Eulerian and Lagrangian chaos we consider two-dimensional flows, where the incompressibility constraint $\boldsymbol{\nabla} \cdot \boldsymbol{u} = 0$ is satisfied taking $u_1 = \partial\psi/\partial x_2$, $u_2 = -\partial\psi/\partial x_1$. The *stream function* $\psi(\boldsymbol{x}, t)$ plays the role of the Hamiltonian for the coordinates (x_1, x_2) of a tracer whose dynamics is given by

$$\frac{\mathrm{d}x_1}{\mathrm{d}t} = \frac{\partial\psi}{\partial x_2}, \qquad \frac{\mathrm{d}x_2}{\mathrm{d}t} = -\frac{\partial\psi}{\partial x_1},$$

(x_1, x_2) are thus canonical variables.

[25]In two-dimensions it is enough to have a time periodic flow and in three the velocity can even be stationary, see Sec. 2.3

In a real fluid, the velocity $\boldsymbol{u}$ is ruled by partial differential equations (PDE) such as the Navier-Stokes equations. However, in weakly turbulent situations, an approximate evolution can be obtained by using the Galerkin approach i.e. writing the velocity field in terms of suitable functions, usually a Fourier series expansion as $\boldsymbol{u}(\boldsymbol{x},t) = \sum_k \boldsymbol{Q}_k(t)\exp(i\boldsymbol{k}\cdot\boldsymbol{x})$, and reducing the Eulerian PDE to a (low dimensional) system of F ODEs (see also Sec. 13.3.2).[26] The motion of a fluid particle is then determined by the $(d+F)$-dimensional system

$$\frac{\mathrm{d}\boldsymbol{Q}}{\mathrm{d}t} = \boldsymbol{f}(\boldsymbol{Q},t) \qquad \text{with } \boldsymbol{Q},\, \boldsymbol{f}(\boldsymbol{Q},t) \in \mathbb{R}^F \tag{11.8}$$

$$\frac{\mathrm{d}\boldsymbol{x}}{\mathrm{d}t} = \boldsymbol{u}(\boldsymbol{x},\boldsymbol{Q}) \qquad \text{with } \boldsymbol{x},\, \boldsymbol{u}(\boldsymbol{x},\boldsymbol{Q}) \in \mathbb{R}^d \tag{11.9}$$

d being the space dimensionality ($d=2$ in the case under consideration) and $\boldsymbol{Q} = (Q_1, ...Q_F)$ the F variables (typically normal modes) representing the velocity field $\boldsymbol{u}$. Notice that Eq. (11.8) describes the Eulerian dynamics that is independent of the Lagrangian one (11.9). Therefore we have a "skew system" of equations where Eq. (11.8) can be solved independently of (11.9).

An interesting example of the above procedure was employed by Boldrighini and Franceschini (1979) and Lee (1987) to study the two-dimensional Navier-Stokes equations with periodic boundary conditions at low Reynolds numbers. The idea is to expand the stream function ψ in Fourier series retaining only the first F terms

$$\psi = -i\sum_{j=1}^{F} \frac{Q_j}{k_j}\mathrm{e}^{i\boldsymbol{k}_j\boldsymbol{x}} + \text{c.c.}\,, \tag{11.10}$$

where c.c. indicates the complex conjugate term. After an appropriate time rescaling, the original PDEs equations can be reduced to a set of F ODEs of the form

$$\frac{\mathrm{d}Q_j}{\mathrm{d}t} = -k_j^2\,Q_j + \sum_{l,m} A_{jlm}Q_lQ_m + f_j, \tag{11.11}$$

where A_{jlm} accounts for the nonlinear interaction among triads of Fourier modes, f_j represents an external forcing, and the linear term is related to dissipation.

Given the skew structure of the system (11.8)-(11.9), three different Lyapunov exponents characterize its chaotic properties [Falcioni *et al.* (1988)]:
λ_{E} for the Eulerian part (11.8), quantifying the growth of infinitesimal uncertainties on the velocity (i.e. on $\boldsymbol{Q}$, independently of the Lagrangian motion);
λ_{L} for the Lagrangian part (11.9), quantifying the separation growth of two initially close tracers evolving in the same flow (same $\boldsymbol{Q}(t)$), assumed to be known;
λ_{T} for the total system of $d+F$ equations, giving the growth rate of separation of initially close particle pairs, when the velocity field is not known with certainty.
These Lyapunov exponents can be measured as [Crisanti *et al.* (1991)]

$$\lambda_{\mathrm{E,L,T}} = \lim_{t\to\infty} \frac{1}{t} \ln \frac{|\boldsymbol{z}(t)^{(\mathrm{E,L,T})}|}{|\boldsymbol{z}(0)^{(\mathrm{E,L,T})}|}$$

[26]This procedure can be performed with mathematical rigor [Lumley and Berkooz (1996)].

where the tangent vector $\boldsymbol{z}^{(\mathrm{E,L,T})}$ evolution is given by the linearization of the Eulerian, the Lagrangian and the total dynamics.[27]

Due to the conservative nature of the Lagrangian dynamics (11.9) there can be coexistence of non-communicating regions with Lagrangian Lyapunov exponents depending on the initial condition (Sec. 3.3). This observation suggests that there should not be any general relation between λ_E and λ_L, as the examples below will further demonstrate. Moreover, as consequences of the skew structure of (11.8)-(11.9), we have that $\lambda_\mathrm{T} = \max\{\lambda_\mathrm{E}, \lambda_\mathrm{L}\}$ [Crisanti *et al.* (1991)].

Some of the above considerations can be illustrated by studying the system (11.8)–(11.9) with the dynamics for $\boldsymbol{Q}$ given by Eq. (11.11). We start briefly recalling the numerical results of Boldrighini and Franceschini (1979) and Lee (1987) about the transition to chaos of the Eulerian problem (11.11) for $F = 5$ and $F = 7$, with the forcing restricted to the third mode $f_j = \mathrm{Re}\,\delta_{j,3}$, Re is the Reynolds number of the flow, controlling the nonlinear terms. For $F = 5$ and $\mathrm{Re} < \mathrm{Re}_1$, there are four stable stationary solutions, say $\widehat{\boldsymbol{Q}}$. At $\mathrm{Re} = \mathrm{Re}_1$, these solutions become unstable, via a Hopf bifurcation [Marsden and McCracken (1976)]. Thus, for $\mathrm{Re}_1 < \mathrm{Re} < \mathrm{Re}_2$, stable limit cycles of the form

$$\boldsymbol{Q}(t) = \widehat{\boldsymbol{Q}} + (\mathrm{Re} - \mathrm{Re}_1)^{1/2}\delta\boldsymbol{Q}(t) + \mathcal{O}(\mathrm{Re} - \mathrm{Re}_1)$$

occur, where $\delta\boldsymbol{Q}(t)$ is periodic with period $T(\mathrm{Re}) = T_0 + \mathcal{O}(\mathrm{Re}-\mathrm{Re}_1)$. At $\mathrm{Re} = \mathrm{Re}_2$, the limit cycles lose stability and Eulerian chaos finally appears through a period doubling transition (Sec. 6.2).

The scenario for fluid tracers evolving in the above flow is as follows. For $\mathrm{Re} < \mathrm{Re}_1$, the stream function is asymptotically stationary, $\psi(\boldsymbol{x}, t) \to \widehat{\psi}(\boldsymbol{x})$ hence, as typical for time-independent one-degree of freedom Hamiltonian systems, Lagrangian trajectories are regular. For $\mathrm{Re} = \mathrm{Re}_1 + \epsilon$, ψ becomes time dependent

$$\psi(\boldsymbol{x}, t) = \widehat{\psi}(\boldsymbol{x}) + \sqrt{\epsilon}\,\delta\psi(\boldsymbol{x}, t) + \mathcal{O}(\epsilon),$$

where $\widehat{\psi}(\boldsymbol{x})$ is given by $\widehat{\boldsymbol{Q}}$ and $\delta\psi$ is periodic in $\boldsymbol{x}$ and in t with period T. As generic in periodically perturbed one-degree of freedom Hamiltonian systems, the region adjacent to a separatrix, being sensitive to perturbations, gives rise to chaotic layers. Unfortunately, the structure of the separatrices (Fig. 11.6 left), and the analytical complications make very difficult the use of Melnikov method (Sec. 7.5) to prove the existence of such chaotic layer. However, already for small $\epsilon = \mathrm{Re}_1 - \mathrm{Re}$, numerical analysis clearly reveals the appearance of layers of Lagrangian chaotic motion (Fig. 11.6 right).

---

[27] In formulae, linearized equations are $\frac{\mathrm{d}z_i^{(\mathrm{E})}}{\mathrm{d}t} = \sum_{j=1}^{F} \frac{\partial f_i}{\partial Q_j}\Big|_{\boldsymbol{Q}(t)} z_j{}^{(\mathrm{E})}$ with $\boldsymbol{z}(t)^{(\mathrm{E})} \in \mathrm{I\!R}^F$, $\frac{\mathrm{d}z_i^{(\mathrm{L})}}{\mathrm{d}t} = \sum_{j=1}^{d} \frac{\partial v_i}{\partial x_j}\Big|_{\boldsymbol{x}(t)} z_j{}^{(\mathrm{L})}$ with $\boldsymbol{z}(t)^{(\mathrm{L})} \in \mathrm{I\!R}^d$ and, finally, $\frac{\mathrm{d}z_i^{(\mathrm{T})}}{\mathrm{d}t} = \sum_{j=1}^{d+F} \frac{\partial G_i}{\partial y_j}\Big|_{\boldsymbol{y}(t)} z_j{}^{(\mathrm{T})}$ with $\boldsymbol{z}(t)^{(\mathrm{T})} \in \mathrm{I\!R}^{F+d}$, where $\boldsymbol{y} = (Q_1, \ldots, Q_F, x_1, \ldots, x_d)$ and $\boldsymbol{G} = (f_1, \ldots, f_F, v_1, \ldots, v_d)$.

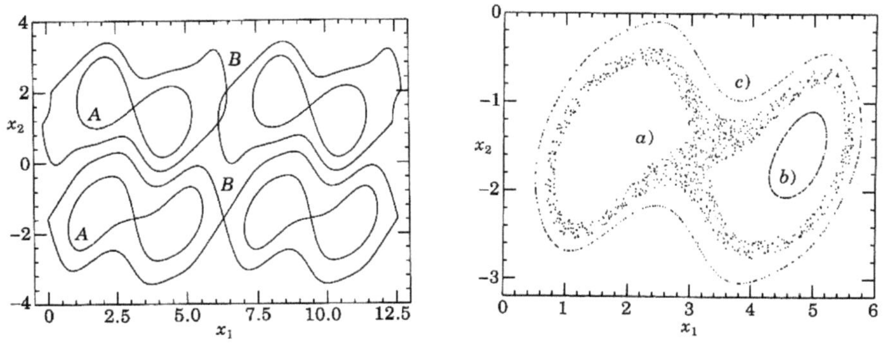

Fig. 11.6 (left) Structure of the separatrices of the Hamiltonian Eq. (11.10) with $F = 5$ and $\mathrm{Re} = \mathrm{Re}_1 - 0.05$. (right) Stroboscopic map displaying the position of three trajectories, at $\mathrm{Re} = \mathrm{Re}_1 + 0.05$, with initial conditions selected close to a separatrix a) or far from it b) and c). The positions are shown at each period of the Eulerian limit cycle (see Falcioni *et al.* (1988) for details.)

From a fluid dynamics point of view, we observe that for these small values of ϵ the separatrices still constitute barriers[28] to the transport of particles in distant regions. Increasing ϵ (as for the standard map, see Chap. 7), the size of the stochastic layers rapidly increase until, at a critical value $\epsilon_c \approx 0.7$, they overlap according to the resonance overlap mechanism (Box B.14). It is then practically impossible to distinguish regular and chaotic zones, and large scale diffusion is finally possible.

The above investigated model illustrated the, somehow expected, possibility of Lagrangian Chaos in the absence of Eulerian Chaos. Next example will show the, less expected, fact that Eulerian Chaos does not always imply Lagrangian Chaos.

11.2.1.2 *Lagrangian chaos in point-vortex systems*

We now consider another example of two-dimensional flow, namely the velocity field obtained by point vortices (Box B.25), which are a special kind of solution of the two-dimensional Euler equation. Point vortices correspond to an idealized case in which the velocity field is generated by N point-like vortices, where the vorticity[29] field is singular and given by $\omega(\boldsymbol{r},t) = \boldsymbol{\nabla} \times \boldsymbol{u}(\boldsymbol{r},t) = \sum_{i=1}^{N} \Gamma_i \delta(\boldsymbol{r} - \boldsymbol{r}_i(t))$, where Γ_i is the circulation of the i-th vortices and $\boldsymbol{r}_i(t)$ its position on the plane at time t. The stream function can be written as

$$\psi(\boldsymbol{r},t) = -\frac{1}{4\pi} \sum_{i=1}^{N} \Gamma_i \ln |\boldsymbol{r} - \boldsymbol{r}_i(t)| \,, \tag{11.12}$$

[28]The presence, detection and study of barriers to transport are important in many geophysical issues [Bower *et al.* (1985); d'Ovidio *et al.* (2009)] (see e.g. Sec. 11.2.2.1) as well as, e.g., in Tokamaks, where devising flow structures able to confine hot plasmas is crucial [Strait *et al.* (1995)].

[29]Note that in $d = 2$ the vorticity perpendicular to the plane where the flow takes place, and thus can be represented as a scalar.

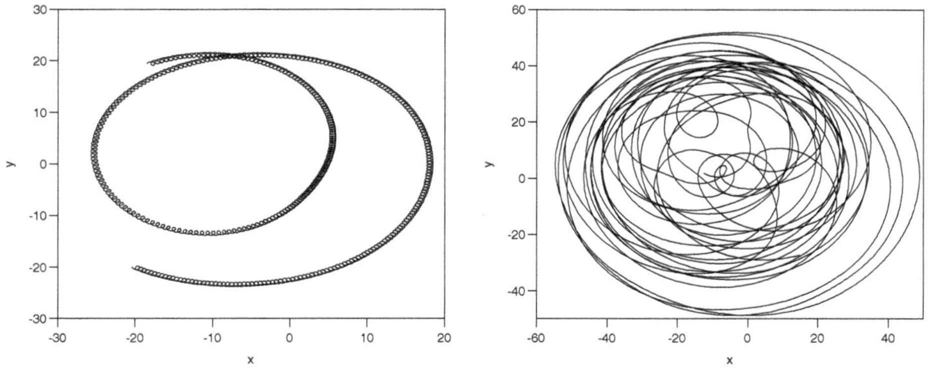

Fig. 11.7 Lagrangian trajectories in the four-vortex system: (left) a regular trajectory around a chaotic vortex; (right) a chaotic trajectory in the background flow.

from which we can derive the dynamics of a tracer particle[30]

$$\frac{\mathrm{d}x}{\mathrm{d}t} = -\sum_i \frac{\Gamma_i}{2\pi} \frac{y - y_i}{|\boldsymbol{r} - \boldsymbol{r}_i(t)|^2} \qquad \frac{\mathrm{d}y}{\mathrm{d}t} = \sum_i \frac{\Gamma_i}{2\pi} \frac{x - x_i}{|\boldsymbol{r} - \boldsymbol{r}_i(t)|^2}, \tag{11.13}$$

where $\boldsymbol{r} = (x, y)$ denotes the tracer position. Of course, Eq. (11.13) represents the dynamics (11.9), which needs to be supplemented with the Eulerian dynamics, i.e. the equations ruling the motion of the point vortices as described in Box B.25.

Aref (1983) has shown that, due to the presence of extra conservation laws, the $N = 3$ vortices problem is integrable while for $N \geq 4$ is not (Box B.25). Therefore, going from $N = 3$ to $N \geq 4$, test particles pass from evolving in a non-chaotic Eulerian field to moving in a chaotic Eulerian environment.[31]

With $N = 3$, three point vortices plus a tracer, even if the Eulerian dynamics is integrable — the stream function (11.12) is time-periodic — the advected particles may display chaotic behavior. In particular, Babiano *et al.* (1994) observed that particles initially released close to a vortex rotate around it with a regular trajectory, i.e. $\lambda_L = 0$, while those released in the background flow (far from vortices) are characterized by irregular trajectories with $\lambda_L > 0$. Thus, again, Eulerian regularity does not imply Lagrangian regularity. Remarkably, this difference between particles which start close to a vortex or in the background flow remains also in the presence of Eulerian chaos (see Fig. 11.7), i.e. with $N \geq 4$, yielding a seemingly paradoxical situation. The motion of vortices is chaotic so that a particle which started close to it displays an unpredictable behavior, as it rotates around the vortex position which moves chaotically. Nevertheless, if we assume the vortex positions to be known and

[30] Notice that the problem of a tracer advected by N vortices is formally equivalent to the case of $N + 1$ vortices where $\Gamma_{N+1} = 0$.

[31] The N-vortex problem resemble the $(N-1)$-body problem of celestial mechanics. In particular, $N = 3$ vortices plus a test particles is analogous to the restricted three-body problem: the test particle corresponds to a chaotic asteroid in the gravitational problem.

consider infinitesimally close particles around the vortex the two particles around the vortex remain close to each other and to the vortex, i.e. $\lambda_L = 0$ even if $\lambda_E > 0$.[32] Therefore, Eulerian chaos does not imply Lagrangian chaos.

It is interesting to quote that also real vortices (with a finite core), as those characterizing two-dimensional turbulence, produce a similar scenario for particle advection with regular trajectories close to the vortex core and chaotic behavior in the background flow [Babiano *et al.* (1994)]. Vortices are thus another example of barrier to transport. One can argue that, in real flows, molecular diffusivity will, sooner or later, let the particles to escape. However, diffusive process responsible for particle escaping is typically very slow, e.g. persistent vortical structures in the Mediterranean sea are able to trap floating buoys up to a month [Rio *et al.* (2007)].

Box B.25: Point vortices and the two-dimensional Euler equation

Two-dimensional ideal flows are ruled by Euler equation that, in terms of the vorticity $\omega\hat{\boldsymbol{z}} = \boldsymbol{\nabla} \times \boldsymbol{u}$ (which is perpendicular to the plane of the flow), reads

$$\partial_t \omega + \boldsymbol{u} \cdot \boldsymbol{\nabla}\omega = 0\,, \tag{B.25.1}$$

expressing the conservation of vorticity along fluid-element paths. Writing the velocity in terms of the stream function, $\boldsymbol{u} = \boldsymbol{\nabla}^{\perp}\psi = (\partial_y, -\partial_x)\psi$, the vorticity is given by $\omega = -\Delta\psi$. Therefore, the velocity can be expressed in terms of ω as [Chorin (1994)],

$$\boldsymbol{u}(\boldsymbol{r},t) = -\boldsymbol{\nabla}^{\perp}\!\int \mathrm{d}\boldsymbol{r}'\, \mathcal{G}(\boldsymbol{r},\boldsymbol{r}')\, \omega(\boldsymbol{r}',t)\,.$$

where $\mathcal{G}(\boldsymbol{r},\boldsymbol{r}')$ is the Green function of the Laplacian operator Δ, e.g. in the infinite plane $\mathcal{G}(\boldsymbol{r},\boldsymbol{r}') = -1/(2\pi)\ln|\boldsymbol{r}-\boldsymbol{r}'|$. Consider now, at $t=0$, the vorticity to be localized on N point-vortices $\omega(\boldsymbol{r},0) = \sum_{i=1}^{N}\Gamma_i\delta(\boldsymbol{r}-\boldsymbol{r}_i(0))$, where Γ_i is the circulation of the i–th vortex. Equation (B.25.1) ensures that the vorticity remains localized, with $\omega(\boldsymbol{r},t) = \sum_{i=1}^{N}\Gamma_i\delta(\boldsymbol{r}-\boldsymbol{r}_i(t))$, which plugged in Eq. (B.25.1) implies that the vortex positions $\boldsymbol{r}_i = (x_i, y_i)$ evolve, e.g. in the infinite plane, as

$$\frac{\mathrm{d}x_i}{\mathrm{d}t} = \frac{1}{\Gamma_i}\frac{\partial H}{\partial y_i} \qquad \frac{\mathrm{d}y_i}{\mathrm{d}t} = -\frac{1}{\Gamma_i}\frac{\partial H}{\partial x_i} \tag{B.25.2}$$

with

$$H = -\frac{1}{4\pi}\sum_{i\neq j}\Gamma_i\Gamma_j \ln r_{ij}$$

where $r_{ij} = |\boldsymbol{r}_i - \boldsymbol{r}_j|$. In other words, N point vortices constitute a N-degree of freedom Hamiltonian system with canonical coordinates $(x_i, \Gamma_i y_i)$. In an infinite plane, Eq. (B.25.2)

[32] It should however remarked that using the methods of time series analysis from a unique long Lagrangian trajectory it is not possible to separate Lagrangian and Eulerian properties. For instance, standard nonlinear analysis tool (Chap. 10) would not give the Lagrangian Lyapunov exponent λ_L, but the total one λ_T. Therefore, in the case under exam one recovers the Eulerian exponent as $\lambda_T = \max(\lambda_E, \lambda_L) = \lambda_E$.

conserves quantities: $Q = \sum_i \Gamma_i x_i$, $P = \sum_i \Gamma_i y_i$, $I = \sum_i \Gamma_i(x_i^2 + y_i^2)$ and, of course, H. Among these only three are in involution (Box B.1), namely $Q^2 + P^2$, H and I as it can be easily verified computing the Poisson brackets (B.1.8) between H and either Q, P or I, and noticing that $\{I, Q^2 + P^2\} = 0$. The existence of these conserved quantities makes thus a system of $N = 3$ vortices integrable, i.e. with periodic or quasi-periodic trajectories.[33] For $N \geq 4$, the system is non-integrable and numerical studies show, apart from non generic initial conditions and/or values of the parameters Γ_i, the presence of chaos [Aref (1983)]. At varying N and the geometry, a rich variety of behaviors, relevant to different contests from geophysics to plasmas [Newton (2001)], can be observed. Moreover, the limit $N \to \infty$ and $\Gamma_i \to 0$ taken in a suitable way can be shown to reproduce the $2D$ Euler equation [Chorin (1994); Marchioro and Pulvirenti (1994)] (see Chap. 13).

11.2.1.3 *Lagrangian Chaos in the ABC flow*

The two-dimensional examples discussed before have been used not only for easing the visualization, but because of their relevance in geophysical fluids, where bidimensionality is often a good approximation (see Dritschell and Legras (1993) and references therein) thanks to the Earth rotation and density stratification, due to temperature in the atmosphere or to temperature and salinity in the oceans. It is however worthy, also for historical reasons, to conclude this overview on Lagrangian Chaos with a three-dimensional example.

In particular, we reproduce here the elegant argument employed by Arnold[34] (1965) to show that Lagrangian Chaos should be present in the ABC flow

$$\boldsymbol{u} = (A \sin z + C \cos y,\ B \sin x + A \cos z,\ C \sin y + B \cos x) \tag{11.14}$$

(where A, B and C are non-zero real parameters), as later confirmed by the numerical experiments of Hénon (1966). Note that in $d = 3$ Lagrangian chaos can appear even if the flow is time-independent.

First we must notice that the flow (11.14) is an exact steady solution of Euler's incompressible equations which, for $\rho = 1$, read $\partial_t \boldsymbol{u} + \boldsymbol{u} \cdot \boldsymbol{\nabla} \boldsymbol{u} = -\boldsymbol{\nabla} p$. In particular, the flow (11.14) is characterized by the fact that the vorticity vector $\boldsymbol{\omega} = \boldsymbol{\nabla} \times \boldsymbol{u}$ is parallel to the velocity vector in all points of the space.[35] In particular, being a steady state solution, we have

$$\boldsymbol{u} \times (\boldsymbol{\nabla} \times \boldsymbol{u}) = \nabla \alpha \ , \qquad\qquad \alpha = p + \boldsymbol{u}^2/2 \ ,$$

where, as a consequence of Bernoulli theorem, $\alpha(\boldsymbol{x}) = p + \boldsymbol{u}^2/2$ is constant along any Lagrangian trajectory $\boldsymbol{x}(t)$. As argued by Arnold, chaotic motion can appear only if $\alpha(\boldsymbol{x})$ is constant (i.e. $\boldsymbol{\nabla}\alpha(\boldsymbol{x}) = 0$) in a finite region of the space, otherwise the trajectory would be confined on the two-dimensional surface $\alpha(\boldsymbol{x}) =$ constant,

[33] In different geometries the system is integrable for $N \leq N^*$, for instance in a half- plane or inside a circular boundary $N^* = 2$, for generic domains one expects $N^* = 1$ [Aref (1983)].

[34] Who introducing such flow predicted *it is probable that such flows have trajectories with complicated topology. Such complications occur in celestial mechanics.*

[35] In real fluids, the flow would decay because of the viscosity [Dombre *et al.* (1986)].

where the motion must be regular as prescribed by the Bendixon-Poincaré theorem. The request $\boldsymbol{\nabla}\alpha(\boldsymbol{x}) = 0$ is satisfied by flows having the Beltrami property $\nabla \times \boldsymbol{u} = \gamma(\boldsymbol{x})\,\boldsymbol{u}$, which is verified by the ABC flow (11.14) with $\gamma(\boldsymbol{x})$ constant.

We conclude noticing that, in spite of the fact that the equation $\mathrm{d}\boldsymbol{x}/\mathrm{d}t = \boldsymbol{u}$ with $\boldsymbol{u}$ given by (11.14) preserves volumes without being Hamiltonian, the phenomenology for the appearence of chaos is not very different from that characterizing Hamiltonian systems (Chap. 7). For instance, Feingold *et al.* (1988) studied a discrete-time version of the ABC flow, and showed that KAM-like features are present, although the range of possible behaviors is richer.

11.2.2 *Chaos and diffusion in laminar flows*

In the previous subsection we have seen the importance of Lagrangian Chaos in enhancing the mixing properties. Here we briefly discuss the role of chaos in the long distance and long time transport properties.

In particular, we consider two examples of transport which underline two effects of chaos, namely the destruction of barriers to transport and the decorrelation of tracer trajectories, which is responsible for large scale diffusion.

11.2.2.1 *Transport in a model of the Gulf Stream*

Western boundary current extensions typically exhibit a meandering jet-like flow pattern, paradigmatic examples are the meanders of the Gulf Stream extension [Halliwell and Mooers (1983)]. These strong currents often separate very different regions of the oceans, characterized by water masses which are quite different in terms of their physical and bio-geochemical characteristics. Consequently, they are associated with very sharp and localized property gradients; this makes the study of mixing processes across them particularly relevant also for interdisciplinary investigations [Bower *et al.* (1985)].

The mixing properties of the Gulf Stream have been studied in a variety of settings to understand the main mechanism responsible for the North-South (and vice versa) transport. In particular, Bower (1991) proposed a kinematic model where the large-scale velocity field is represented by an assigned flow whose spatial and temporal characteristics mimic those observed in the ocean. In a reference frame moving eastward, the Gulf-Stream model reduces to the following stream function

$$\psi = -\tanh\left[\frac{y - B\cos(ky)}{\sqrt{1 + k^2B^2\sin^2(kx)}}\right] + cy \ . \tag{11.15}$$

consisting of a spatially periodic streamline pattern (with k being the spatial wave number, and c being the retrograde velocity of the "far field") forming an meandering (westerly) current of amplitude B with recirculations along its boundaries (see Fig. 11.8 left).

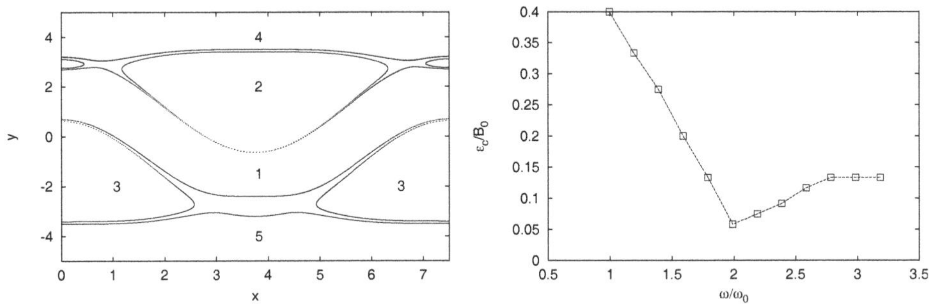

Fig. 11.8 (left) Basic pattern of the meandering jet flow (11.15), as identified by the separatrices. Region 1 is the jet (the Gulf stream), 2 and 3 the Northern and Southern recirculating regions, respectively. Finally, region 4 and 5 are the far field. (right) Critical values of the periodic perturbation amplitude for observing the overlap of the resonances, ϵ_c/B_0 vs ω/ω_0, for the stream function (11.15) with $B_0 = 1.2$, $c = 0.12$ and $\omega_0 = 0.25$. The critical values have been estimated following, up to 500 periods, a cloud of 100 particles initially located between the 1 and 2.

Despite its somehow artificial character, this simplified model enables to focus on very basic mixing mechanisms. In particular, Samelson (1992) introduced several time dependent modifications of the basic flow (11.15): by superposing a time-dependent meridional velocity or a propagating plane wave and also a time oscillation of the meander amplitude

$$B = B_0 + \epsilon \cos(\omega t + \phi)$$

where ω and ϕ are the frequency and phase of the oscillations. In the following we focus on the latter.

Clearly, across-jet particle transport can be obtained either considering the presence of molecular diffusion [Dutkiewicz *et al.* (1993)] (but the process is very slow for low diffusivities) or thanks to chaotic advection as originally expected by Samelson (1992). However, the latter mechanism can generate across-jet transport only in the presence of overlap of resonances otherwise the jet itself constitutes a barrier to transport. In other words we need perturbations strong enough to make the regions 2 and 3 in the left panel of Fig. 11.8 able to communicate after particle sojourns in the jet, region 1. A shown in Cencini *et al.* (1999b), overlap of resonances can be realized for $\epsilon > \epsilon_c(\omega)$ (Fig. 11.8 right): for $\epsilon < \epsilon_c(\omega)$ chaos is "localized" in the chaotic layers, while for $\epsilon > \epsilon_c(\omega)$ vertical transport occurs.

Since in the real ocean the two above mixing mechanisms, chaotic advection and diffusion, are simultaneously present, particle exchange can be studied through the progression from periodic to stochastic disturbances. We end remarking that choosing the model of the parameters on the basis of observations, the model can be shown to be in the condition of overlap of the resonances [Cencini *et al.* (1999b)].

11.2.2.2 *Standard and Anomalous diffusion in a chaotic model of transport*

An important large scale transport phenomenon is the diffusive motion of particle tracers revealed by the long time behavior of particle displacement

$$\langle\, (x_i(t) - x_i(0))(x_j(t) - x_j(0))\,\rangle \simeq 2D^E_{ij}t\,, \tag{11.16}$$

where $x_i(t)$ (with $i = 1, \dots, d$) denotes the particle position.[36]

Typically when studying large scale motion of tracers, the full Langevin equation (11.6) is considered, and D^E_{ij} indicates the eddy diffusivity tensor [Majda and Kramer (1999)], which is typically much larger than the molecular diffusivity D. However, the diffusive behavior (11.16) can be obtained also in the absence of molecular diffusion, i.e. considering the dynamics (11.7). In fact, provided we have a mechanism able to avoid particle entrapment (e.g. molecular noise or overlap of resonances), for diffusion to be present it is enough that the particle velocity decorrelates in the time course as one can realize noticing that

$$\langle (x_i(t) - x_i(0))^2 \rangle = \int_0^t \mathrm{d}s \int_0^t \mathrm{d}s' \,\langle u_i(\boldsymbol{x}(s))\, u_i(\boldsymbol{x}(s'))\rangle \simeq 2\,t \int_0^t \mathrm{d}\tau\; C_{ii}(\tau)\,, \tag{11.17}$$

where $C_{ij}(\tau) = \langle v_i(\tau) v_j(0)\rangle$ is the correlation function of the Lagrangian velocity, $\boldsymbol{v}(t) = \boldsymbol{u}(\boldsymbol{x}(t), t)$. It is then clear that if the correlation decays in time fast enough for the integral $\int_0^\infty \mathrm{d}\tau\; C_{ii}(\tau)$ to be finite, we have a diffusive motion with

$$D^E_{ii} = \lim_{t\to\infty} \frac{1}{2\,t} \langle (x_i(t) - x_i(0))^2 \rangle = \int_0^\infty \mathrm{d}\tau\; C_{ii}(\tau)\,. \tag{11.18}$$

Decay of Lagrangian velocity correlation functions is typically ensured either by molecular noise or by chaos, however anomalously slow decay of the correlation functions can, sometimes, give rise to *anomalous diffusion* (superdiffusion), with $\langle (x_i(t) - x_i(0))^2 \rangle \sim t^{2\nu}$ with $\nu > 1/2$ [Bouchaud and Georges (1990)].

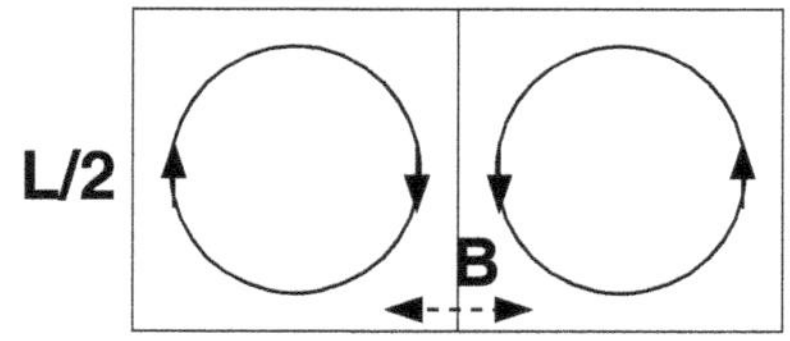

Fig. 11.9 Sketch of the basic cell in the cellular flow (11.19). The double arrow indicates the horizontal oscillation of the separatrix with amplitude B.

---

[36] Notice that, Eq. (11.16) has an important consequence on the transport of a scalar field $\theta(\boldsymbol{x}, t)$, as it implies that the coarse-grained concentration $\langle\theta\rangle$ (where the average is over a volume of linear dimension larger than the typical velocity length scale) obeys Fick equation:

$$\partial_t \langle\theta\rangle = D^E_{ij} \partial_{x_i} \partial_{x_j} \langle\theta\rangle \qquad i, j = 1, \dots, d\,.$$

Often, the goal of transport studies it to compute D^E given the velocity field, for which there are now well established techniques (see, e.g. Majda and Kramer (1999)).

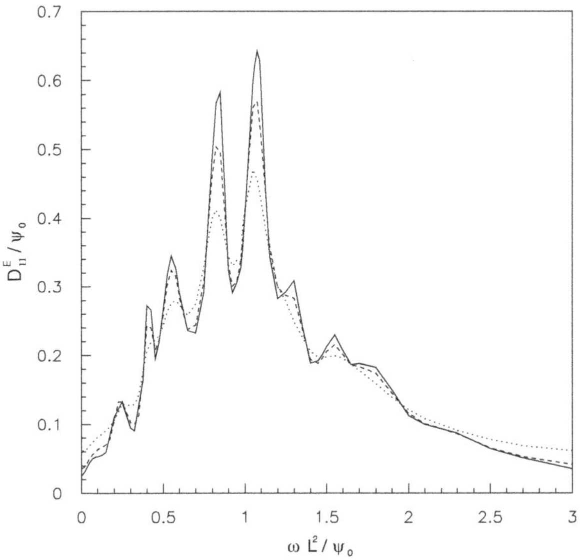

Fig. 11.10 D^E_{11}/ψ_0 *vs* $\omega L^2/\psi_0$ for different values of the molecular diffusivity D/ψ_0. $D/\psi_0 = 3\times10^{-3}$ (dotted curve); $D/\psi_0 = 1\times10^{-3}$(broken curve); $D/\psi_0 = 5\times10^{-4}$ (full curve).

Instead of presenting a complete theoretical treatment (for which the reader can refer to, e.g., Bouchaud and Georges (1990); Bohr *et al.* (1998); Majda and Kramer (1999)), here we discuss a simple example illustrating the richness of behaviors which may arise in the transport properties of a system with Lagrangian chaos. In particular, we consider a cellular flow mimicking Rayleigh-Bénard convection (Box B.4) which is described by the stream function [Solomon and Gollub (1988)]:

$$\psi(x,y,t) = \psi_0 \sin\left[\frac{2\pi}{L}(x + B\sin(\omega t))\right]\sin\left(\frac{2\pi}{L}y\right). \qquad (11.19)$$

The resulting velocity field, $\boldsymbol{u} = (\partial_y\psi, -\partial_x\psi)$, consists of a spatially periodic array of counter-rotating, square vortices of side $L/2$, L being the periodicity of the cell (Fig. 11.9). Choosing $\psi_0 = UL/2\pi$, U sets the velocity intensity. For $B \neq 0$, the time-periodic perturbation mimics the even oscillatory instability of the Rayleigh–Bénard convective cell causing the lateral oscillation of the rolls [Solomon and Gollub (1988)]. Essentially the term $B\sin(\omega t)$ is responsible for the horizontal oscillation of the separatrices (see Fig. 11.9). Therefore, for fixed B, the control parameter of particle transport is $\omega L^2/\psi_0$, i.e. the ratio between the lateral roll oscillation frequency ω and the characteristic circulation frequency ψ_0/L^2 inside the cell.

We consider here the full problem which includes the periodic oscillation of the separatrices and the presence of molecular diffusion, namely the Langevin dynamics (11.6) with velocity $\boldsymbol{u} = (\partial_y\psi, -\partial_x\psi)$ and ψ given by Eq. (11.19), at varying the molecular diffusivity coefficient D. Figure 11.10 illustrates the rich structure of the eddy diffusivity D^E_{11} as a function of the normalized oscillation frequency $\omega L^2/\psi_0$, at varying the diffusivity. We can identify two main features represented by the peaks and off-peaks regions, respectively which are characterized by the following properties [Castiglione *et al.* (1998)].

At decreasing D, the off-peaks regions become independent of D, suggesting that the limit $D \to 0$ is well defined. Therefore, standard diffusion can be realized even in the absence of molecular diffusivity because oscillations of the separatrices provide a mechanism for particles to jump from one cell to another. Moreover, chaos is strong enough to rapidly decorrelate the Lagrangian velocity and thus Eq. (11.18) applies.

On the contrary, the peaks become more and more pronounced and sharp as D decreases, suggesting the development of singularities in the pure advection limit, $D \to 0$, for specific values of the oscillation frequency. Actually, as shown in Castiglione *et al.* (1998, 1999), for $D \to 0$ anomalous superdiffusion sets in a narrow window of frequencies around the peaks, meaning that[37]

$$\langle (x(t) - x(0))^2 \rangle \propto t^{2\nu} \qquad \text{with} \qquad \nu > 1/2\,.$$

Superdiffusion is due to the slow decay of the Lagrangian velocity correlation function making $\int_0^\infty \mathrm{d}\tau\, C_{ii}(\tau) \to \infty$ and thus violating Eq. (11.18). The slow decay is not caused by the failure of chaos in decorrelating Lagrangian motion but by the establishment of a sort of synchronization between the tracer circulation in the cells and their global oscillation that enhances the coherence of the jumps from cell to cell, allowing particles to persist in the direction of jump for long periods.

Even if the cellular flow discussed here has many peculiarities (for instance, the mechanism responsible for anomalous diffusion is highly non-generic), it constitutes an interesting example as it contains part of the richness of behaviors which can be effectively encountered in Lagrangian transport. Although with different mechanisms in respect to the cellular flow, anomalous diffusion is generically found in intermittent maps [Geisel and Thomae (1984)], where the anomalous exponent ν can be computed with powerful methods [Artuso *et al.* (1993)].

It is worth concluding with some general considerations. Equation (11.17) implies that superdiffusion can occur only if one of, or both, the conditions

(I) finite variance of the velocity: $\langle v^2 \rangle < \infty$,

(II) fast decay of Lagrangian velocities correlation function: $\int_0^t d\tau\, C_{ii}(\tau) < \infty$,

are violated, while when both I) and II) are verified standard diffusion takes place with effective diffusion coefficients given by Eq. (11.18).

While violations of condition I) are actually rather unphysical, as an infinite velocity variance is hardly realized in nature, violation of II) are possible. A possibility to violate II) is realized by the examined cellular flow, but it requires to consider the limit of vanishing diffusivity. Indeed for any $D > 0$ the strong coherence in the direction of jumps between cells, necessary to have anomalous diffusion, will sooner or later be destroyed by the decorrelating effect of the molecular noise term

[37]Actually, as discussed in Castiglione *et al.* (1999), studying moments of the displacement, i.e. $\langle |x(t) - x(0)|^q \rangle$, the anomalous behavior displays other nontrivial features.

of Eq. (11.6). In order to observe anomalous diffusion with $D > 0$ in incompressible velocity fields the velocity $\boldsymbol{u}$ should possess strong spatial correlations [Avellaneda and Majda (1991); Avellaneda and Vergassola (1995)], as e.g. in random shear flows [Bouchaud and Georges (1990)].

We conclude mentioning that in velocity fields with multiscale properties, as in turbulence, superdiffusion can arise for the relative motion between two particles $\boldsymbol{x}_1$ and $\boldsymbol{x}_2$. In particular, in turbulence, we have $\langle |\boldsymbol{x}_1 - \boldsymbol{x}_2|^2 \rangle \propto t^3$ (see Box B.26), as discovered by Richardson (1926).

Box B.26: Relative dispersion in turbulence

Velocity properties at different length-scales determine two-particle separation, $\boldsymbol{R}(t) = \boldsymbol{x}_2(t) - \boldsymbol{x}_1(t)$, indeed

$$\frac{\mathrm{d}\boldsymbol{R}}{\mathrm{d}t} = \delta_R \boldsymbol{u} = \boldsymbol{u}(\boldsymbol{x}_1(t) + \boldsymbol{R}(t), t) - \boldsymbol{u}(\boldsymbol{x}_1(t), t)\,. \tag{B.26.1}$$

Here, we briefly discuss the case of turbulent flows (see Chap. 13 and, in particular, Sec. 13.2.3), which possess a rich multiscale structure and are ubiquitous in nature [Frisch (1995)]. Very crudely, a turbulent flow is characterized by two length-scales: a small scale ℓ below which dissipation is dominating, and a large scale $\mathcal{L}$ representing the size of the largest flow structures, where energy is injected. We can thus identify three regimes, reflecting in different dynamics for the particle separation: for $r \ll \ell$ dissipation dominates, and $\boldsymbol{u}$ is smooth; in the so-called inertial range, $\ell \ll r \ll \mathcal{L}$, the velocity differences display a non-smooth behavior,[38] $\delta_r u \propto r^{1/3}$; for $r \gg \mathcal{L}$ the velocity field is uncorrelated.

At small separations, $R \ll \ell$, and hence short times (until $R(t) \lesssim \ell$) the velocity difference in (B.26.1) is well approximated by a linear expansion in R, and chaos with exponential growth of the separation, $\langle \ln R(t) \rangle \simeq \ln R(0) + \lambda t$, is observed ($\lambda$ being the Lagrangian Lyapunov exponent). In the other asymptotics of long times and large separations, $R \gg \mathcal{L}$, particles evolve with uncorrelated velocities and the separation grows diffusively, $\langle R^2(t) \rangle \simeq 4D^E\, t$; the factor 4 stems from the asymptotic independence of the two particles.

Between these two asymptotics, we have $\delta_R v \sim R^{1/3}$ violating the Liptchiz condition — non-smooth dynamical systems — and from Sec. 2.1 we know that the solution of Eq. (B.26.1) is, in general, not unique. The basic physics can be understood assuming $\ell \to 0$ and considering the one-dimensional version of Eq. (B.26.1) $\mathrm{d}R/\mathrm{d}t = \delta_R v \propto R^{1/3}$ and $R(0) = R_0$. For $R_0 > 0$, the solution is given by

$$R(t) = \left(R_0^{2/3} + 2t/3\right)^{3/2}. \tag{B.26.2}$$

If $R_0 = 0$ two solutions are allowed (non-uniqueness of trajectories): $R(t) = [2t/3]^{3/2}$ and the trivial one $R(t) = 0$. Physically speaking, this means that for $R_0 \neq 0$ the solution becomes independent of the initial separation R_0, provided t is large enough. As easily

[38] Actually, the scaling $\delta_r u \propto r^{1/3}$ is only approximately correct due to intermittency [Frisch (1995)] (Box B.31), here neglected. See Boffetta and Sokolov (2002) for an insight on the role of intermittency in Richardson diffusion.

derived from (B.26.2), the separation grows anomalously

$$\langle R^2(t)\rangle \sim t^3$$

which is the well known Richardson (1926) law for relative dispersion. The mechanism underlying this "anomalous" diffusive behavior is, analogously to the absolute dispersion case, the violation of the condition II), i.e. the persistence of correlations in the Lagrangian velocity differences for separations within the inertial range [Falkovich *et al.* (2001)].

11.2.3 *Advection of inertial particles*

So far we considered particle tracers that, having the same density of the carrier fluid and very small size, can be approximated as point-like particles having the same velocity of the fluid at the position of the particle, i.e. $\boldsymbol{v}(t) = \boldsymbol{u}(\boldsymbol{x}(t), t)$, with the phase space coinciding with the particle-position space. However, typical impurities have a non-negligible size and density different from the fluid one as, e.g., water droplets in air or air bubbles in water. Therefore, the tracer approximation cannot be used, and the dynamics has to account for all the forces acting on a particle such as drag, gravity, lift etc [Maxey and Riley (1983)]. In particular, drag forces causes inertia — hence the name *inertial particles* — which makes the dynamics of such impurities dissipative as that of tracers in compressible flows. Dissipative dynamics implies that particle trajectories asymptotically evolve on a dynamical[39] attractor in phase space, now determined by both the position ($\boldsymbol{x}$) and velocity ($\boldsymbol{v}$) space, as particle velocity differs from the fluid one (i.e. $\boldsymbol{v}(t) \neq \boldsymbol{u}(\boldsymbol{x}(t), t)$). Consequently, even if the flow is incompressible, the impurities can eventually distribute very inhomogeneously (Fig. 11.11a), similarly to tracers in compressible flows [Sommerer and Ott (1993); Cressman *et al.* (2004)].

Nowadays, inertial particles constitute an active, cross-disciplinary subject relevant to fundamental and applied contexts encompassing engineering [Crowe *et al.* (1998)], cloud physics [Pruppacher and Klett (1996)] and planetology [de Pater and Lissauer (2001)]. It is thus useful to briefly discuss some of their main features.

We consider here a simple model, where the impurity is point-like with a velocity dynamics[40] accounting for viscous and added mass forces, due to the density contrast with the fluid, i.e.

$$\frac{\mathrm{d}\boldsymbol{x}}{\mathrm{d}t} = \boldsymbol{v}(t)\,, \qquad \frac{\mathrm{d}\boldsymbol{v}}{\mathrm{d}t} = \frac{\boldsymbol{u}(\boldsymbol{x}(t), t) - \boldsymbol{v}(t)}{\tau_p} + \beta D_t \boldsymbol{u}\,. \tag{11.20}$$

The difference between the particle ρ_p and fluid ρ_f density is measured by $\beta = 3\rho_f/(2\rho_p + \rho_f)$ (notice that $\beta \in [0:3]$; $\beta = 0$ and 3 correspond to particles much heavier and lighter than the fluid, respectively), while $\tau_p = a^2/(2\beta\nu)$ is the Stokes

[39] If the flow is stationary the attractor is a fixed set of the space, as in the Lorenz system, in non-autonomous systems the attractor is dynamically evolving.

[40] More refined models require to account for other forces, for a detailed treatment see Maxey and Riley (1983), who wrote the complete equations for small, rigid spherical particles.

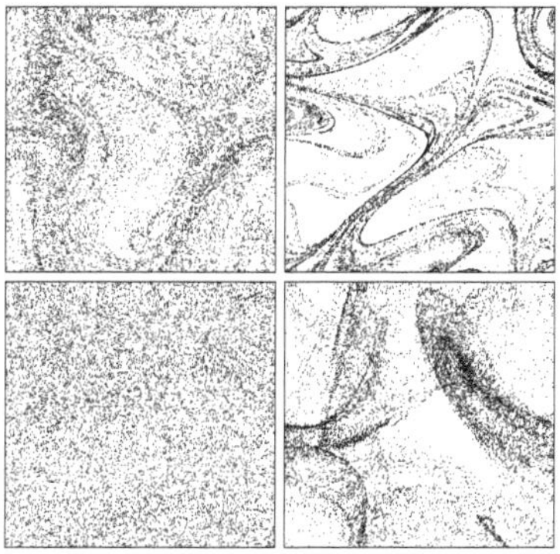

(a)

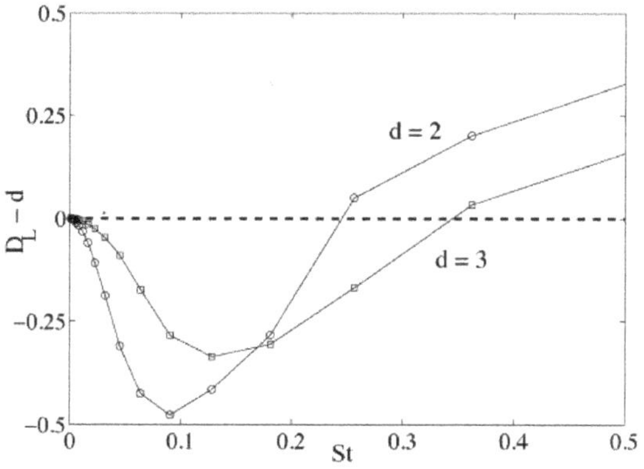

(b)

Fig. 11.11 (a) Snapshots of 10^4 particles heavier than the fluid ($\beta = 0$) at (proceeding clockwise from the top left corner) small, intermediate, order one, and larger than one Stokes number values. (b) Difference between Lyapunov D_L and space dimension in $d = 2$ and $d = 3$ computed in a random laminar flow. [Courtesy of J. Bec]

response time, which is proportional to the square of the particle radius a and inversely proportional to the fluid viscosity ν. In real situations, the fluid velocity field $\boldsymbol{u}(\boldsymbol{x}, t)$ dynamically evolves with the Navier-Stokes equation:

$$D_t \boldsymbol{u} = \partial_t \boldsymbol{u} + \boldsymbol{u} \cdot \boldsymbol{\nabla} \boldsymbol{u} = \nu \Delta \boldsymbol{u} - \boldsymbol{\nabla} p / \rho_f + \boldsymbol{f} \,, \qquad \text{with} \quad \boldsymbol{\nabla} \cdot \boldsymbol{u} = 0 \,,$$

where $D_t \boldsymbol{u}$ denotes the convective derivative, p the pressure and $\boldsymbol{f}$ an external stirring acting on the fluid. Of course, Eq. (11.20) can also be studied with simple flow field models [Benczik *et al.* (2002); Bec (2003)].[41] Within this model inertial particle dynamics depends on two dimensionless control parameters: the constant β, and the Stokes number $St = \tau_p / \tau_u$, measuring the ratio between the particle response time and the smallest characteristic time of the flow τ_u (for instance, the correlation time in a laminar flow, or the Kolmogorov time in turbulence, see Chap. 13).

Both from a theoretical and an applied point of view, the two most interesting features emerging in inertial particles are the appearance of strongly inhomogeneous distributions — *particle clustering* — and the possibility, especially at large Stokes number, to have close particles having large velocity differences.[42] Indeed both these properties are responsible for an enhanced probability of chemical, biological or physical interaction depending on the context. For instance, these properties are crucial to the time scales of rain [Falkovich *et al.* (2002)] and planetesimals

[41] In these cases $D_t \boldsymbol{u}$ is substituted with the derivative along the particle path.

[42] These features are absent for tracers in incompressible flows where dynamics is conservative, and particle distribution soon becomes uniform thanks to chaos-induced mixing. Moreover, particles at distance $r = |\boldsymbol{x}_1 - \boldsymbol{x}_2|$ have small velocity differences as a consequence of the smoothness of the underlying flow, i.e. $|\boldsymbol{v}_1(t) - \boldsymbol{v}_2(t)| = |\boldsymbol{u}(\boldsymbol{x}_1, t) - \boldsymbol{u}(\boldsymbol{x}_2, t)| \propto r$.

formation in the early Solar System [Bracco *et al.* (1999)]. In the following, we briefly discuss the issue of particle clustering.

The phenomenology of particle clustering can be understood as follows [Bec (2003)]. First, notice that the system (11.20) can be rewritten as the ODE

$$\frac{\mathrm{d}\boldsymbol{z}}{\mathrm{d}t} = \boldsymbol{F}(\boldsymbol{z}, t)\,,$$

with $\boldsymbol{z} = (\boldsymbol{x}, \boldsymbol{v})$ and $\boldsymbol{F} = [\boldsymbol{v}, (\boldsymbol{u} - \boldsymbol{v})/\tau_p + \beta D_t \boldsymbol{u}]$, making explicit that, in a d-dimensional flow, inertial particles actually leave in a $(2 \times d)$-phase space. Therefore, the attractor will generically have a fractal dimension $D_F < 2d$, with D_F being a function of both β and St. Second, we observe that $\boldsymbol{\nabla} \cdot \boldsymbol{F} = -d/\tau_p$, i.e. phase-space volumes are uniformly contracted (Sec. 2.1.1) at a rate $-d/\tau_p$. In particular, for $\tau_p \to 0$ (*viz.* $St \to 0$) the contraction rate is infinite, which physically means that particle dynamics reduces to that of tracers, and the $(2 \times d)$-dimensional phase space contracts to the d-dimensional one, i.e. we recover the conservative dynamics of tracers in position space. In this limit, $D_F = d$, i.e. the fractal dimension of the attractor coincides with the dimensionality of the coordinate space and, consequently no clustering of particles can be observed. In the opposite asymptotics of extremely large response times, $St \to \infty$, the phase-space contraction rate goes to zero, indicating a conservative dynamics in the full $(2 \times d)$-dimensional position-velocity phase space. Physically, this limit corresponds to a gas of particles, essentially unaffected by the presence of the flow, so that the attractor is nothing but the full phase space with $D_F = 2d$. Also in this case no clustering in position space is observed. Between these two asymptotics, it may occur that $D_F < d$, so that looking at particle positions we can observe clustered distributions. These qualitative features are well reproduced in Fig. 11.11a. Following Bec (2003), the fractal dimension can be estimated through the Kaplan-Yorke or Lyapunov dimension D_L (Sec. 5.3.4) at varying the particle response time, in a simple model laminar flow. The results for $\beta = 0$ are shown in Fig. 11.11b: in a range of (intermediate) response time values $D_L - d < 0$, indicating that $D_F < d$ and thus clustering as indeed observed.

The above phenomenological picture well describes what happens in realistic flows obtained simulating the Navier-Stokes equation [Bec *et al.* (2006, 2007); Calzavarini *et al.* (2008)] (see Fig. 11.12) and also in experiments [Eaton and Fessler (1994); Saw *et al.* (2008)]. For Navier-Stokes flows the fluid mechanical origin of particle clustering can be understood by a simple argument based on the perturbative expansion of Eq. (11.20) for $\tau_p \to 0$, giving [Balkovsky *et al.* (2001)]

$$\boldsymbol{v} = \boldsymbol{u} + \tau_p(\beta - 1) D_t \boldsymbol{u} = \boldsymbol{u} + \tau_p(\beta - 1)(\partial_t \boldsymbol{u} + \boldsymbol{u} \cdot \boldsymbol{\nabla} \boldsymbol{u}) \tag{11.21}$$

which correctly reproduces the tracer limit $\boldsymbol{v} = \boldsymbol{u}$ for $\tau_p = 0$ or $\beta = 1$, i.e. $\rho_p = \rho_f$. Within this approximation, similarly to tracers, the phase space reduces to the position space and inertia is accounted by the particle velocity field $\boldsymbol{v}(\boldsymbol{x}, t)$ which is now compressible even if the fluid is incompressible. Indeed, from Eq. (11.21) it follows that

$$\boldsymbol{\nabla} \cdot \boldsymbol{v} = \tau_p(1 - \beta) \boldsymbol{\nabla} \cdot (\boldsymbol{u} \cdot \boldsymbol{\nabla} \boldsymbol{u}) \neq 0\,.$$

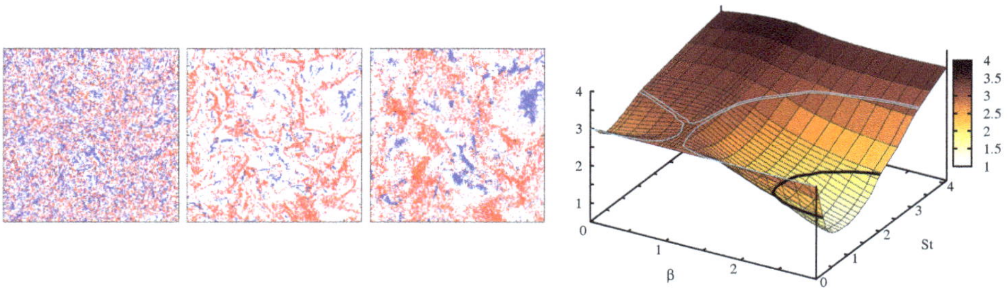

(a) (b)

Fig. 11.12 (a) Heavy β=0 (red) and light β=3 (blue) particle positions, in a slice of a three dimensional turbulent flow at moderately high Reynolds number for three different Stokes number values, from left to right, St= 0.1, 1, 4.1. (b) Lyapunov dimension D_L as function of both β and St. The cyan curves display the $D_L = 2.9, 3.1$ isolines, while the black one displays the $D_L = 2$ isoline. Data refer to simulations from Calzavarini *et al.* (2008).

Making explicit the r.h.s. of the above equation in terms of the symmetric and antisymmetric part of the stress tensor [Chong *et al.* (1990)], i.e. $\mathbb{S}_{ij} = (\partial_i u_j + \partial_j u_i)/2$ and $\mathbb{W}_{ij} = (\partial_i u_j - \partial_j u_i)/2$, respectively we have $\boldsymbol{\nabla} \cdot \boldsymbol{v} \propto \tau_p(\beta - 1)(\mathbb{S}^2 - \mathbb{W}^2)$ from which we see that: heavy particles, $\beta < 1$, have negative (positive) compressibility for $\mathbb{S}^2 > \mathbb{W}^2$ ($\mathbb{S}^2 < \mathbb{W}^2$) meaning that tend to accumulate in strain dominated regions and escape from vorticity dominated ones; for light particles $\beta > 1$ the opposite is realized and thus tend to get trapped into high vorticity regions. Therefore, at least for $St \ll 1$, we can trace back the origin of particle clustering to the *preferential concentration* of particles in or out of high vorticity regions depending on their density.

It is well known that three-dimensional turbulent flows are characterized by vortex filaments (almost one-dimensional inter-winded lines of vorticity) which can be visualized seeding the fluid flow (water in this case) with air bubbles [Tritton (1988)]. On the contrary, particles heavier than the fluid escape from vortex filaments generating sponge like structures. These phenomenological features find their quantitative counterpart in the fractal dimension of the aggregates that they generate. For instance, in Fig. 11.12 we show the Lyapunov dimension of inertial particles as obtained by using (11.20) for several values of β and St. As expected, light particles ($\beta > 1$) are characterized by fractal dimensions considerably smaller (approaching $D_L = 1$ — the signature of vortex filaments — for $St \approx 1$, value at which clustering is most effective) than those of heavy ($\beta < 1$) particles.

11.3 Chaos in population biology and chemistry

In this section we mainly discuss two basic problems concerning population biology and reaction kinetics, namely the Lotka-Volterra predator-prey model [Lotka (1910); Volterra (1926b,a)] and the Belousov-Zhabotinsky chemical reaction [Belousov (1959); Zhabotinsky (1991)], that constitute two milestones of nonlinear dynamics theory.

Mathematical biology is a branch of applied mathematics which studies the changes in the composition of populations. Historically its origins can be traced back to the demographic analysis by Malthus and Verlhust (Sec. 3.1), but, during the years, the development of mathematical biology has greatly expanded till embodying ecology, genetics and immunology. In population biology, we are generally interested in the time variation of the number of individuals of certain species. Species compete, evolve and disperse to seek resources for sustaining their struggle for the existence. Depending on the specific environment and settings, often the interplay among individuals involves a sort of loss-win mechanisms that can be exemplified to the form of predator-prey interactions. In this context, the role of chaos is still a controversial issue, and the common wisdom suggests that a chaotic behavior is the exceptional event rather than a rule. The typical "incorrect" argument raised is the stability of systems that would make chaos improbable. Accordingly, populations are expected to undergo cyclical fluctuations mostly triggered by living cycles, seasonal or climate changes. On the other hand, the alternative line of reasoning recognizes in the extreme variability and in the poor long-term predictability of several complex biological phenomena a fingerprint of nonlinear laws characterized by sensitive dependence on initial conditions.

In Chemistry, where the rate equations have the same structure of those of population dynamics, we have a similar phenomenology with the concentration of reagents involved in chemical reactions in place of the individuals. Rate equations, written on the basis of the elementary chemical rules, can generate very complex behaviors in spite of their simplicity as shown in the sequel with the example of the Belousov-Zhabotinsky reaction [Zhabotinsky (1991)]. We stress that in all of the examples discussed in this section we assume spatial homogeneity, this entails that the phenomena we consider can be represented by ODE of the state variables, the role of inhomogeneity will be postponed to the next Chapter.

11.3.1 *Population biology: Lotka-Volterra systems*

Species sharing the same ecosystem are typically in strong interaction. At a raw level of details, the effects exerted by a species on another can be re-conducted to three main possibilities: *predation*, *competition* or *cooperation* also termed mutualism. In the former two cases, a species subtracts individuals or resources to another one, whose population tends to decrease. In the latter, two or more species take mutual benefit from the respective existence and the interaction promotes their simultaneous growth. These simple principles define systems whose evolution in general is supposed to reach stationary or periodic states. Lotka-Volterra equations, also known as *predator-prey system*, are historically one of the first attempt to construct a mathematical theory of a simple biological phenomena. They consist in a pair of nonlinear ODE describing the interactions of two species, one acting as predator and the other as prey. Possible realistic examples of predator-prey systems are: resource-consumer, plant-herbivore, parasite-host, tumor cells (virus)-immune system, susceptible-infectious interactions, etc. These equations were proposed independently by Lotka (1910) and Volterra (1926b,a)[43]

$$\frac{dx}{dt} = r_1 x - \gamma_1 xy \qquad (11.22)$$

$$\frac{dy}{dt} = -r_2 y + \gamma_2 xy \qquad (11.23)$$

where x is the number of some prey (say, rabbits); y is the number of predators (wolves); r_1, γ_1, r_2 and γ_2 are positive parameters embodying the interaction between the two species. The assumptions of LV-model are the following. In the absence of predators, prey-population grows indefinitely at rate r_1. Thus, in principle, preys have infinite food resources at disposal and the only limitation to their increment stems from predation represented by the term $-\gamma_1 xy$. The fate of predators in absence of preys is "extinction" at rate r_2, condition prevented by the positive term $\gamma_2 xy$, describing hunting.

The dynamics of the model is rather simple and can be discussed conveniently by looking at the phase portrait. There are two fixed points $P_0 = (0,0)$ and $P_1 = (r_2/\gamma_2, r_1/\gamma_1)$, the first corresponds to extinction of both species while the second refers to an equilibrium characterized by constant populations. Linear stability matrices (Sec. 2.4) computed at the two points are

$$\mathbb{L}_0 = \begin{pmatrix} r_1 & 0 \\ 0 & -r_2 \end{pmatrix} \qquad \text{and} \qquad \mathbb{L}_1 = \begin{pmatrix} 0 & -r_2\frac{\gamma_1}{\gamma_2} \\ r_1\frac{\gamma_2}{\gamma_1} & 0 \end{pmatrix}.$$

Therefore P_0 admits eigenvalues $\lambda_1 = r_1$ and $\lambda_2 = -r_2$, hence is a saddle, while P_1 has pure imaginary eigenvalues $\lambda_{1,2} = \pm\sqrt{r_1 r_2}$. In the small oscillation approximation around the fixed point P_1, one can easily check that the solutions of linearized LV-equations (11.22)-(11.23) evolve with a period $T = 2\pi/\sqrt{r_1 r_2}$. An important property of LV-model is the existence of the integral of motion,

$$H(x,y) = r_2 \ln x + r_1 \ln y - \gamma_2 x - \gamma_1 y, \qquad (11.24)$$

as a consequence, the system exhibits periodic orbits coinciding with isolines of the functions $H(x,y) = H_0$ (Fig. 11.13a), where the value of H_0 is fixed by the initial

[43] Volterra formulated the problem stimulated by the observation of his son in law, the Italian biologist D'Ancona, who discovered a puzzling fact. During the first World War, the Adriatic sea was a dangerous place, so that large-scale fishing effectively stopped. Upon studying the statistics of the fish markets, D'Ancona noticed that the proportion of predators was higher during the war than in the years before and after. The same equations were also derived independently by Lotka (1910) some years before as a possible model for oscillating chemical reactions.

conditions $x(0) = x_0$ and $y(0) = y_0$.[44] Therefore, as shown in Fig. 11.13b, the time evolution consists of cyclic fluctuations of the two populations, for which predator population follows the variation of the preys with a certain dephasing, known as the *law of periodic fluctuations*. The biological origin of oscillations is clear: abundance of hunters implies large killing of preys, that, on the long term, means shortage of food for predators thus their decline. This decrease, in turn, causes the increase of preys and so on, in cyclical alternates.

Another interesting property of LV-model concerns the average over a cycle of number of prey/predator populations that, independently of initial conditions, reads

$$\langle x \rangle = r_2/\gamma_2 \,, \qquad \langle y \rangle = r_1/\gamma_1 \,. \tag{11.25}$$

This result, known as *law of averages*, can be derived writing, e.g., Eq. (11.22) in logarithmic form and averaging it on a period T

$$\frac{\mathrm{d}\ln x}{\mathrm{d}t} = r_1 - \gamma_1 y \qquad\qquad \frac{1}{T}\int_0^T \mathrm{d}t\, \frac{\mathrm{d}\ln x}{\mathrm{d}t} = r_1 - \gamma_1 \langle y \rangle \,.$$

The periodicity of $x(t)$ makes the left hand side vanishing and thus $\langle y \rangle = r_1/\gamma_1$. The law of averages has the paradoxical consequence that, if the birth rate of preys decreases $r_1 \to r_1 - \epsilon_1$ and, simultaneously, the predator extinction rate increases $r_2 \to r_2 + \epsilon_2$, the average populations vary as $\langle x \rangle \to \langle x \rangle + \epsilon_2/\gamma_2$ and $\langle y \rangle \to \langle y \rangle - \epsilon_1/\gamma_1$, respectively (*law of perturbations of the averages*). This property, also referred to as *Volterra's paradox*, implies that a simultaneous changes of the rates, which causes a partial extinction of both species, favors on average the preys. In other words, if the individuals of the two species are removed from the system by an external action, the average of preys tends to increase. Even though this model is usually considered inadequate for representing realistic ecosystems because too qualitative, it remains one of the simplest example of a pair of nonlinear ODE sustaining cyclical fluctuations. For this reason, it is often taken as an elementary building block when modeling more complex food-webs.

The main criticism that can be raised to LV-model is its structural instability due to the presence of a conservation law $H(x, y) = H_0$ conferring the system an Hamiltonian character. A generic perturbation, destroying the integral of motion where orbits lie, changes dramatically LV-behavior. Several variants have been proposed to generalize LV-model to realistic biological situations, and can be expressed as

$$\begin{aligned} \frac{\mathrm{d}x}{\mathrm{d}t} &= F(x,y)x \\ \frac{\mathrm{d}y}{\mathrm{d}t} &= G(x,y)y \,, \end{aligned} \tag{11.26}$$

---

[44]The existence of integral of motion H can be shown by writing the Eqs. (11.22,11.23) in a Hamiltonian form through the change of variables $\xi = \ln x, \eta = \ln y$

$$\frac{\mathrm{d}\xi}{\mathrm{d}t} = r_1 - \gamma_1 \mathrm{e}^{\eta} = \frac{\partial \mathcal{H}}{\partial \eta} \qquad \frac{\mathrm{d}\eta}{\mathrm{d}t} = -r_2 + \gamma_2 \mathrm{e}^{\xi} = -\frac{\partial \mathcal{H}}{\partial \xi} \,,$$

where the conserved Hamiltonian reads $\mathcal{H}(\xi, \eta) = r_2\xi - \gamma_2 \mathrm{e}^{\xi} + r_1\eta - \gamma_1 \mathrm{e}^{\eta}$, that in terms of original variables x, y gives the constant $H(x, y)$.

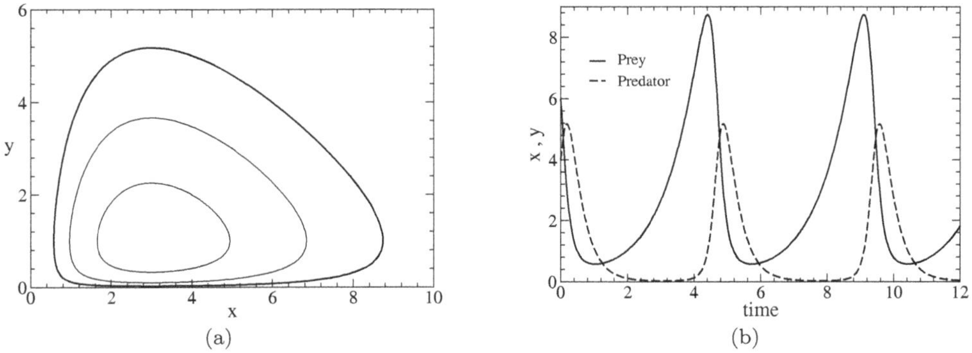

Fig. 11.13 (a) Phase-space portrait of LV-system described by the isolines of $H(x,y)$ (11.24). (b) Oscillating behavior in prey-predator populations of LV-equation for $r_1 = 1.0$, $r_2 = 3.0$, $\gamma_{1,2} = 1.0$.

where G and F are the rates at which prey/predator populations change. Following Verhulst, the first improvement can be introduced by considering a logistic growth (Sec. 3.1) of preys in absence of hunting:

$$F(x,y) = r_1 \left(1 - \frac{x}{K}\right) - \gamma_1 y$$

where K represents the *carrying capacity*: the maximal number of individuals an environment can support. More in general, the hunting rate, γ_1, is supposed to contain a saturation effect in predation term, with respect to the standard LV-model. As typical choices of $\gamma_1(x)$, we can mention [Holling (1965)],

$$\frac{a}{b+x}, \qquad \frac{ax}{b^2+x^2}, \qquad \frac{a[1-\exp(-bx)]}{x},$$

that when plugged into Eq. (11.26)) make the rate bounded. Also the rate $G(x,y)$ is certainly amenable to more realistic generalizations by preferring, e.g., a logistic growth to the simple form of Eq. (11.23). In this context, it is worth mentioning *Kolmogorov's predator-prey model.* Kolmogorov (1936) argued that the term $\gamma_2 xy$ is too simplistic, as it implies that the growth rate of predators can increase indefinitely with prey abundance, while it should saturate to the maximum reproductive rate of predators. Accordingly, he suggested the modified model

$$\frac{\mathrm{d}x}{\mathrm{d}t} = r(x)x - \gamma(x)y$$
$$\frac{\mathrm{d}y}{\mathrm{d}t} = q(x)y$$

where $r(x)$, $\gamma(x)$ and $q(x)$ are suitable functions of the prey abundance and predators are naturally "slaved" to preys. He made no specific hypothesis on the functional form of $r(x)$, $\gamma(x)$ and $q(x)$ requiring only that:

(a) In the absence of predators, the birth rate of preys $r(x)$ decreases when the population increases, becoming at a certain point negative. This means that a sort of inter-specific competition among preys is taken into account.

(b) The birth rate of predators $q(x)$ increases with prey population, going from negative (food shortage) to positive (food abundance).
(c) The function $\gamma(x)$ is such that: $\gamma(0) = 0$ and $\gamma(x) > 0$ for $x > 0$.

With these three conditions, Kolmogorov obtained a complete phase diagram, showing that a two-species predator-prey competition may lead to, extinction of predators, stable coexistence of preys and predators or, finally, oscillating cycles. He also generalized the differential equation to more than two species,[45] introducing most of the classification nowadays used in population dynamics. Moreover, Kolmogorov pointed to the strong character of the assumptions behind an approach based on differential equations. In particular, he argued that populations are composed of individuals and statistical fluctuations may not be negligible, especially for small populations. In practice, there exists a fourth scenario: at the minimum of a large oscillation, fluctuations can extinguish the prey population, thereby causing the extinction of predators too. In this remark Kolmogorov underscored the importance of discreetness in population dynamics becoming the precursor of what nowadays is termed "agent based formulation" of population biology, where individuals are "particles" of the system interacting with other individuals via effective couplings. An interesting discussion on this subject can be found in Durrett and Levin (1994).

11.3.2 *Chaos in generalized Lotka-Volterra systems*

According to Poincaré-Bendixon theorem (Sec. 2.3), the original Lotka-Volterra model and its two-dimensional autonomous variants as well cannot sustain chaotic behaviors. To observe chaos, it is necessary to increase the number of interacting species to $N \geq 3$. Searching for multispecies models generating complex behaviors is a necessary step to take into account the wealth of phenomenology commonly observed in Nature, which cannot be reduced to a simple 2-species context. However, the increase of N in LV-models does not necessarily imply chaos, therefore it is natural to wonder "under which conditions do LV-models entail structurally stable chaotic attractors?". Answering such a question is a piece of rigorous mathematics applied to population biology that we cannot fully detail in this book. We limit to mention the contribution by Smale (1976), who formulated the following theorem on a system with N competing populations x_i

$$\frac{\mathrm{d}x_i}{\mathrm{d}t} = x_i M_i(x_1, \ldots, x_N) \qquad\qquad i = 1, \ldots, N\,.$$

He proved that the above ODE, with $N \geq 5$, can exhibit any asymptotic behavior, including chaos, under the following conditions on the functions $M_i(\boldsymbol{x})$: 1) $M_i(\boldsymbol{x})$ is infinitely differentiable; 2) for all pairs i and j, $\partial M_i(\boldsymbol{x})/\partial x_j < 0$, meaning that only species with positive intrinsic rate $M_i(0)$ can survive; 3) there exist a constant C such that, for $|\boldsymbol{x}| > C$ then $M_i(\boldsymbol{x}) < 0$ for all i. The latter constraint corresponds

[45] See for instance Murray (2002); the generalized version is sometimes referred to as *Kolmogorov model*.

to bounded resources and environments. Ecosystems satisfying conditions 1-3 are said to belong to Smale's class.

A LV-model involving N species, also termed a food web in biological contexts, assumes, in analogy, that the evolution of population i receives contributions from isolated species i, and from its interactions with a generic species j which on the average is proportional to the rate at which individuals from i and j encounter each other. In this case, populations and their growth rates form N dimensional vectors $\boldsymbol{x} = (x_1, \ldots, x_N)$, $\boldsymbol{r} = (r_1, \ldots, r_N)$ and the interactions define a $N \times N$-matrix $\mathbb{J}$ often termed *community matrix*. Then the equation for the i-th species becomes

$$\frac{\mathrm{d}x_i}{\mathrm{d}t} = r_i x_i \left(1 - \sum_j \mathbb{J}_{ij} x_j\right) \qquad i = 1, \ldots, N \, . \tag{11.27}$$

r_i is the positive/negative growth rate of the i-the isolated species. The entries of the coupling matrix $\mathbb{J}_{ij}$ model the interaction between species i and j, while the diagonal elements $\mathbb{J}_{ii}$ incorporate intra-specific competitions. For instance, $r_i\mathbb{J}_{ij} > 0$ indicates that the encounter of i and j will lead to the increase of x_i, while, when $r_i\mathbb{J}_{ij} < 0$, their encounter will cause a decrease of individuals belonging to species i.[46] Arnèodo *et al.* (1982) have shown that a typical chaotic behavior can arise in a three species model like Eq. (11.27), for instance by choosing the following values of parameters

$$\boldsymbol{r} = \begin{pmatrix} 1.1 \\ -0.5 \\ 1.75 \end{pmatrix} \qquad \mathbb{J} = \begin{pmatrix} 0.5 & 0.5 & 0.1 \\ -0.5 & -0.1 & 0.1 \\ 1.55 & 0.1 & 0.1 \end{pmatrix} .$$

The attractor and the chaotic evolution of the trajectories are shown in Fig. 11.14, where we observe aperiodic oscillations qualitatively similar to that produced by the Lorenz system. Thus we can say that presence of chaos does not destroy the structure of LV-cycles (Fig. 11.13) but rather it disorders their regular alternance and changes randomly their amplitude.

We conclude this short overview on the theoretical and numerical results on LV-systems by mentioning a special N-dimensional version investigated by Goel *et al.* (1971)

$$\frac{\mathrm{d}x_i}{\mathrm{d}t} = r_i x_i + \frac{1}{\beta_i} \sum_{j=1}^{N} a_{ij} x_i x_j \qquad \text{with} \qquad a_{ij} = -a_{ji} \tag{11.28}$$

where the positive coefficients β_i^{-1} are named "equivalence" number by Volterra. The difference with respect to the generic system (11.27) lies in the antisymmetry

[46] Equation (11.27) can be also interpreted as a second order expansion in the populations of more complex models.

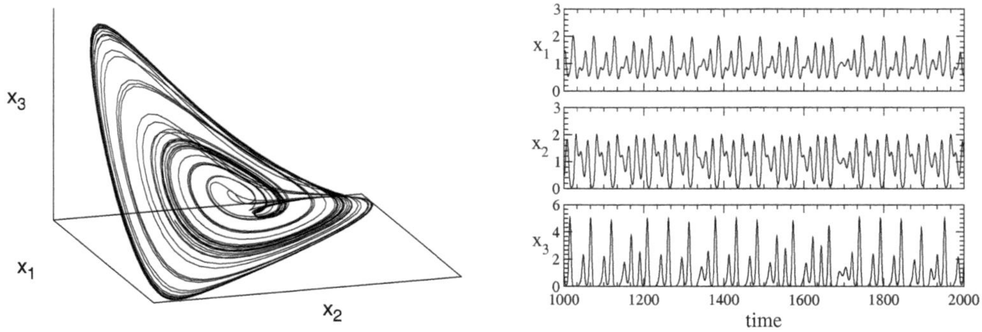

(a) (b)

Fig. 11.14 (a) 3-dimensional attractor generated by the LV-system Eqs. (11.27) with initial conditions: $x_1(0) = 1.28887$, $x_2(0) = 1.18983$, and $x_3(0) = 0.819691$. (b) Snapshot of separated trajectories of the three species corresponding to the attractor. The three patterns consist in a irregular series of aperiodic oscillations that might recall vaguely the cycles of Fig. 11.13.

properties of couplings. The non-trivial fixed point $(q_1, ...q_N)$ satisfies the linear equation

$$r_i\beta_i + \sum_{j=1}^{N} a_{ij}q_j = 0 \ ,$$

of course $(0, 0, ...0)$ is the trivial fixed point. We can introduce the new variables:

$$u_i = \ln\left(\frac{x_i}{q_i}\right)$$

that remain bounded quantities since all the x_i's remain positive if their initial values are positive. The quantity

$$G(\boldsymbol{u}) = \sum_i q_i\beta_i[\exp(u_i) - u_i] = \sum_i q_i\beta_i\left[\frac{x_i}{q_i} - \ln\left(\frac{x_i}{q_i}\right)\right]$$

is invariant under the time evolution, in analogy with the two-species model. In addition, Liouville theorem holds for the variables $\{u_i\}$,

$$\sum_{i=1}^{N} \frac{\partial}{\partial u_i}\left(\frac{du_i}{dt}\right) = 0 \ .$$

The two above properties can be used, in the limit $N \gg 1$, to build up a formal statistical mechanical approach to the system (11.28) with asymmetric couplings [Goel *et al.* (1971)]. We do not enter here the details of the approach, it is however important to stress that numerical studies [Goel *et al.* (1971)] have shown that, for the above system, chaos can take place when $N \geq 4$. Moreover, via the same computation carried out for the original LV-model, one can prove that the population averages $\langle x_i \rangle = q_i$ coincide with the fixed point, in analogy with Eqs. (11.25).[47]

[47]The demonstration is identical to that of the 2 species models, which works even in the absence of periodic solution, provided each $x_i(t)$ is a bounded function of t

The presence of chaotic behaviors in ecological systems, like food webs, has been conjectured on the basis of theoretical models generalizing Lotka-Volterra approach, where the concomitant interactions of species in competition and predation can generate chaos as actually observed in computer simulations [May (1974)]. On the experimental side, however, the poor reproducibility affecting biological observations and the lack of a large and robust amount of data have often limited the possibility of a neat detection of chaos in web of interacting species. Despite the relevance of the issue, poor attention has been devoted to experiments under laboratory controlled conditions able to provide clear evidences for chaos in biology. Only recently, a laboratory experience has been conceived with the purpose to detect long-term chaos in plankton food webs [Benincà *et al.* (2008)]. A plankton community isolated from the Baltic Sea has been studied for more than eight years. The experiment was maintained under constant external conditions and the development of the different plankton species was monitored twice per week. This simple food web never settled to equilibrium state and species abundances continued to vary wildly. Mathematical techniques based on nonlinear data analysis methods (Chap. 10) give a evidence for the presence of chaos in this system, where fluctuations, caused by competition and predation, give rise to a dynamics with none of the species prevailing on the others. These findings show that, in this specific food web, species abundances are essentially unpredictable in the long term. Although short-term prediction seems possible, on the long-term one can only indicate the range where species-population will fluctuate.

11.3.3 *Kinetics of chemical reactions: Belousov-Zhabotinsky*

The dynamical features encountered in population biology pertain also to chemical reactions, where the dynamical states now represent concentrations of chemical species. At a first glance, according to the principles of thermodynamics and chemical kinetics, complex behaviors seem to be extraneous to most of the chemical reactions as they are expected to reach quickly and monotonically homogeneous equilibrium states. However, complex behaviors and chaos can emerge also in chemical systems, providing they are kept in appropriate out-of-equilibrium conditions. In the literature, the class of chaotic phenomena pertaining to chemical contexts is known as *chemical chaos*.

Let us consider a generic chemical reaction such as

$$\alpha A + \beta B \underset{k_{-1}}{\overset{k}{\rightleftharpoons}} \gamma C + \delta D \,, \tag{11.29}$$

where A, B are reagents and C, D products, with the stoichiometric coefficients $\alpha, \beta, \gamma, \delta$, and where the dimensional coefficients k and k_{-1} are the forward and reverse reaction constants, respectively. Chemical equilibrium for the reaction (11.29) is determined by the *law of mass action* stating that, for a balanced chemical equation at a certain temperature T and pressure p, the equilibrium reaction rate, defined by the ratio

$$\frac{[C]^\gamma [D]^\delta}{[A]^\alpha [B]^\beta} = K_{eq}(T,p)\,, \tag{11.30}$$

is constant and it only depends on T and p [Atkins and Jones (2004)]. The square brackets indicate the concentration of chemical species. When the reaction (11.29) is far from equilibrium, the reaction rate characterizing how concentration of substances change in time is formally defined as[48]

$$R = -\frac{1}{\alpha}\frac{\mathrm{d}[A]}{\mathrm{d}t} = -\frac{1}{\beta}\frac{\mathrm{d}[B]}{\mathrm{d}t} = \frac{1}{\gamma}\frac{\mathrm{d}[C]}{\mathrm{d}t} = \frac{1}{\delta}\frac{\mathrm{d}[D]}{\mathrm{d}t}\,. \tag{11.31}$$

The phenomenological quantity R depends on the peculiar reaction mechanism and, more specifically, on the concentrations of reactants (and often of products too). Moreover, it is affected by stoichometric coefficients, pressure and temperature and by the presence of catalysts and inhibitor. In the simple example (11.29), we expect that $R = R(\alpha, [A], \ldots, \delta, [D], T, p)$. The dependence on concentrations is generally unknown a priori[Atkins and Jones (2004)] and has to be determined only through careful experimental measurements. For instance if from an experiment on the simple reaction $A + B \to C$, we discover that the formation rate of product C depends on the 3-rd power of $[A]$ and on first power of $[B]$, then we are allowed to write $\mathrm{d}[C]/\mathrm{d}t \propto [A]^3[B]$. The powers 3 and 1 are said *order of the reaction* with respect to species A and B respectively.

A reasonable assumption often done is that R satisfies the mass action also outside the equilibrium, thus it depends on concentrations raised to the corresponding stoichiometric coefficient, in formulae

$$R = k_1[A]^\alpha[B]^\beta - k_{-1}[C]^\gamma[D]^\delta. \tag{11.32}$$

The above expression is obtained as a "in-out" equation between a forward process where reactants A and B disappear at the rate $k_1[A]^\alpha[B]^\beta$ and reverse process involving the increase of products at the rate $k_{-1}[C]^\gamma[D]^\delta$. According to Eqs. (11.31) and (11.32) the variation of concentration with time for e.g. substances A and D is governed by the ODEs

$$\begin{aligned}\frac{\mathrm{d}[A]}{\mathrm{d}t} &= -\alpha(k_1[A]^\alpha[B]^\beta - k_{-1}[C]^\gamma[D]^\delta)\\ \frac{\mathrm{d}[D]}{\mathrm{d}t} &= \delta(k_1[A]^\alpha[B]^\beta - k_{-1}[C]^\gamma[D]^\delta)\,.\end{aligned}$$

At equilibrium, the rate R vanishes recovering the mass action law Eq. (11.30) with $K_{eq}(T) = k_1/k_{-1}$. This is the formulation of the detailed balance principle in the context of chemical reactions.

[48] In the definition of rates, the reason why the time derivative of concentrations are normalized by the corresponding stoichiometric coefficient is clear by considering the case $A+2B \to C$, where for every mole of A, two moles of B are consumed. Therefore, the consumption velocity of B is twice that of A. Moreover, as reagent are consumed, their rate is negative while products that are generated have a positive derivative, this is sign convention usually adopted.

Although the mass action applies to a wide class of reactions, including enzymatic activity (for an example, see Box B.27), kinetics mechanisms do not generally follow it. In particular, the powers do not coincide with the values prescribed by the reaction stoichiometry [Atkins and Jones (2004)].

A chemical reaction that represented and still constitutes an important step in dynamical system theory is the one studied by Belousov and later by Zhabotinsky. In the '50s, Belousov incidentally discovered that a reaction generated by a certain mix of reactants, in appropriate concentrations, caused the solution to perform surprising reproducible long-lived oscillations between yellow and a colorless state. The history of this discovery and its publication on scientific journals was weird. Indeed, Belousov made two attempts to publish his findings, but the paper was rejected, with the objection that the result explanation was unclear. The work was finally published in a minor journal without peer-review [Belousov (1959)]. Later, in 1961, Zhabotinsky, at that time a graduate student, rediscovered and improved the Belousov reaction continuing to study the process [Zhabotinsky (1991)]. The results, however, remained unknown to the Western scientific community until 1968, year in which they were presented to a conference held in Prague. Since then, BZ-reaction became probably the most studied oscillating reaction both theoretically and experimentally. Although it was not certainly the first known oscillating reaction, it was no more considered just as a curiosity, becoming soon the paradigm of oscillatory phenomenology in chemistry [Hudson and Rössler (1986)].[49] Before this discovery, most of chemists were convinced that chemical reactions were immune from stationary oscillations, rather, according to intuition from thermodynamics, reaction were expected to proceed spontaneously and unidirectionally towards the compatible thermodynamical equilibrium. In the literature, the oscillating behavior is called a chemical clock, it is typical of systems characterized by *bistability*, a regime for which a system visits cyclically two stable states. Bistable mechanisms are considered important also to biology because they often represent the prototypes of basic biochemical processes occurring in living organisms (see next section).

The chemistry of BZ-reaction is, in principle, rather simple and corresponds to oxidation (in acid medium) of an organic acid by bromate ions in the presence of a metal ion catalyst. Different preparations are possible, the original Belousov's experiment consists of *sulphuric acid* H_2SO_4, medium in which it is dissolved: *malonic acid* $CH_2(C00H)_2$, *potassium bromate* $KBr0_3$, *cerium sulfate* $Ce_2(S0_4)_3$ as a catalyst. The reaction, with small adjustments, works also when Cerrum ions are replaced by Ferrum ions as catalysts. If the reactants are well mixed in a beaker by stirring the solution, then one observes oscillations lasting several minutes in the system, characterized by a solution changing alternately between a yellow color and colorless state. The yellow color is due to the abundance of ions Ce^{4+}, while the colorless state corresponds to the preponderance of Ce^{3+} ions. When BZ-reaction

[49]The Briggs-Rauscher reaction is another well known oscillating chemical reaction, easier to be produced than BZ-reaction. Its color changes from amber to a very dark blue are clearly visible.

occurs in non-homogeneous media, it generates spiraling patterns produced by the combined effect of a local nonlinear chemical process and diffusion (see Fig. 12.1 in the next Chapter).

The original explanation of the oscillation proposed by Field, Köros and Noyes (1972) employs the same kinetics and thermodynamics principles governing standard reactions. It is known as FKN mechanism, and involves 18 reactions (steps) and 21 reactants. However, in 1974, the same authors proposed a simplified set of reactions, called "Oregonator",[50] able to capture the very essence of the BZ-mechanism. The set of chemical reactions can be summarized as follows:

$$\begin{aligned} A+Y &\xrightarrow{k_1} X+P \\ A+X &\xrightarrow{k_2} 2X+2Z \\ X+Y &\xrightarrow{k_3} 2P \\ 2X &\xrightarrow{k_4} A+P \\ B+Z &\xrightarrow{k_5} (f/2)Y+.... \end{aligned}$$

where $A = BrO_3^-$, $B = CH_2(C00H)_2$, $P = HBrO$, $X = HBrO_2$, $Y = Br^-$, $Z = Ce^{4+}$ and dots indicate other products that are inessential to the mechanism explanation. Since some stoichiometry of the reaction is still uncertain, a tunable parameter f is considered to fit the data. The second reaction is autocatalytic as it involves compound X both as reagent and product, this is crucial to generate oscillations. With the help of mass action law, the kinetic differential equations of the system can be written as

$$\begin{aligned} \frac{d[X]}{dt} &= k_1[A]\cdot[Y] + k_2[A]\cdot[X] - k_3[X]\cdot[Y] - 2k_4[X]^2 \\ \frac{d[Y]}{dt} &= -k_1[A]\cdot[Y] - k_3[X]\cdot[Y] + k_5 f/2[B]\cdot[X] \\ \frac{d[Z]}{dt} &= 2k_2[A]\cdot[X] - k_5[B]\cdot[Z] \,. \end{aligned}$$

After a transient, the dynamics of the system sets on limit cycle oscillations, and this behavior depends crucially on the values of rates $(k_1, \dots, k_5)$, coefficient f and initial concentrations. Numerical integration, as in Fig. 11.15, shows that model (11.33) could successfully reproduce oscillations and bistability. Moreover, it is able to explain and predict most experimental results on the BZ-reaction, however it cannot exhibit irregular oscillations and chaos.

The importance of BZ-reaction to dynamical system theory relies on the fact that it represents a laboratory system, relatively easy to master, showing a rich phenomenology. Furthermore, chaotic behaviors have been observed in certain recent experiments [Schmitz *et al.* (1977); Hudson and Mankin (1981); Roux (1983)] where the BZ-reactions were kept in continuous stirring by a CSTR reactor.[51] Chemical

[50] The name was chosen in honor to the University of Oregon where the research was carried out.
[51] The acronym CSTR stands for Continuous-flow Stirred Tank Reactor is most frequently used

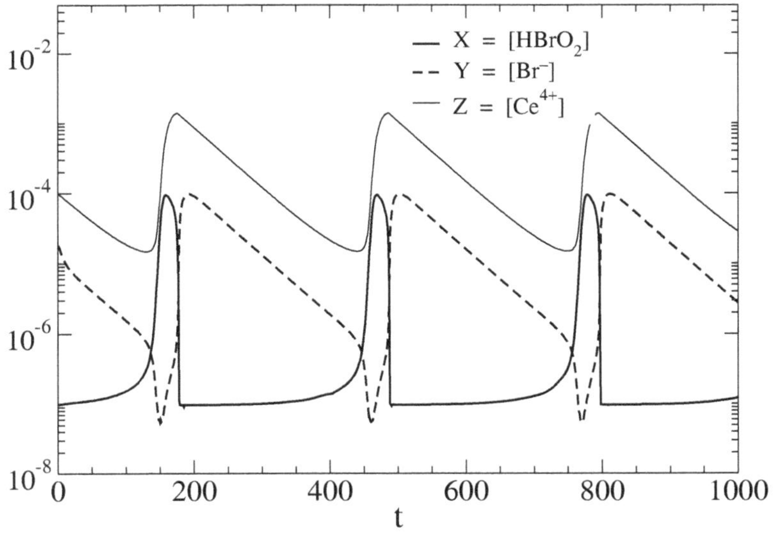

Fig. 11.15 Oscillatory behavior of concentrations $[X]$,$[Y]$ and $[Z]$ in the Oregonator system (11.33) with parameters: $k_1 = 1.2$, $k_2 = 8.0$, $k_3 = 8 \times 10^{-5}$, $k_4 = 2 \times 10^{-5}$, $k_5 = 1.0$, $f = 1.5$. Concentrations of A and B are kept constant and set to $[A] = 0.06$ and $[B] = 0.02$.

chaos shows up as chemical concentrations which neither remain constant nor oscillate periodically, rather they increase and decrease irregularly making their evolution unpredictable for long times. Several theoretical attempts have been suggested to explain the presence of chaotic regimes in BZ-like experiments. One of the simplest approach can be derived by reducing "Oregonator"-like systems to a three components model with the identification and elimination of fast variables [Zhang *et al.* (1993)]. Such models have the same nonlinearities already encountered in Lorenz and Lotka-Volterra systems, its chaotic behaviors are thus characterized by aperiodic fluctuations as for Lorenz's attractor (Sec. 3.2).

Box B.27: Michaelis-Menten law of simple enzymatic reaction

As an important application of mass action law, we can mention the Michaelis-Menten law governing the kinetics of elementary enzymatic reactions [Leskovac (2003); Murray (2002)]. Enzymes are molecules that increase (i.e. catalyze) the rate of a reaction even of many orders of magnitude. Michaelis-Menten law describes the rate at which an enzyme E interacts with a substrate S in order to generate a product P. The simplest catalytic

experimental setup to study chemical reaction maintained out of equilibrium. In a typical CSTR experiment, fresh reagents are continuously pumped into the reactor tank while an equal volume of the solution is removed to work at constant volume. A vigorous stirring guarantees, in a good approximation, the instantaneous homogeneous mixing of the chemicals inside the reactor vessel. The flow, feedstream and stirring rates, as well as temperature are all control parameters of the experiments.

reaction can be represented as

$$E + S \underset{k_{-1}}{\overset{k}{\rightleftharpoons}} ES \xrightarrow{k_p} E + P \, .$$

The last stage is assumed irreversible, thus the product P does not re-bind to the enzyme E and the first process is considered so fast that it reaches the equilibrium sooner than products. The chemistry of the reaction is characterized by the four concentrations, $[S]$, $[E]$, $[ES]$ and $[P]$. However, the constraint that the total amount of E (in complex or free) is conserved, $[E]_0 = [ES] + [E] = \text{const}$, allows the elimination of a variable between $[E]$ and $[ES]$. According to mass-action law, we can write the three equations

$$\begin{aligned}
\frac{\mathrm{d}[S]}{\mathrm{d}t} &= k_{-1}[ES] - k_1[E][S] \\
\frac{\mathrm{d}[ES]}{\mathrm{d}t} &= -(k_p - k_{-1})[ES] + k_1[E][S] \\
\frac{\mathrm{d}[P]}{\mathrm{d}t} &= k_p[ES]
\end{aligned}$$

where $[E] = [E]_0 - [ES]$. Notice that the last equation is trivial as it couples P and ES concentrations only. The quasi-steady state condition of the first step of the reaction $\mathrm{d}[ES]/\mathrm{d}t = 0$ leads to the relation

$$-k_p[ES] + k_1([E]_0 - [ES])[S] - k_{-1}[ES] = 0$$

then the concentration of the complex ES is

$$[ES] = \frac{[E]_0[S]}{(k_p + k_{-1})/k_1 + [S]} = \frac{[E]_0[S]}{K_M + [S]}$$

where $(k_p + k_{-1})/k_1 = K_M$ is the well-known Michaelis-Menten constant. The final result about the P-production rate recasts as

$$\frac{\mathrm{d}[P]}{\mathrm{d}t} = V_{max}\frac{[S]}{K_M + [S]}, \qquad V_{max} = k_p[E]_0 \tag{B.27.1}$$

indicating that such a rate is the product of a maximal rate and V_{max} and a function $\phi([S]) = [S]/(K_M + [S])$ of the substrate concentration only. Michaelis-Menten kinetics, like other classical kinetic theories, is a simple application of mass action law which relies on free diffusion and thermal-driven collisions. However, many biochemical or cellular processes deviate significantly from such conditions.

11.3.4 *Chemical clocks*

Cyclical and rhythmic behaviors similar to those observed in the Belousov-Zhabotinsky reaction are among the peculiar features of living organism [Winfree (1980)]. Chemical oscillating behaviors are usually called *chemical clocks*, and characterize systems which operate periodically among different stable states. Chemical clocks can be found at every level of the biological activity, well known examples are *circadian rhythms* which correspond to the 24-hour periodicity of several physiological processes of plants and animals. Cyclical phenomena with different clocks can be also found in a multitude of metabolic and genetic networks concurring to perform and regulate the complex stages of cell life. Typical oscillations in metabolic pathways are associated either to regulation of protein synthesis, *epigenetic oscillations* (with periods of hours) or to regulation of enzyme activity, *metabolic oscillations* (with periods of minutes). Although cell biochemistry is extremely complex, researchers have been able to identify some elementary but fundamental oscillatory processes, we can mention the well known: *glycolytic oscillations* observed in muscles and yeast, *oscillations in cytosolic calcium* Ca^{+2}, a response of cells to mechanical or chemical stimulation, presumably in order to communicate with one another and coordinate their activity over larger regions; *pulsatile inter-cellular communications* and *periodic synthesis of the cyclic Adenosin mono-phosphate* (cAMP) controlling cell differentiation and chemotaxis. For a review on the subject see Goldbeter (1996).

From a mathematical point of view these oscillating behaviors can be explained in terms of attracting limit cycles in the chemical rate equations governing the biochemistry of cellular processes. In this context, dynamical system theory plays a fundamental role in identifying the conditions for the onset and stability of cyclical behaviors. At the same time, it is also of interest understanding whether the presence of instabilities and perturbations may give rise to bursts of chaotic regimes able to take the system outside its steady cyclical state. Enzymatic activity is commonly at the basis of metabolic reactions in the cells, which are brightest examples of natural occurring oscillating reactions. Several experiments have shown that cell cultures generally show periodic increase in the enzyme concentrations during cellular division. The periodic change in enzyme synthesis implies the presence of metabolic regulatory mechanism with some kind of feedback control [Tyson (1983)]. To explain theoretically oscillations in enzyme activities and possible bifurcation towards chaotic states Decroly and Goldbeter (1982) considered a model of two coupled elementary enzymatic reactions, sketched in Figure 11.16, involving

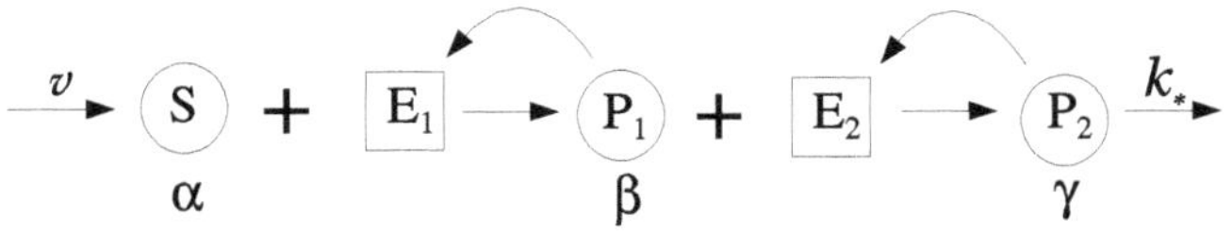

Fig. 11.16 Cascade of enzymatic reactions proposed as elementary metabolic system showing simple and chaotic oscillations in the rescaled concentrations α, β and γ. Note the presence of a positive feedback due to the re-bind of products P_1 and P_2 to the corresponding enzymes E_1 and E_2. (see Box. B.28).

the allosteric enzymes E_1 and E_2 a substrate S synthesized at constant rate v. S is transformed into the product P_1 by the catalytic reaction driven by E_1, which in turn, is activated by P_1. A second enzymatic reaction transforms P_1 into P_2 by the catalysis of a second allosteric enzyme E_2, finally the product P_2 disappears at a

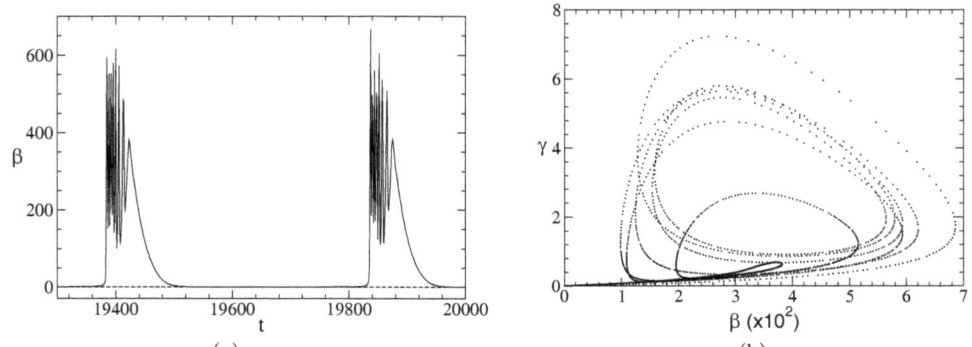

(a) (b)

Fig. 11.17 Complex oscillations in the reaction sketched in Fig. 11.16: (a) normalized concentration $\beta = [P_1]/k_1$ versus time and (b) projection on the (β, γ)-plane. The details on the model are discussed in Box B.28, specifically data are obtained by using Eq. (B.28.1) with parameters: $\sigma_1 = \sigma_2 = 10s^{-1}$, $q_1 = 50$, $q_2 = 0.02$, $L_1 = 5 \times 10^8$, $L_2 = 10^2$ and $d = 0$, $v = 0.44s^{-1}$, $k_* = 3s^{-1}$.

rate $k_*[P_2]$. The Box B.28 reports the three coupled dynamical equations for concentrations $[S]$, $[P_1]$, $[P_2]$ and summarizes the principal chemical basis of the model. However a deeper discussion would require a complex chemical reasoning which is far from the purpose of this section, the interested reader can refer to [Goldbeter and Lefever (1972)]. Here we can mention the conclusions of the work by Decroly and Goldbeter (1982). The model exhibits a rich bifurcation pattern upon changing the parameters v and k_*. The plot of the steady values of the normalized substrate concentration $\alpha \propto [S]$ versus k_* revealed a variety of steady-state behaviors, ranging from: limit-cycle oscillations, excitability, bi-rhythmicity (coexistence of two periodic regimes) and period doubling route to chaos, with bursting of complex oscillations (Fig. 11.17). It is interesting to observe that chaos in the model occurs only in a small parameter range, suggesting that regimes of regular oscillations are largely overwhelming. Such a parameter fine-tuning to observe chaos was judged by the authors in agreement with what observed in Nature, where all known metabolic processes show (regular) cyclical behaviors.

Box B.28: A model for biochemical oscillations

In this box we briefly illustrate a model of a simple metabolic reaction obtained by coupling two enzymatic activities, as proposed by Decroly and Goldbeter (1982) and sketched in Figure 11.16. The process, represented with more details in Fig. B28.1, involves an enzyme E_1 and a substrate S that react generating a product P_1, which in turns acts a substrate for a second enzyme E_2 which catalyzes the production of P_2. Further assumptions on the model are (Fig. B28.1):
(a) the enzymes E_1, E_2 are both dimers existing in two allosteric forms: active "R" or inactive "T", which interconvert one another via the equilibrium reaction $R_0 \leftrightarrow T_0$ of rate

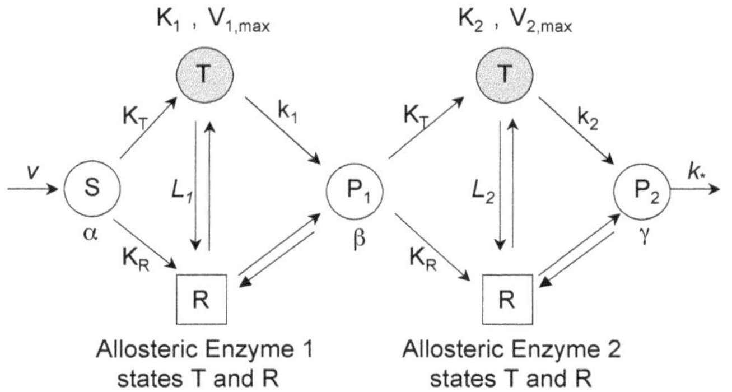

Fig. B28.1 Cartoon of the coupled enzymatic reactions considered by Decroly and Goldbeter (1982). Enzymes E_1 and E_2 are allosteric and can assume two conformations R and T which interconvert one another, with equilibrium constant L. Both forms can bind the substrate S injected at constant rate v. The product P_1 from the first catalytic reaction acts a substrate for the second reaction, and binds only to R forms of both enzymes. The final product P_2 is extracted at constant rate k_*.

L, the index 0 refers to the free enzyme (not bound to S). In other words this enzyme kinetics is assumed to follow the cooperative model of Monod, Wyman and Chateaux (1965) (MWC) of allostery;[52]
(b) the substrate binds to both forms, while the product acting as an effector binds only to the active form R;
(c) the form R carrying the substrate decays irreversibly yielding the product.
The evolution equations for the metabolites are

$$\begin{aligned}\frac{d\alpha}{dt} &= \frac{v}{K_1} - \sigma_1\phi(\alpha,\beta)\\ \frac{d\beta}{dt} &= q_1\sigma_1\phi(\alpha,\beta) - \sigma_2\psi(\beta,\gamma)\\ \frac{d\gamma}{dt} &= q_2\sigma_2\psi(\beta,\gamma) - k_s\gamma\end{aligned} \tag{B.28.1}$$

where coefficients: $\sigma_{1,2}$, $q_{1,2}$ and k_s depends on the enzymatic properties of the reaction defined by K_1, K_2 and $V_{1,max}, V_{2,max}$, the Michaelis-Menten constants and maximal rates

[52]Monod-Wyman-Changeux hypothesized that in an allosteric enzyme E carrying n binding sites, each subunit called a promoter can exist in two different conformational states "R" (relaxed) or "T" (tense) states (Fig. B28.1). In any enzyme molecule E, all promoters must be in the same state i.e. all subunits must be in either the R or T state. The R state has a higher affinity for the ligand S than T, thus the binding of S, will shift the equilibrium of the reaction in favor of the R conformation. The equations that characterize, the fractional occupancy of the ligand binding-site and the fraction of enzyme molecules in the R state are:

$$\overline{Y} = \frac{\alpha(1+\alpha)^{n-1} + Lc\alpha(1+c\alpha)^{n-1}}{(1+\alpha)^n + L(1+c\alpha)^n}, \qquad \overline{R} = \frac{(1+\alpha)^n}{(1+\alpha)^n + L(1+c\alpha)^n}$$

where, $\alpha = [S]/K_R$ is the normalized concentration of ligand S, $L = [T_0]/[R_0]$ is the allosteric constant, the ratio of proteins in the "T" and "R" states free of ligands, $c = K_R/K_T$ is the ratio of the affinities of R and T states for the ligand. This model explains sigmoid binding properties as change in concentration of ligand over a small range will lead to a large a high association of the ligand to the enzyme.

of enzymes E_1, E_2 respectively (see Eq. (B.27.1)) and by k_1 and k_2 characterizing the dissociation constants of product P_1 for E_1 and P_2 for E_2 (Fig. B28.1). The rates v and k_* describe the injection of substrate S and the removal of P_2 respectively. The variables of the system are the rescaled concentrations $\alpha = [S]/K_1$, $\beta = [P_1]/k_1$, $\gamma = [P_2]/k_2$, and accordingly the coefficients in Eq. (B.28.1) correspond to $\sigma_i = V_{i,max}/K_i$ $(i = 1, 2)$, $q_1 = K_1/k_1$, $q_2 = k_1/k_2$, $k_s = k_*/k_2$. The functions ϕ, ψ, due to allostery and cooperativity, are no longer obtained via Michaelis-Menten theory (Box B.27), but are derived from a MWC-like model which describes properly cooperative binding in the presence of allosteric behavior

$$\phi = \frac{\alpha(1+\alpha)(1+\beta)^2}{L_1 + (1+\alpha)^2(1+\beta)^2} \qquad \psi = \frac{\beta(1+d\beta)(1+\gamma)^2}{L_2 + (1+d\beta)^2(1+\gamma)^2}$$

where L_1 and L_2 are the constants of the MWC-model referred to each enzyme, $d = k_1/K_2$. The specific form of ϕ and ψ reflect the assumption that enzymes E_1 and E_2 are both dimers with exclusive binding of ligands to R state (more active state).

11.4 Synchronization of chaotic systems

Synchronization (from Greek σύν: syn = the same, common and χρόνος: chronos = time) is a common phenomenon in nonlinear dynamical systems discovered at the beginning of modern science by Huygens (1673) who, while performing experiments with two pendulum clocks (also invented by him), observed

> *... It is quite worths noting that when we suspended two clocks so constructed from two hooks imbedded in the same wooden beam, the motions of each pendulum in opposite swings*[53] *were so much in agreement that they never receded the least bit from each other and the sound of each was always heard simultaneously. Further, if this agreement was disturbed by some interference, it reestablished itself in a short time. For a long time I was amazed at this unexpected result, but after a careful examination finally found that the cause of this is due to the motion of the beam, even though this is hardly perceptible.*

Huygens's observation qualitatively explains the phenomenon in terms of the imperceptible motion of the beam: in modern language we understand clocks synchronization as the result of the coupling induced by the beam.

Despite the early discovery, synchronization was systematically investigated and theoretically understood only in the XX-th century by Appleton (1922) and van der Pol (1927) who worked with triode electronic generators. Nowadays the fact that different parts of a system or two coupled systems (not only oscillators) can synchronize is widely recognized and found numerous applications in electric and mechanical engineering. Synchronization is also exploited in optics with the synchronization of lasers [Simonet *et al.* (1994)] and is now emerging as a very fertile research field in biological sciences where synchrony plays many functional roles in circadian rhythms, firing of neurons, adjustment of heart rate with respiration

[53] He observed an anti-phase synchronization of the two pendula.

or locomotion, etc. (the interested reader can find a plenty of examples and an exhaustive treatment of the subject in the book by Pikovsky *et al.* (2001)).

Intriguingly, also chaotic systems can synchronize. The following two sections present *phase synchronization* in chaotic oscillators[54] [Rosenblum *et al.* (1996)], which generalizes that of regular oscillators like Huygens' pendula, and *complete synchronization* of two coupled, identical chaotic systems [Fujisaka and Yamada (1983); Pecora and Carroll (1990)]. The latter example is particularly interesting as it displays a new phenomenon known as *on-off intermittency* [Fujisaka and Yamada (1985, 1986); Heagy *et al.* (1994)] and allows us to think back a few concepts introduced in the first part, namely the definition of attractor and the Lyapunov exponents. *Generalized synchronization* of non-identical systems or *lag/anticipated synchronization* are not considered here, for them the reader can refer to Pikovsky *et al.* (2001).

For the sake of self-consistency next section summarizes some basic facts about the synchronization of regular oscillators.

11.4.1 *Synchronization of regular oscillators*

Stable limit cycles, such as those arising via Hopf's bifurcation (Box B.11) or those characterizing the van der Pol oscillator (Box B.12), constitute the typical example of periodic, self-sustained oscillators encountered in dynamical systems. In limit cycles of period T_0, it is always possible to introduce a phase variable ϕ evolving with constant angular velocity,[55]

$$\frac{\mathrm{d}\phi}{\mathrm{d}t} = \omega_0 \,, \tag{11.33}$$

with $\omega_0 = 2\pi/T_0$. Notice that displacements along the limit cycle do not evolve as corresponds to a phase shift and the dynamics (11.33) has a zero Lyapunov exponent, the amplitude of the oscillations is characterized by a contracting dynamics (not shown) and thus by a negative Lyapunov exponent.

In the following, we consider two examples of synchronization, an oscillator driven by a periodic forcing and two coupled oscillators, which can be treated in a unified framework. For weak distortion of the limit cycle, the former problem can be reduced to the equation

$$\frac{\mathrm{d}\phi}{\mathrm{d}t} = \omega_0 + \epsilon q(\phi - \omega t),$$

where $q(\theta + 2\pi) = q(\theta)$ with, in general, $\omega \neq \omega_0$ and ϵ, which should be small, controls the strength of the external driving [Pikovsky *et al.* (2001)]. The phase difference between the oscillator and the external force is given by $\psi = \phi - \omega t$ and

[54]See below the Rössler model.

[55]Indeed, for non-uniform rotation the phase φ dynamics is given by $\mathrm{d}\varphi/\mathrm{d}t = \varpi(\varphi)$ where $\varpi(\varphi + 2\pi) = \varpi(\varphi)$, then an uniformly rotating phase ϕ can be obtained via the transformation $\phi = \omega_0 \int_0^{\varphi} \mathrm{d}\theta\, [\varpi(\theta)]^{-1}$ (where ω_0 is determined by the condition $2\pi = \omega_0 \int_0^{2\pi} \mathrm{d}\theta\, [\varpi(\theta)]^{-1}$).

can be regarded as a slow variable moving in a rotating frame. Synchronization means here phase-locking, i.e. $\psi = \text{const}$ (e.g. for Huygens' pendula $\psi = \pi$ as they were oscillating in anti-phase). Denoting with $\nu = \omega_0 - \omega$ the frequency mismatch (detuning),[56] the equation for ψ reads

$$\frac{d\psi}{dt} = \nu + \epsilon q(\psi)\,. \tag{11.34}$$

The above equation also describes the synchronization of two coupled oscillators

$$\begin{aligned}\frac{d\phi_1}{dt} &= \omega_1 + \epsilon g_1(\phi_1, \phi_2)\\ \frac{d\phi_2}{dt} &= \omega_2 + \epsilon g_2(\phi_1, \phi_2)\,,\end{aligned} \tag{11.35}$$

where $g_{1,2}$ are 2π-periodic functions of their arguments and ϵ tunes the coupling strength. To highlight the similarity with the previous case, it is useful to expand $g_{1,2}$ in Fourier series $g_i(\phi_1, \phi_2) = \sum_{p,q} a_i^{(p,q)} e^{ip\phi_1 + iq\phi_2}$, and notice that $\phi_i(t) = \omega_i t$ at the zero-th order of approximation. Consequently, averaging over a period, all terms of the expansion vanish except those satisfying the resonance condition $p\omega_1 + q\omega_2 \approx 0$. Now, assuming $n\omega_1 \approx m\omega_2$, all terms $p = nk$ and $q = -mk$ are resonant and thus contribute, so that, defining $q_{1,2}(n\phi_1 - m\phi_2) = \sum_k a_{1,2}^{(nk,-mk)} e^{ik(n\phi_1 - m\phi_2)}$, the system (11.35) reads

$$\begin{aligned}\frac{d\phi_1}{dt} &= \omega_1 + \epsilon q_1(n\phi_1 - m\phi_2)\\ \frac{d\phi_2}{dt} &= \omega_2 + \epsilon q_2(n\phi_1 - m\phi_2)\,,\end{aligned}$$

which can also be reduced to Eq. (11.34) with $\psi = n\phi_1 - m\phi_2$, $\nu = n\omega_1 - m\omega_2$, and $q = q_1 - q_2$. The case $n = m = 1$ corresponds to the simplest instance of synchronization.

Now we briefly discuss phase synchronization in terms of Eq. (11.34) with the simplifying choice $q(\psi) = \sin\psi$ leading to the Adler equation [Adler (1946)]

$$\frac{d\psi}{dt} = \nu + \epsilon \sin\psi\,. \tag{11.36}$$

The qualitative features of this equation are easily understood by noticing that it describes a particle on the real line $\psi \in [-\infty : \infty]$ (here ψ is not restricted to $[0 : 2\pi]$) that moves in an inclined potential $V(\psi) = -\nu\psi + \epsilon\cos(\psi)$ in the overdamped limit.[57] The potential is characterized by a repetition of minima and maxima, where $dV/d\psi = 0$ when $\epsilon < |\nu|$ (Fig. 11.18a), which disappear for $\epsilon \geq |\nu|$ (Fig. 11.18b). The dynamics (11.36) is such to drive the particle towards the potential minima (when it exists), which corresponds to the phase-locked state $d\psi/dt = 0$ characterized by a constant phase difference ψ_0. Synchronous oscillations

[56] As clear from below, by defining $\nu = m\omega_0 - n\omega$ and $\psi = m\phi - n\omega t$ the same equation can be used to study more general forms of phase locking.

[57] The equation of a particle x in a potential is $md^2x/dt + \gamma dx/dt + dV/dx = 0$. The overdamped limit corresponds to the limit $\gamma \to \infty$ or, equivalently, $m \to 0$.

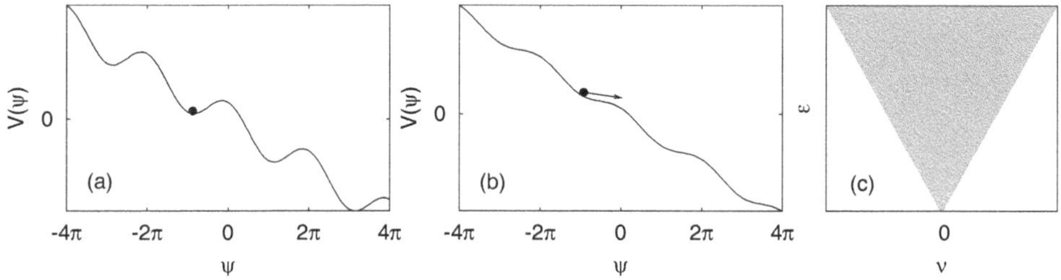

Fig. 11.18 Phase interpreted as a particle moving in an inclined potential $V(\psi)$: (a) $\epsilon > |\nu|$, synchronization is possible as particles fall in a potential minimum, (b) $\epsilon < |\nu|$, outside the synchronization region the minima disappear and the synchronization is no more possible. Panel (c) shows the synchronization region (shaded area) in the (ν, ϵ) plane. [After Pikovsky *et al.* (2001)]

are thus possible only in the triangular-shaped synchronization region illustrated in Fig. 11.18c.[58] For two coupled oscillators, the sign of ϵ being positive or negative determines a repulsive or attractive interaction, respectively leading to an "anti-phase" or "in-phase" synchronization (clearly in Huygens pendula $\epsilon > 0$).

For generic functions $q(\psi)$ the corresponding potential $V(\psi)$ may display a richer landscape of minima and maxima complicating the synchronization features.

We conclude this brief overview mentioning the case of oscillators subjected to external noise, where the phase evolves with the Langevin equation $\mathrm{d}\phi/\mathrm{d}t = \omega_0 + \eta$, η being some noise term. In this case, phase dynamics is akin to that of a Brownian motion with drift and $\langle(\phi(t) - \phi(0) - \omega_0 t)^2\rangle \simeq 2Dt$, where D is the diffusion coefficient. When such noisy oscillators are subjected to a periodic external driving or are coupled, the phase difference is controlled by a noisy Adler equation $\mathrm{d}\psi/\mathrm{d}t = -\nu + \epsilon \sin\psi + \eta$. The synchronization is still possible but, as suggested by the mechanical analogy (Fig. 11.18a,b), a new phenomenon known as *phase slip* may appear [Pikovsky *et al.* (2001)]: due to the presence of noise the particle fluctuates around the minimum (imperfect synchronization) and, from time to time, can jump to another minimum changing the phase difference by $\pm 2\pi$.

11.4.2 *Phase synchronization of chaotic oscillators*

Chaotic systems often display an irregular oscillatory motion for some variables and can be regarded as chaotic oscillators like, for instance, the Rössler (1976) system

$$\begin{aligned}
\frac{\mathrm{d}x}{\mathrm{d}t} &= -y - z \\
\frac{\mathrm{d}y}{\mathrm{d}t} &= x + ay \\
\frac{\mathrm{d}z}{\mathrm{d}t} &= b + z(x - c)\,,
\end{aligned} \tag{11.37}$$

[58]Such regions exist also for $m : n$ locking states and define the so-called *Arnold tongues.*

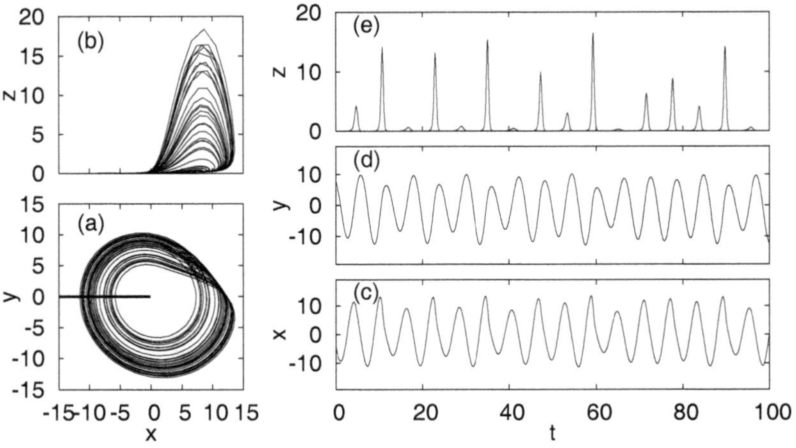

Fig. 11.19 Rössler system for $a=0.15$, $b=0.4$ and $c=8.5$. (a) Projection of the attractor on the (x, y)-plane, the black thick line indicate the Poincaré section explained in the text. (b) Projection of the attractor on the (x, z)-plane. Temporal signals $x(t)$ (c), $y(t)$ (d) and $z(t)$ (e).

whose dynamics is shown in Fig. 11.19. The time signals $x(t)$ and $y(t)$ (Fig. 11.19c,d) display chaotically modulated oscillations, and the projection on the (x, y)-plane (Fig. 11.19a) looks like a well defined rotation around the origin, suggesting to define the oscillation amplitude A and phase ϕ as $A = \sqrt{x^2+y^2}$ and $\phi = \mathrm{atan}(y/x)$, respectively. However, different non-equivalent definitions of phase are possible [Rosenblum *et al.* (1996)]. For instance, we can consider the Poincaré map obtained by the intersections of the orbit with the semi-plane $y = 0$ and $x < 0$. At each crossing, which happens at times t_n, the phase change by 2π so that we have [Pikovsky *et al.* (2001)]

$$\phi(t) = 2\pi \frac{t - t_n}{t_{n+1} - t_n} + 2\pi n \,, \qquad t_n \le t \le t_{n+1} \,.$$

Unlike limit cycles, the phase of a chaotic oscillators cannot be reduced to a uniform rotation. In general, we expect that the dynamics can be described, in the interval $t_n \le t < t_{n+1}$ as

$$\frac{\mathrm{d}\phi}{\mathrm{d}t} = \omega(A_n) = \omega_0 + F(A_n)$$

where A_n is the amplitude at time t_n, whose dynamics can be considered discrete as we are looking at the variables defined by the Poincaré map $A_{n+1} = G(A_n)$. The amplitude evolves chaotically and we can regard the phase dynamics as that of an oscillator with average frequency ω_0 with superimposed a "chaotic noise" $F(A_n)$.[59]

[59] Recalling the discussion at the end of Sec. 11.4.1 we can see that the phase of a chaotic oscillator evolves similarly to that of a noisy regular oscillator, where chaotic fluctuations play the role of noise. Indeed it can be shown that, similarly to noise oscillators, the phase dynamics of a single chaotic oscillator is diffusive. We remark that replacing, for analytical estimates or for conceptual reasoning, deterministic chaotic signals with noise is a customary practice which very often produce not only qualitatively correct expectations but also quantitatively good estimates.

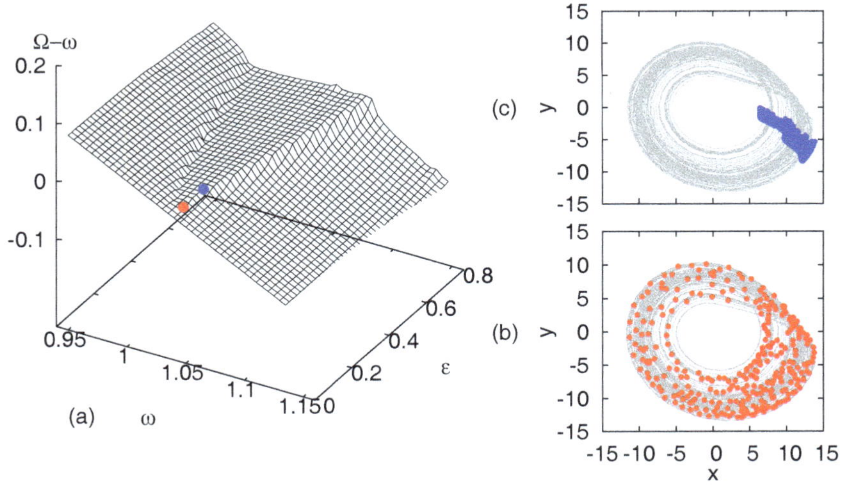

Fig. 11.20 (a) $\Omega-\omega$ as a function of ω and ϵ in the periodically forced Rössler model. Stroboscopic map: (b) outside the synchronization region, for $\omega = 1.04$ and $\epsilon = 0.05$ (red spot in (a)); (c) inside the synchronization region, for $\omega = 1.04$ and $\epsilon = 0.16$ (blue spot in (a)). The projection of the Rössler attractor is shown in gray. [After Pikovsky *et al.* (2001)]

Following Pikovsky *et al.* (2000) we illustrate the synchronization of the phase of Rössler system with a periodic driving force of frequency ω and intensity ϵ, the idea is to add in the r.h.s. of the first equation of the system (11.37) a term $\epsilon \cos(\omega t)$. Synchronization means that the average frequency of rotation $\Omega = \lim_{t\to\infty} \phi/t$ (with the phase ϕ defined, e.g., by the Poincaré map) should become equal to that of the forcing ω when the driving intensity ϵ is large enough, exactly as it happens in the case of regular oscillators. In Fig. 11.20a we show the frequency difference $\Omega - \omega$ as a function of ω and ϵ, we can clearly detect a triangular-shaped plateau where $\Omega = \omega$ (similarly to the regular case, see Fig. 11.18c), which defines the synchronization region. The matching between the two frequencies and thus phase locking can be visualized by the stroboscopic map shown in Fig. 11.20b,c, where we plot the position of the trajectory $\boldsymbol{r}_k = [x(\tau_k), y(\tau_k)]$ on the (x, y)-plane at each forcing period, $\tau_k = 2k\pi/\omega$. When ϵ and ω are inside the synchronization region, the points concentrate in phase and disperse in amplitude signaling phase-locking (Fig. 11.20b to be compared with Fig. 11.20c).

It is now interesting to see how the forcing modifies the Lyapunov spectrum and how the presence of synchronization influences the spectrum. In the absence of driving, the Lyapunov exponents of Rössler system are such that $\lambda_1 > 0$, $\lambda_2 = 0$ and $\lambda_3 < 0$. If the driving is not too intense the system remains chaotic ($\lambda_1 > 0$) but the second Lyapunov exponent passes from zero to negative values when the synchronization takes place. The signature of the synchronization is thus present in the Lyapunov spectrum, as well illustrated by the next example taken by Rosenblum *et al.* (1996).

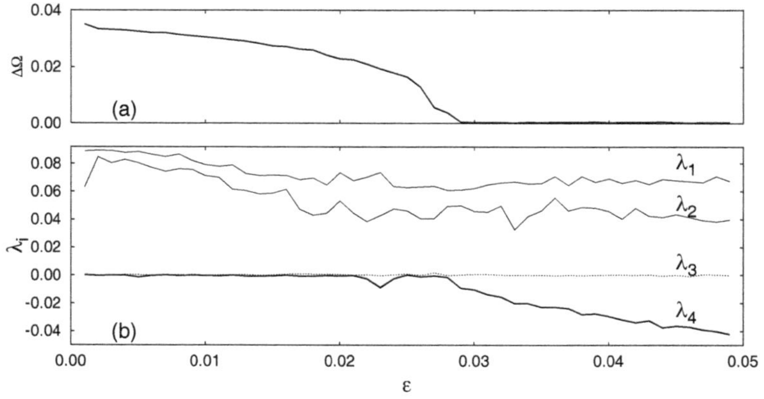

Fig. 11.21 (a) Observed frequency difference $\Delta\Omega$ as a function of the coupling ϵ for the coupled Rössler system with $\nu = 0.015$, $a = 0.15$ $b = 0.2$ and $c = 10$. (b) The first 4 Lyapunov exponents λ_i $(i = 1, 4)$ for the same system. Notice that $\lambda_4 < 0$ in correspondence of the phase synchronization.[After Pikovsky *et al.* (2001)]

We now consider two coupled Rössler oscillators:

$$\frac{\mathrm{d}x_1}{\mathrm{d}t} = -(1+\nu)y_1 - z_1 + \epsilon(x_2 - x_1) \qquad \frac{\mathrm{d}x_2}{\mathrm{d}t} = -(1-\nu)y_2 - z_2 + \epsilon(x_1 - x_2)$$
$$\frac{\mathrm{d}y_1}{\mathrm{d}t} = (1+\nu)x_1 + ay_1 \qquad \frac{\mathrm{d}y_2}{\mathrm{d}t} = (1-\nu)x_2 + ay_2$$
$$\frac{\mathrm{d}z_1}{\mathrm{d}t} = b + z_1(x_1 - c) \qquad \frac{\mathrm{d}z_2}{\mathrm{d}t} = b + z_2(x_2 - c)\,,$$

where $\nu \neq 0$ sets the frequency mismatch between the two systems and ϵ tunes the coupling intensity. Above a critical coupling strength $\epsilon > \epsilon_c$, the observed frequency mismatch $\Delta\Omega = \lim_{t\to\infty} |\phi_1(t) - \phi_2(t)|/t$ goes to zero signaling the phase synchronization between the two oscillators (Fig. 11.21a). The signature of the transition is well evident looking at the Lyapunov spectrum of the two coupled models. Now we have 6 Lyapunov exponents and the spectrum is degenerate for $\epsilon = 0$, as the systems decouple. For any ϵ, λ_1 and λ_2 remain positive meaning that the two amplitudes are always chaotic. It is interesting to note that in the asynchronous regime $\lambda_3 \approx \lambda_4 \approx 0$ meaning that the phases of the oscillators a nearly independent despite the coupling, while as synchronization sets in λ_3 remains close to zero and λ_4 becomes negative signaling the locking of the phases (Fig. 11.21b). Essentially, the number of non-negative Lyapunov exponents estimate the effective number of variables necessary to describe the system. Synchronization leads to a reduction of the effective dimensionality of the system, two or more variables become identical, and this reflect in the Lyapunov spectrum as discussed above.

Phase synchronization has been experimentally observed in electronic circuits, and lasers (see Kurths (2000)).

11.4.3 Complete synchronization of chaotic systems

Complete synchronization means that all the variables of two identical, coupled chaotic systems evolve in synchrony, so that the two systems behave as a unique. The most interesting aspect of such a phenomenon is the transition from asynchronous to synchronous dynamics. We illustrate the basic features of the problem mostly following Pikovsky and Grassberger (1991) (see also Fujisaka and Yamada (1983, 1985, 1986); Glendinning (2001)) and consider the simple case of two identical, symmetrically coupled maps, i.e. the system:

$$\begin{aligned} x(t+1) &= (1-\epsilon)f(x(t)) + \epsilon f(y(t)) \\ y(t+1) &= \epsilon f(x(t)) + (1-\epsilon)f(y(t)), \end{aligned} \tag{11.38}$$

where $f(x)$ can be any of the map we considered so far. The above system admits two limiting cases: for $\epsilon = 0$, x and y are independent and uncorrelated; for $\epsilon = 1/2$, independently of the initial condition, one step is enough to have trivial synchronization $x = y$. It is then natural to wonder whether a critical coupling strength $0 < \epsilon_c < 1/2$ exists such that complete synchronization can happen (i.e. $x(t) = y(t)$ for any t larger than a time t^*), and to characterize the dynamics in the neighborhood of the transition.

In the following we shall discuss the complete synchronization of the skew tent map:

$$f(x) = \begin{cases} x/p & 0 \le x \le p \\ (1-x)/(1-p) & p < x \le 1 \end{cases} \qquad 0 < p < 1 \tag{11.39}$$

which for $p = 1/2$ recovers the usual tent map. This map presents the advantages of being chaotic for any p and generic, as the growth rate of perturbation can fluctuate for $p \neq 1/2$.[60] Furthermore, it is reasonably simple to allow for some analytical computation, for instance, when uncoupled it is characterized by a uniform invariant density, $\rho^{inv}(x) = 1$, and we know the Lyapunov exponent:

$$\lambda = \int \mathrm{d}x \, \rho^{inv}(x) \, \ln|f'(x)| = \langle \ln|f'(x)| \rangle = -p\ln p - (1-p)\ln(1-p). \tag{11.40}$$

Let us now consider the coupled system (11.38) with f the skew tent map (11.39). As clear from Fig. 11.22, the increase of the coupling strength ϵ enhances the correlation between $x(t)$ and $y(t)$ until they become identical, i.e. synchronous. Interestingly, this happens for $\epsilon < 1/2$.

Before studying the properties of the transition, we need to determine the critical point ϵ_c at which the synchronous state becomes stable: this is rather simple but yet instructive. We can introduce two new variables

$$u = \frac{1}{2}(x+y) \qquad \text{and} \qquad v = \frac{1}{2}(x-y).$$

[60] Though non-generic, it is worthy remarking that in the case of discontinuous maps nontrivial effects can be induced by finite size instabilities as those observed in Fig. 9.9 for the Bernoulli map [Cencini and Torcini (2005)].

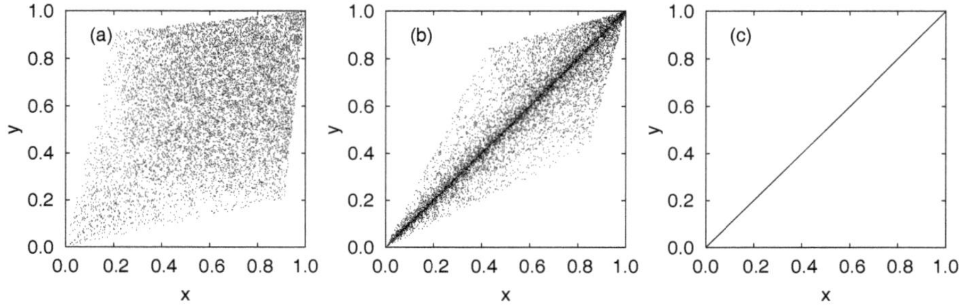

Fig. 11.22 Attractor of two symmetrically coupled skewed tent map for $a = 0.7$ and (a) $\epsilon = 0.1$, (b) $\epsilon = 0.21$ and (c) $\epsilon = 0.25$, which represent coupling values below, close and above the critical coupling value $\epsilon_c = 0.228\ldots$ for observing synchronization. [After Pikovsky and Grassberger (1991)]

Geometrically speaking, u and v correspond to a rotation by 45° degrees of the original variables, so that synchronized dynamics lies on the diagonal ($x = y$) and transverse dynamics out of it. In terms of these new variables Eq. (11.38) reads:

$$\begin{aligned} u(t+1) &= \frac{1}{2}\left(f[u(t)+v(t)] + f[u(t)-v(t)]\right) \\ v(t+1) &= \left(\frac{1}{2} - \epsilon\right)\left(f[u(t)+v(t)] - f[u(t)-v(t)]\right) , \end{aligned} \tag{11.41}$$

admitting the synchronous solution $v(t) = 0$ and $u(t+1) = f(u(t))$ for all ϵ. The stability of such a solution can be explored linearizing the system (11.41) around the solution $u(t)$ of the uncoupled dynamics $u(t+1) = f(u(t))$:

$$w_u(t+1) = f'(u(t))\, w_u(t) \tag{11.42}$$

$$w_v(t+1) = (1-2\epsilon) f'(u(t))\, w_v(t) , \tag{11.43}$$

being (w_u, w_v) the tangent vectors. We can treat the dynamics of w_u and w_v separately as they result uncoupled. The map $u(t+1) = f(u(t))$ is chaotic so that, recalling the definition of Lyapunov exponents (Sec. 5.3), we have

$$\lambda_u = \lim_{t\to\infty} \frac{1}{t} \ln \left| \frac{w_u(t)}{w_u(0)} \right| = \langle \ln |f'(u(t))| \rangle = \lambda \tag{11.44}$$

$$\lambda_v = \lim_{t\to\infty} \frac{1}{t} \ln \left| \frac{w_v(t)}{w_v(0)} \right| = \ln|1-2\epsilon| + \langle \ln |f'(u(t))| \rangle = \ln|1-2\epsilon| + \lambda \tag{11.45}$$

λ being the LE of the uncoupled map, which coincides with the diagonal exponent λ_u while the *transverse* one $\lambda_\perp \equiv \lambda_v$ depends on the coupling

$$\lambda_\perp(\epsilon) = \ln|1-2\epsilon| + \lambda \,. \tag{11.46}$$

If $\lambda_\perp > 0$ (resp. < 0) — *on average* — transverse perturbations grow (resp. decay), so that the criterion for the stability of the synchronized state is

$$\lambda_\perp(\epsilon_c) = 0 \quad \Longrightarrow \quad \epsilon_c = \frac{1 - e^{-\lambda}}{2} ,$$

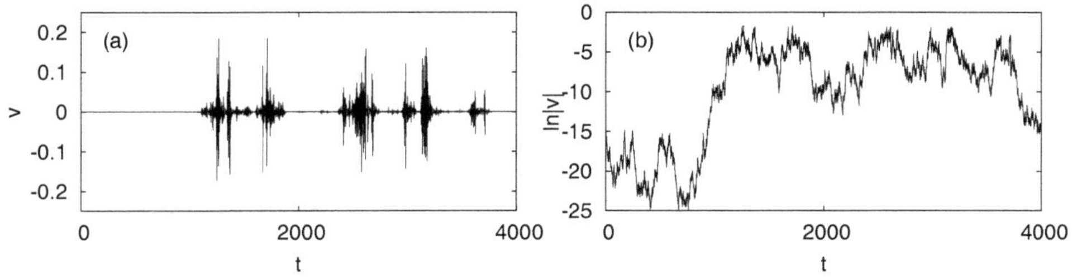

Fig. 11.23 (a) Time evolution of the transverse variable $v = (x - y)/2$ for the skewed tent map with $a = 0.7$ close to the critical coupling, i.e. $\epsilon = \epsilon_c - 10^{-3}$. Note the large excursions interrupted by long quiescent periods. (b) The logarithm $\ln|v|$ displays a diffusive-like behavior (see text).[After Pikovsky and Grassberger (1991)]

e.g., for the skew tent map at $p = 0.7$ (as in Fig. 11.22) we have $\epsilon_c = 0.228\ldots$, as one can verify numerically.[61] Remarkably, we can determine the Lyapunov exponent λ of an unknown chaotic system by finding ϵ_c inverting the above relation [Schuster *et al.* (1986)].

However, $\lambda_\perp$ characterizes only the average behavior and close to the transition, large and irregular fluctuations in the transverse direction can be observed, Fig. 11.23a,b. The transverse dynamics v is actually very intermittent with wild $\mathcal{O}(1)$ burst and essentially synchronized $v \approx 0$ phases. This kind of intermittent dynamics is however rather different from that encountered while discussing the transition to chaos (Sec. 6.3), and is termed *on-off* or *modulational intermittency* [Fujisaka and Yamada (1985, 1986)]. The origin of this behavior can be easily identified by reconsidering the tangent, transverse dynamics (11.43), which can be rewritten for the variable $z = \ln|w_v(t)|$ as

$$z(t+1) = z(t) + \lambda_\perp + \zeta(u(t)) \qquad \text{with} \qquad \zeta(u(t)) = \ln|f'(u(t))| - \lambda\,, \tag{11.47}$$

where again the dynamics of u is independent from that of z and given by the chaotic map $u(t+1) = f(u(t))$. Due to the irregularity of chaotic trajectories we can consider $\zeta(u(t))$ as a random term,[62] and thus interpret the motion of z as a random walk with bias given by $\lambda_\perp$.

Above the transition, $\lambda_\perp > 0$ then the transverse variable grows on average, whereas below the transition $\lambda_\perp < 0$, the transverse variable contracts on average. At the transition ($\lambda_\perp = 0$) the random walk is unbiased. This interpretation is consistent with the observed behavior of $\ln|v|$ shown in Fig. 11.23b, and is at the origin of the observed intermittent behavior of the transverse dynamics. The

[61] In this respect we remark that in evaluating the critical point, typically the synchronization is observed at $\epsilon \lesssim \epsilon_c$. This is due to the finite precision of numerical computations, e.g. in double precision $|x - y| < \mathcal{O}(10^{-16})$ will be treated as zero. To avoid this trouble one can introduce an infinitesimal $\mathcal{O}(10^{-16})$ mismatch in the parameters of the map, so to make the two system not perfectly identical.

[62] Notice that this is possible because we have chosen the generic case of a map with fluctuating growth rate, as the skew tent map for $p \neq 0.5$.

random walk interpretation of the transverse dynamics can also be used to predict the statistics of the time duration τ of the laminar events (Fig. 11.23a) which takes the form $p(\tau) \propto \tau^{-2/3}e^{-\alpha\tau}$, where $\alpha \propto \lambda_\perp^2$ [Heagy *et al.* (1994); Pikovsky *et al.* (2001)].

In general, $\zeta(u(t))$ may have non-trivial time correlations so that the treatment of the random walk (11.47) should be carefully performed, this problem is absent for the skew tent map (11.39) where we have that $\zeta = -\ln p - \lambda$ with probability p and $\zeta = -\ln(1-p) - \lambda$ with probability $1-p$, and λ as in Eq. (11.40), so that the analytical treatment can be done explicitly [Pikovsky and Grassberger (1991); Pikovsky *et al.* (2001)].

The random walk dynamics above introduced can be characterized in terms of the diffusion constant related to the variance of $\zeta(u(t))$. It is now interesting to notice that the latter is determined by the finite time Lyapunov exponent statistics of the map, indeed we can solve Eq. (11.47) as

$$z(T) = z(0) + \lambda_\perp T + \sum_{t=0}^{T-1} \zeta(u(t)) = z(0) + \lambda_\perp T + T(\gamma(T) - \lambda)$$

where $\gamma(T) = 1/T \sum_{t=0}^{T-1} \ln|f'(u(t))| = 1/T \ln(w_v(T)/w_v(0))$ is the finite time LE, notice that $\gamma(T) - \lambda \to 0$ for $T \to \infty$ but, in general, it fluctuates for any finite time. As discussed in Sec. 5.3.3, the distribution of $\gamma(T)$ is controlled by the Cramer function so that $P_T(\gamma) \sim \exp(-S(\gamma)T)$, which can be approximated, near the minimum in $\gamma = \lambda$, by a parabola $S(\gamma) = (\gamma-\lambda)^2/(2\sigma^2)$ and we have $\langle(z(T)-z(0)-\lambda_\perp T)^2\rangle \propto \sigma^2 T$, i.e. σ^2 gives the diffusion constant for the transverse perturbation dynamics.

Close to the transition, where the bias disappear $\lambda_\perp = 0$, $\gamma(T) - \lambda$ can still display positive fluctuations which are responsible for the intermittent dynamics displayed in Fig. 11.23a. For $\epsilon > \epsilon_c$ the steps of the random walk are shifted by $\zeta(u(t)) - |\lambda_\perp|$ so that for large enough ϵ the fluctuations may all become negative. Usually when this happen we speak of *strong synchronization* in contrast with the *weak synchronization* regime, in which fluctuations of the synchronized state are still possible. In particular, for the strong synchronization to establish we need $\lambda_\perp + \gamma_{max} - \lambda = \ln|1 - 2\epsilon_c| + \gamma_{max}$ to become negative, i.e. the coupling should exceed $\epsilon_{max} = (1 - e^{-\gamma_{max}})/2$, γ_{max} being the maximal expanding rate (i.e. the supremum of the support of $S(\gamma)$), which typically coincides with the most unstable periodic orbit of the map.

For instance, for the skew tent map with $p > 1/2$ the Lyapunov exponent is $\lambda = -p\ln p - (1-p)\ln(1-p)$ while the maximal rate $\gamma_{max} = -\ln(1-p) > \lambda$ is associated to the unstable fixed point $x^* = 1/(2-p)$. Therefore, strong synchronization sets in for $\epsilon > \epsilon_{max} = p/2 > \epsilon_c$. Similarly one can define an $\epsilon_{min} < \epsilon_c$ at which the less unstable periodic orbits start to become synchronized, so that we can identify the four regimes displayed in Fig. 11.24.

We remark that in the strongly synchronized regime the diagonal constitutes the attractor of the dynamics as all points of the (x, y)-plane collapse on it remaining

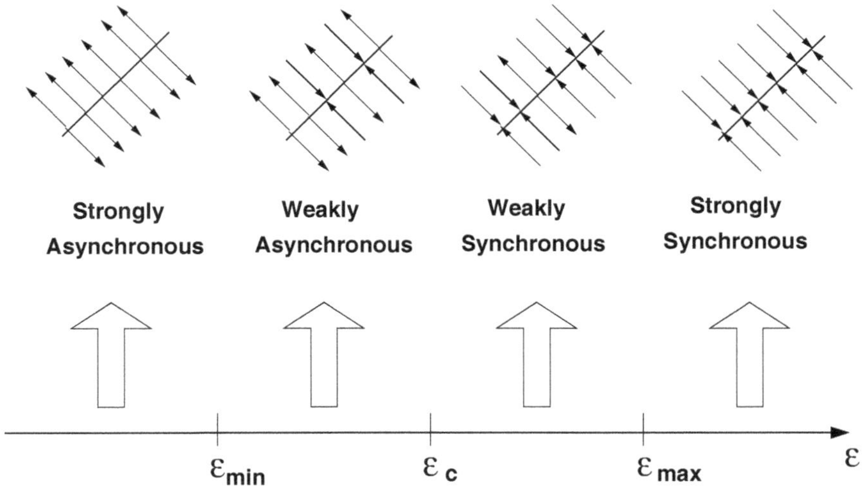

Fig. 11.24 The different regimes: $\epsilon < \epsilon_{min}$ strongly asynchronous all symmetric trajectories are unstable; $\epsilon_{min} < \epsilon < \epsilon_c$ weakly asynchronous an increasing number of symmetric trajectories become stable; $\epsilon_c < \epsilon < \epsilon_{max}$ weakly synchronous, most of symmetric trajectories are stable with few unstable; $\epsilon > \epsilon_{max}$ strongly synchronous, all symmetric trajectories are stable. With symmetric trajectory we mean $x(t) = y(t)$, i.e. a synchronized state. See Pikovsky and Grassberger (1991); Glendinning (2001). [After Pikovsky *et al.* (2001)]

there for ever. In the weakly synchronized case, the dynamics bring the trajectories on the diagonal, but each time a trajectory come close to an unstable orbit with associated Lyapunov exponent larger than the typical one λ, it can escape for a while before coming to it. In this case the attractor is said to be a *Milnor* (probabilistic) *attractor* [Milnor (1985)].

We conclude mentioning that complete synchronization has been experimentally observed in electronic circuits [Schuster *et al.* (1986)] and in lasers [Roy and Thornburg (1994)].

Chapter 12

Spatiotemporal Chaos

> Mathematics is a part of physics. Physics is an experimental science, a part of natural science. Mathematics is the part of physics where experiments are cheap.
>
> *Vladimir Igorevich Arnold*

At variance with low-dimensional systems, many degrees of freedom chaotic systems exhibit behaviors which cannot be easily subsumed under a unified theoretical framework. Without attempting an exhaustive review, this Chapter surveys a few phenomena emerging in high-dimensional, spatially extended, chaotic systems, together with the tools worked up for their characterization.

12.1 Systems and models for spatiotemporal chaos

Many natural systems require a field description in terms of partial differential equations (PDE) or can be modeled in terms of many coupled, discrete elements with their own (chaotic) dynamics. In such high-dimensional systems, unlike low-dimensional ones, chaos becomes apparent not only through the occurrence of temporally unpredictable dynamics but also through unpredictable spatial patterns. For instance, think of a fluid at high Reynolds number, chemical reactions taking place in a tank, or competing populations in a spatial environment, where diffusion, advection or other mechanisms inducing movement are present. There are also high dimensional systems for which the notion of space has no-meaning, but still the complexity of emerging behaviors cannot be reduced to the irregular temporal dynamics only, as e.g. neural networks in the brain.

In the following we briefly introduce the systems and phenomena of interest with emphasis on their modeling. Above we touched upon classes of high dimensional systems described by PDEs or coupled ODEs, we should also mention coupled maps (arranged on a lattice or on a network) or discrete-state models as Cellular Atomata[1]

[1]Even if CA can give rise to very interesting conceptual questions and are characterized by a rich spectrum of behaviors, we will not consider them in this book. The interested reader may consult

Table 12.1 Classification of typical high dimensional models.

Space	**Time**	**State**	**Model**
continuous	continuous	continuous	partial differential equations (PDE)
discrete	continuous	continuous	coupled ordinary differential equations (ODE)
discrete	discrete	continuous	coupled map lattices (CML)
discrete	discrete	discrete	cellular automata (CA)

which have been successfully employed to model chemical or biological interacting units. Table 12.1 summarizes the possible descriptions.

In this Chapter we mostly focus on nonlinear macroscopic systems with a spatial structure and a number of degrees of freedom extensive in the system size. These *extended* dynamical systems can display complex temporal and spatial evolution — *spatiotemporal chaos.* Hohenberg and Shraiman (1989) provided the following definition of the above terms (see also Cross and Hohenberg (1993)). For a generic system of size L, we can define three characteristic scales: the *dissipation* length ℓ_D (scales smaller than ℓ_D are essentially inactive), the *excitation* length ℓ_E (where energy is produced by an external force or by internal instabilities) and a suitably defined *correlation* length ξ. Two limiting cases can be considered:

- **1.** When the characteristic lengths are of the same order ($\ell_D \sim \ell_E \sim \xi \sim \mathcal{O}(L)$) distant portions of the system behave coherently. Consequently, the spatial extension of the system is unimportant and can be disregarded in favor of low-dimensional descriptions, as the Lorenz model is for Rayleigh-Bénard convection (Box B.4). Under these conditions, we only have *temporal* chaos.
- **2.** When $L \geq \ell_E \gg \xi \gg \ell_D$ distant regions are weakly correlated and the number of (active) degrees of freedom, the number of positive Lyapunov exponents, the Kolmogorov-Sinai entropy and the attractor dimension are extensive in the system size [Ruelle (1982); Grassberger (1989)]. Space is thus crucial and spatiotemporal unpredictability may take place as, for instance, in Rayleigh-Bérnad convection for large aspect ratio [Manneville (1990)].

12.1.1 *Overview of spatiotemporal chaotic systems*

The above picture is approximate but broad enough to include systems ranging from fluid dynamics and nonlinear optics to biology and chemistry, some of which will be illustrated in the following.

12.1.1.1 *Reaction diffusion systems*

In Sec. 11.3 we mentioned examples of chemical reactions and population dynamics, taking place in a homogeneous environment, that generate temporal chaos.

Wolfram (1986); Badii and Politi (1997); Jost (2005).

In the real world, concentration of chemical or biological species are not space-homogeneous, and can diffuse in space, thus a proper description of these Reaction-Diffusion (RD) systems requires to consider PDEs such as

$$\partial_t \boldsymbol{c}(\boldsymbol{x},t) = \mathbb{D}\Delta \boldsymbol{c}(\boldsymbol{x},t) + \boldsymbol{f}(\boldsymbol{c}(\boldsymbol{x},t)) \tag{12.1}$$

where $\boldsymbol{c} = (c_1, \ldots, c_N)$ represents the concentration of N reagents, $\boldsymbol{x} = (x_1, \ldots, x_d)$ the coordinates in a d-dimensional space, $\mathbb{D}_{ij}$ the $(N \times N)$-diffusive matrix and, finally, $\boldsymbol{f} = (f_1, \ldots, f_N)$. the chemical kinetics or the biological interactions.

It is well established that PDEs like Eq. (12.1) may give rise to complex spatiotemporal evolutions from travelling patterns to spatiotemporal chaos. For instance, traveling-front solutions characterize FKPP-like dynamics, from Fisher (1937) and Kolmogorov, Petrovskii and Piskunov 1937, which in $d = 1$ is obtained taking $f(c) \propto c(1 - c)$, i.e. a sort of spatial logistic equation. Since Turing (1953), the competition between reaction and diffusion is known to generate nontrivial patterns. See, for instance, the variety of patterns arising in the Belousov-Zhabotinsky reaction (Fig. 12.1). Nowadays, many mechanisms for the generation of patterns have been found — *pattern formation* [Cross and Hohenberg (1993)] — relevant to many chemical [Kuramoto (1984); Kapral and Showalter (1995)] and biological [Murray (2003)] problems.

Patterns arising in RD-systems may be stationary, periodic and temporally [Tam and Swinney (1990); Vastano *et al.* (1990)] or spatiotemporally chaotic. For instance, pattern disruption by defects provides a mechanism for spatiotemporally unpredictable behaviors, see for instance the various ways spatiotemporal chaos may emerge from spiral patterns, [Ouyang and Swinney (1991); Ouyang and Flesselles (1996); Ouyang *et al.* (2000); Zhan and Kapral (2006)]. Thus, RD-systems constitute a typical experimental and theoretical framework to study spatiotemporal chaos.

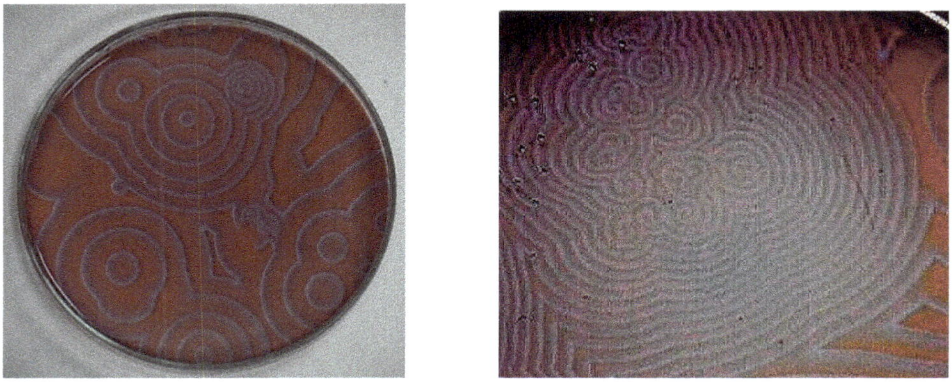

Fig. 12.1 Patterns generated by the reagents of the Belousov-Zhabotinsky reaction taking in a "capsula-Petri", without stirring. We can recognize target patterns (left), spirals and broken spirals (right). [Courtesy of C. López and E. Hernández-García.]

12.1.1.2 *Rayleigh Bénard convection*

Rayleigh-Bénard convection (see also Box B.4) has long been a paradigm for pattern formation and spatiotemporal chaos [Cross and Hohenberg (1993)]. The system consists of a horizontal layer of fluid heated from below and is characterized by three dimensionless parameters: the Rayleigh number Ra (which is proportional to the temperature difference, see Eq. (B.4.2)) and the Prandtl number Pr (the ratio between fluid viscosity and thermal diffusivity) specify the fluid properties; while the system geometry is controlled by the aspect ratio $\Gamma \equiv L/d$, where L and d are the horizontal and vertical size of the system, respectively.

Different dynamical regimes are observed at varying the control parameters. When Ra is larger than the critical value for the stability of the conduction state, but the aspect ratio is small, the system organizes in a regular pattern of convective rolls where chaos can manifest in the temporal dynamics [Maurer and Libchaber (1979, 1980); Gollub and Benson (1980); Giglio *et al.* (1981); Ciliberto and Rubio (1987)], well captured by low-dimensional models (Box B.4).

At increasing the aspect ratio, but keeping Ra small, the ordered patterns of convective rolls destabilize and organize similarly to the patterns observed in RD-systems [Meyer *et al.* (1987)]. Different types of pattern may compete creating defects [Ciliberto *et al.* (1991); Hu *et al.* (1995)]. For example, Fig. 12.2 illustrates an experiment where spirals and striped patterns compete, this regime is usually dubbed spiral defect chaos and is one of the possible ways spatiotemporal chaos manifests in Rayleigh-Bénard convection [Cakmur *et al.* (1997)]. Typically, defects constitute the seeds of spatiotemporal disordered evolutions [Ahlers (1998)], which, when fully developed, are characterized by a number of positive Lyapunov exponents that increases with the system size [Paul *et al.* (2007)]. For a review of experiments in various conditions see Bodenschatz *et al.* (2000).

Fig. 12.2 Competition and coexistence of spiral defect chaos and striped patterns in Rayleigh-Bénard convection. [Courtesy of E. Bodenschatz]

12.1.1.3 *Complex Ginzburg-Landau and Kuramoto-Sivanshisky equations*

When close to an instability or a bifurcation point (e.g. a Hopf bifurcation) the dynamics of spatially extended systems can be decomposed in slow and fast components. By applying standard perturbative approaches [Kuramoto (1984); Cross and Hohenberg (1993)], universal equations for the slow component can be derived. These PDEs, often called amplitude equations, provide a mathematical lab for studying many spatiotemporal phenomena. Among these equations, an important position is occupied by the Complex Ginzburg-Landau equation (CGLE) which is relevant to nonlinear optics, liquid crystals, superconductors and other systems, see Aranson and Kramer (2002) and references therein. The CGLE is usually written as

$$\partial_t A = A + (1 + ib)\Delta A - (1 + ic)|A|^2 A\,, \tag{12.2}$$

where $A(\boldsymbol{x}, t) = |A(\boldsymbol{x}, t)|e^{i\phi(\boldsymbol{x},t)}$ is a complex field, Δ the Laplacian and b and c two parameters depending on the original system. Already in one spatial dimension, depending on the values of b and c, a variety of behaviors is observed from periodic waves and patterns to spatiotemporal chaotic states of various nature:
Phase chaos (Fig. 12.3a) characterized by a chaotic evolution in which $|A| \neq 0$ everywhere, so that the well-defined phase drives the spatiotemporal evolution;
Defect (or *amplitude*) *chaos* (Fig. 12.3b) in which defects, i.e. places where $|A| = 0$ and thus the phase is indeterminate, are present (for the transition from phase to defect chaos see Shraiman *et al.* (1992); Brusch *et al.* (2000));
Spatiotemporal intermittency (Fig. 12.3c) characterized by a disordered alternation of space-time patches having $|A| \approx 0$ or $|A| \sim \mathcal{O}(1)$ [Chaté (1994)].

In two dimensions, the CGLE displays many of the spatiotemporal behaviors observed in RD-systems and Rayleigh-Bénard convection [Chaté and Manneville

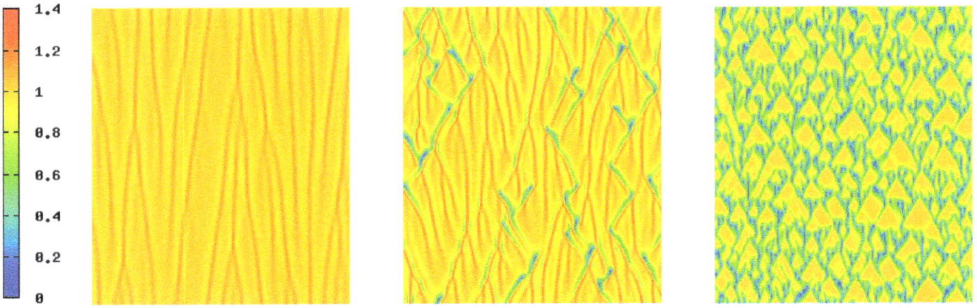

(a) (b) (c)

Fig. 12.3 Spatiotemporal evolution of the amplitude modulus $|A|$ according to the CGLE: regime of phase turbulence (a), amplitude or defect turbulence (b) and spatiotemporal intermittency (c). Time runs vertically and space horizontally. Data have been obtained by means of the algorithm described in Torcini *et al.* (1997). [Courtesy of A. Torcini]

Fig. 12.4 Spiral breakup in the two-dimensional CGLE. [Courtesy of A. Torcini]

(1996)], and spatiotemporal chaos is typically accompanied and caused by the breaking of basic patterns (Fig. 12.4).

As phase chaos involves the phase $\phi(x,t)$ dynamics only, Eq. (12.2) can be simplified in an equation for the gradient of the phase, which in $d = 1$ reads ($u = \partial_x \phi$) [Kuramoto (1984)]

$$\partial_t u = -\partial_x^2 u - \partial_x^4 u + u \partial_x u \,,$$

whose unique control parameter is the system size L. This is the Nepomnyashchy-Kuramoto-Sivashinsky equation,[2] as it was independently derived by Nepomnyashchy (1974), for describing the free surface of a fluid falling down an inclined plane, by Kuramoto and Tsuzuki (1976), in the framework of chemical reactions, and Sivashinsky (1977), in the context combustion flame propagation. For $L \gg 1$, the Nepomnyashchy-Kuramoto-Sivashinsky equation displays spatiotemporal chaos, whose basic phenomenology can be understood in Fourier space, i.e., using $\hat{u}(k,t) = (1/L) \int_0^L \mathrm{d}x \; u(x,t) \exp(-ikx)$, so that it becomes:

$$\frac{\mathrm{d}\hat{u}(k,t)}{\mathrm{d}t} = k^2(1-k^2)\hat{u}(k,t) - i \sum_p p\hat{u}(k-p,t)\hat{u}(p,t) \,.$$

We can readily see that at $k < 1$ the linear term on the r.h.s. becomes positive leading to a large-scale instability while small scales ($k > 1$) are damped by diffusion and thus regularized. The nonlinear term preserves the total energy $\int_0^L \mathrm{d}x \; |u(x,t)|^2$ and redistributes it among the modes. We have thus an internal instability driving the large scales, dissipation at small scales and nonlinear/chaotic redistribution of energy among the different modes [Kuramoto (1984)].

[2]Typically, in the literature it is known as the Kuramoto-Sivashinsky, indeed the contribution of Nepomnyashchy was recognized only much later.

12.1.1.4 *Coupled map lattices*

We can think of spatiotemporally chaotic systems as an ensemble of weakly coupled chaotic systems distributed in space. In this perspective, at the beginning of '80s, Kuznetsov (1983), Kaneko (1984) and Kapral (1985) introduced coupled map lattice (CML) models (see Kaneko (1993); Chazottes and Fernandez (2004)), which can be considered a prototype for chaotic extended systems.

In a nutshell, CMLs consist in a (regular) lattice, say in one spatial dimension ($d = 1$), with L sites, $i = 1, \dots, L$, to each of which is associated a discrete-time state-vector $\boldsymbol{u}_i(t) = (u_i^1, \dots, u_i^D)$, where D is the number of variables necessary to describe the state. On this $(D \times L)$-dimensional phase space, we can then define a dynamics such as

$$\boldsymbol{u}_i(t+1) = \sum_j \epsilon_{ij} \boldsymbol{f}(\boldsymbol{u}_j(t)) \,, \tag{12.3}$$

where $\boldsymbol{f}$ is a D-dimensional map, e.g. the logistic ($D=1$) or Hénon ($D=2$) map, and ϵ_{ij} is the coupling matrix among different sites (chaotic units). One of the most common choice is nearest-neighbor (diffusive) coupling ($\epsilon_{ij} = 0$ for $|i-j| > 1$) that, with symmetric coupling and for $D = 1$, can be written as:

$$u_i(t+1) = (1-\epsilon) f(u_i(t)) + \frac{\epsilon}{2} \left[f(u_{i-1}(t)) + f(u_{i+1}(t)) \right] \,. \tag{12.4}$$

This equation is in the form of a discrete Laplacian mimicking, in discrete time and space, a Reaction-Diffusion system. Of course, other symmetrical, non-symmetrical or non-nearest-neighbor types of coupling can be chosen to model a variety of physical situations, see the examples presented in the collection edited by Kaneko (1993).

Tuning the coupling strength ϵ and the nonlinear parameters of f, a variety of behaviors is observed (Fig. 12.5): from space-time patterns to spatiotemporal chaos similar to those found in PDEs. Discrete-time models are easier to study and surely easier to simulate on a computer, therefore in the following sections we shall discuss spatiotemporal chaos mostly relying on CMLs.

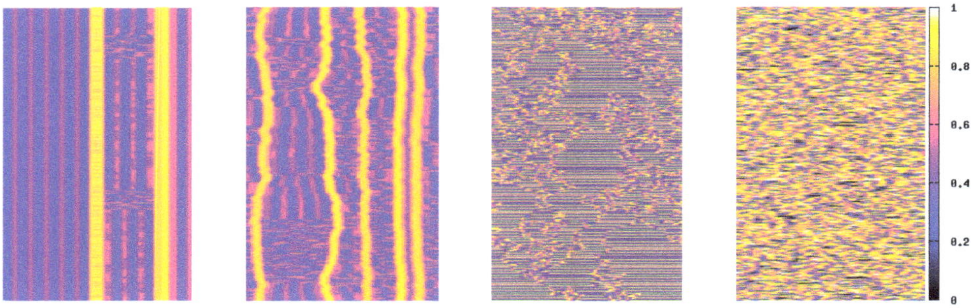

(a) (b) (c) (d)

Fig. 12.5 Typical spatiotemporal evolutions of CML (12.4) with $L=100$ logistic maps, $f(x) = rx(1-x)$: (a) spatially frozen regions of chaotic activity ($r=3.6457$ and $\epsilon=0.4$); (b) spatiotemporal irregular wandering patterns ($r=3.6457$ and $\epsilon=0.5$); (c) spatiotemporal intermittency ($r=3.829$ and $\epsilon=0.001$); (d) fully developed spatiotemporal chaos ($r=3.988$ $\epsilon=1/3$).

12.1.1.5 *Nonlinear lattices: the Fermi-Pasta-Ulam model*

Chaotic spatiotemporal evolutions characterize also high-dimensional conservative systems, which are important also for statistical mechanics. For instance, a solid can be modeled in terms of a lattice of weakly coupled, slightly non-harmonic oscillators. In such models, the goal is to derive thermodynamics and (heat) transport properties from the nonlinear microscopic dynamics, as in the much studied FPU model, from the seminal work of Fermi, Pasta and Ulam (1955), defined by the Hamiltonian

$$H(\boldsymbol{q},\boldsymbol{p}) = \sum_{i=1}^{L} \left[\frac{p_i^2}{2} + V(q_i - q_{i-1}) \, , \right]$$

where p_i, q_i indicate the momentum and coordinate of the mass $m = 1$ oscillator at each lattice site $i = 1, \ldots, L$ and $V(x) = x^2/2 + \beta x^4/4$ is a nonlinear deformation of the harmonic potential. For $\beta \neq 0$, the normal modes are coupled by the nonlinearity and we can wonder about the spatiotemporal propagation of energy [Lepri *et al.* (2003)]. Despite its simplicity, the FPU-model presents many interesting and still poorly understood features [Gallavotti (2007)], which are connected to important aspects of equilibrium and non-equilibrium statistical mechanics. For this reason, we postpone its discussion to Chapter 14. Here, we just note that this extended system is made of coupled ODEs and mention that conservative high dimensional systems include Molecular Dynamics models [Ciccotti and Hoover (1986)], which stand at the basis of microscopical (computational and theoretical) approaches to fluids and gases, within classical mechanics.

12.1.1.6 *Fully developed turbulence*

Perhaps, the most interesting and studied instance of high-dimensional chaotic system is constituted by the Navier-Stokes equation which rules the evolution of fluid velocity fields. We have already seen that increasing the control parameter of the system, namely the Reynolds number Re, the motion undergoes a series of bifurcations with increasingly disordered temporal behaviors ending in an unpredictable spatiotemporal evolution for Re $\gg 1$, which is termed *fully developed turbulence* [Frisch (1995)]. Although fully developed turbulence well fit the definition of spatiotemporal chaos given at the beginning of the section, we prefer to treat it separately from the kind of models considered above, part for a tradition based on current literature and part for the specificity of turbulence and its relevance to fluid mechanics. The next Chapter is devoted to this important problem.

12.1.1.7 *Delayed ordinary differential equations*

Another interesting class of systems that generate "spatio"-temporal behaviors is represented by time-delayed differential equations, such as

$$\frac{\mathrm{d}\boldsymbol{x}}{\mathrm{d}t} = \boldsymbol{f}(\boldsymbol{x}(t), \boldsymbol{x}(t-\tau)) \, , \tag{12.5}$$

where the evolution depends on both the current state and the state at some past time $t-\tau$. Equations of this type occur, for instance, in nonlinear optics when considering lasers with delayed feedback mechanisms [Ikeda and Matsumoto (1987); Arecchi *et al.* (1991, 1992)], or in modeling biological processes such as hemopoiesis or respiration [Mackey and Glass (1977)]. Equation (12.5) defines an infinite dimensional dynamical system, indeed the evolution depends on the state vectors in the time-window $[t-\tau : t]$, and time is continuous. An explicit integration scheme for Eq. (12.5) spotlights the connection with spatiotemporal chaotic systems. For instance, consider the simple Euler integration scheme with time step dt, in terms of which Eq. (12.5) is approximated by the finite difference equation $\boldsymbol{x}(t+\mathrm{d}t) = \boldsymbol{x}(t) + \boldsymbol{f}(\boldsymbol{x}(t), \boldsymbol{x}(t-\tau))\mathrm{d}t$. By reabsorbing $\mathrm{d}t$ in the definition of $\boldsymbol{f}$, such an equation can be rewritten as the N-dimensional mapping [Farmer (1982)]

$$\begin{aligned} \boldsymbol{x}_1(k+1) &= \boldsymbol{x}_N(k) + \boldsymbol{f}(\boldsymbol{x}_N(k), \boldsymbol{x}_1(k)) \\ &\ \vdots \\ \boldsymbol{x}_j(k+1) &= \boldsymbol{x}_{j-1}(k) + \boldsymbol{f}(\boldsymbol{x}_{j-1}(k+1), \boldsymbol{x}_j(k)) \end{aligned} \tag{12.6}$$

where $j = 2, \ldots, N$, $\mathrm{d}t = \tau/(N-1)$ and the generic term $\boldsymbol{x}_i(k)$ corresponds to $\boldsymbol{x}(t = i\mathrm{d}t + k\tau)$, with $i = 1, \ldots, N$. The system (12.6) is an asynchronously updated, one-dimensional CML of size N, where the non-local in time interaction has been converted into a local in (fictitious) space coupling. The tight connection with spatiotemporal chaotic systems has been pushed forward in several studies [Arecchi *et al.* (1992); Giacomelli and Politi (1996); Szendro and López (2005)].

12.1.2 *Networks of chaotic systems*

For many high-dimensional chaotic systems the notion of space cannot be properly defined. For example, consider the CML (12.3) with $\epsilon_{ij} = 1/L$, this can be seen either as a mean field formulation of the diffusive model (12.4) or as a new class of non-spatial model. Similar mean field models can be also constructed with ODEs, as in the coupled oscillators Kuramoto (1984) model, much studied for synchronization problems. One can also consider models in which the coupling matrix ϵ_{ij} has non-zero entries for arbitrary distances between sites $|i-j|$, making appropriate a description in terms of a network of chaotic elements.

A part from Sec. 12.5.2, in the sequel we only discuss systems where the notion of space is preponderating. However, we mention that nonlinear coupled systems in network topology are nowadays very much studied in important contexts such as neurophysiology. In fact, the active unit of the brain — the neurons — are modeled in terms of nonlinear single units (of various complexity from simple integrate and fire models [Abbott (1999)] to Hodgkin and Huxley (1952) or complex compartmental models [Bower and Beeman (1995)]) organized in complex highly connected networks. It is indeed estimated that in a human brain there are $\mathcal{O}(10^9)$ neurons and each of them is coupled to many other neurons through $\mathcal{O}(10^4)$ synapses.

12.2 The thermodynamic limit

Lyapunov exponents, attractor dimensions and KS-entropies can be defined (and, at least the LEs, numerically computed) also for extended systems. An issue of particular interest is to understand the behavior of such quantities as the system size L increases. To illustrate the basic ideas we consider here the simplest setting as provided by a diffusively coupled one-dimensional lattice of maps

$$u_i(t+1) = (1-\epsilon) f(u_i(t)) + \frac{\epsilon}{2}\left[f(u_{i-1}(t)) + f(u_{i+1}(t))\right] , \qquad (12.7)$$

with $i = 1, \ldots, L$, and periodic boundary conditions, $u_{L+1} = u_1$ and $u_{-1} = u_L$.

For $L < \infty$, the system has a finite dimensional phase space and LEs,[3] KS-entropy and fractal dimensions are well defined. However, to build a statistical description of spatiotemporal chaos, as pointed out by Ruelle (1982), we should require that the phenomenology of these systems does not depend on their size L and thus the existence of the thermodynamic limit for the Lyapunov spectrum

$$\lim_{L\to\infty} \lambda_i(L) = \Lambda(x = i/L) \qquad x \in [0:1] ,$$

with $\Lambda(x)$ a non-increasing function, defining the *density* of Lyapunov exponents.

The density $\Lambda(x)$ can be analytically computed in a simple, pedagogical example. Consider the evolution of L tangent vectors associated to the CML (12.7)

$$w_j(t+1) = (1-\epsilon) f'(u_j(t)) w_j(t) + \frac{\epsilon}{2}\left[f'(u_{j-1}(t)) w_{j-1}(t) + f'(u_{j+1}(t)) w_{j+1}(t)\right] . \quad (12.8)$$

For the generalized shift map, $f(x) = rx \mod 1$, as $f'(u) = r$, the tangent evolution simplifies in

$$w_j(t+1) = r\left[(1-\epsilon) w_j(t) + \frac{\epsilon}{2}\left(w_{j-1}(t) + w_{j+1}(t)\right)\right] ,$$

which can be solved in terms of plane-waves, $w_j(t) = \exp(\lambda_p t + i k_p j)$ with $k_p = (p-1)\, 2\pi/L$ and $p = 1, \ldots, L$, obtaining the Lyapunov density

$$\Lambda(x) = \ln r + \ln|1 - \epsilon + \epsilon\cos(2\pi x)| .$$

In this specific example the tangent vectors are plane waves and thus are homogeneously spread in space (see also Isola *et al.* (1990)), in general this is not the case and spatial localization of the tangent vectors may take place, we shall come back this phenomenon later.

For generic maps, the solution of Eq. (12.8) cannot be analytically obtained and numerical simulations are mandatory. For example, in Fig. 12.6a we show $\Lambda(x)$ for a CML of logistic maps, $f(x) = rx(1-x)$, with parameters leading to a well-defined thermodynamic limit, to be contrasted with Fig. 12.6c, where it does not, due to the frozen chaos phenomenon [Kaneko (1993)]. In the latter case, chaos localizes in a specific region of the space, and is not extensive in the system size: the step-like structure of $\Lambda(x)$ relates to the presence of frozen regular regions (Fig. 12.5a).

[3] For $L \to \infty$, Lyapunov exponents may depend on the chosen norm [Kolmogorov and Fomin (1999)]. We shall see in Sec. 13.4.3 that this is not just a subtle mathematical problem.

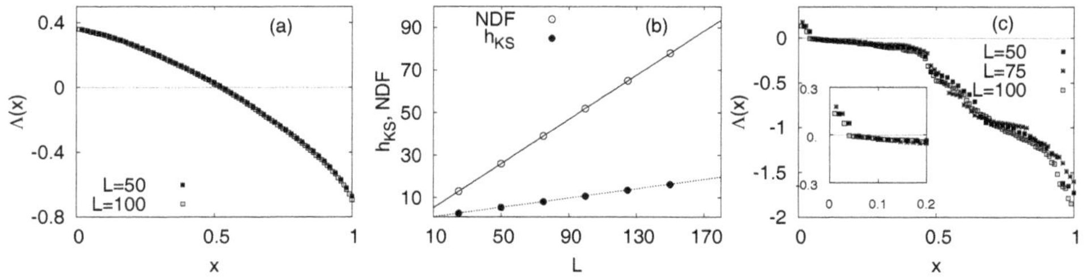

Fig. 12.6 (a) Density of Lyapunov exponents $\Lambda(x = i/L)$ for a CML of logistic maps with $r = 3.988$ and $\epsilon = 0.3$, notice the collapse on a L-independent function. This is the typical behavior of $\Lambda(x)$ in fully developed spatiotemporal chaos (Fig. 12.5d). (b) For the same system as in (a) Linear scaling of the number of degrees of freedom (NDF), i.e. number of positive Lyapunov exponents, and of the Kolmogorov-Sinai entropy h_{KS}, computed through the Pesin relation (8.23), with the system size. (c) Same as (a) for $r = 3.6457$ and $\epsilon = 0.4$, corresponding to the regime of frozen chaos Fig. 12.5a. Inset: zoom of the region close to the origin.

Once the existence of a Lyapunov density is proved, some results of low dimensional systems, such as the Kaplan and Yorke conjecture (Sec. 5.3.4) and the Pesin relation (Sec. 8.4.2), can be easily generalized to spatially extended chaotic systems [Ruelle (1982); Grassberger (1989); Bunimovich and Sinai (1993)]. It is rather straightforward to see that the Pesin relation (see Eq. (8.23)) can be written as

$$\mathcal{H}_{KS} = \lim_{L\to\infty} \frac{h_{KS}}{L} = \int_0^1 \mathrm{d}x\, \Lambda(x)\, \Theta(\Lambda(x))$$

$\Theta(x)$ being the step function. In other words we expect the number of positive LEs and the KS-entropy to be linear growing functions of L as shown in Fig. 12.6b. In the same way, the *dimension density* $\mathcal{D}_F = \lim_{L\to\infty} D_F/L$ can be obtained through the Kaplan and Yorke conjecture (Sec. 5.3.4) which reads

$$\int_0^{\mathcal{D}_F} \mathrm{d}x\, \Lambda(x) = 0\,.$$

The existence of a good thermodynamic limit is supported by numerical simulations [Kaneko (1986); Livi *et al.* (1986)] and exact results [Sinai (1996)]. For instance, Collet and Eckmann (1999) proved the existence of a density of degrees of freedom in the CGLE (12.2), as observed in numerical simulations [Egolf and Greenside (1994)].

We conclude the section with a couple of remarks. Figure 12.7 shows the behavior of the maximal LE for a CML of logistic maps as a function of the nonlinear parameter, r. The resulting curve is rather smooth, indicating a regular dependence on r, which contrasts with the non-smooth dependence observed in the uncoupled map. Thus the presence of many degrees of freedom has a "regularization" effect, so that large systems are usually structurally more stable than low dimensional ones [Kaneko (1993)] (see also Fig. 5.15 in Sec. 5.3 and the related discussion). Another aspect worth of notice is that, as for low-dimensional chaotic dynamics, temporal auto-correlation functions $\langle u_i(t)u_i(t+\tau)\rangle$ of the state $u_i(t)$ in a generic

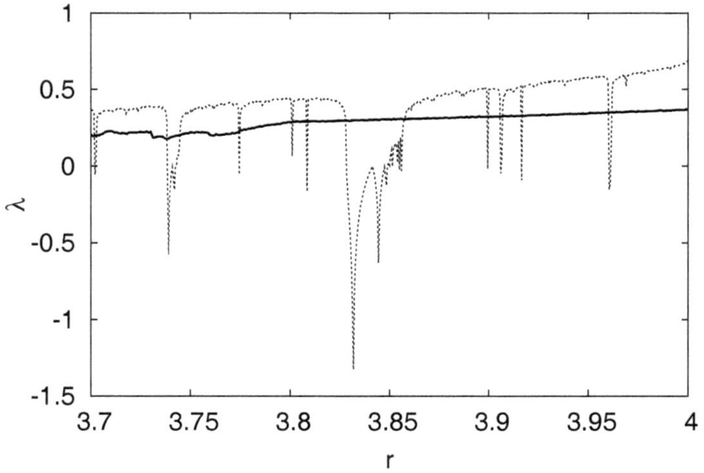

Fig. 12.7 λ_1 vs r for a CML, with $L = 100$ and $\epsilon = 0.2$, of logistic maps (solid line) compared with the same quantity for the single logistic map (dotted line).

site i of an extended chaotic system typically decay, indicating memory-loss of the initial condition, and thus a well-defined temporal chaoticity. Similarly, in order to have a well-defined spatiotemporal chaoticity, also spatial correlation functions, $\langle u_i(t)u_{i+k}(t)\rangle$, must decay [Bunimovich and Sinai (1993)].

12.3 Growth and propagation of space-time perturbations

12.3.1 *An overview*

In low dimensional systems, no matter how the initial (infinitesimal) disturbance is chosen, after a — usually short — relaxation time, T_R, the eigendirection associated to the maximal growth rate dominates for almost all initial conditions (Sec. 5.3) [Goldhirsch *et al.* (1987)]. On the contrary, in high-dimensional systems this is not necessarily true: when many degrees of freedom are present, different choices for the initial perturbation are possible (e.g., localized or homogeneous in space), and it is not obvious that the time T_R the tangent vectors take to align along the maximally expanding direction is the same for all of them [Paladin and Vulpiani (1994)].

In general, the phenomenology can be very complicated. For instance, also for homogeneous disturbances, the tangent-space dynamics may lead to localized tangent vectors [Kaneko (1986); Falcioni *et al.* (1991)], by mechanisms similar to Anderson localization of the wave function in disordered potential [Isola *et al.* (1990); Giacomelli and Politi (1991)], or to wandering weakly localized structures (see Sec. 12.4.1) [Pikovsky and Kurths (1994a)]. Of course, this severely affects the prediction of the future evolution leading to the coexistence of regions characterized by long or short predictability times [Primo *et al.* (2007)]. For instance, for initially localized disturbances, the main contribution to the predictability time comes from the time the perturbation takes to propagate through the system or to align along the maximally expanding direction, which can be of the order of the system size [Paladin and Vulpiani (1994)].

Standard Lyapunov exponents are typically inadequate to account for perturbations with particular space-time shapes. Thus a number of new indicators have been introduced such as: *temporal* or *specific* [Politi and Torcini (1992)] and *spatial* LEs [Giacomelli and Politi (1991)], characterizing perturbations exponentially shaped in space and time, respectively; the *comoving* LE [Kaneko (1986); Deissler (1987)] accounting for the spatiotemporal evolution of localized perturbations; (local or) *boundary* LE [Pikovsky (1993); Falcioni *et al.* (1999)] which is particularly relevant to asymmetric systems where convective instabilities can present [Deissler (1987); Deissler and Kaneko (1987); Aranson *et al.* (1988)]. Some of these indicators are connected and can be presented in a unified formulation [Lepri *et al.* (1996, 1997)].

Extended systems are often characterized by the presence of long-lived coherent structures, which move maintaining their shape for rather long times. Although predicting their evolution can be very important, e.g. think of cyclonic/anti-cyclonic structures in the atmosphere, it is not clear how to do it, especially due to the dependence of the predictability time on the chosen norm. Hence, often it is necessary to adopt *ad hoc* treatments, based on physical intuition (see, e.g., Sec. 13.4.3).

12.3.2 *"Spatial" and "Temporal" Lyapunov exponents*

Let us consider a generic CML such as

$$u_i(t+1) = f\left((1-\epsilon)u_i(t) + \frac{\epsilon}{2}(u_{i-1}(t) + u_{i+1}(t))\right) , \tag{12.9}$$

with periodic boundary conditions in space.[4] Following Giacomelli and Politi (1991); Politi and Torcini (1992) (see also Lepri *et al.* (1996, 1997)) we now consider generic perturbations with an exponential profile both in space and in time, i.e.[5]

$$|\delta u_i(t)| \propto e^{\mu i + \lambda t} ,$$

the profile being identified by the spatial μ and temporal λ rates. Studying the growth rate of such kind of perturbations requires to introduce *temporal* (or *specific*) and *spatial* Lyapunov exponents. For the sake of simplicity, we treat spatial and temporal profiles separately.

We start considering infinitesimal, exponentially shaped (in space) perturbations $\delta u_i(t) = \Phi_i(t)\exp(\mu i)$, with modified boundary condition in tangent space, i.e.

[4] Equation (12.9) is equivalent to Eq. (12.7) through the change of variables $v_i(t) = f(u_i(t))$

[5] Through this and the next sections with some abuse of notation we shall denote different generalizations of the Lyapunov exponents with the same symbol $\lambda(\cdot)$, whose meaning should be clear from the argument. Notice that also the symbol λ without any argument appears here, this should be interpreted as the imposed exponential rate in the temporal profile of the perturbation.

$\delta u_{L+1}(t)=\mathrm{e}^{\mu L}\delta u_0(t)$. The evolution of $\Phi_i(t)$ then reads [Politi and Torcini (1992)]

$$\Phi_i(t+1) = m_i(t)\left[\frac{\epsilon}{2}\mathrm{e}^{-\mu}\Phi_{i-1}(t) + (1-\epsilon)\Phi_i(t) + \frac{\epsilon}{2}\mathrm{e}^{\mu}\Phi_{i+1}(t)\right],$$

with $m_i(t) = f'[(1-\epsilon)u_i(t) + \epsilon/2(u_{i-1}(t)+u_{i+1}(t))]$. For each value of μ, we can compute the *temporal* or *specific* Lyapunov spectrum $\lambda_i(\mu)$ with $i = 1,\ldots,L$. A typical perturbation with an exponential profile of rate μ is amplified/decreased as

$$|\delta u_i(t)| \approx |\delta u_i(0)|\mathrm{e}^{\lambda_1(\mu)t}, \tag{12.10}$$

where $\lambda_1(\mu)$ is the maximal specific LE associated to the perturbation with spatial rate μ. In well behaving extended systems, a density of such exponents can be defined in the thermodynamic limit [Lepri *et al.* (1996, 1997)],

$$\lambda(\mu, n_\lambda) = \lambda_j(\mu) \quad \text{with} \quad n_\lambda = j/L. \tag{12.11}$$

Notice that for $\mu = 0$ the usual Lyapunov spectrum is recovered. For an extension to PDEs see Torcini *et al.* (1997).

Now we consider perturbations exponentially decaying in time, $\delta u_i(t) = \exp(\lambda t)\Psi_i(t)$, whose evolution is characterized by the *spatial* LE [Giacomelli and Politi (1991)]. In this case the tangent dynamics reads

$$\Psi_i(t+1) = \mathrm{e}^{-\lambda}m_i(t)\left[\frac{\epsilon}{2}\Psi_{i-1}(t) + (1-\epsilon)\Psi_i(t) + \frac{\epsilon}{2}\Psi_{i+1}(t)\right],$$

and can be formally solved via a transfer matrix approach applied to the equivalent spatio-temporal recursion [Lepri *et al.* (1996)]

$$\theta_{i+1}(t)=\Psi_i(t) \text{ and } \Psi_{i+1}(t)=-2\frac{(1-\epsilon)}{\epsilon}\Psi_i(t)+\frac{2\mathrm{e}^{\lambda}}{\epsilon m_i(t)}\Psi_i(t+1) - \theta_i(t). \tag{12.12}$$

The computation of the spatial LE is rather delicate and requires prior knowledge of the multipliers $m_i(t)$ along the whole trajectory. Essentially the problem is that to limit the memory storage resources, one is forced to consider only trajectories with a finite length T and to impose periodic boundary conditions in the time axis, $\theta_i(T+1) = \theta_i(1)$ and $\Psi_i(T+1) = \Psi_i(1)$. We refer to Giacomelli and Politi (1991) (see also Lepri *et al.* (1996)) for details. Similarly to the specific LE, for $T \to \infty$ a density spatial Lyapunov exponent can be defined

$$\mu(\lambda, n_\mu) = \mu_j(\lambda) \quad \text{with} \quad n_\mu = \frac{j-1/2}{T} - 1, \tag{12.13}$$

where $(j-1/2)/T - 1$ ensures $n_\mu = 0$ to be a symmetry center independently of T.

Expressing Eqs. (12.11) and (12.13) in terms of the densities $n_\lambda(\mu,\lambda)$ and $n_\mu(\mu,\lambda)$ the (μ,λ)-plane is completely characterized. Moreover, these densities can be connected through an entropy potential, posing the basis of a thermodynamic approach to spatiotemporal chaos [Lepri *et al.* (1996, 1997)].

We conclude the section mentioning that spatial LE can be used to the characterize the spatial localization of tangent vectors. We sketch the idea in the following. The localization phenomenon is well understood for wave functions in disordered systems [Anderson (1958)]. Such problem is exemplified by the discrete version of the two-dimensional (infinite strip) Schrödinger equation with random potential [Crisanti *et al.* (1993b)], in the tight-binding approximation,

$$\psi_{n,m+1} + \psi_{n,m-1} + \psi_{n-1,m} + \psi_{n+1,m} = (E - V_{n,m})\psi_{n,m} \,, \tag{12.14}$$

where $n = 1, 2 \ldots, \infty$ and $m = 1, \ldots, D$ (D being the vertical size of the strip, with imposed periodic boundary conditions), $\psi_{m,n}$ and $V_{m,n}$ are the wave function and random potential, respectively. Theory predicts the existence of exponentially localized wave functions for arbitrary small disorder [Crisanti *et al.* (1993b)]. The idea is now to establish an analogy between Eq. (12.14) and (12.12), noticing that, in the latter, time can be seen as a space index and assuming that chaotic fluctuations of $m_i(t)$ play the role of the random potential $V_{n,m}$. With this analogy $\mu(\lambda, 0)$ can be interpreted as the inverse of the localization length of the tangent vector associated to λ, provided λ belongs to the LE spectrum [Giacomelli and Politi (1991); Lepri *et al.* (1996)].

12.3.3 *The comoving Lyapunov exponent*

In spatiotemporal chaotic systems, generic perturbations not only grow in time but also propagate in space [Grassberger (1989)]. A quantitative characterization of such propagation can be obtained in terms of the *comoving* Lyapunov exponent,[6] generalizing the usual LE to a non stationary frame of reference [Kaneko (1986); Deissler (1987); Deissler and Kaneko (1987)].

We consider CMLs (extension to the continuous time and space being straightforward [Deissler and Kaneko (1987)]) with an infinitesimally small perturbation localized on a single site of the lattice.[7] The perturbation evolution along the line $i(t) = [vt]$ ($[\cdot]$ denoting the integer part) in the space-time plane defined by is expected to behave as

$$|\delta u_i(t)| \approx |\delta u_0(0)| e^{\lambda(v=i/t)t} \,, \tag{12.15}$$

where the perturbation is initially at the origin $i = 0$. The exponent $\lambda(v)$ is the largest comoving Lyapunov exponent, i.e.

$$\lambda(v) = \lim_{t\to\infty} \lim_{L\to\infty} \lim_{|\delta u_0(0)|\to 0} \frac{1}{t} \ln \left| \frac{\delta u_{[vt]}(t)}{\delta u_0(0)} \right| \,,$$

where the order of the limits is important to avoid boundary effects and that $\lambda(v = 0) = \lambda_{max}$. The spectrum of comoving LEs can be, in principle, obtained

[6] Another interesting quantity for this purpose is represented by (space-time) correlation functions, which for scalar fields $u(\boldsymbol{x}, t)$ can be written as

$$C(\boldsymbol{x}, \boldsymbol{x}'; t, t') = \langle u(\boldsymbol{x}, t) u(\boldsymbol{x}', t') \rangle - \langle u(\boldsymbol{x}, t) \rangle \langle u(\boldsymbol{x}', t') \rangle \,,$$

where $\langle [\ldots] \rangle$ indicates an *ensemble* average. For statistically stationary and translation invariant systems one has $C(\boldsymbol{x}, \boldsymbol{x}'; t, t') = C_{i,j}(\boldsymbol{x} - \boldsymbol{x}'; |t - t'|)$. When the dynamics yields to propagation phenomena, the propagation velocity can be inferred by looking at the peaks of C, which are located at $||\boldsymbol{x} - \boldsymbol{x}'|| \approx V_p |t - t'|$, being V_p the propagation velocity.

[7] An alternative and equivalent definition initializes the perturbation on $W \ll L$ sites.

using the comoving Jacobian matrix $\mathbb{J}_{ij}(v, \boldsymbol{u}(t)) = \partial u_{i+[v(t+1)]}(t+1)/\partial u_{j+[vt]}(t)$. However, as the limit $L \to \infty$ is implicitly required, the meaning of this spectrum is questionable, hence we focus only on the maximum comoving Lyapunov exponent.

There exists an interesting connection between comoving and specific LE. First we notice that Eq. (12.15) implies that the perturbation is locally exponential in space,[8] i.e. $|\delta u_i(t)| \sim \exp(\mu i)$ with $\mu = \ln(|\delta u_{i+1}(t)|/|\delta u_i(t)|) = \mathrm{d}\lambda(v)/\mathrm{d}v$. Then using Eq. (12.10) with $|\delta u_i(0)| = |\delta u_0(0)|\mathrm{e}^{\mu i}$, posing $i = vt$ and comparing with Eq. (12.15) we have

$$\lambda(\mu) = \lambda(v) - v\frac{d\lambda(v)}{\mathrm{d}v}, \tag{12.16}$$

which is a Legendre transform connecting $(\lambda(\mu), \mu)$ and $(\lambda(v), v)$. By inverting the transformation we also have that $v = \mathrm{d}\lambda(\mu)/\mu$.

In closed systems with symmetric coupling one can show that $\lambda(v) = \lambda(-v)$. Moreover, $\lambda(v)$ can be approximated assuming that the perturbation grows exponentially at a rate λ_{max} and diffuses in space, due to the coupling Eq. (12.9), with diffusion coefficient $\epsilon/2$. This leads to [Deissler and Kaneko (1987)]

$$|\delta u_i(t)| \sim \frac{|\delta u_0(0)|}{\sqrt{2\pi\epsilon t}} \exp\left(\lambda_{max} t - \frac{i^2}{2\epsilon t}\right).$$

Comparing with (12.15), we have that

$$\lambda(v) = \lambda_{max} - \frac{v^2}{2\epsilon}, \tag{12.17}$$

which usually well approximates the behavior close to $v = 0$, although deviations from a purely diffusive behavior may be present as a result of the discretization [van de Water and Bohr (1993); Cencini and Torcini (2001)].

In open and generically asymmetric systems $\lambda(v) \neq \lambda(-v)$ and further t the maximal growth rate may be realized for $v \neq 0$. In particular, there are cases in which $\lambda(0) < 0$ and $\lambda(v) > 0$ for $v \neq 0$, these are called convectively unstable systems (see Sec. 12.3.5).

12.3.4 *Propagation of perturbations*

Figure 12.8a shows the spatiotemporal evolution of a perturbation initially localized in the middle of a one-dimensional lattice of locally coupled tent maps. As clear from the figure, the perturbation grows in amplitude and linearly propagates in space with a velocity V_p [Kaneko (1986)]. Such a velocity can be measured by following the left and right edges of the disturbance within a preassigned threshold. Simulations show that V_p is independent of both the amplitude of the initial perturbation and of the threshold value, so that it is a well-defined quantity [Kaneko (1986)] (see also Politi and Torcini (1992); Torcini *et al.* (1995); Cencini and Torcini (2001)).

[8] Notice that $|\delta u_{i+1}(t)| \sim |\delta u_0(0)| \exp(\lambda((i+1)/t)t)$ which, for large t, can be expanded in $\lambda((i+1)/t) \simeq \lambda(v) + (\mathrm{d}\lambda(v)/\mathrm{d}v)/t$ [Politi and Torcini (1992)].

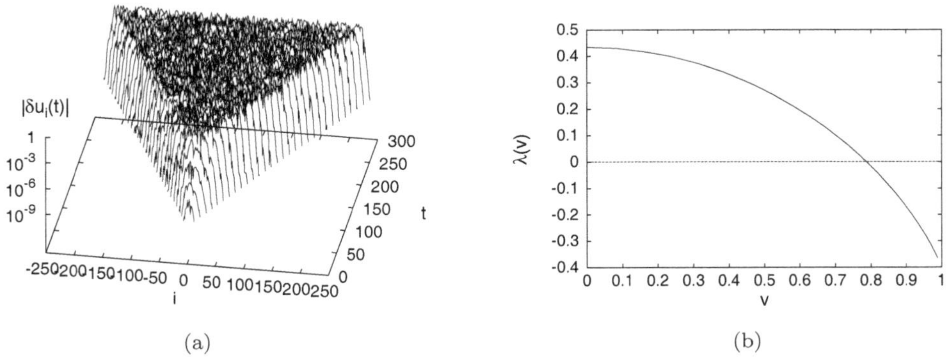

Fig. 12.8 (a) Space-time evolution of $|\delta u_i(t)|$ for an initially localized perturbation on the middle of the lattice with amplitude 10^{-8}. Tent maps, $f(x) = a(1/2 - |x - 1/2|)$, have been used with $a = 2$, $\epsilon = 2/3$ and $L = 1001$. (b) Comoving LE, $\lambda(v)$, for $v > 0$ for the same system.The condition $\lambda(V_P) = 0$ identifies the perturbation velocity $V_P \approx 0.78$.

Clearly, in the frame of refrence comoving with the perturbation, it neither grows nor decreases, meaning that V_p solves the equation

$$\lambda(V_p) = 0 \,. \tag{12.18}$$

Therefore, the growth and propagation of an infinitesimal perturbation is completely characterized by the comoving LE (see Fig. 12.8b). In this respect it is interesting to notice that the prediction (12.18) coincides with the velocity measured when the perturbation is no more infinitesimal indeed, as shown in Fig. 12.8a, the velocity does not change when the perturbation acquires a finite amplitude. The reason for such a numerical coincidence is that, as shown below, what does matter for the evolution of the perturbation is the edge of the perturbation, which is always infinitesimal.

In order to better appreciate the above observation, it is instructive to draw an analogy between perturbation propagation and front propagation in Reaction-Diffusion systems [Torcini *et al.* (1995)]. For instance, consider the FKPP equation $\partial_t c = D\Delta c + f(c)$ with $f(c) \propto c(1-c)$ [Kolmogorov *et al.* (1937); Fisher (1937)] (see also Sec. 12.1.1.1). If $f'(0) > 0$, the state $c = 0$ is unstable while $c = 1$ is stable, hence a localized perturbation of the state $c(x) = 0$ will evolve generating a propagating front with the stable state $c = 1$ invading the stable one. In this rather simple equation, the propagation velocity can be analytically computed to be $V_F = 2\sqrt{f'(0)D}$ [Fisher (1937); Kolmogorov *et al.* (1937); van Saarloos (1988, 1989)]. We can now interpret the perturbation propagation in CML as a front establishing from a chaotic (fluctuating) state and an unstable state. From Eq. (12.17), it is easy to derive $V_p \approx 2\sqrt{\lambda_{max}\epsilon/2}$ which is equivalent to the FKPP result once we identify $D \to \epsilon/2$ and $f'(0) \to \lambda_{max}$. Although the approximation (12.17) and thus the expression for V_p are not always valid [Cencini and Torcini (2001)], the similarity between front propagation in FKPP-like systems and perturbation propagation in spatially extended chaotic systems is rather evident, in particular we can identify $c(x,t)$ with $|\delta u_i(t)|$.

The above analogy can be tightened showing that the propagation velocity V_p is selected by the dynamics as the minimum allowed velocity, like in FKPP [van Saarloos (1988, 1989)]. Similarly to FKPP, the leading edge of the perturbation (i.e. $|\delta u_i| \approx 0$) is typically characterized by an exponential decaying profile

$$\delta u_i \sim \exp(-\mu i)\,.$$

As a consequence from Eq. (12.11) we have that

$$\delta u_i(t) \sim \exp(\lambda(\mu)t - \mu i)\,,$$

meaning that the front edge is exponential with a spatial rate μ and that propagates with velocity $V(\mu) = \lambda(\mu)/\mu$. Since generic, localized perturbations always give rise to the same propagation speed V_p, it means that the leading edge should be characterized by a specific $\mu^\star$ with $V_p = V(\mu^\star)$. In particular, if the analogy with FKPP is working, the exponent $\mu^\star$ selected by the dynamics should be such that $V_p = V(\mu^\star)$ is the minimal allowed value [van Saarloos (1988, 1989); Torcini *et al.* (1995)]. To test such an hypothesis we must show that $\mathrm{d}V/\mathrm{d}\mu|_{\mu^\star} = 0$ while $\mathrm{d}^2V/\mathrm{d}^2\mu|_{\mu^\star} < 0$. From $V(\mu) = \lambda(\mu)/\mu$, and inverting Eq. (12.16),[9] we have that [Torcini *et al.* (1995)]

$$\frac{\mathrm{d}V}{\mathrm{d}\mu} = \frac{1}{\mu}\left(\frac{\mathrm{d}\lambda}{\mathrm{d}\mu} - \frac{\lambda}{\mu}\right) = -\frac{\lambda(v)}{\mu^2}\,,$$

which implies that $\mathrm{d}V/\mathrm{d}\mu|_{\mu^\star} = 0$ if $V_p = V(\mu = \mu^\star)$ as, from Eq. (12.18), $\lambda(V_p) = 0$. Being a Legendre transform, $\lambda(\mu)$ is convex, thus the minimum is unique and

$$V_p = \frac{\lambda(\mu^\star)}{\mu^\star} = \left.\frac{\mathrm{d}\lambda(\mu)}{\mathrm{d}\mu}\right|_{\mu=\mu^\star}\,. \tag{12.19}$$

Therefore, independently of the perturbation amplitude in the core of the front, the dynamics is driven by the infinitesimal edges, where the above theory applies.

Equation (12.19) generalizes the so-called marginal stability criterion [van Saarloos (1988, 1989)] for propagating FKPP-like fronts, which are characterized by a reaction kinetics $f(c)$ such that $\max_c\{f(c)/c\}$ is realized at $c = 0$ and coincides with $f'(0)$. For non FKPP-like[10] reaction kinetics when $\max_c\{f(c)/c\} > f'(0)$ for some $c > 0$, front propagation is no-more controlled by the leading edge ($c \approx 0$) dynamics as the stronger instability is realized at some finite c-value. We thus speak of front *pushed* (from the interior) instead of *pulled* (by the edge) (for a simplified description see Cencini *et al.* (2003)). It is thus natural to seek for an analogous phenomenology in the case of perturbation propagation in CMLs.

Figure 12.9a is obtained as Fig. 12.8a by using for the local dynamics the generalized shift map $f(x) = rx \bmod 1$. In the course of time, two regimes appear:

[9] Notice that here as the perturbation is taken with the minus sign we have $\mu = -\mathrm{d}\lambda(v)/\mathrm{d}v$.

[10] As e.g. the kinetics $f(c) = (1-c)\mathrm{e}^{-A/c}$ (A being a positive constant) which appears in some combustion problems.

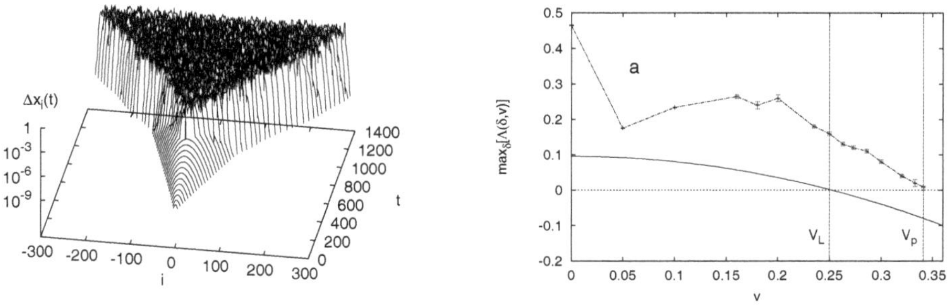

(a) (b)

Fig. 12.9 (a) Space-time evolution of $|\delta u_i(t)|$ for an initially localized perturbation on the middle of the lattice with amplitude 10^{-8}, for a CML of generalized shift maps, $f(x) = rx \mod 1$, with $r = 1.1$, $\epsilon_0 = 1/3$ and $L = 1001$. (b) $\max_\delta[\lambda(\delta, v)]$ (dashed line with points) versus v, compared with $\lambda(v)$ (continuous line). The two vertical lines indicates the velocity obtained by (12.18) which is about 0.250 and the directly measured one $V_p \approx 0.342$. Note that $\max_\delta[\lambda(\delta, v)]$ approaches zero exactly at V_p.

until the amplitude of the perturbation remains "infinitesimal", it propagates similarly to the CML of tent maps with a velocity V_L well predicted by the linearized dynamics and thus obeying the equation $\lambda(V_L) = 0$; at later times, when the perturbation amplitude becomes large enough, a different propagation speed $V_p(> V_L)$ is selected.

Recalling Fig. 9.9 and related discussion (Sec. 9.4.1), it is tempting to attribute the above phenomenology to the presence of strong nonlinear instabilities, which were characterized by means of the Finite Size Lyapunov Exponent (FSLE). To account for such an effect, Cencini and Torcini (2001) introduced the finite size comoving Lyapunov exponent, $\lambda(\delta, v)$ which generalize the comoving LE to finite perturbations. As shown in Fig. 12.9b, $\max_\delta\{\lambda(\delta, v)\}$ vanishes exactly at the measured propagation velocity $V_p > V_L$ suggesting to generalize Eq. (12.18) in

$$\max_\delta \{\lambda(\delta, V_p)\} = 0.$$

Numerical simulations indicate that deviations from the linear prediction given by (12.18) and (12.19) should be expected whenever $\lambda(\delta, v = 0) > \lambda(0, 0) = \lambda_{max}$.

In some CMLs, it may happen that $\lambda(0, 0) < 0$ but $\lambda(\delta, 0) > 0$ for some $\delta > 0$ [Cencini and Torcini (2001)], so that spatial propagation of disturbances can still be observed even if the system has a negative LE [Torcini *et al.* (1995)], i.e. when we are in the presence of the so-called *stable chaos* phenomenon. Stable chaos, first discovered by Politi *et al.* (1993), manifests with unpredictable behaviors contrasting with the fact that the LE is negative (see Box B.29 for details).

Summarizing, for short range coupling, the propagation speed is finite and fully determines the spatiotemporal evolution of the perturbation. We mention that for long-range coupling, as e.g. Eq. (12.3) with $\epsilon_{ij} \propto |i - j|^{-\alpha}$, the velocity of propagation is unbounded [Paladin and Vulpiani (1994)], but it can still be characterized by generalizing the specific LE to power-law (instead of exponentially) shaped perturbations [Torcini and Lepri (1997)].

Box B.29: Stable chaos and supertransients

Sometime, in extended systems, Lyapunov analysis alone is unable to characterize the wealth of observed dynamical behaviors and, remarkably, "Lyapunov stability" is not necessarily synonymous of "regular motion". A striking example is represented by the complex dynamical regimes observed in certain extended systems despite a fully negative Lyapunov spectrum [Politi *et al.* (1993); Cecconi *et al.* (1998)]. This apparently paradoxical phenomenon has been explained by noticing that, in finite-size systems, unpredictable evolutions persist only as transient regimes, until the dynamics falls onto the stable attractors as prescribed by the negative LE. However, transient lifetimes scale exponentially with the system size L. Consequently, in the thermodynamic limit $L \to \infty$, the supertransients become relevant (disordered) stationary regimes, while regular attractors become inaccessible. In this perspective, it makes sense to speak of "Lyapunov stable chaotic regimes" or "Stable Chaos" (SC) [Politi *et al.* (1993)] (see also Politi and Torcini (2009) for a recent review on this subject). SC has been shown by computer simulations to be a robust phenomenon observed in certain CML and also in chains of Duffing oscillators [Bonaccini and Politi (1997)]. Moreover, SC appears to be, to some extent, structurally stable [Ershov and Potapov (1992); Politi and Torcini (1994)].

We emphasize that SC must not be confused with Transient Chaos [Tél and Lai (2008)] which can appear also in low dimensional systems, and is a truly chaotic regime characterized by a positive Lyapunov exponent, that become negative only when the dynamics reaches the stable attractor. In high-dimensional systems, besides transient chaos, one can also have chaotic supertransients [Crutchfield and Kaneko (1988)] (see also Tél and Lai (2008)) characterized by exponentially (in the system size) long chaotic transients, and stable (trivial) attractors. Also in these systems, in thermodynamic limit, attractors are irrelevant, as provocatively stated in the title of Crutchfield and Kaneko (1988) work: *Are attractors relevant to turbulence?* In this respect, SC phenomena are a non-chaotic counterpart of the chaotic supertransient, although theoretically more interesting as the LE is negative during the transient but yet the dynamics is disordered and unpredictable.

We now illustrate the basic features if SC systems with the CML

$$x_i(t+1) = (1-2\sigma)f(x_i(t)) + \sigma f(x_{i-1}(t)) + \sigma f(x_{i+1}(t)). \tag{B.29.1}$$

where

$$f(x) = \begin{cases} bx & 0 \le x < 1/b \\ a + c(x - 1/b) & 1/b \le x \le 1\,, \end{cases}$$

with $a = 0.07$, $b = 2.70$, $c = 0.10$. For such parameters' choice, the map $f(x)$ is linearly expanding in $0 \le x < 1/b$ and contracting in $1/b \le x \le 1$. The discontinuity at $x = 1/b$ determines a point-like nonlinearity. The isolated map dynamics is globally attracted to a cycle of period-3, with LE $\lambda_0 = \log(b^2c)/3 \simeq -0.105$. As the diffusive coupling maintains the stability, it might be naively expected to generate a simple relaxation to periodic solutions. On the contrary, simulations show that, for a broad range of coupling values,

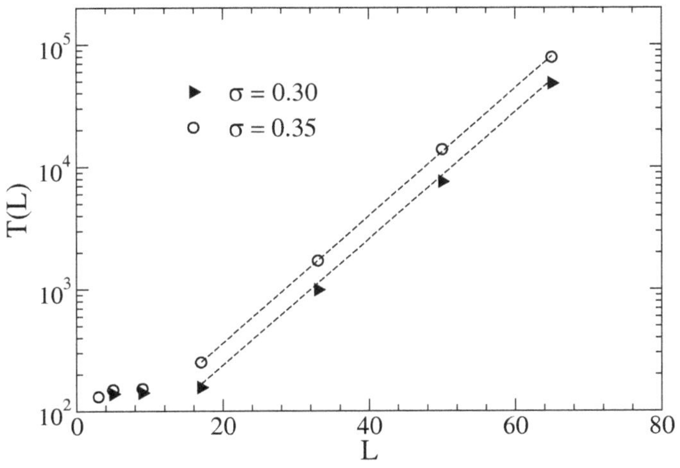

Fig. B29.1 Transient time T versus system-size L for SC-CML (B.29.1), for two different values of σ. Straight lines indicate the scaling $T(L) \sim \exp(\alpha L)$.

the system displays complex spatiotemporal patterns, akin to those generated by genuine chaotic CML, characterized by correlations which decay fast both in time and space [Politi *et al.* (1993); Cecconi *et al.* (1998)]. As discussed by Politi *et al.* (1993); Bunimovich *et al.* (1992); Badii and Politi (1997) a criterion for distinguishing disordered from ordered regimes is provided by the scaling properties of transient duration with the chain length L. The transient time is defined as the minimal number of iterations necessary to observe a recurrence. The study of short chains at different sizes shows that the transient regimes actually last for a time increasing exponentially with L: $T(L) \sim \exp(\alpha L)$ (Fig. B29.1). For the CML Eq. (B.29.1), it is interesting also to consider how, upon changing σ, the dynamics makes an order-disorder transition by passing from periodic to chaotic-like regimes. We know that as Lyapunov instability cannot operate, the transition is only controlled by the transport mechanisms of disturbance propagation [Grassberger (1989)]. Similarly to Cellular Automata [Wolfram (1986)], such transport can be numerically measured by means of *damage spreading* experiments. They consist in evolving to replicas of the system (B.29.1) differing, by a finite amount, only in a small central region R_0, of size $\ell(0)$. The region R_0 represents the initial disturbance, and its spreading $l(t)$ during the evolution allows the disturbance propagation velocity to be measured as

$$V_p = \lim_{t\to\infty} \frac{\langle \ell(t) \rangle}{t}.$$

Positive values of V_p as a function of σ locate the coupling values where the system behavior is "unstable" to finite perturbations. Again, it is important to remark as discussed in the section the difference with truly chaotic CML. As seen in Fig. 12.9, in the chaotic case, perturbations when remain infinitesimal undergo a propagation controlled by Lyapunov instabilities (linear regime), then when they are no longer infinitesimal, nonlinear effects may change the propagation mechanisms selecting a different velocity (nonlinear regimes). In SC models, the first (linear) regime is absent and the perturbation transport is a fully nonlinear phenomenon, and can be characterized through the FSLE [Torcini *et al.* (1995); Cencini and Torcini (2001)].

We can conclude that SC phenomenology provides a convincing evidence that, in extended systems with a significant number of elements, strange attractors and exponential sensitivity to initial data are not necessary conditions to observe complex chaotic-like behaviors.

12.3.5 *Convective chaos and sensitivity to boundary conditions*

In this section we consider systems with a privileged direction as found in a variety of physical contexts, such as turbulent jets, boundary layers and thermal convection. In the presence of an asymmetry, the propagation of fluctuations proceeds preferably along a given direction, think of turbulent spots swept by a constant wind, and we usually speak of *open-flow systems* or simply *flow systems* [Aranson *et al.* (1988); Jensen (1989); Bohr and Rand (1991)]. A minimal model is represented by a chain of maps with unidirectional coupling [Aranson *et al.* (1988); Pikovsky (1989); Jensen (1989)]:

$$u_i(t+1) = (1-\epsilon)f(u_i(t)) + \epsilon f(u_{i-1}(t)) \,. \tag{12.20}$$

Typically, the system (12.20) is excited (driven) through an imposed boundary condition at the beginning of the chain ($i=0$), while it is open at end ($i=L$). Excitations propagate from left to right boundary, where exit from the system. Therefore, for any finite L, interesting dynamical aspects of the problem are transient. Different kinds of boundary condition $x_0(t)$, corresponding to different driving mechanisms, can be considered. For instance, we can choose $x_0(t)=x^*$, with x^* being an unstable fixed point of the map $f(x)$, or generic time-dependent boundary conditions where $x_0(t)$ is a periodic, quasiperiodic or chaotic function of time [Pikovsky (1989, 1992); Vergni *et al.* (1997); Falcioni *et al.* (1999)].

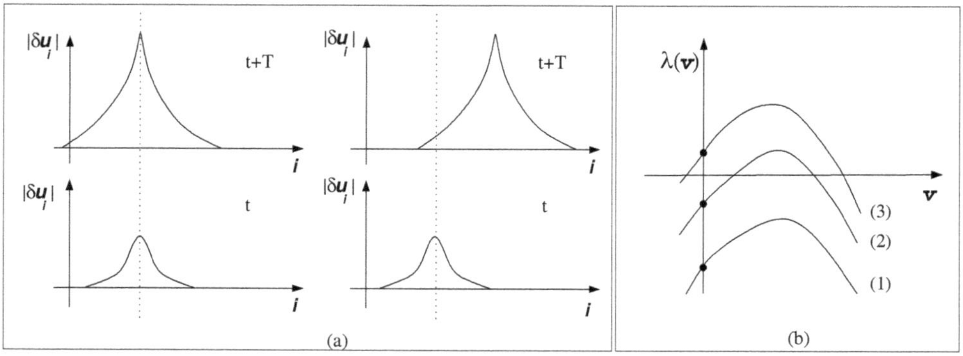

Fig. 12.10 Convective instability. (a) Sketch of perturbation growth, at two instant of time, for an absolutely (left) and convectively (right) unstable system. (b) Sketch of the behavior of $\lambda(v)$ for (1) an absolutely and convectively stable flow, (2) absolutely stable but convectively unstable flow, and (3) absolutely unstable flow.

In the presence of an asymmetry as, in particular, an unidirectional coupling as in the above model, it may happen that a perturbation grows exponentially along the flow but vanishes locally (Fig. 12.10a right) and we speak of *convective instability* in contrast to absolute instability (Fig. 12.10a left). Such instabilities can be quantitatively characterized in terms of the comoving Lyapunov exponent (Fig. 12.10b). Being the system asymmetric, assuming without loss of generality an unidirectional coupling toward the right, we can restrict to positive velocities. As sketched in Fig. 12.10b, we have: (1) absolute stability when $\lambda(v) < 0$ for all $v \geq 0$; (2) convective instability if $\lambda_{max} = \lambda(v=0) < 0$ and $\lambda(v) > 0$ for some velocities $v > 0$; (3) standard chaos (absolute instability) whenever $\lambda_{max} = \lambda(v=0) > 0$. In spite of the negative largest LE, convective unstable systems usually display unpredictable behaviors. In Box B.29 we discussed the phenomenon of stable chaos which is also an example of unpredictability with negative LE, but for convective instabilities the mechanisms leading to unpredictability are rather different.

Unpredictable behaviors observed in convectively unstable systems are linked to the sensitivity to small perturbations of the boundary conditions (at the beginning of the chain), which are always present in physical systems. These are amplified by the convective instability, so that perturbations exponentially grow while propagating along the flow. We thus need to quantify the degree of sensitivity to boundary conditions. This can be achieved in several ways [Pikovsky (1993); Vergni *et al.* (1997)]. Below we follow Vergni *et al.* (1997) (see also Falcioni *et al.* (1999)) who linked the sensitivity to boundary conditions to the comoving LE.

We restrict the analysis to infinitesimal perturbations. The uncertainty $\delta u_i(t)$, on the determination of the variable at time t and site i, can be written as the superposition of the uncertainty on the boundary condition at previous times $\delta u_0(t-\tau)$ with $\tau = i/v$:

$$\delta u_i(t) \sim \int \mathrm{d}v\, \delta u_0(t-\tau) \mathrm{e}^{\lambda(v)\tau} \approx \delta_0 \int \mathrm{d}v\, \mathrm{e}^{[\lambda(v)/v]i}\,, \tag{12.21}$$

where, without loss of generality, we assumed $|\delta u_0(t)| = \delta_0 \ll 1$ for any t. Being interested in the asymptotic spatial behavior, $i \to \infty$, we can write:

$$|\delta u_i(t)| \sim \delta_0 \mathrm{e}^{\Gamma^* i}, \qquad \text{with} \qquad \Gamma^* = \lim_{n\to\infty} \frac{1}{n} \left\langle \ln\left(\frac{|\delta u_n|}{\delta_0}\right)\right\rangle,$$

which defines a spatial-complexity index, where brackets mean time averages. A steepest-descent estimate of Eq. (12.21) gives

$$\Gamma^* = \max_v \left[\frac{\lambda(v)}{v}\right],$$

establishing a link between the comoving and the "spatial" complexity index Γ^*, i.e. between the convective instability and sensitivity to boundary conditions.

The above expression, however, does not properly account for the growth-rate fluctuations (Sec. 5.3.3), which can be significant as perturbations reside in the system only for a finite time. Fluctuations can be taken into account in terms of

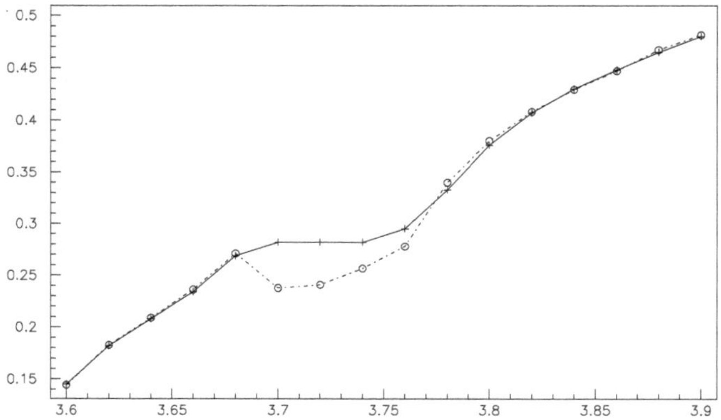

Fig. 12.11 Γ (+) and Γ^* ($\bigcirc$) vs. r for a flow system (12.20) of logistic maps, $f(x) = rx(1-x)$, with $\epsilon = 0.7$ and quasi-periodic boundary conditions (the system is convectively unstable for all the considered values of the parameters). The region where Γ and Γ^* differs is where fluctuations are important, see text.

the effective comoving Lyapunov exponent $\tilde{\gamma}_t(v)$ that gives the exponential changing rate of a perturbation, in the frame of reference moving with velocity v, on a finite time interval t. In particular, Equation (12.21) should be replaced by

$$\delta u_i(t) \sim \delta_0 \int \mathrm{d}v \, \mathrm{e}^{[\tilde{\gamma}_t(v)/v]i},$$

and, as a consequence,

$$\Gamma = \lim_{i\to\infty} \frac{1}{i} \left\langle \ln \left| \frac{\delta u_i}{\delta_0} \right| \right\rangle = \left\langle \max_v \left[\frac{\tilde{\gamma}_t(v)}{v} \right] \right\rangle \geq \max_v \left[\frac{\langle \tilde{\gamma}_t(v) \rangle}{v} \right] \equiv \Gamma^*$$

where, as for the standard LE, $\lambda(v) = \langle \tilde{\gamma}_t(v) \rangle$ (see Sec. 5.3.3). This means that, in the presence of fluctuations, Γ cannot be expressed in terms of $\lambda(v)$. However, Γ^* is often a good approximation of Γ and, in general, a lower bound [Vergni *et al.* (1997)], as shown in Fig. 12.11.

12.4 Non-equilibrium phenomena and spatiotemporal chaos

In this section we discuss the evolution of the tangent vector in generic CML, the complete synchronization of chaotic extended systems and the phenomenon of spatiotemporal intermittency. Despite the differences, these three problems share the link with peculiar non-equilibrium critical phenomena. In particular, we will see that tangent vector dynamics is akin to roughening transition in disordered interfaces [Kardar *et al.* (1986); Halpin-Healy and Zhang (1995)], while both synchronization and spatiotemporal intermittency fit the broad class of non-equilibrium phase transitions to an adsorbing state [Hinrichsen (2000); Ódor (2004); Muñoz (2004)], represented by the synchronized and the quiescent state, respectively. For the sake of self-consistency, in Box B.30 relevant facts about the non-equilibrium processes of interest are summarized.

Box B.30: Non-equilibrium phase transitions

The theory of critical phenomena is well established in the context of equilibrium statistical mechanics [Kadanoff (1999)] while its extension to non-equilibrium processes is still under development. However, it is noteworthy that most of the fundamental concepts of equilibrium models — such as phase transitions, universality classes and scaling — remain valid, to some extent, in non-equilibrium processes [Hinrichsen (2000); Ódor (2004)]. Due to the wealth of systems and phenomena, it is impossible to properly summarize the subject in a short Box. Therefore, here we focus on specific non-equilibrium processes which, as discussed in the main text, are relevant to some phenomena encountered in extended dynamical systems. In particular, we consider non-equilibrium processes characterized by the presence of an adsorbing state, i.e. a somehow trivial state from which the system cannot escape. Typical examples of adsorbing state are the infected phase in epidemic spreading, the empty state in certain reaction diffusion. The main issue in this context is to characterize the system properties when the transition to the adsorbing state is critical, with scaling laws and universal behaviors. We will examine two particular classes of such transitions characterized by distinguishing features. In addition, we will briefly summarize some known results of the roughening of interfaces, which is relevant to many non-equilibrium processes [Halpin-Healy and Zhang (1995)].

Directed Percolation

The Directed Percolation (DP) introduced by Broadbent and Hammersley (1957) is an anisotropic generalization of the percolation problem, characterized by the presence of a preferred direction of percolation, e.g. that of time t. One of the simplest models exhibiting such a kind of transition is the Domany and Kinzel (1984) (DK) probabilistic cellular automata in $1+1$ dimension (one dimension is represented by time). DK model can be illustrated as a two dimensional lattice, the vertical direction coinciding with time t-axis and the horizontal one the space i. On each lattice site we define a discrete variable $s_{i,t}$ which can assume two values 0, unoccupied or inactive, and 1, occupied or active. The system starts randomly assigning the variables at time 0, i.e. $s_{i,0}$, then the evolution is determined by a Markovian probabilistic rule based on the conditional probabilities $P(s_{i,t+1}|s_{i-1,t}, s_{i+1,t})$ which are chosen as follows

$$
\begin{aligned}
&P(0|0,0) = 1 \quad \text{and} \quad P(1|0,0) = 0 \\
&P(1|1,0) = P(1|0,1) = p_1 \\
&P(1|1,1) = p_2 \,,
\end{aligned}
$$

p_1, p_2 being the control parameters. The first rule ensures that the inactive configuration ($s_i = 0$ for any i) is the adsorbing state. The natural order parameter for such an automaton is the density $\rho(t)$ of active sites. In terms of $\rho(t)$ it is possible to construct the two-dimensional phase diagram of the model (i.e. the square $p_1, p_2 \in [0:1]$) which shows a second order transition line (in the $p_1 > 1/2$ region) separating the active from the adsorbing (inactive) phase. In the active phase, an occupied site can eventually propagate through the lattice, forming a percolating cluster. Numerical studies provided strong evidences that the whole transition line, with an exception at $p_2 = 1$, share the same critical properties, being characterized by the same set of critical exponents. These exponents are indeed common to other models and we speak of the DP universality class (see the reviews Hinrichsen (2000); Ódor (2004)).

A possible way to characterize the critical scaling is, for example, to fix $p_2 = \text{const}$ and vary $p_1 = p$. Then there will be a critical value p_c below which the system is asymptotically adsorbed to the inactive state, and above which an active cluster of sites percolates. Close to the critical point, i.e. for $|p - p_c| \ll 1$, the following scaling behaviors are observed

$$\langle \rho(t) \rangle_t \sim |p - p_c|^{\beta}\,, \quad \ell_c \sim |p - p_c|^{-\nu_\perp}\,, \quad \tau_c \sim |p - p_c|^{-\nu_\parallel}\,,$$

where $\langle \rho(t) \rangle_t$ denotes the time average, while ℓ_c and τ_c are the correlation length and time, respectively. Further, at $p = p_c$, the density decreases in time as a power law

$$\rho(t) \sim t^{-\delta}\,.$$

It should be also remarked that the active phase is stable only in the infinite size limit $L \to \infty$, so that the adsorbing state is always reached for finite systems ($L < \infty$) in a time $\tau \sim L^z$, where z is the so-called dynamic exponent. Actually, DP transition is scale invariant and only three exponents are independent [Hinrichsen (2000); Ódor (2004)], in particular

$$\delta = \frac{\beta}{\nu_\parallel} \quad \text{and} \quad z = \frac{\nu_\parallel}{\nu_\perp}\,.$$

A field theoretic description of DP is feasible in terms of the Langevin equation[11]

$$\partial \rho(x,t) = \Delta \rho(x,t) + a\rho(x,t) - b\rho^2(x,t) + \sqrt{\rho(x,t)}\,\eta(x,t) \qquad \text{(B.30.1)}$$

for the local (in space-time) density of active sites $\rho(x,t)$, where a plays the role of p in the DK-model and b has to be positive to ensure that the pdf of $\rho(x,t)$ is integrable. The noise term $\eta(x,t)$ represents a Gaussian process δ-correlated both in time and in space, the important fact is that the noise term is multiplied by $\sqrt{\rho(x,t)}$, which ensures that the state $\rho = 0$ is actually adsorbing, in the sense that once entered it cannot be escaped, see Hinrichsen (2000); Ódor (2004) for details. This field description allows perturbative strategies to be devised for computing the critical exponents. However, it turns out that such perturbative techniques failed in predicting the values of the exponents in the $1 + 1$ dimensional case as fluctuations are very strong, so that we have to trust the best numerical estimates

$$\beta = 0.276486(8) \qquad \delta = 0.159464(6) \qquad z = 1.580745(1).$$

Multiplicative Noise

We now briefly consider another interesting class of adsorbing phase transitions, called Multiplicative Noise[12] (MN) [Grinstein *et al.* (1996); Muñoz (2004)], described by a Langevin

[11] If a multiplicative noise (i.e. depending on the system state) is present in the Langevin equation, it is necessary to specify the adopted convention for the stochastic calculus [Gardiner (1982)]. Here, and in the following, we assume the Ito rule.

[12] We remark that in the mathematical terminology the term multiplicative noise generically denotes stochastic differential equations where the effect of noise is not merely additive term, as in

equation analogous to that of DP but with important differences,

$$\partial\rho(x,t) = \Delta\rho(x,t) + a\rho(x,t) - b\rho^2(x,t) - c\rho^3(x,t) + \rho(x,t)\eta(x,t) \tag{B.30.2}$$

where a and b are free parameters, a being the control parameter, c has to be positive to ensure the integrability of the pdf of ρ. As for DP, $\eta(x,t)$ is a Gaussian process δ-correlated both in time and in space but, unlike DP, now multiplied by $\rho(x,t)$ instead of $\sqrt{\rho(x,t)}$. This is enough to change the nature of the adsorbing state and, consequently, the universality class of the resulting transition. For $b < 0$ the transition is discontinuous, while for $b > 0$ is DP-like but with different exponents

$$\beta = 1.70(5) \qquad \delta = 1.10(5) \qquad z = 1.53(7)\,.$$

It is worth noticing that MN displays several analogies with non-equilibrium processes described by the Kardar, Parisi and Zhang (1986) (KPZ) equation, whose characteristics are briefly discussed below.

Kardar-Parisi-Zhang equation and surface roughening

As discussed in Sec. 12.4.1, the dynamics of tangent vectors share interesting connections with the roughening of disordered interfaces described by the KPZ equation [Kardar *et al.* (1986)] that, in one space dimension, reads

$$\partial_t h(x,t) = \nu\Delta h(x,t) - \lambda(\partial_x h(x,t))^2 + v + \eta(x,t)\,, \tag{B.30.3}$$

where ν, λ and v are free parameters and $\eta(x,t)$ is a δ-correlated in time and space Gaussian process having zero mean. The field $h(x,t)$ can be interpreted as the profile of an interface, which has a deterministic growth speed v and is subjected to noise and nonlinear distortion, the latter controlled by λ, while is locally smoothed by the diffusive term, controlled by ν.

Interfaces ruled by the KPZ dynamics exhibit critical roughening properties which can be characterized in terms of proper critical exponents. In this case, the proper order parameter is the width or roughness of the interface

$$W(L,t) = \left[\langle(h(x,t) - \langle h(x,t)\rangle)^2\rangle\right]^{1/2}$$

which depends on the system size L and on the time t, brackets indicate spatial averages. For $L \to \infty$, the roughness degree W of the interface grows in time as

$$W(L \to \infty, t) \sim t^\beta$$

while in finite systems, after a time $\tau(L) \sim L^z$, saturate to a L-dependent value

$$W_{sat}(L, t \to \infty) \sim L^\alpha\,.$$

Interestingly, interfaces driven by KPZ display scale invariance implying that the above exponent are not independent, and in particular $z = \alpha/\beta$, another consequence of the scale

Eq. (B.30.3), but the noise itself depend on the system state. In this respect, both Eq. (B.30.1) and Eq. (B.30.2) are equations with multiplicative noise in mathematical jargon. The term Multiplicative Noise as used to denote Eq. (B.30.2) just refers to the designation usually found in the physical literature to indicate this specific equation.

invariance is also the following finite-size scaling relation [Krug and Meakin (1990)]

$$W(L,t) = L^\alpha g(t/L^z)\,,$$

g being a proper scaling function. Unlike DP and MN, for the KPZ exact renormalization group computations are available which predict the critical exponents to be

$$\alpha = 1/2 \quad \beta = 1/3 \quad z = 3/2\,.$$

The demanding reader is refereed to the review by Halpin-Healy and Zhang (1995) and the original work by Kardar, Parisi and Zhang (1986).

12.4.1 *Spatiotemporal perturbations and interfaces roughening*

Tangent vectors, i.e. the infinitesimal perturbations, $w_i(t) = \delta u_i(t)$, can either localize with mechanisms similar to Anderson localization (Sec. 12.3.2) or give rise to *dynamical localization*, which is a more generic but weaker form of localization characterized by the slow wandering of localized structures [Pikovsky and Kurths (1994a); Pikovsky and Politi (1998, 2001)], as discussed in the sequel.

Consider the CML

$$u_i(t+1) = f\left(\tilde{u}_i(t)\right)) \,, \qquad i = 1, \ldots, L$$

with $\tilde{u}_i(t) = (1-\epsilon)u_i(t) + (\epsilon/2)(u_{i-1}(t) + u_{i+1}(t))$. Take as local dynamics the logistic map, $f(x) = rx(1-x)$, in the fully developed spatiotemporal chaos regime (e.g. $r = 4$ and $\epsilon = 2/3$) where the state of the lattice is statistically homogeneous (Fig. 12.12a). A generic infinitesimal perturbation evolves in tangent space as

$$w_i(t+1) = m_i(t)\left((1-\epsilon)w_i(t) + \frac{\epsilon}{2}(w_{i-1}(t) + w_{i+1}(t))\right) \tag{12.22}$$

with $m_i(t) = f'(\tilde{u}_i(t)) = r(1 - 2\tilde{u}_i(t))$. Even if the initial condition of the tangent vector is chosen statistically homogeneous in space, at later times, it usually localizes in a small portion of the chain (Fig. 12.12b) with its logarithm $h_i(t) = \ln|w_i(t)|$ resembling a disordered interface (Fig. 12.12c). Unlike Anderson localization, however, the places where the vector is maximal in modulus slowly wander in space (Fig. 12.12d), so that we speak of dynamic localization [Pikovsky and Politi (1998)].

We can thus wonder about the origin of such a phenomenon, which is common to discrete and continuous as well as conservative and dissipative systems [Pikovsky and Politi (1998, 2001)], and has important consequence for the predictability problem in realistic models of atmosphere circulation [Primo *et al.* (2005, 2007)]. After Pikovsky and Kurths (1994a) we know that the origin of dynamical localization can be traced back to the dynamics of $h_i(t) = \ln|w_i(t)|$, which behaves as an interface undergoing a *roughening process* described by the KPZ equation [Kardar *et al.* (1986)] (Box B.30). In the following we sketch the main idea and some of the consequences.

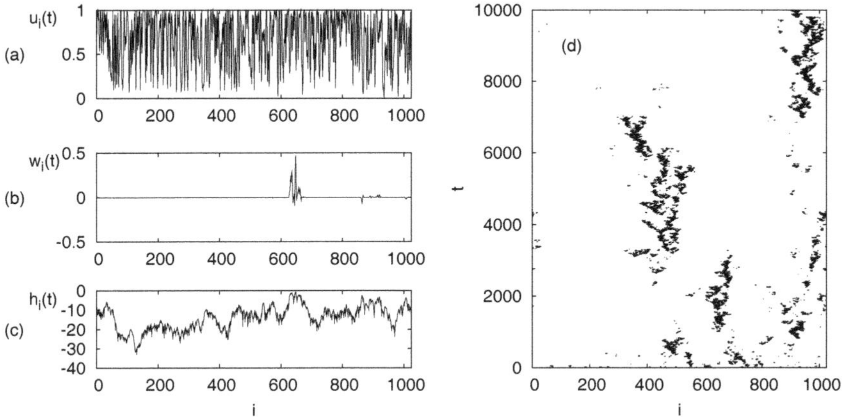

Fig. 12.12 Tangent vector evolution in a one-dimensional lattice of logistic maps at $r = 4$ and democratic coupling $\epsilon = 2/3$. (a) Typical state of the chain $u_i(t)$ at stationarity. (b) Tangent vector $w_i(t)$ at the same time. (c) Logarithm of the tangent vector $h_i(t) = \ln |w_i(t)|$. (c) Space-time locations of the sites where $|w_i(t)|$ exceeds a preassigned threshold value. [After Pikovsky and Politi (1998)]

From Equation (12.22), it is easily derived the evolution rule

$$h_i(t+1) = \ln m_i(t) + h_i(t) + \ln \left[(1-\epsilon) + \frac{\epsilon}{2}\left(e^{\Delta_+ h_i(t)} + e^{\Delta_- h_i(t)}\right)\right]$$

with $\Delta_\pm h_i(t) = h_{i\pm1}(t) - h_i(t)$. Approximating $h_i(t)$ with a time and space continuous field $h(x,t)$, in the small-coupling ($\epsilon \ll 1$) limit, the equation for h reduces to the KPZ equation (see Box B.30)

$$\frac{\partial h}{\partial t} = \frac{\epsilon}{2}\frac{\partial^2 h}{\partial x \partial x} + \frac{\epsilon}{2}\left(\frac{\partial h}{\partial x}\right)^2 + \xi(x,t) \tag{12.23}$$

with both the diffusion coefficient and nonlinear parameter equal to $\epsilon/2$. The noise term $\xi(x,t)$ models the chaotic fluctuations of the multipliers $m(x,t)$.

Even if the above analytic derivation cannot be rigorously handled in generic systems[13] the emerging dynamics is rather universal as numerically tested in several systems. The KPZ equation (Box B.30) describes the evolution of an initially flat profile advancing with a constant speed

$$v = \lim_{t\to\infty} \lim_{L\to\infty} \langle \partial_t h(x,t) \rangle \,,$$

which is nothing but the Lyapunov exponent, and growing width or roughness $W(L,t) = \langle (h(x,t) - \langle h(x,t) \rangle)^2 \rangle^{1/2} \sim t^\beta$ with $\beta = 1/3$, in one spatial dimension. In finite systems $L < \infty$, the latter saturates to a L-dependent value $W(L,\infty) \sim L^\alpha$ with $\alpha = 1/2$. Extensive numerical simulations [Pikovsky and Kurths (1994a); Pikovsky and Politi (1998, 2001)] have shown that the exponents β and α always

[13]Indeed the derivation is meaningful only if $m_i(t) > 0$, in addition, typically, ξ and h are not independent variables.

match the KPZ-values in a variety of models, supporting the claim that the KPZ universality class well describes the tangent vector dynamics (see also Primo *et al.* (2005, 2007)).

Remarkably, the link with KPZ allows the finite-size corrections to the maximal LE to be estimated using to the scaling law (derived from the equivalent one in KPZ [Krug and Meakin (1990)], see Box B.30)

$$\lambda(T, L) = L^{-1} g(T/L^z) \tag{12.24}$$

where L is the system size, T the time employed to measure the LE (mathematically speaking this should be infinite), $g(x)$ is a universal function and $z = 3/2$ the dynamic exponent. In principle, we would like to have access to the thermodynamic quantity, i.e. $\lambda(\infty, \infty)$, which is impossible numerically. However, from the scaling law (12.24) it is immediate to derive that

$$\lambda(\infty, L) - \lambda(\infty, \infty) \sim L^{-1} \qquad \text{and} \qquad \lambda(T, \infty) - \lambda(\infty, \infty) \sim T^{-2/3}\,.$$

These two finite-size relationships can then be used to estimate the asymptotic value $\lambda(\infty, \infty)$ [Pikovsky and Politi (1998)].

We finally observe that KPZ equation is related to other problems in statistical mechanics such as directed polymers in random media, for which analytical techniques exist for estimating the partition function, opening the gate for analytically estimating the maximal LE in some limits [Cecconi and Politi (1997)].

12.4.2 *Synchronization of extended chaotic systems*

Coupled extended chaotic systems can synchronize as low-dimensional ones (Sec. 11.4.3) but the presence of spatial degrees of freedom adds new features to the synchronization transition, configuring it as a non-equilibrium phase transition to an adsorbing state (Box B.30), i.e. the synchronized state [Grassberger (1999); Pikovsky *et al.* (2001)].

Consider two locally coupled replicas of a generic CML with diffusive coupling

$$\begin{aligned} u_i^{(1)}(t) &= (1-\gamma) f(\tilde{u}_i^{(1)}(t)) + \gamma f(\tilde{u}_i^{(2)}(t)) \\ u_i^{(2)}(t) &= (1-\gamma) f(\tilde{u}_i^{(2)}(t)) + \gamma f(\tilde{u}_i^{(1)}(t)) \end{aligned} \tag{12.25}$$

where γ tunes the coupling strength between replicas, and $\tilde{u}_i^{(\alpha)}(t) = (1-\epsilon)u_i^{(\alpha)}(t) + (\epsilon/2)(u_{i-1}^{(\alpha)}(t) + u_{i-1}^{(\alpha)}(t))$ with $\alpha = 1, 2$ being the replica index, and ϵ the diffusive-coupling strength within each replica. Following the steps of Sec. 11.4.3, i.e. linearizing the dynamics close to the synchronized state, the transverse dynamics reduces to Eq. (12.22) with a multiplicative factor $(1-2\gamma)$, so that the transverse Lyapunov exponent is obtained as

$$\lambda_\perp(\gamma) = \lambda + \ln(1 - 2\gamma)$$

which is analogous to Eq. (11.46) with λ denoting now the maximal LE of the single CML. It is then natural to expect that for $\gamma > \gamma_c = [1 - \exp(-\lambda)]/2$ the system

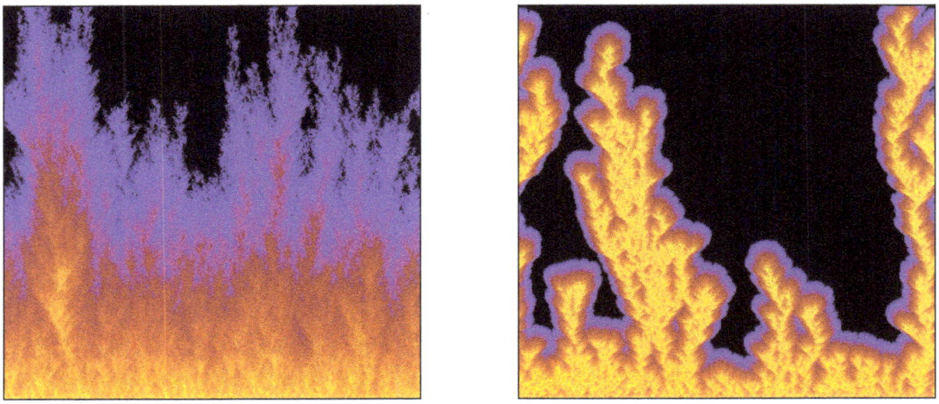

Fig. 12.13 Spatiotemporal evolution of the logarithm of the synchronization error $\ln(\mathcal{W}_i(t))$ for (left) tent maps $f(x) = 2(1 - |1/2 - x|)$ and (right) Bernoulli shift maps $f(x) = 2x \mod 1$, for both the system size is $L = 1024$ and the diffusive coupling strength $\epsilon = 2/3$. The coupling among replicas is chosen slightly above the critical one which is $\gamma_c \approx 0.17605$ for the tent map and $\gamma_c \approx 0.28752$ for the shift map. Notice that for the tent map $\lambda_\perp(\gamma_c) \approx 0$ while for the shift one $\lambda_\perp(\gamma_c) < 0$. Colors code the intensity of $\ln(\mathcal{W}_i(t))$ black means synchronization.

synchronizes, i.e. the synchronization error $\mathcal{W}_i(t) = |u_i^{(1)}(t) - u_i^{(2)}(t)|$ asymptotically vanishes. Unlike synchronization in low-dimensional systems, now $\mathcal{W}_i(t)$ is a field evolving in space-time, thus it is interesting not only to understand when it goes to zero but also the way it does.

Figure 12.13 shows the spatiotemporal evolution of $\mathcal{W}_i(t)$ for a CML with local dynamics given by the tent (left) and Bernoulli shift (right) map, slightly above the critical coupling for the synchronization γ_c. Two observations are in order.

First, the spatiotemporal evolution of $\mathcal{W}_i(t)$ is rather different for the two maps suggesting that the synchronization transitions are different in the two models [Baroni *et al.* (2001); Ahlers and Pikovsky (2002)] i.e., in the statistical mechanics jargon, that they belong to different universality classes.

Second, as explained in the figure caption, for the tent map $\mathcal{W}_i(t)$ goes to zero together with $\lambda_\perp$ at $\gamma_c = [1 - \exp(-\lambda)]/2$, while for the Bernoulli map synchronization takes place at $\gamma_c \approx 0.28752$ even though $\lambda_\perp$ vanishes at $\gamma = 0.25$. Therefore, in the latter case the synchronized state is characterized by a negative transverse LE, implying that the synchronized state is a "true" absorbing state, once it is reached on a site it stays there for ever. A different scenario characterizes the tent map where the synchronized state is marginal, and fluctuations may locally desynchronize the system. This difference originates from the presence of strong nonlinear instabilities in the Bernoulli map [Baroni *et al.* (2001); Ginelli *et al.* (2003)] (see Sec. 9.4.1), which are also responsible for the anomalous propagation properties seen in Sec. 12.3.4.[14]

[14]It should be remarked the importance of the combined effect of nonlinear instabilities and of the presence of many coupled degrees of freedom, indeed for two Bernoulli maps the synchronization

Analogously with the non-equilibrium phase transitions described in Box B.30, a quantitative characterization can be achieved by means of the spatially averaged synchronization error $\rho_\gamma(t) = \langle \mathcal{W}_i(t) \rangle$ which is the order parameter. Below threshold ($\gamma < \gamma_c$), as the two replicas are not synchronized, $\rho_\gamma(t)$ asymptotically saturates to a finite value $\rho^*(\gamma)$ depending on the distance from the critical point $\gamma_c - \gamma$. Above threshold ($\gamma > \gamma_c$), the replicas synchronize in a finite time so that $\rho_\gamma(t) \to 0$. Therefore, it is interesting to look at the time behavior exactly at γ_c. In both cases, as the synchronization transition is a critical phenomenon, we expect power-law behaviors. In fact, extensive numerical simulations [Baroni *et al.* (2001); Ahlers and Pikovsky (2002); Ginelli *et al.* (2003); Droz and Lipowski (2003); Cencini *et al.* (2008)] of different tent-like and Bernoulli-like maps have shown that

$$\rho_{\gamma_c}(t) \sim t^{-\delta} \quad \text{and} \quad \rho^*(\gamma) \sim (\gamma_c - \gamma)^\beta$$

with different values for the exponents δ and β, see below.

The spatiotemporal evolution of the synchronization error for the Bernoulli map Fig. 12.13 (right) reveals the typical features of *Directed Percolation* (DP), a universality class common to many non-equilibrium phenomena with adsorbing states [Grassberger (1997); Hinrichsen (2000); Ódor (2004)] (Box B.30). This naive observation is confirmed by the values of critical exponents, $\delta \approx 0.16$ and $\beta \approx 0.27$, which agree with the best known estimates for DP [Hinrichsen (2000)]. The fact that Bernoulli-like maps belong to the DP universality class finds its root in the existence of a well defined adsorbing state (the synchronized state), the transverse LE being negative at the transition, and in the peculiar propagation properties of this map (Sec. 12.3.4) [Grassberger (1997)]. Actually chaos is not a necessary condition to have this type of synchronization transition which has been found also in maps with stable chaos [Bagnoli and Cecconi (2000)] where the LE is negative (Box B.29) and cellular automata [Grassberger (1999)]. Unfortunately, the nonlinear nature of this phenomenon makes difficult the mapping of Eq. (12.25) onto the field equation for DP (Box B.30), this was possible only for a stochastic generalization of Eq. (12.25) [Ginelli *et al.* (2003)].

On the contrary, for tent-like maps, when $\mathcal{W}_i(t) \to 0$ we can reasonably expect the dynamics of the synchronization error to be given by Eq. (12.22) (with a $(1-2\gamma)$ multiplicative factor). Thus, in this case, the synchronization transition should be connected with the KPZ phenomenology [Pikovsky and Kurths (1994a); Grassberger (1997, 1999)]. Indeed a refined analysis [Ahlers and Pikovsky (2002)] shows that for this class of maps the dynamics of the synchronization error can be mapped to the KPZ equation with the addition of a saturation term, e.g. proportional to $-p|\mathcal{W}_i(t)|^2\, \mathcal{W}_i(t)$ (p being a free parameter that controls the strength of the nonlinear saturation term), preventing its unbounded growth. Therefore, similarly to the previous section, denoting with h the logarithm of the synchronization error,

transition is still determined by the vanishing of the transverse LE, while nonlinear instabilities manifest by looking at other observables [Cencini and Torcini (2005)].

Eq. (12.23) generalizes to [Ahlers and Pikovsky (2002)]

$$\partial_t h = -p e^{2h(x,t)} + \frac{\epsilon}{2}\partial_x^2 h + \frac{\epsilon}{2}(\partial_x h)^2 + \xi(x,t) + a\,,$$

where a is related to the distance from the critical point. In this picture, synchronization corresponds to an interface moving towards $h = -\infty$, while the exponential saturation term (the first on the r.h.s.) prevents the interface from unbounded positive growth, hence the name *bounded KPZ* (BKPZ) for this transition, which is essentially coincident with the Multiplicative Noise (MN) universality class discussed in Box B.30. The (negative) average interface velocity is nothing but transverse Lyapunov exponent. The critical properties of BKPZ transition has been studied by means of renormalization group method and numerical simulations [Grinstein *et al.* (1996); Muñoz (2004)]: the critical exponents are in reasonable agreement with those found for tent-like maps, i.e. $\delta \approx 1.24$ and $\beta \approx 0.7$, confirming the MN universality class.[15]

We conclude mentioning that a unifying field theoretic framework able to describe the synchronization transition in extended systems has been proposed by [Muñoz and Pastor-Satorras (2003)] and that, thanks to the mapping discussed in Sec. 12.1.1.7, exactly the same synchronization properties can be observed in delayed systems [Szendro and López (2005)].

12.4.3 *Spatiotemporal intermittency*

Among the many phenomena appearing in extended systems, a special place is occupied by spatiotemporal intermittency (STI), a term designating all situations in which a spatially extended system presents intermittency both in its spatial structures and temporal evolutions [Bohr *et al.* (1998)]. STI characterizes many systems such as fluids [Ciliberto and Bigazzi (1988)], where in some conditions sparsely turbulent spots may be separated by laminar regions of various sizes, liquid crystals [Takeuchi *et al.* (2007)], the Complex Ginzburg-Landau equation (see Fig. 12.3c) [Chaté (1994)] (and thus all the phenomena described by it, see Sec. 12.1.1.3) and model systems as coupled map lattices (see Fig. 12.5c). In spite of its ubiquity many features are still unclear. Although many numerical evidences indicate that STI belongs to the Directed Percolation universality class (Box B.30), as conjectured by Pomeau (1986) on the basis of arguments resting on earlier works by Janssen (1981) and Grassberger (1982), it is still debated whether a unique universality class is able to account for all the observed features of STI [Grassberger and Schreiber (1991); Bohr *et al.* (2001)].

A minimal model able to catch the main features of STI was introduced in a seminal paper by Chaté and Manneville (1988), this amounts to a usual CML with

[15]Actually while β and ζ are in a reasonable good agreement, δ appears to be slightly larger pinpointing the need of refined analysis [Cencini *et al.* (2008)].

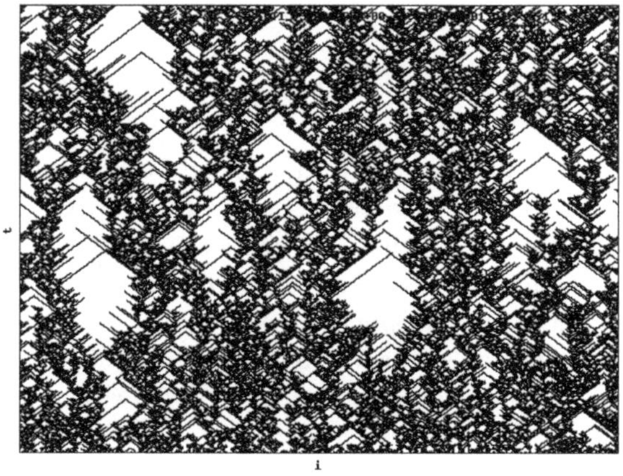

Fig. 12.14 Spatiotemporal evolution of the Chaté-Manneville model for STI with $a = 3$ and $\epsilon = 0.361$: turbulent states ($u_i(t) < 1$) are in black while laminar ones ($u_i(t) > 1$) in white.

diffusive coupling and local dynamics given by the map

$$f(x) = \begin{cases} a(1/2 - |x - 1/2|) & \text{for} \quad x < 1 \\ x & \text{for} \quad x > 1\,. \end{cases} \tag{12.26}$$

This map, if uncoupled, for $a > 2$ produces a quite trivial dynamics: a chaotic transient — the *turbulent state* — till $x < 1$ where the map evolves as in a usual tent map, followed by a fixed point — the *laminar state*[16] — as soon as $x > 1$. The latter is an adsorbing state, meaning that a laminar state cannot become turbulent again. When diffusively coupled on a lattice, the map (12.26) gives rise to a rather nontrivial dynamics: if the coupling is weak $\epsilon < \epsilon_c$, after a transient it settles down to a quiescent laminar state; while, above ϵ_c, a persistent chaotic motion with the typical features of STI is observed (Fig. 12.14).

In STI the natural order parameter is the density of active (turbulent) sites $\rho_\epsilon(t) = 1/L \sum_i \Theta(1 - u_i(t))$, where $\Theta(x)$ denotes the Heaviside step function, in terms of which the critical region close to ϵ_c can be examined. Several numerical studies, in a variety of models, have found contradictory values for the critical exponents: some in agreement with DP universality class, some different and, in certain cases, the transition displays discontinuous features akin to first order phase transitions (see Chaté and Manneville (1988); Grassberger and Schreiber (1991); Bohr *et al.* (1998, 2001) and reference therein). The state-of-the-art of the present understanding of STI mostly relies on the observations made by Grassberger and Schreiber (1991) and Bohr *et al.* (2001), who by investigating generalization of Chaté-Manneville model were able to highlight the importance of long-lived "soliton-like" structures. However, while Grassberger and Schreiber (1991) expecta-

[16]Notice that we speak of STI also when the laminar state is not a fixed point but, e.g., a periodic state as for the logistic map in Fig. 12.5c [Grassberger (1989)].

tion is that such solitons simply lead to long crossovers before finally recovering DP properties, Bohr *et al.* (2001) have rather convincingly shown that STI phenomena cannot be reduced to a unique universality class.

12.5 Coarse-grained description of high dimensional chaos

We close this Chapter by briefly discussing some issues related to the problem of system description and modeling. In the context of low dimensional systems, we have already seen (Chap. 9 and Chap. 10) that changing the level of description or, more precisely, the (scale) resolution at which we observe the signal casts light on many aspects allowing the establishment of more efficient representation/models of the system. In fact, coarser descriptions typically lead to a certain freedom in modeling. For instance, even if a system is stochastic at some scale it may be effectively described as a deterministic one or viceversa (see Sec. 10.3). Yet another example is when from the huge number of molecules which compose a fluid we derive the hydrodynamic description in terms of the Navier-Stokes equation. In the following two subsections we discuss two examples emphasizing some important aspects of the problem.

12.5.1 *Scale-dependent description of high-dimensional systems*

The first example, taken from Olbrich *et al.* (1998), well illustrates that high-dimensional systems are able to display non-trivial behaviors at varying the scale of the magnifying glass used to observe them. In particular, we focus on a flow system [Aranson *et al.* (1988)] described by the unidirectional coupled map chain

$$u_j(t+1) = (1-\sigma)f(u_j(t)) + \sigma u_{j+1}(t) \tag{12.27}$$

where, as usual, $j(=1,\ldots,L)$ denotes spatial index of the chain having length L, t the discrete time, and σ the coupling strength.

It is now interesting to wonder whether, by looking at a long time record of a single scalar observable, such as the state variable on a site e.g. $u_1(t)$, we can recognize the fact that the system is high dimensional. This is obviously important both for testing the possibilities of nonlinear time series analysis and to understand which would be the best strategy of modeling if we want to mimic the behavior of a single element of the system.

The natural way to proceed is to apply the embedding method discussed in Sec. 10.2 to compute, for instance, the correlation integral $C_m^q(\varepsilon)$ (10.8), where m and ε indicate the embedding dimension and the observation scale, respectively. From $C_m^q(\varepsilon)$ we can obtain quantities such as the correlation dimension at varying the resolution $D_m^{(2)}(\varepsilon)$ (10.9) and the correlation ε-entropy $h^{(2)}(\varepsilon)$ (10.10).

Olbrich *et al.* (1998) performed a detailed numerical study (also supported by analytical arguments) of both $h_m^{(2)}(\varepsilon)$ and $D_m^{(2)}(\varepsilon)$ at varying ε and m. In the limit

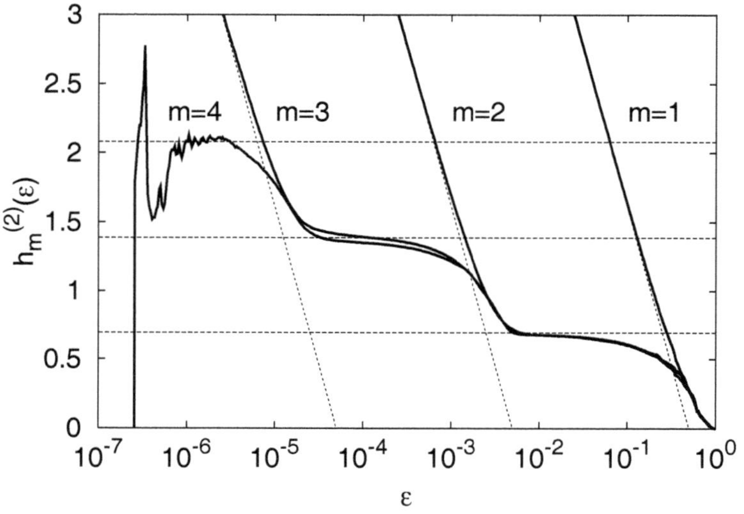

Fig. 12.15 $h_m^{(2)}(\varepsilon)$ for $m = 1, \ldots, 4$, computed with the Grassberger-Procaccia method, for the system (12.27) using the tent map $f(x) = 2|1/2 - |x - 1/2||$ and coupling strength $\sigma = 0.01$. Horizontal lines indicate the entropy steps which appear at decreasing ε, while the oblique (dashed) lines indicate $\ln(1/\varepsilon)+C_m$, where C_m depends on the embedding dimension, which is the behavior expected for noise. For $m \geq 4$ the number of data does not allow to explore the small ε range. [Courtesy of H. Kantz and E. Olbrich]

of small coupling $\sigma \to 0$, the following scale-dependent scenario emerges (Fig. 12.15)

for $1 \geq \varepsilon \geq \sigma$ and $m \geq 1$, $h_m^{(2)}(\varepsilon) \simeq \lambda_s$, where λ_s is the Lyapunov exponent of the single (uncoupled) map $x(t+1) = f(x(t))$, and $D_m^{(2)}(\varepsilon) \simeq 1$;
for $\sigma \geq \varepsilon \geq \sigma^2$ and $m \geq 2$, $h_m^{(2)}(\varepsilon) \simeq 2\lambda_s$ and $D_m^{(2)}(\varepsilon) \simeq 2$;
⋮
for $\sigma^{n-1} \geq \varepsilon \geq \sigma^n$ and $m \geq n$ $h_m^{(2)}(\varepsilon) \simeq n\lambda_s$ and $D_m^{(2)}(\varepsilon) \simeq n$;

Of course, while reducing the observation scale, it is necessary to increase the embedding dimension, otherwise one simply has $h_m^{(2)}(\varepsilon) \sim \ln(1/\varepsilon)$ as for noise (Fig. 12.15).

The above scenario suggests that we can understand the dynamics at different scales as ruled by a hierarchy of low-dimensional systems whose "effective" dimension $n_{eff}(\varepsilon)$ increases as ε decreases [Olbrich *et al.* (1998)]:

$$n_{eff}(\varepsilon) \sim \left[\frac{\ln(1/\varepsilon)}{\ln(1/\sigma)}\right],$$

where $[\ldots]$ indicates the integer part. Therefore, the high-dimensionality of the system becomes obvious only for smaller and smaller ε while taking larger and larger m embedding dimensions. In fact, only for $\varepsilon \leq \sigma^N$, we can realize deterministic and high-dimensional character of the system, signaled by the plateau $h^{(2)}(\varepsilon) \simeq N\lambda_s$.

It is interesting to observe that, given the resolution ε, a suitable (relatively) low dimensional noisy system can be found, which is able to mimic the evolution of, e.g. $u_1(t)$ given by Eq. (12.27). For instance, if we limit the resolution of our magnifying glass to, say, $\varepsilon \geq \sigma$ we can mimic the evolution of $u_1(t)$ by using a one-dimensional stochastic maps as

$$u(t+1) = (1-\sigma)f(u(t)) + \sigma\xi(t)\,,$$

provided the noise $\xi(t)$ has a probability distribution not too far from the typical one of one element of the original system [Olbrich *et al.* (1998)]. Analogously, for $\varepsilon \geq \sigma^n$ with $n \ll L$ the system can be by an n-dimensional deterministic system, i.e. a chain of maps with n elements, plus noise.

Summarizing, adopting a scale-dependent description of high dimensional systems gives us some freedom in modeling them in terms of low dimensional systems with the addition of noise. Thus, this example renforces what was observed in Sec. 10.3, namely the fact that changing the point of view (the observation scale) may change the "character" of the observed system.

12.5.2 *Macroscopic chaos: low dimensional dynamics embedded in high dimensional chaos*

High dimensional systems are able to generate nontrivial collective behaviors. A particularly interesting one is *macroscopic chaos* [Losson *et al.* (1998); Shibata and Kaneko (1998); Cencini *et al.* (1999a)] arising in globally coupled map (GCM)

$$u_n(t+1) = (1-\sigma)f(u_n(t)) + \frac{\sigma}{N}\sum_{i=1}^{N} f(u_i(t)), \tag{12.28}$$

N being the total number of elements. GCM can be seen as a mean field version of the standard CML though, strictly speaking, no notion of space can be defined, sites are all equivalent.

Collective behaviors can be detected by looking at a macroscopic variable; in Eq. (12.28) an obvious one is the mean field

$$m(t) = \frac{1}{N}\sum_{i=1}^{N} u_i(t)\,.$$

Upon varying the coupling σ and the nonlinear parameter of the maps $f(x)$, $m(t)$ displays different behaviors:

(a) *Standard Chaos*: $m(t)$ follows a Gaussian statistics with a definite mean and standard deviation $\sigma_N = \sqrt{\langle m^2(t)\rangle - \langle m^2(t)\rangle} \sim N^{-1/2}$;
(b) *Macroscopic Periodicity*: $m(t)$ displays a superposition of a periodic function and small fluctuations $\mathcal{O}(N^{-1/2})$;
(c) *Macroscopic Chaos*: $m(t)$ may also display an irregular motion as evident from the return plot of $m(t)$ vs. $m(t-1)$ in Fig. 12.16, that appears as a structured function (with thickness $\mathcal{O}(N^{-1/2})$), suggesting a chaotic collective dynamics.

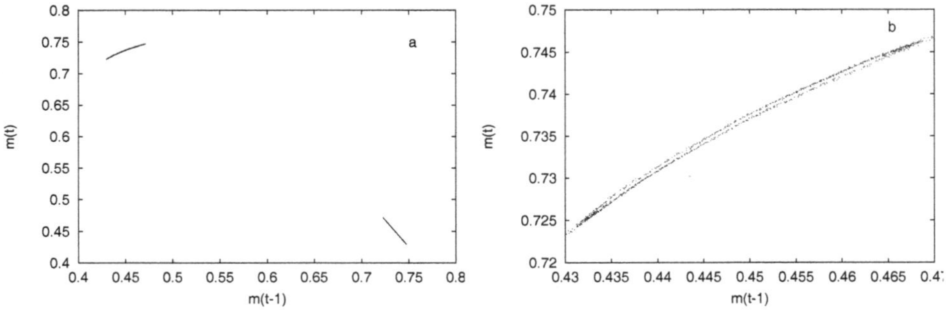

Fig. 12.16 Return map of the mean field: (a) $m(t)$ versus $m(t-1)$ with local dynamics given by the tent map $f(x) = a(1 - |1/2 - x|)$ with $a = 1.7$ $\sigma = 0.3$ and $N = 10^6$; (b) is an enlargement of (a). From Cencini *et al.* (1999a).

Phenomena (a) and (b) have been also observed in CML with local coupling in high dimensional lattices [Chaté and Manneville (1992)], for case (c), as far as we know, there is not a direct evidence in finite dimensional CMLs. We also remark that (a) is a rather natural behavior in the presence of chaos. Essentially $m(t)$ amounts to the sum of random (more precisely, chaotically wandering) variables so that a sort of central limit theorem and law of large numbers can be expected. Behaviors such as (b) and (c) are, in this respect, more interesting as reveal the presence of non trivial correlations even when many positive LE are present.

Intuition may suggest that the mean field evolves on times longer than those of the full dynamics (i.e. the microscopic dynamics), which are basically set by $1/\lambda_{max}$, the inverse of the maximal LE of the full system — which we can call the *microscopic Lyapunov exponent* λ_{micro}. At least conceptually, macroscopic chaos for GCM resembles hydrodynamical chaos emerging from molecular motion. There, in spite of the huge microscopic Lyapunov exponent ($\lambda_1 \sim 1/\tau_c \sim 10^{11}$ s^{-1}, τ_c being the collision time), rather different behaviors may appear at the hydrodynamical (coarse-grained) level: regular motion (with $\lambda_{hydro} \leq 0$), as for laminar fluids, or chaotic (with $0 < \lambda_{hydro} \ll \lambda_1$), as in moderately turbulent flows. In principle, knowledge of hydrodynamic equations makes possible to characterize the macroscopic behavior by means of standard dynamical system techniques. However, in generic CML there are no systematic methods to build up the macroscopic equations, apart from particular cases, where macroscopic chaos can be characterized also by means of a self-consistent Perron-Frobenius nonlinear operator [Perez and Cerdeira (1992); Pikovsky and Kurths (1994b); Kaneko (1995)], see also Cencini *et al.* (1999a) for a discussion of this aspect.

The microscopic Lyapunov exponent cannot be expected to account for the macroscopic motion, because related to infinitesimal scales, where as seen in the previous section the high dimensionality of the system is at play. A possible strategy, independently proposed by Shibata and Kaneko (1998) and Cencini *et al.* (1999a),

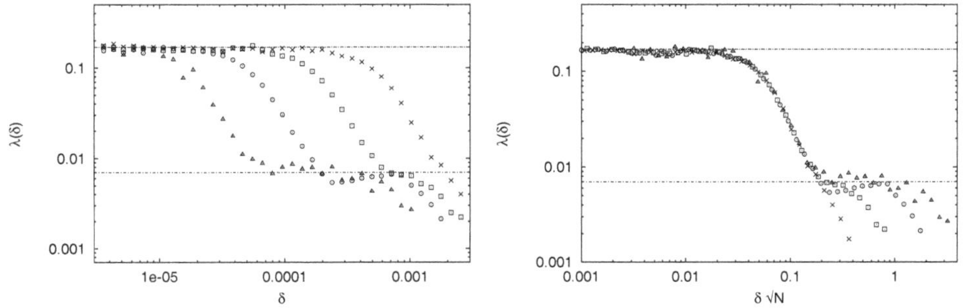

Fig. 12.17 $\lambda(\delta)$ versus δ for a GCM (12.28) of tent maps $f(x) = a(1 - |1/2 - x|)$ with $a = 1.7$, $\sigma = 0.3$ and $N = 10^4$ ($\times$), $N = 10^5$ ($\square$), $N = 10^6$ ($\odot$) and $N = 10^7$ ($\triangle$) . The two horizontal lines indicate the microscopic LE $\lambda_{micro} \approx 0.17$ and the macroscopic LE $\lambda_{Macro} \approx 0.007$. The average is over $2 \cdot 10^3$ realizations for $N = 10^4$, 10^5, 10^6 and 250 realizations for $N = 10^7$. (b) The same as (a) rescaling the δ–axis with $\sqrt{N}$. From Cencini *et al.* (1999a).

is to use the Finite Size Lyapunov Exponent[17] (FSLE) (see Sec. 9.4). In the limit of infinitesimal perturbations $\delta \to 0$, $\lambda(\delta) \to \lambda_{max} \equiv \lambda_{micro}$; while, for finite δ, the δ-dependence of $\lambda(\delta)$ may provide information on the characteristic time-scales governing the macroscopic motion.

Figure 12.17a shows $\lambda(\delta)$ versus δ in the case of macroscopic chaos [Cencini *et al.* (1999a)]. Two plateaus can be detected: at small values of δ ($\delta \leq \delta_1$), as expected from general considerations, $\lambda(\delta) = \lambda_{micro}$; while for $\delta \geq \delta_2$ another plateau $\lambda(\delta) = \lambda_{Macro}$ defines the "macroscopic" Lyapunov exponent. Moreover, δ_1 and δ_2 decrease at increasing N as $\delta_1,\ \delta_2 \sim 1/\sqrt{N}$ (Fig. 12.17b). It is important to observe that the macroscopic plateau, almost non-existent for $N = 10^4$, becomes more and more resolved and extended on large values of $\delta\sqrt{N}$ at increasing N up to $N = 10^7$. We can thus argue that the macroscopic motion is well defined in the thermodynamics limit $N \to \infty$.

In conclusion, we can summarize the main outcomes as follows:

- at small δ ($\ll 1/\sqrt{N}$) the "microscopic" Lyapunov exponent is recovered, i.e. $\lambda(\delta) \approx \lambda_{micro}$
- at large δ ($\gg 1/\sqrt{N}$), $\lambda(\delta) \approx \lambda_{Macro}$ which can be much smaller than the microscopic one.

The emerging scenario is that at a coarse-grained level, i.e. $\delta \gg 1/\sqrt{N}$, the system can be described by an "effective" hydro-dynamical equation (which in some cases can be low-dimensional), while the "true" high-dimensional character appears only at very high resolution, i.e. $\delta \leq \mathcal{O}(N^{1/2})$, providing further support to the picture which emerged with the example analyzed in the previous subsection.

[17] A way to measure it is by means of the algorithm described in Sec. 9.4 applied to the evolution of $|\delta m(t)|$, initialized at $\delta m(t) = \delta_{min} \ll 1$ by shifting all the elements of the unperturbed system by the quantity δ_{min} (i.e. $u_i'(0) = u_i(0) + \delta_{min}$), for each realization.

Chapter 13

Turbulence as a Dynamical System Problem

All exact science is dominated by the idea of approximation.
Bertrand Russell (1872–1970)

This Chapter discusses some aspects of fluid motion and, in particular, turbulence under a dynamical systems perspective. Despite the Navier-Stokes equation, ruling the evolution of fluid flows, has been introduced almost two centuries ago, its understanding is still a challenging open issue, posing fluid dynamics research as an active field in mathematics, physics and applied sciences. For instance, a rigorous proof for the existence of the solution, at any time, of the three-dimensional Navier-Stokes equation is still missing [Doering and Gibbon (1995); Doering (2009)], and the search for such a proof is currently on the list of the millennium problems at the Clay Mathematics Institute (see `http://www.claymath.org/millennium`). The much less ambitious purpose of this Chapter is to overview some aspects of turbulence relevant to dynamical systems, such as the problem of reduction of the degrees of freedom and the characterization of predictability. For the sake of self-consistency, it is also summarized the current phenomenological understanding of turbulence in both two and three dimensions and briefly sketched the statistical mechanics description of ideal fluids.

13.1 Fluids as dynamical systems

Likely, the most interesting instance of high-dimensional chaotic system is the Navier-Stokes equation (NSE)

$$\partial_t \boldsymbol{v} + \boldsymbol{v} \cdot \boldsymbol{\nabla} \boldsymbol{v} = -\frac{1}{\rho} \boldsymbol{\nabla} p + \nu \Delta \boldsymbol{v} + \boldsymbol{f}\,, \qquad \boldsymbol{\nabla} \cdot \boldsymbol{v} = 0\,,$$

which is the Newton's second law ruling an incompressible fluid velocity field $\boldsymbol{v}$ of density ρ and viscosity ν; p being the pressure and $\boldsymbol{f}$ an external driving force. When other fields, such as the temperature or the magnetic field, interact with the fluid velocity $\boldsymbol{v}$, it is necessary to modify the NSE and add new equations, for instance thermal convection is described by Boussinesq's equation (Box B.4). Here and in the following, however, we focus on the NSE.

NSE can be studied in two (2D) or three (3D) space dimensions. While the 3D case is of unequivocal importance, we stress that 2D turbulence is not a mere academical problem but is important and relevant to applications. Indeed, thanks to Earth rotation and density stratification, both the atmosphere and oceans dynamics are well approximated by the two-dimensional NSE, at least as far as large scale motions are concerned [Dritschell and Legras (1993); Monin and Yaglom (1975)].

It is worth remarking from the outset that two-dimensional fluids are quite different from three-dimensional ones, as evident by re-writing the NSE in terms of the vorticity $\boldsymbol{\omega} = \boldsymbol{\nabla} \times \boldsymbol{v}$. In 2D vorticity is a scalar $\omega(\boldsymbol{x}, t)$ (or, more precisely, a vector perpendicular to the plane identified by the fluid, i.e. $\boldsymbol{\omega} = (0, 0, \omega)$) which, neglecting external forces, obeys the equation

$$\partial_t \omega + (\boldsymbol{v} \cdot \boldsymbol{\nabla})\, \omega = \nu \Delta \omega\,,$$

while in 3D $\boldsymbol{\omega}(\boldsymbol{x}, t)$ is a vector field ruled by

$$\partial_t \boldsymbol{\omega} + (\boldsymbol{v} \cdot \boldsymbol{\nabla})\, \boldsymbol{\omega} = (\boldsymbol{\omega} \cdot \boldsymbol{\nabla})\, \boldsymbol{v} + \nu \Delta \boldsymbol{\omega}\,.$$

The 2D equation is formally identical to the transport equation for a passive scalar field (see Eq. (11.5)) so that, in the inviscid limit ($\nu = 0$), vorticity is conserved along the motion of each fluid element. Such a property stands at the basis of a theorem for the existence of a regular solution of the 2D NSE, valid at any time and for arbitrary ν. On the contrary, in 3D the term $(\boldsymbol{\omega} \cdot \boldsymbol{\nabla})\boldsymbol{v}$, which is at the origin of vorticity stretching, constitutes the core of the difficulties in proving the existence and uniqness of a solution, at any time t and arbitrary ν (see Doering (2009) for a recent review with emphasis on both mathematical and physical aspects of the problem). Currently, only the existence for $t \lesssim 1/\sup_{\boldsymbol{x}} |\boldsymbol{\omega}(\boldsymbol{x}, 0)|$ can be rigorously proved [Rose and Sulem (1978)]. For $\nu = 0$, i.e. the 3D Euler equation, the vorticity stretching term relates to the problem of finite-time singularities, see Frisch *et al.* (2004) for a nice introduction and review on such problem.

Surely, *Fully developed turbulence* (FDT) is the most interesting regime of fluid motion and among the most important high-dimensional chaotic systems. In order to illustrate FDT, we can consider a classical fluid dynamics experiment: in a wind tunnel, an air mass conveyed by a large fan impinges an obstacle, which significantly perturbs the downstream fluid velocity. In principle, the flow features may depend on the fluid viscosity ν, the size L and shape of obstacles, mean wind velocity U, and so on. Remarkably, dimensional analysis reveals that, once the geometry of the problem is assigned, the NSE is controlled by a dimensionless combination U, L and ν, namely the Reynolds number

$$\mathrm{Re} = UL/\nu\,.$$

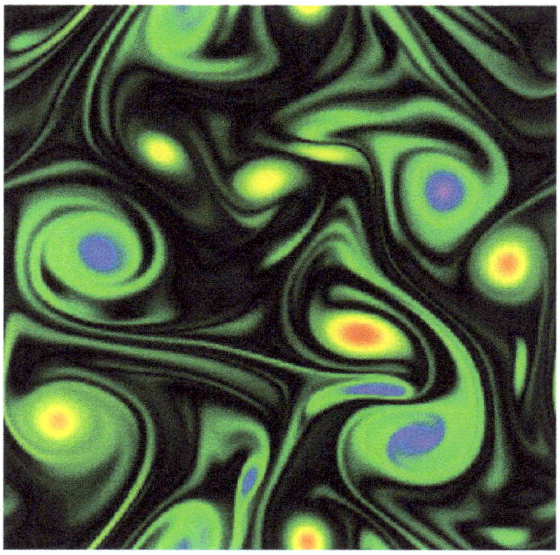

Fig. 13.1 Typical snapshot of the intensity of the vorticity field in two-dimensional turbulent flows.

Increasing the Reynolds number, fluid motion passes through a series of bifurcations, with more and more disordered temporal behaviors, ending in an unpredictable spatiotemporal chaotic behavior, when Re $\gg 1$, characterized by the appearance of large and small whirls. In this regime, all the scales of motion, from that of

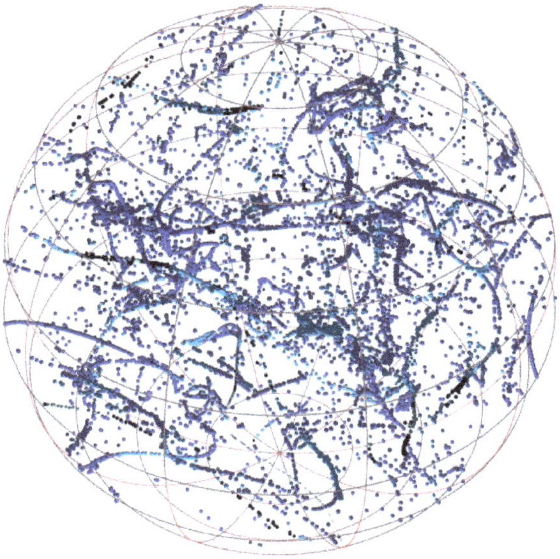

Fig. 13.2 Vorticity filaments in 3D turbulence visualized through the positions of bubbles, colors code the Laplacian of the fluid pressure. [Courtesy of E. Calzavarini]

the obstacle to the very small ones where dissipation takes place, are excited and we speak of Fully Developed Turbulence [Frisch (1995)]. Besides the relevance to applications in engineering or geophysics, the fundamental physical interest in this regime is motivated by the existence, at sufficiently small scales, of universal statistical properties, independent of the geometry, detailed forcing mechanisms and fluid properties [Monin and Yaglom (1975); Frisch (1995)].

Strictly speaking, fully developed turbulence well fits the definition of spatiotemporal chaos given in the previous Chapter. However, we preferred a separate discussion for a tradition based on current literature in the field, and for the main connotative trait of FDT, contrasting with typical spatiotemporal chaotic systems, namely the presence of many active spatial and temporal scales. Such a feature, indeed, makes turbulence somehow similar to critical phenomena [Eyink and Goldenfeld (1994)].

After a brief introduction to the statistical features of perfect fluids and the phenomenology of turbulence, this Chapter will focus on two aspects of turbulence.

The first topic is a general problem we face when studying any partial differential equations. Something we have overlooked before is that each time we use a PDE we are actually coping with an infinite dimensional dynamical system. It is thus relevant understanding whether and how to reduce the description of the problem to a finite number (small or large depending on the specific case) of degrees of freedom, e.g. by passing from a PDE to a finite set of ODEs. For instance, in the context of spatiotemporal chaotic models discussed in Chapter 12, the spontaneous formation of patterns suggests the possibility of a reduced descriptions in terms of the defects (Fig. 12.2). Similar ideas apply also to turbulence where the role of defects is played by coherent structures, such as vortices in two-dimensions (Fig. 13.1) or vortex filaments in three-dimensions (Fig. 13.2). Actually, the dichotomy between descriptions in terms of statistical or coherent-structures approaches is one of the oldest and still unsolved issues of turbulence and high dimensional systems in general [Frisch (1995); Bohr *et al.* (1998)]. Of course, other strategies to reduce the number of degrees of freedom are possible. As we shall see, some of these approaches can be carried out with mathematical rigor, which sometimes can hide the physics of the problem, while some others have a strong physical motivation, but may lack of mathematical rigor.

The second aspect touched by this Chapter concerns the predictability problem in turbulent systems: thinking of the atmosphere, it is clear the great interest and importance of understanding the limits of our possibility to forecast the weather. However, the presence of many time and spatial scales makes often standard tools, like Lyapunov exponents or Kolmogorov-Sinai entropy, inadequate. Moreover, the duality between coherent structures or statistical theories presents itself again when trying to develop a theory of predictability in turbulence. We will brief describe the problems and some attempts in this direction at the end of the Chapter.

13.2 Statistical mechanics of ideal fluids and turbulence phenomenology

Before introducing the phenomenology of fully developed turbulence it is instructive to discuss the basic aspects of the statistical mechanics of three and two dimensional ideal fluids.

13.2.1 *Three dimensional ideal fluids*

Incompressible ideal fluids are ruled by the Euler equation

$$\partial_t \boldsymbol{v} + \boldsymbol{v} \cdot \boldsymbol{\nabla} \boldsymbol{v} = -\frac{1}{\rho} \boldsymbol{\nabla} p, \qquad \boldsymbol{\nabla} \cdot \boldsymbol{v} = 0\,, \tag{13.1}$$

which is nothing but NSE for an inviscid ($\nu = 0$) and unforced fluid. In spite of the fact that it is not Hamiltonian, as briefly sketched below, it is possible to develop an equilibrium statistical mechanical treatment for the Euler equation [Kraichnan (1958); Kraichnan and Montgomery (1980)], in perfect analogy with the micro-canonical formalism used in standard Hamiltonian systems [Huang (1987)].

Consider a fluid contained in a cube of side L and assume periodic boundary conditions, so that the velocity field can be expressed by the Fourier series

$$\boldsymbol{v}(\boldsymbol{x}, t) = \frac{1}{L^{3/2}} \sum_{\boldsymbol{k}} \boldsymbol{u}(\boldsymbol{k}, t)\, \mathrm{e}^{i\boldsymbol{k}\cdot\boldsymbol{x}} \tag{13.2}$$

with $\boldsymbol{k} = 2\pi\boldsymbol{n}/L$ ($\boldsymbol{n} = (n_1, n_2, n_3)$ with n_j integer) denoting the wave-vector. Plugging expression (13.2) into Euler equation (13.1), and imposing an ultraviolet cutoff, $\boldsymbol{u}(\boldsymbol{k}) = 0$, for $k = |\boldsymbol{k}| > k_{max}$, the original PDE is converted in a finite set of ODEs. Then exploiting the incompressibility condition $\boldsymbol{u}(\boldsymbol{k}) \cdot \boldsymbol{k} = 0$, after some algebra, it is possible to identify a subset of independent amplitudes $\{Y_a\}$ from the Fourier coefficients $\{\boldsymbol{u}(\boldsymbol{k}, t)\}$, in term of which the set of ODEs reads

$$\frac{\mathrm{d}Y_a}{\mathrm{d}t} = \sum_{b,c=1}^{N} A_{abc} Y_b Y_c \tag{13.3}$$

where $a = 1, 2, \ldots, N$, with $N \propto k^3_{max}$ being the total number of degrees of freedom considered. In particular, the coefficients A_{abc} have the properties $A_{abc} = A_{acb}$ and $A_{abc} + A_{bca} + A_{cab} = 0$. The latter property, inherited by the nonlinear advection and pressure terms of the Euler equation, ensures energy conservation[1]

$$\frac{1}{2} \sum_{a=1}^{N} Y_a^2 = E = \mathrm{const}\,,$$

while incompressibility ensures the validity of Liouville theorem

$$\sum_{a=1}^{N} \frac{\partial}{\partial Y_a} \frac{\mathrm{d}Y_a}{\mathrm{d}t} = 0\,.$$

[1]Beyond energy, also helicity $H = \int \mathrm{d}\boldsymbol{x}\, (\boldsymbol{\nabla} \times \boldsymbol{v}(\boldsymbol{x}, t)) \cdot \boldsymbol{v}(\boldsymbol{x}, t) = i \sum (\boldsymbol{k} \times \boldsymbol{u}(\boldsymbol{k}, t)) \cdot \boldsymbol{u}(\boldsymbol{k}, t)$ is conserved. However, the sign of H being not well defined it plays no role for the statistical mechanics treatment of Euler equation, and it is thus ignored in the following.

Recalling how equilibrium statistical mechanics of Hamiltonian systems is obtained [Huang (1987)], it is easily recognized that energy conservation and Liouville theorem suffice to derive the microcanonical distribution on the constant energy surface $\frac{1}{2}\sum_a Y_a^2 = E$, the symplectic structure of Hamiltonian system playing no role. In particular, for large N, the invariant probability density of $\{Y_a\}$ is given by

$$P_{inv}(\{Y_a\}) \propto \mathrm{e}^{-\frac{\beta}{2}\sum_{a=1}^{N} Y_a^2},$$

$\beta = 1/T$ being the inverse temperature.

Therefore, 3D Euler equation is well captured by standard equilibrium statistical mechanics with the Gaussian-Gibbs measure. The degrees of freedom are coupled through the nonlinear terms which preserve energy, redistributing it among the Fourier modes so to recover energy equipartition $\langle Y_a^2 \rangle = 2\frac{E}{N} = \beta^{-1}$ among the degrees of freedom.

13.2.2 *Two dimensional ideal fluids*

The statistical mechanics treatment of two dimensional ideal fluids is more delicate as, in principle, there exist an infinite number of conserved quantities (the vorticity of each fluid element is conserved) [Kraichnan and Montgomery (1980)]. However, a generic truncation,[2] necessary to a statistical approach, preserves only two positive quadratic quantities,[3] namely energy $E = \frac{1}{2}\int \mathrm{d}\boldsymbol{x}\, |\boldsymbol{v}(\boldsymbol{x},t)|^2$ and enstrophy $\Omega = \frac{1}{2}\int \mathrm{d}\boldsymbol{x}\, |\boldsymbol{\nabla}\times\boldsymbol{v}(\boldsymbol{x},t)|^2$, which in Fourier space reads

$$E = \frac{1}{2}\sum_a Y_a^2 = \text{const} \quad \text{and} \quad \Omega = \frac{1}{2}\sum_a k_a^2 Y_a^2 = \text{const}\,. \tag{13.4}$$

The presence of an additional constant of the motion has important consequences for the statistical features.

The procedure for deriving equilibrium statistical mechanics is similar to that of 3D fluids and we obtain a set of ODEs as Eq. (13.3) with the additional constraint $k_a^2 A_{abc} + k_b^2 A_{bca} + k_c^2 A_{cab} = 0$ on coefficients A_{abc}, that ensures enstrophy conservation (13.4). Now the microcanonical distribution should be build on the surface where both energy and enstrophy are constant, i.e. $\frac{1}{2}\sum_a Y_a^2 = E$ and $\frac{1}{2}\sum_a k_a^2 Y_a^2 = \Omega$. Therefore, in the large N limit, we have the distribution [Kraichnan and Montgomery (1980)]

$$P_{inv}(\{Y_a\}) \propto \mathrm{e}^{-\frac{1}{2}(\beta_1 \sum_{a=1}^{N} Y_a^2 + \beta_2 \sum_{a=1}^{N} k_a^2 Y_a^2)} \tag{13.5}$$

where the Lagrange multipliers β_1 and β_2 are determined by E and Ω, and

$$\langle Y_a^2 \rangle = \frac{1}{\beta_1 + \beta_2 k_a^2}\,.$$

The above procedure is meaningful only when the system is truncated, $k_{min} \le k_a \le k_{max}$. As $\langle Y_a^2 \rangle$ must be positive, the unique constraint is $\beta_1 + \beta_2 k_a^2 > 0$, which if

[2] For example, setting to zero all modes $k > k_{max}$.

[3] We mention that, as showed by Hald (1976), an "ad hoc" truncation may preserve other constants of motion in addition to energy and enstrophy, but this is not important for what follows.

$k_{min} > 0$ implies that β_1 can be also negative. Therefore, at varying the values of E and Ω, as first recognized by Onsager (1949), both positive and negative temperature are possible in contrast with typical Hamiltonian statistical mechanics (see also Kraichnan and Montgomery (1980)).[4] Roughly speaking states with negative temperature correspond to configurations where energy mainly concentrates in the infrared region, i.e. on large scale structures [Eyink and Spohn (1993)]. Negative temperature states are not an artifact due to the truncated Fourier series expansion of the velocity field, and are present also in the point vortex representation of the 2D Euler equation, see below.

This unconventional property is present also in other fluid dynamical systems such magnetohydrodynamics and geostrophic systems where Eq. (13.5) generalizes to

$$P_{inv}(\{Y_a\}) \propto \mathrm{e}^{-\frac{1}{2}\sum_{ab}\alpha_{a,b}Y_aY_b}\,, \tag{13.6}$$

$\{\alpha_{ab}\}$ being a positive matrix, with entries that depend on both the specific form of the invariants and the values of the Lagrange multipliers.

Numerical results show that systems described by inviscid cut-offed ODEs as Eq. (13.3), with quadratic invariants and Liouville theorem are ergodic and mixing if N is large enough [Orszag and Patterson Jr (1972); Kells and Orszag (1978)], and arbitrary initial distributions of $\{Y_a\}$ evolve towards the Gaussian (13.6).

13.2.3 *Phenomenology of three dimensional turbulence*

Fully developed turbulence corresponds to the limit $\mathrm{Re} = UL/\nu \to \infty$ which, holding the characteristic scale L and velocity U fixed, can be realized for $\nu \to 0$. Therefore, at first glance, we may be tempted to think that FDT can be understood from the equilibrium statistical mechanics of perfect fluids. The actual scenario is completely different.

We start analyzing the various terms of the NSE

$$\partial_t \boldsymbol{v} + \boldsymbol{v}\cdot\boldsymbol{\nabla}\boldsymbol{v} = -\frac{1}{\rho}\boldsymbol{\nabla}p + \nu\Delta\boldsymbol{v} + \boldsymbol{f}\,. \tag{13.7}$$

The forcing term, acting on characteristic scale L, injects energy at an average rate $\langle \boldsymbol{f}\cdot\boldsymbol{v}\rangle = \bar{\epsilon}$, here and hereafter the brackets indicate the average over space and time. As discussed previously, the nonlinear terms ($\boldsymbol{v}\cdot\boldsymbol{\nabla}\boldsymbol{v}$ and $\boldsymbol{\nabla}p$) preserve the total energy and thus simply redistribute it among the modes, i.e. the different scales. Finally, the viscous term, which is mostly acting at small scales,[5] dissipates energy at an average rate $\nu\sum_{i,j}\langle(\partial_j v_i)^2\rangle$.

No matter how large the Reynolds number is, upon waiting long enough, experiments show that a statistically stationary turbulent state settles on. The very

[4] Two dimensional Euler is not the only system where negative temperatures may appear, see Ramsey (1956) for a general discussion of such an issue.

[5] Notice that the dissipation term is proportional to $(\partial_i v_j)^2$ which in Fourier space means a term proportional to k^2, which becomes important at large k's and thus at very small scales.

existence of such a stationary state means that the rate of energy dissipation always balances the input rate [Frisch (1995)]

$$\nu\langle(\partial_j v_i)^2\rangle \approx \overline{\epsilon} = \mathcal{O}(U^3/L)\,, \tag{13.8}$$

where the latter equality stems from dimensional analysis. From this important result we deduce that the limit $\nu \to 0$ is singular, and thus that Euler equation ($\nu = 0$) is different from NSE at high Reynolds number. As a consequence, the statistical mechanics of an inviscid fluids is essentially irrelevant for turbulence [Rose and Sulem (1978); Frisch (1995)]. The non vanishing of the limit $\lim_{\nu\to 0}\nu\langle(\partial_j v_i)^2\rangle = \overline{\epsilon} = \mathcal{O}(U^3/L)$ is technically called *dissipative anomaly* and is at the core of the difficulties in building a theory of turbulence. Noticing that $\nu\sum_{i,j}\langle(\partial_j v_i)^2\rangle = \nu\langle|\boldsymbol{\omega}|^2\rangle$ it is not difficult to realize that the dissipative anomaly is also connected with the mathematical problem of demonstrating the existence, at any time, of the solution of NSE for arbitrary ν.

The action of the various terms in Eq. (13.7) suggests a phenomenological description in terms of the so-called Richardson's *energy cascade* (Fig. 13.3). In this phenomenological framework, forcing acts as a source of excitations generating *eddies* at the scale of energy injection, i.e. patches of fluid correlated over a scale L. Such eddies, thanks to the nonlinear terms, undergo a process of destabilization that "breaks" them in smaller and smaller eddies, generating a cascade of energy (fluctuations of the velocity field) toward smaller and smaller scales. This energy cascade process, depicted in Fig. 13.3, is then arrested when eddies reach a scale ℓ_D small enough for dissipation to be the dominating mechanism. In the range of scales $\ell_D \ll \ell \ll L$, the main contribution comes from the nonlinear (inertial) terms and thus is called *inertial range*. Such a range of scales bears the very authentic nonlinear effects of NSE and thus constitutes the central subject of turbulence research, at least from the theoretical point of view.

Besides the finite energy dissipation, another important and long known experimental result is about the velocity power spectrum $E(k)$[6] which closely follows a power law decay $E(k) \propto k^{-5/3}$ over the inertial range [Monin and Yaglom (1975); Frisch (1995)]. The important fact is that the exponent $-5/3$ seems to be universal, being independent of the fluid and the detailed geometry or forcing. Actually, as discussed below and in Box B.31, a small correction to the $5/3$ value is present, but this seems to be universal. At larger wave-number the spectrum falls off with an exponential-like behavior, whereas the small-k behavior (i.e. at large scales) depends on the mechanism of forcing and/or boundary conditions. A typical turbulence spectrum is sketched in Fig. 13.4. The two crossovers refer to the two characteristic scales of the problem: the excitation scale $L \sim k_L^{-1}$, associated with the energy containing eddies, and the dissipation scale $\ell_D \sim k_D^{-1}$, related to the smallest active eddies. The presence of a power law behavior in between these two extremes unveils that no other characteristic scale is involved.

[6] $E(k)\mathrm{d}k$ is the contribution to the kinetic energy of the Fourier modes in an infinitesimal shell of wave-numbers $[k\!:k+\mathrm{d}k]$.

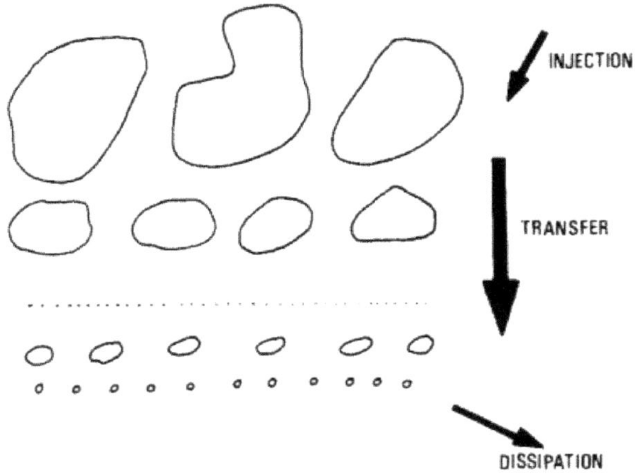

Fig. 13.3 Cartoon illustrating Richardson's cascade of energy in three-dimensional turbulence, with the three basic processes of energy injection, transfer and dissipation.

Besides the power spectrum $E(k)$, central quantities for developing a theoretical understanding of turbulence are the structure functions of the velocity field

$$S_p(\ell) = \langle [(\boldsymbol{v}(\boldsymbol{x}+\boldsymbol{\ell},t) - \boldsymbol{v}(\boldsymbol{x},t)) \cdot \hat{\boldsymbol{\ell}}]^p \rangle \,,$$

which are the p-moments of the velocity difference over a distance $\ell = |\boldsymbol{\ell}|$ projected in the direction of the displacement $\hat{\boldsymbol{\ell}} = \boldsymbol{\ell}/\ell$ (these are more precisely called longitudinal structure functions). We used as unique argument the distance ℓ because we assumed homogeneity (independence of the position $\boldsymbol{x}$), stationarity (independence of t) and isotropy (no dependence on the direction of the displacement $\hat{\boldsymbol{\ell}}$). Unless specified these three properties will always be assumed in the following.

The second order structure functions ($p = 2$) can be written in terms of the spatial correlation function $C_2(\ell)$ as $S_2(\ell) = 2[C_2(0) - C_2(\ell)]$. As, thanks the Wiener-Khinchin theorem, $C_2(\ell)$ is nothing but the Fourier transform of the power spectrum, it is easily obtained that the $5/3$ exponent of the spectrum translates in the power law behavior $S_2(\ell) \sim (\ell/L)^{2/3}$, see Monin and Yaglom (1975) or Frisch (1995) for details. For $p > 2$, we can thus explore higher order statistical quantities than power spectrum. In the following, as we mostly consider dimensional analysis, we shall often disregard the vectorial nature of the velocity field and indicate with $\delta v(\ell)$ a generic velocity difference over a scale ℓ, and with $\delta_{\|} v(\ell)$ the longitudinal difference, $\delta_{\|} v(\ell) = [(\boldsymbol{v}(\boldsymbol{x}+\boldsymbol{\ell},t) - \boldsymbol{v}(\boldsymbol{x},t)) \cdot \hat{\boldsymbol{\ell}}]$.

A simple and elegant explanation of the experimental findings on the energy spectrum is due to Kolmogorov (1941) (K41). In a nutshell, K41 theory assumes the Richardson cascade process (Fig. 13.4) and focuses on the inertial range, where we can safely assume that neither injection nor dissipation play any role. Thus, in the inertial range, the only relevant quantity is the injection or, equivalently (via

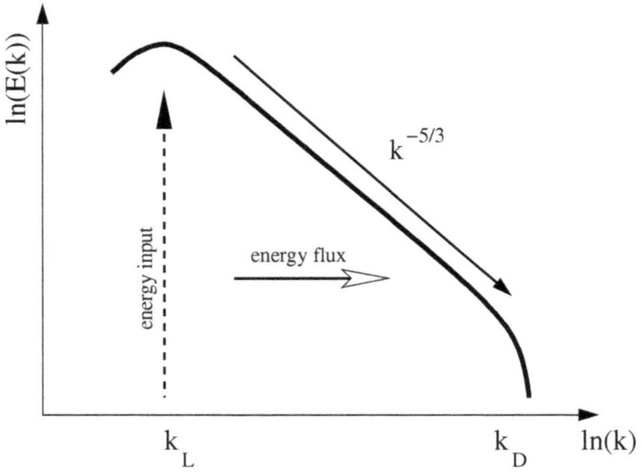

Fig. 13.4 Sketch of a typical turbulent energy spectrum, $L \approx k_L^{-1}$ is the energy containing integral scale and $\ell_D \approx k_D^{-1}$ the dissipative Kolmogorov scale.

Eq. (13.8)), the dissipation rate $\bar{\epsilon}$. This means that the statistical properties of the velocity field should only depend on $\bar{\epsilon}$ and the scale ℓ. The unique dimensional combination of the two leads to the K41 scaling law

$$\delta v(\ell) \sim (\bar{\epsilon}\ell)^{1/3} \sim U(\ell/L)^{1/3}, \tag{13.9}$$

which also yields the result that the energy transfer rate at scale ℓ, which can estimated to be $\delta v^3(\ell)/\ell$,[7] is constant and equal to the dissipation rate, $\delta v^3(\ell)/\ell \approx \bar{\epsilon}$. Notice that Eq. (13.9) implied that, in the inertial range, the velocity field is only Hölder continuous, i.e. non-differentiable, with Hölder exponent $h = 1/3$. Neglecting the small correction to the spectrum exponent (discussed in Box B.31), this dimensional result explains the power spectrum behavior as it predicts

$$E(k) = C_K \bar{\epsilon}^{2/3} k^{-5/3},$$

where C_K is a constant, whose possible universality should be tested experimentally as dimensional arguments provide no access to its value. Moreover, the scaling (13.9) agrees with an exact result, again derived by Kolmogorov in 1941 from the Navier-Stokes equation, known as "4/5 law" stating that [Frisch (1995)]

$$\langle \delta v_{||}^3(\ell) \rangle = -\frac{4}{5}\bar{\epsilon}\ell. \tag{13.10}$$

Assuming that the scaling (13.9) holds down to the dissipative scale ℓ_D (called the Kolmogorov length), setting to order unity the "local Reynolds numbers" $\ell_D \delta v(\ell_D)/\nu = \mathcal{O}(1)$, we can estimate how ℓ_D changes with Re

$$\ell_D \sim L \mathrm{Re}^{-3/4}.$$

[7] i.e. given by the ratio between energy fluctuation at that scale $\delta v^2(\ell)$ and the characteristic time at scale ℓ, dimensionally given by $\ell/\delta v(\ell)$.

A natural extension of K41 theory to higher order structure functions leads to

$$S_p(\ell) \sim \left(\frac{\ell}{L}\right)^{\zeta_p}$$

with $\zeta_p = p/3$. Even based on a phenomenological ground, the above result would provide a rather complete understanding of the statistical properties of turbulence, if confirmed by experiments. Actually, experimental and numerical results, [Anselmet *et al.* (1984); Arnèodo *et al.* (1996)] have shown that K41 scaling $\zeta_p = p/3$ is not exact. Indeed the exponent ζ_p is a nonlinear function (Box B.31), with $\zeta_3 = 1$ as a consequence of the "4/5 law". Such nonlinear behavior as a function of p indicates a breakdown of the perfect self-similarity characterizing Kolmogorov-Richardson energy cascade. Larger and larger deviations from mean values are observed as smaller and smaller scales are sampled: a phenomenon going under the name of intermittency [Frisch (1995); Bohr *et al.* (1998)].

Until the '90s there was an alive debate on whether such deviations from K41 scaling were just a finite Reynolds effect, disappearing at very high Reynolds numbers, or a genuine Re $\to \infty$ behavior. Nowadays, thanks to accurate experiments and numerical simulations, a general consensus has been reached on the fact that intermittency in turbulence is a genuine phenomenon [Frisch (1995)], whose first-principle theoretical explanation is still lacking. Some steps towards its understanding have been advanced in the simpler, but still important, case of passive scalar transport (Sec. 11.2) in turbulent flows, where the mechanisms of intermittency for the scalar field have been unveiled [Falkovich *et al.* (2001)]. Nevertheless, as far as intermittency in fluid turbulence is concerned, a rather powerful phenomenological theory has been developed over the years which is able to account for many aspects of the problem. This is customarily known as the multifractal model of turbulence which was introduced by Parisi and Frisch (1985) (see Box B.31 and Boffetta *et al.* (2008) for a recent review).

Box B.31: Intermittency in three-dimensional turbulence: the multifractal model

This Box summarizes the main aspects of the multifractal model of turbulence. First introduced by Parisi and Frisch (1985), this phenomenological model had and important role in statistical physics, disordered systems and chaos. Among its merits there is the recognition of the inexactness of the original idea, inherited by critical phenomena, that just few scaling exponents are relevant to turbulence (and more generally in complex systems). Nowadays, is indeed widely accepted that an infinite set of exponents is necessary for characterizing the scaling properties of 3D turbulent flows. As already underlined in Sec. 5.2.3, from a technical point of view the multifractal model is basically a large deviation theory (Box B.8). We start noticing that the Navier-Stokes equation is formally invariant under the scaling transformation:

$$\boldsymbol{x} \to \chi \boldsymbol{x} \quad \boldsymbol{v} \to \chi^h \boldsymbol{v}, \quad t \to \chi^{1-h} t, \quad \nu \to \chi^{h+1} \nu,$$

with $\chi > 0$. Notice also that such a transformation leaves the Reynolds number invariant. Symmetry considerations cannot determine the exponent h, which at this level is a free parameter. K41 theory corresponds to global invariance with $h = 1/3$, which is in disagreement with experiments and simulations [Anselmet *et al.* (1984); Arnèodo *et al.* (1996)], which provides convincing evidence that the exponent of the structure functions ζ_p is a nonlinear function of p (Fig. B31.1), implying a breakdown of global invariance in the turbulent cascade. It can be shown that K41 theory corresponds to assume energy dissipation to occur homogeneously in the full three-dimensional space [Paladin and Vulpiani (1987); Frisch (1995)], which somehow disagrees with the sparse vorticity structures observed in high Re flows (Fig. 13.2)[8]. A simple extension of K41 thus consists in assuming the energy dissipation uniformly distributed on a homogeneous fractal with dimension $D_F < 3$. In this simplified view, the active eddies of size ℓ contributing to energy flux do not fill the whole space but only a fraction $\propto \ell^{3-D_F}$. As the energy flux is dimensionally given by $\delta v^3(\ell)/\ell$, and it is on average constant and equal to $\bar{\epsilon}$, i.e. $\ell^{3-D_F}\delta v^3(\ell)/\ell \approx \bar{\epsilon}$, assuming the scaling $\delta v(\ell) \sim \ell^h$ we have $h = 1/3 - (3 - D_F)/3$, which recovers K41 for $D_F = 3$. This assumption (called absolute curdling or β-model [Frisch (1995)]) allows for a small correction to K41, but still in the framework of global scale invariance. In particular, it predicts $\zeta_p = (D_F - 2)p/3 + (3 - D_F)$ which, for $D_F \simeq 2.83$ is in fair agreement with the experimental data for $p \lesssim 6 - 7$, but it fails in describing the large p behavior, which is clearly nonlinear in p.

Gathering up experimental observations, the multifractal model assumes local scaling invariance for the velocity field, meaning that the exponent h is not unique for the whole space. The idea is to think the space as parted in many fractal sets each with fractal dimension $\mathcal{D}(h)$, where $\delta v(\ell) \sim \ell^h$ [Frisch (1995); Benzi *et al.* (1984)]. More formally, it is assumed that in the inertial range

$$\delta v_{\boldsymbol{x}}(\ell) \sim \ell^h ,$$

if $\boldsymbol{x} \in S_h$, where S_h is a fractal set having dimension $\mathcal{D}(h)$ with h belonging to a certain interval of values $h_{min} < h < h_{max}$. In this way the probability to observe a given scaling exponent h at scale ℓ is $P_h(\ell) \sim (\ell/L)^{3-\mathcal{D}(h)}$, and the scaling exponents of the structure function can be computed as

$$S_p(\ell) = \langle |\delta v(\ell)^p \rangle \sim \int_{h_{min}}^{h_{max}} \mathrm{d}h \, \left(\frac{\ell}{L}\right)^{hp+(3-\mathcal{D}(h))} \sim \left(\frac{\ell}{L}\right)^{\zeta_p} . \qquad \text{(B.31.1)}$$

As $\ell/L \ll 1$, the integral in Eq. (B.31.1) can be approximated by the steepest descent method, which gives

$$\zeta_p = \min_h \{hp + 3 - \mathcal{D}(h)\} ,$$

so that $\mathcal{D}(h)$ and ζ_p are related by a Legendre transform. The "4/5 law", $\zeta_3 = 1$, imposes

$$\mathcal{D}(h) \leq 3h + 2 . \qquad \text{(B.31.2)}$$

K41 corresponds to the case of a unique singularity exponent $h = 1/3$ with $\mathcal{D}(h = 1/3) = 3$; similarly, for β-model $h = (D_F - 2)/3$ with $\mathcal{D}(h = (D_F - 2)/3) = D_F$.

Unfortunately, no method is known to directly compute $\mathcal{D}(h)$, or equivalently ζ_p, from NSE. Therefore, we should resort to phenomenological models. A first step in this direction is a represented by a simple multiplicative processes known as random β-model [Benzi

[8] We recall that energy dissipation is proportional to enstrophy, i.e. the square vorticity.

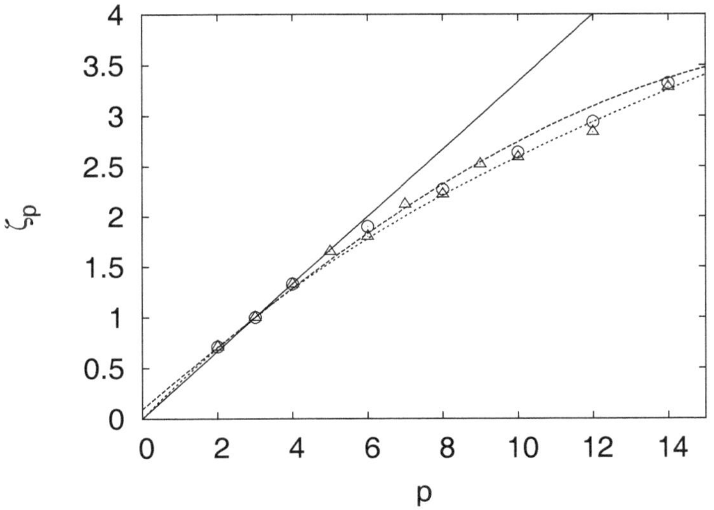

Fig. B31.1 Structure function scaling exponents ζ_p plotted vs p. Circles and triangles correspond to the data of Anselmet *et al.* (1984). The solid line corresponds to Kolmogorov scaling $p/3$; the dashed line is the random beta model prediction (B.31.3) with $B = 1/2$ and $x = 7/8$; the dotted line is the She and Lévêque (1994) prediction (B.31.4) with $\beta = 2/3$.

et al. (1984)]. It describes the energy cascade through eddies of size $\ell_n = 2^{-n}L$, L being the energy injection length. At the n-th step of the cascade, a mother eddy of size ℓ_n splits into daughter eddies of size ℓ_{n+1}, and the daughter eddies cover a fraction $\beta_j \leq 1$ of the mother volume. As the energy flux is constant throughout the scales, $v_n = \delta v(\ell_n)$ receives contributions only on a fraction of volume $\prod_{j=1}^{n} \beta_j$, so that $v_n = v_0 \ell_n^{1/3} \prod_{j=1}^{n} \beta_j^{-1/3}$ where the β_j's are independent, identically distributed random variables. A reasonable phenomenological assumption is to imagine a turbulent flow as composed by laminar and singular structures. This can be modeled by taking $\beta_j = 1$ with probability x and $\beta_j = B = 2^{-(1-3h_{min})}$ with probability $1 - x$, h_{min} setting the most singular structures of the flow. The above multiplicative process generates a two-scale Cantor set (Sec. 5.2.3) with a fractal dimension spectrum

$$\mathcal{D}(h) = 3 + (3h - 1)\left[1 + \log_2\left(\frac{1-3h}{1-x}\right)\right] + 3h\log_2\left(\frac{x}{3h}\right),$$

while the structure functions exponents are

$$\zeta_p = p/3 - \log_2[x + (1-x)B^{1-p/3}]. \qquad \text{(B.31.3)}$$

Two limit cases are $x = 1$, corresponding to K41, and $x = 0$ which is the β-model with $D_F = 2 + 3h_{min}$. Setting $x = 7/8$ and $h_{min} = 0$ (i.e. $B = 1/2$), Eq. (B.31.3) provides a fairly good fit of the experimental exponents (Fig. B31.1).

In principle, as we have the freedom to choose the function $\mathcal{D}(h)$, i.e. an infinite number of free parameter, the fit can be made as good as desired. The nice aspect of the random β-model is to have reduced this infinite set of parameters to a few ones, chosen on phenomenological ground. Another popular choice is She and Lévêque (1994) model which gives

$$\zeta_p = (2\beta - 1)p/3 + 2(1 - \beta^{p/3}) , \tag{B.31.4}$$

in good agreement with experimental data for $\beta = 2/3$.

Although far from being a first principles model, the multifractal model allows for predicting other nontrivial statistical features [Frisch (1995)], such as the pdf of the velocity gradient [Benzi *et al.* (1991)], the existence of an intermediate dissipative range [Frisch and Vergassola (1991); Biferale *et al.* (1999)] and precise scaling predictions for Lagrangian quantities [Arnèodo *et al.* (2008)]. Once $\mathcal{D}(h)$ is obtained by fitting the experimental data, then all the predictions obtained in the multifractal model framework must be checked without additional free parameters [Boffetta *et al.* (2008)].

The multifractal model for turbulence links to the $f(\alpha)$ vs α description of the singular measures in chaotic attractors presented in Sec. 5.2.3. In order to show this connection let us recall Kolmogorov (1962) (K62) revised theory [Frisch (1995)] stating that velocity increments $\delta v(\ell)$ scale as $(\epsilon_\ell \ell)^{1/3}$ where ϵ_ℓ is the energy dissipation space-averaged over a cube of side ℓ. Let us introduce the measure $\mu(\boldsymbol{x}) = \epsilon(\boldsymbol{x})/\bar{\epsilon}$, a partition of non overlapping cells of size ℓ and the coarse-graining probability $P_i(\ell) = \int_{\Lambda_\ell(\boldsymbol{x})} d\mu(\boldsymbol{y})$, where $\Lambda_l(\boldsymbol{x}_i)$ is a side-ℓ cube centered in $\boldsymbol{x}_i$, of course $\epsilon_\ell \sim \ell^{-3} P(\ell)$. Following the notation of Sec. 5.2.3, denote with α the scaling exponent of $P(\ell)$ and with $f(\alpha)$ the fractal dimension of the sub-fractal having scaling exponent α, we can introduce the generalized dimensions $D(p)$:

$$\sum_i P_i(\ell)^p \sim \ell^{(p-1)D(p)} \quad \text{with} \quad (p-1)D(p) = \min_\alpha [p\alpha - f(\alpha)] \ .$$

Noting that $\langle \epsilon_\ell^p \rangle = \ell^3 \sum \epsilon_\ell^p$, we have $\langle \epsilon_\ell^p \rangle \sim \ell^{(p-1)(D(p)-3)}$; therefore the correspondences

$$h \leftrightarrow \frac{\alpha - 2}{3} \ , \ \ \mathcal{D}(h) \leftrightarrow f(\alpha) \ , \ \ \zeta_p = \frac{p}{3} + \left(\frac{p}{3} - 1\right)\left(D(p/3) - 3\right)$$

can be established. Notice that having assumed $\delta v(\ell) \sim (\epsilon_\ell \ell)^{1/3}$ the result $\zeta_3 = 1$ holds independently of the choice for $f(\alpha)$. We conclude, noticing that the lognormal theory K62, where $\zeta_p = p/3 + \mu p(3 - p)/18$, is a special case of the multifractal model with $\mathcal{D}(h)$ being a parabola having maximum at $D_F = 3$, while the parameter μ is determined by the fluctuation of $\ln(\epsilon_\ell)$ [Frisch (1995)].

13.2.4 *Phenomenology of two dimensional turbulence*

In 2D, the phenomenology of turbulence is rather different. The major source of difference comes from the fact that Euler equation ($\nu = 0$) in two dimension preserves vorticity of each fluid elements. For $\nu \neq 0$ this conservation entails the absence of dissipative anomaly, meaning that

$$\nu \sum_{i,j} \langle (\partial_i v_j)^2 \rangle = \nu \langle \omega^2 \rangle = \nu \langle \Omega \rangle = \mathcal{O}(\nu) \ .$$

Under these circumstances the energy cascade scenario *à la* Richardson, with a constant flux of energy from the injection to the dissipative scale (Fig. 13.3), does not hold anymore. The energy cascade towards the small scales with a constant energy flux would indeed lead to an unbounded growth of enstrophy $\langle \Omega \rangle \to \infty$, which in

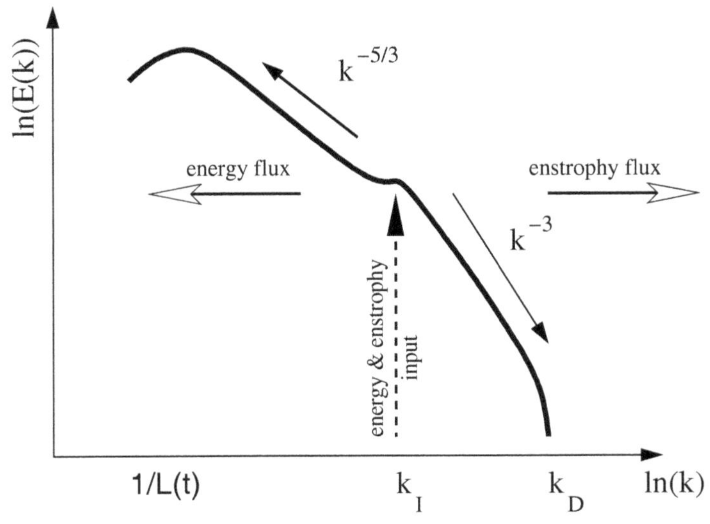

Fig. 13.5 Sketch of the energy spectrum $E(k)$ of two-dimensional turbulence.

the unforced case is conserved. The regularity of the limit $\nu \to 0$ means that, unlike 3D turbulence, energy is not dissipated any longer when $\mathrm{Re} \to \infty$. Therefore, the system cannot establish a statistically steady state. These observations pose thus a conundrum on the fate of energy and enstrophy in 2D turbulence.

In a seminal work Kraichnan (1967) was able to compose this puzzle and to build a theory of two-dimensional turbulence, incorporating the above discussed observations. The idea is as follows. Due to the the forcing term, energy and enstrophy are injected on scale L_I (wave-number $k_I \sim 1/L_I$) at a rate $\overline{\epsilon} = \langle \boldsymbol{f} \cdot \boldsymbol{v} \rangle$ and $\overline{\eta} = \langle (\boldsymbol{\nabla} \times \boldsymbol{f})\omega \rangle$, respectively. Then, a double cascade establishes thanks to the nonlinear transfer of energy and enstrophy among the modes (scales): energy flows towards the large scales ($\ell > L_I$) while enstrophy towards the small scales ($\ell < L_I$). In the following we analyze separately these two processes and their consequences on the energy spectrum, which is sketched in Fig. 13.5.

As time proceeds, the *inverse*[9] energy cascade establishes generating a velocity field correlated on a time-growing scale $L(t)$. In the range of scales $L_I \ll \ell \ll L(t)$, analogously to K41 theory of 3D turbulence, the statistical features of the velocity should only depend on the energy flux $\overline{\epsilon}$ and the scale, so that by dimensional reasoning we have

$$\delta v(\ell) \sim (\overline{\epsilon}\ell)^{1/3} .$$

In other terms, in the range of wave numbers $1/L(t) \ll k \ll k_I \approx 1/L_I$, the power spectrum behaves as in 3D turbulence (Fig. 13.5)

$$E(k) \sim \overline{\epsilon}^{2/3} k^{-5/3} ,$$

[9]In contrast with the *direct* (toward the large wave-numbers, i.e. small scales) energy cascade of 3D turbulence.

and K41 scaling $\zeta_p = p/3$ is expected for the structure functions, also in agreement with the 2D equivalent of the "4/5" law which is the "3/2" law [Yakhot (1999)]

$$\langle \delta v_{||}^3(\ell) \rangle = \frac{3}{2}\bar{\varepsilon}\ell \,.$$

It is noteworthy that the r.h.s. of the above equation has an opposite sign with respect to Eq. (13.10), this is the signature of the cascade being directed towards the large scales. In bounded systems the large scale $L(t)$ cannot grow arbitrarily, the inverse cascade will be sooner or later stopped when the largest available scale is reached, causing the condensation of energy [Smith and Yakhot (1993)]. The latter phenomenon is often eliminated by the presence of a large scale energy dissipation mechanism, due to friction of the fluid with the bottom or top surface which can be modeled adding to the r.h.s. of NSE a term of form $-\alpha \boldsymbol{v}$ (known as Ekman friction). This extra dissipative mechanism is usually able to stop the cascade at a scale larger than that of injection, $L_\alpha > L_I$, but smaller than the domain size.

At scales $\ell < L_I$, the energy transfer contribution is negligible and a direct cascade of enstrophy takes place, where the rate of enstrophy dissipation $\overline{\eta}$ plays the role of $\overline{\epsilon}$. Physical arguments similar to K41 theory suggest that the statistical features of velocity differences should only depend on the scale ℓ and the enstrophy flux $\overline{\eta}$. It is then easily checked that there exists a single possible dimensional combination giving

$$\delta v(\ell) \sim \overline{\eta}\ell$$

for scales comprised in between the injection L_I and dissipative scale $\ell_D \sim L\mathrm{Re}^{-1}$ (where viscous forces dominate the dynamics). The above scaling implies that the velocity field is smooth (differentiable) with spectrum (Fig. 13.5)

$$E(k) \sim \overline{\eta}^{2/3} k^{-3} \quad ,$$

for $k_I < k < k_D (\approx \ell_D^{-1})$. Actually, a refined treatment lead Kraichnan and Montgomery (1980) to a slightly different spectrum $E(k) \sim k^{-3}[\ln(k/k_D)]^{-1/3}$, which would be more consistent with some of the assumptions of the theory, see Rose and Sulem (1978) for a detailed discussion.

Nowadays, as supported by experimental [Tabeling (2002)] and numerical [Boffetta (2007)] evidences, there is a quite wide consensus on the validity of the double cascade scenario. Moreover, theoretical arguments [Yakhot (1999)] and numerical simulations [Boffetta *et al.* (2000c)] have shown that the inverse cascade is in extremely good agreement with Kraichnan predictions. In particular, no significant deviations from K41 scaling have been detected with the statistics of the velocity increments deviating very mildly from the Gaussian. The situation is much less clear for the enstrophy direct cascade, where deviations from the predicted spectrum are often observed for $k \gg k_I$ and even universality with respect to the forcing has been questioned (see e.g. Frisch (1995)).

It is worth concluding this overview by mentioning that 2D turbulence is characterized by the emergence of coherent structures (typically vortices, see Fig. 13.1) which, especially when considering the decaying (unforced) problem, eventually dominate the dynamics [McWilliams (1984)]. Coherent structures are rather regular weakly-dissipative regions of fluids in the turbulent background flow, whose interactions can be approximately described by a conservative dynamics. We shall reconsider coherent structures in the sequel when discussing point vortices and predictability.

13.3 From partial differential equations to ordinary differential equations

In abstract terms, the Navier-Stokes equation, as any PDE, has infinitely many degrees of freedom. While this is not a big issue when dealing with mathematical approaches, it constitutes a severe limitation for numerical computations. Of course, there are, more or less standard, rigorous ways to reduce the original PDE to a set of ODEs, such as finite differences schemes, discrete Fourier transforms etc., necessary to perform a Direct Numerical Simulation[10] (DNS) of the NSE. However, this may be not enough when the number of degrees of freedom becomes daunting huge. When this happens, clever, though often less rigorous, methods must be employed. Typically, these techniques allow the building up of a set of ODEs with (relatively) few degrees of freedom, which model the original dynamics, and thus need to be triggered by physical hints. The idea is then to make these ODEs able to describe, at least, some specific features of the problem under investigation.

Before illustrating some of these methods, it is important to reckon the degrees of freedom of turbulence, meaning the minimal number of variables necessary to describe a turbulent flow.

13.3.1 *On the number of degrees of freedom of turbulence*

Suppose that we want to discretize space and time to build a numerical scheme for a DNS of the Navier-Stokes equation, how many modes or grid points $\mathcal{N}$ do we need in order to faithfully reproduce the flow features?[11]

Through the 3D turbulence phenomenological theory (Sec. 13.2.3) we can estimate the Kolmogorov length (fixing the border between inertial and dissipative behaviors) as

$$\ell_D \sim L\mathrm{Re}^{-3/4}$$

where L, as usual, denotes the typical large scale. Being ℓ_D the smallest active scale, for an accurate simulation we need to resolve, at least, scales $\lesssim \ell_D$. We must thus

[10] The term direct numerical simulation is typically used to indicate numerical schemes aiming to integrate in details and faithfully a given equation.
[11] The faithful reproducibility can be tested, e.g. by checking that increasing the number of grid points or decreasing the time step do not change significantly the results.

employ a spatial mesh $\Delta x \lesssim \ell_D$ or a maximal Fourier wave-number $k_{max} \gtrsim \ell_D^{-1}$. Therefore, we can roughly estimate the number of degrees of freedom to be

$$\mathcal{N} \sim \left(\frac{L}{\ell_D}\right)^3 \sim \mathrm{Re}^{9/4} .$$

Considering that in laboratory setups and in the atmosphere, Re ranges in $\mathcal{O}(10^4) - \mathcal{O}(10^{18})$, for instance the Reynolds number of a person swimming in the pool is about 4×10^6 while that of a blue whale in the sea is 3×10^8, it is easily realized that $\mathcal{N}$ is typically huge. The above formula is based on K41, taking into account intermittency (Box B.31) minor corrections to the exponent 9/4 should be considered [Paladin and Vulpiani (1987); Bohr *et al.* (1998)].

An additional practical difficulty in DNS of 3D turbulence relates to the necessary time step Δt. Each scale ℓ is characterized by a characteristic time, typically dubbed eddy turnover time, which can be dimensionally estimated as

$$\tau(\ell) \sim \ell/\delta v(\ell) \sim LU^{-1}(\ell/L)^{2/3} , \tag{13.11}$$

meaning that turbulence possesses many characteristic times hierarchically ordered from the slowest $\tau_L = L/U$, associated with the large scales, to the fastest $\tau_D \sim \tau_L \mathrm{Re}^{-1/2}$, pertaining to the Kolmogorov scale. Of course, a faithful and numerically stable computation requires $\Delta t \lesssim \tau_D$. Consequently, the number of time steps necessary for integrating the flow over a time period τ_L grows as $\mathcal{N}_T \sim \mathrm{Re}^{1/2}$, meaning that the total number of operations grows as $\mathcal{N}\mathcal{N}_T \sim \mathrm{Re}^{11/4}$. Such bounds discourage any attempt to simulate turbulent flows with $\mathrm{Re} \gtrsim 10^6$ (roughly enough for a swimming person, far below for a blue whale!). Therefore, in typical geophysical and engineering applications small scales modeling is unavoidable.[12] For an historical and foreseeing discussion about DNS of 3D turbulence see Celani (2007).

In 2D, the situation is much better. The dissipative scale, now called Kraichnan length, behaves as $\ell_D \sim L\mathrm{Re}^{-1/2}$ [Kraichnan and Montgomery (1980)], so that

$$\mathcal{N} \sim \left(\frac{L}{\ell_D}\right)^2 \sim \mathrm{Re} .$$

Therefore, detailed DNS can be generically performed without the necessity of small scale parametrization also for rather large Reynolds numbers. However, when simulating the inverse energy cascade, the slowest time scale associated to the growing length $L(t)$ (see Sec. 13.2.4), which is still growing with $\mathrm{Re}^{1/2}$, may put severe bounds to the total integration time.

More rigorous estimates of the number of degrees of freedom can be obtained in terms of the dimension of the strange attractor characterizing turbulent flows, by

[12] One of the most common approach is the so-called large eddy simulation (LES). It was formulated and used in the late '60s by Smagorinsky to simulate atmospheric air currents. During the '80s and '90s it became widely used in engineering [Moeng (1984)]. In LES the large scale motions of the flow are calculated, while the effect of the smaller universal (so-called sub-grid) scales are suitably modeled.

using the Lyapunov (or Kaplan-Yorke) dimension D_L (Sec. 5.3.4). In particular, Doering and Gibbon (1995) and Robinson (2007) found that in 2D

$$D_L \leq C_1 \mathrm{Re}\,(1 + C_2 \ln \mathrm{Re}) \ ,$$

which is in rather good agreement, but for a logarithmic correction, to the phenomenological prediction. While in 3D they estimated

$$D_L \leq C \left(\frac{L}{\ell_D}\right)^{4.8} \sim \mathrm{Re}^{3.6} \, .$$

The three dimensional bound, as consequences of technical difficulties, is not very strict. Indeed it appears to be much larger than the phenomenologically predicted result $\mathrm{Re}^{9/4}$.

13.3.2 *The Galerkin method*

Knowing the bounds to the minimum number of degrees of freedom necessary to simulate a turbulent flow, we can now discuss a mathematically rigorous technique to pass from the NSE to a set of ODEs, faithfully reproducing it. In particular, we aim describing the Galerkin method, which was briefly mentioned in Box B.4 while deriving the Lorenz model.

The basic idea is to write the velocity field (as well as the pressure and the forcing) in terms of a complete, orthonormal (infinite) set of eigenfunctions $\{\psi_n(\boldsymbol{x})\}$:

$$\boldsymbol{v}(\boldsymbol{x}, t) = \sum_n \boldsymbol{a}_n(t) \psi_n(\boldsymbol{x}) \, . \tag{13.12}$$

Substituted the expansion (13.12) in NSE, the original PDE is transformed in an infinite set of ODEs for the coefficients $\boldsymbol{a}_n(t)$. Considering the infinite sum, the procedure is exact but useless as we still face the problem of working with an infinite number of degrees of freedom. We thus need to approximate the velocity field by truncating (13.12) to a finite N, by imposing $\boldsymbol{a}_n = 0$ for $n > N$, so to obtain a finite set of ODEs. From the previous discussion on the number $\mathcal{N}$ of variables necessary for a turbulent flow, it is clear that, provided the eigenfunctions $\{\psi_n(\boldsymbol{x})\}$ are suitably chosen, such an approximation can be controlled if $N \gtrsim \mathcal{N}(\mathrm{Re})$. This can actually be rigorously proved [Doering and Gibbon (1995)].

The choice of the functions $\{\psi_n(\boldsymbol{x})\}$ depends on the boundary conditions. For instance, in 2D, with periodic boundary conditions on a square of side-length 2π, we can expand the velocity field in Fourier series with a finite number of modes belonging to a set $\mathcal{M}$. Accounting also for incompressibility, the sum reads

$$\boldsymbol{v}(\boldsymbol{x}, t) = \sum_{\boldsymbol{k} \in \mathcal{M}} Q_{\boldsymbol{k}}(t) \frac{\boldsymbol{k}^{\perp}}{|\boldsymbol{k}|} \mathrm{e}^{i\boldsymbol{k}\cdot\boldsymbol{x}} \, , \tag{13.13}$$

where $\boldsymbol{k}^{\perp} = (k_2, -k_1)$. In addition the reality of $\boldsymbol{v}(\boldsymbol{x}, t)$ implies $Q_{\boldsymbol{k}} = -Q^*_{-\boldsymbol{k}}$. Plugging the expansion (13.13) into NSE, the following set of ODEs is obtained

$$\frac{\mathrm{d}Q_{\boldsymbol{k}}}{\mathrm{d}t} = -i \sum_{\boldsymbol{k}+\boldsymbol{k}'+\boldsymbol{k}''=0} \frac{(\boldsymbol{k}')^{\perp} \cdot \boldsymbol{k}''(k''^2 - k'^2)}{2kk'k''} Q^*_{\boldsymbol{k}'} Q^*_{\boldsymbol{k}''} - \nu k^2 Q_{\boldsymbol{k}} + f_{\boldsymbol{k}} \tag{13.14}$$

where $\boldsymbol{k}, \boldsymbol{k}'$ and $\boldsymbol{k}''$ belong to $\mathcal{M}$ and if $\boldsymbol{k} \in \mathcal{M}$ then $-\boldsymbol{k} \in \mathcal{M}$ [Lee (1987)], and $f_{\boldsymbol{k}}$ is the Fourier coefficient of the forcing.

When the Reynolds number is not too large, few modes are sufficient to describe NSE dynamics. As already discussed in Chap. 6, in a series of papers Franceschini and coworkers investigated in details, at varying the Reynolds number, the dynamical features of system (13.14) with a few number of modes ($N = 5 - 7$) for understanding the mechanisms of transition to chaos [Boldrighini and Franceschini (1979); Franceschini and Tebaldi (1979, 1981)]. In particular, for $N = 5$ they observed, for the first time in a system derived from first principles, Feigenbaum period doubling scenario.

Galerkin method, with a few practical modifications, such as the so-called pseudo-spectral method,[13] can be used as a powerful DNS method for NSE both in 2D and 3D, with the already discussed limitation in the Reynolds number that can be reached [Celani (2007)]. Other eigenfunctions often used in DNS are the wavelets [Farge (1992)].

13.3.3 *Point vortices method*

In two-dimensional ideal fluids, the Euler equation can be reduced to a set of ODEs in an exact way for special initial conditions, i.e. when the vorticity at the initial time $t = 0$ is localized on N point-vortices. In such a case, vorticity remains localized and the two-dimensional Euler equation reduces to $2N$ ODEs. We already examined the case of few ($N \sim 2 - 4$) point-vortices systems in the context of transport in fluids (Sec. 11.2.1.2). For moderate values of N, the point-vortex system has been intensively studied in different contexts from geophysics to plasmas [Newton (2001)]. Here, we reconsider the problem when N is large. As shown in Box B.25, in the case of an infinite plane the centers $\{\boldsymbol{r}_i = (x_i, y_i)\}$ of the N vortices evolve according to the dynamics

$$\Gamma_i \frac{\mathrm{d}x_i}{\mathrm{d}t} = \frac{\partial H}{\partial y_i} \qquad \Gamma_i \frac{\mathrm{d}y_i}{\mathrm{d}t} = -\frac{\partial H}{\partial x_i} \tag{13.15}$$

with the Hamiltonian

$$H = -\frac{1}{4\pi} \sum_{i \neq j} \Gamma_i \Gamma_j \ln r_{ij}$$

where $r_{ij}^2 = (x_i - x_j)^2 + (y_i - y_j)^2$.

Remarkably, in the limit $N \to \infty\,, \Gamma_i \to 0\,,$ which can be realized e.g. taking $N^2 |\Gamma_i| \to$ const if $\sum_i \Gamma_i = 0$, or $N|\Gamma_i| \to$ const if $\sum_i \Gamma_i \neq 0$, the system (13.15) can be proved to approximate of the 2D Euler equation [Chorin (1994); Marchioro

[13] The main (smart) trick is to avoid to work directly with Eq. (13.14), this because for the straightforward computation of the terms on the right side of (13.14), or the corresponding equation in 3D, one needs $\mathcal{O}(N^2)$ operations. The pseudo-spectral method, which uses in a systematic way fast Fourier transform and operates both in real space and Fourier space, reduces the number of operations to $\mathcal{O}(N \ln N)$ [Orszag (1969)].

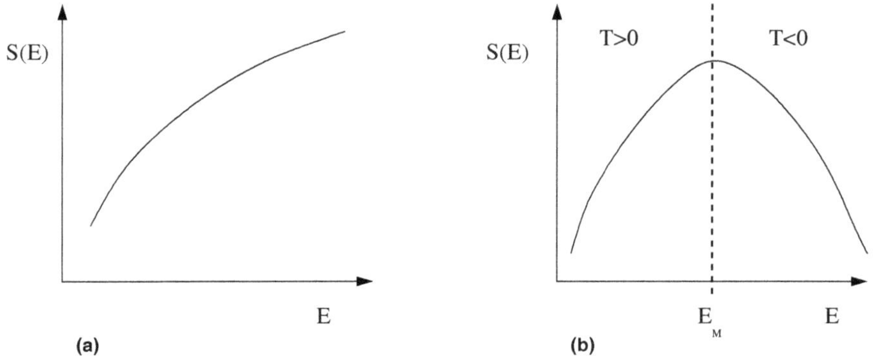

Fig. 13.6 Possible behaviors of entropy $S(E)$: (a) standard case $S(E)$ is a non-decreasing function of E and the temperature is always positive; (b) non-standard behavior in which states at negative temperature are possible. See text for explanation.

and Pulvirenti (1994); Aref and Siggia (1980)]. As a consequence, the point-vortex method can be used for numerical simulations of the inviscid hydrodynamics. Moreover, with suitable modifications, the point vortices can also reproduce some features of the dissipative problem such as the inverse energy cascade of two-dimensional Navier-Stokes turbulence [Siggia and Aref (1981)].

The point-vortex model with large N is also interesting from the point of view of statistical mechanics because, similarly to 2D Euler equation (Sec. 13.2.2), the statistical behavior presents non standard features due to the simultaneous conservation of energy and enstrophy. In a seminal paper at the origin of the modern statistical hydrodynamics, Onsager (1949) studied the statistical properties of a large number of point-vortices. We now summarize the main results. Consider N point vortices confined in a domain of area A. As in the usual microcanonical description of the statistical mechanics [Huang (1987)], introduce the quantity

$$g(E) = \int \dots \int \mathrm{d}x_1 \dots \mathrm{d}x_N \mathrm{d}y_1 \dots \mathrm{d}y_N \, \delta[E - H(x_1, \dots, x_N, y_1, \dots, y_N)]$$

and define the entropy $S(E)$ and the temperature T through the relationships

$$S(E) = \ln g(E) \,, \qquad \frac{1}{T} = \frac{\mathrm{d}S(E)}{\mathrm{d}E} = \frac{1}{g(E)} \frac{\mathrm{d}g(E)}{\mathrm{d}E} \;.$$

The phase space volume with energy smaller than E is

$$\psi(E) = \int_{-\infty}^{E} \mathrm{d}e \, g(e) \;,$$

note that $\psi(-\infty) = 0$ and $g(E)$ is a positive function, and thus $\psi(E)$ increases monotonically from zero to the maximum value A^N, as $E \to \infty$. This means that $g = \psi'$ must attains its maximum at a certain value E_M, see Fig. 13.6, so that for $E > E_M$ the entropy $S(E)$ is a decreasing function and hence $T(E)$ is negative. The high energy states $E \gg E_M$ are those in which the vortices are crowded. In a system with positive and negative Γ_i, negative temperature states correspond to the presence of well separated large groups of vortices with the same vorticity sign. On the contrary for $E \ll E_M$ (positive temperature) the vortices of opposite Γ_i tend to remain close. Negative temperatures thus correspond to configurations where same sign vortices are organized in clustered structures.

We conclude mentioning the interesting attempts by Robert and Sommeria (1991) and Pasmanter (1994) to describe, in terms of the 2D inviscid equilibrium statistical mechanics, the common and spectacular phenomena of long-lived, large-scale structures which appear in real fluids such as the red spot of Jupiter's atmosphere and other coherent structures in geophysics.

13.3.4 *Proper orthonormal decomposition*

Long-lived coherent structures appearing in both 2D and 3D fluid flows are often the main subject of investigation in systems relevant to applications such as, e.g., the wall region of a turbulent boundary layer, the annular mixing layer and thermal convection [Lumley and Berkooz (1996); Holmes *et al.* (1997)]. In these situations, performing a standard DNS is not the best way to approach the problem. Indeed, as suggested by intuition, the basic features of coherent structures are expected to be, at least in principle, describable in terms of systems with few variables. In these circumstances the main question is how to build reliable low dimensional models.

Remaining in the framework of Galerkin methods (Sec. 13.3.2), the basic idea is to go beyond "obvious" choices, as trigonometric functions or special polynomials, dictated only by the geometry and symmetries of the system, and to use a "clever" complete, orthonormal set of eigenfunctions $\{\phi_n(\boldsymbol{x})\}$, chosen according to the specific dynamical properties of the problem under investigation [Lumley and Berkooz (1996); Holmes *et al.* (1997)]. Such procedure, called proper orthonormal decomposition (POD),[14] allows low-dimensional systems, able to capture the coherent structures, to be determined starting from experimental or numerical data.

The main idea of the method can be described as follows. For the sake of notation simplicity, we consider a scalar field $u(x,t)$ in a one-dimensional space, evolving according to a generic PDE

$$\partial_t u = \mathcal{L}[u] \,, \tag{13.16}$$

where $\mathcal{L}[u]$ is a nonlinear differential operator. The main point of the method is to determine the set $\{\phi_n(x)\}$ in the truncated expansion

$$u^{(N)}(x,t) = \sum_{n=1}^{N} a_n(t)\phi_n(x) \;, \tag{13.17}$$

in such a way to maximize, with respect to a given norm, the projection of the approximating field $u^{(N)}$ on the measured one u. In the case of L_2-norm, it is

[14]POD goes under several names in different disciplines, e.g. Karhunen-Loéve decomposition, principal component analysis and singular value decomposition.

necessary to find $\phi_1, \phi_2, \ldots$ such that the quantity

$$\frac{\langle |(u^{(N)}, u)|^2 \rangle}{|(u^{(N)}, u^{(N)})|^2}$$

is maximal, here (,) denotes the inner product in Hilbert space with L_2-norm and $\langle\ \rangle$ the ensemble average (or, assuming ergodicity, the time average).

From the calculus of variations, the above problem reduces to find the eigenvalues and eigenfunctions of the integral equation

$$\int dx' R(x, x') \phi_n(x') = \lambda_n \phi_n(x) \ ,$$

where the kernel $R(x, x')$ is given by the spatial correlation function $R(x, x') = \langle u(x) u(x') \rangle$. The theory of Hilbert-Schmidt operators [Courant and Hilbert (1989)] guarantees the existence of a complete orthonormal set of eigenfunctions $\{\phi_n(x)\}$ such that $R(x, x') = \sum_n \phi_n(x) \phi_n(x')$. The field u is thus reconstructed using this set of function and using the series (13.17), where the eigenvalues $\{\lambda_k\}$ are ordered in such a way to ensure that the convergence of the series is optimal. This means that for any N the expansion (13.17) is the best approximation (in L_2-norm). Inserting the expansion (13.17) into Eq. (13.16) yields a set of ODEs for the coefficients $\{a_n\}$. Essentially, the POD procedure is a special case of the Galerkin method which captures the maximum amount of "kinetic energy" among all the possible truncations with N fixed.[15]

POD has been successfully used to model different phenomena such as, e.g., jet-annular mixing layer, 2D flow in complex geometries and the Ginzburg-Landau equation [Lumley and Berkooz (1996); Holmes *et al.* (1997)]. One of the nicest application has been developed by Aubry *et al.* (1988) for the wall region in a turbulent boundary layer, where organized structures are experimentally observed. The behavior of these structures is intermittent in space an time, with bursting events corresponding to large fluctuations in the turbulent energy production. The low dimensional model obtained by POD is in good agreement with experiments and DNS, performed with a much larger number of variables.

We conclude stressing that POD is not a straightforward procedure as the norm should be chosen accurately for the specific purpose and problem: the best model is not necessarily obtained by keeping the most energetic modes, e.g. L_2-norm may exclude modes which are essential to the dynamics. Therefore, thoughtful selections of truncation and norms are necessary in the construction of convincing low-dimensional models [Smith *et al.* (2005)].

13.3.5 *Shell models*

The proper orthonormal decomposition works rather well for coherent structures (which are intrinsically low-dimensional), we now discuss another class of (rela-

[15]POD procedure can be formulated also for other inner products, and consequently different norms, in such a way the selected POD modes are optimal for quantities other that kinetic energy.

tively) low-dimensional models which have been proposed for "simulating" the energy cascade process typical of high Reynolds number turbulent flows. The huge number of degrees of freedom of turbulence, growing as $\mathrm{Re}^{9/4}$ (Sec. 13.3.1), seems to rule out the use of low-dimensional models for the energy cascade, unless very crude approximations are used. Clearly, this means that we can just hope to reproduce some qualitative features or, in the best case, provide a caricature of some dynamical and/or statistical features of turbulent flows.

In particular, we are interested in describing a class of dynamical systems, dubbed *shell models* for reasons that will be soon clear, that originated from the works of Gledzer (1973) and Ohkitani and Yamada (1989) (GOY) and further developed by many authors, see Bohr *et al.* (1998) and Biferale (2003) for a throughout discussion of shell models for turbulence and beyond. The basic idea is to implement a dynamical (energy) cascade model with a set of complex variables u_n, $n = 1, \ldots, N$ representing the velocity fluctuation in a shell of wave-numbers $k_n < |\boldsymbol{k}| < k_{n+1}$. The wave-numbers are chosen geometrically spaced $k_n = k_0 2^n$ so to limit the number of variables needed to describe the inertial range physics. In this way, the spatial and vectorial structure of the original problem is completely disregarded. Then, some insights are used to write the equation ruling the set of variables $\{u_n\}$. A basic source of inspiration is NSE written in Fourier space, where the modes interact in triads (see e.g. Eq. (13.14)): only three modes are involved at the same time. As we need to simplify the complexity of the equations, we can think to retain the triad structure by cutting some interactions. Due to the hierarchical organization of the characteristic times associated to the different scales (see Eq. (13.11)), we can assume that only close modes, i.e. variables referring to close scales, can interact. The justification for this is that distant modes (say k_n and k_m with $|m - n| \gg 1$) have so different time-scales that the resulting interaction would be very weak. This assumption is known as hypothesis of the locality of the cascade and can be substantiated with refined analysis of NSE [Rose and Sulem (1978)].[16] In particular, considering triads of the form (k_{n-2}, k_{n-1}, k_n) we obtain the GOY model

$$\left(\frac{\mathrm{d}}{\mathrm{d}t}+\nu k_n^2\right)u_n = i\left[k_n u_{n+1}^* u_{n+2}^* - \delta k_{n-1} u_{n-1}^* u_{n+1}^* - (1-\delta) k_{n-2} u_{n-2}^* u_{n-1}^*\right] + f_n, \quad (13.18)$$

where the term $\nu k_n^2 u_n$ corresponds to dissipation, f_n to the forcing term (typically restricted on the first shells). As one can see the nonlinear terms on the l.h.s. have the same structure of those of the Navier-Stokes equation. Moreover, for $\nu = f_n = 0$, the total energy $E \equiv 1/2 \sum_n |u_n|^2$ and the phase space volume are preserved for any value of the free parameter δ, so that we can derive an equilibrium statistical mechanics similar to that of NSE.

[16]Such a hypothesis is reasonable in 3D, but not in the 2D direct cascade of enstrophy, which is known to be at the border of locality [Rose and Sulem (1978); Frisch (1995)].

The free parameter δ is important in determining the properties of the GOY model. Indeed, besides energy, Eq. (13.18) admits another conserved quantity

$$Q_\delta = \sum_n k_n^\alpha |u_n|^2 \qquad \text{with} \qquad \alpha = -\log_2(\delta - 1) \quad ,$$

whose nature depends on δ. For $\delta > 1$, α is real and Q_δ is an enstrophy-like invariant (see Eq. (13.4)). Actually, for $\delta = 5/4$, we have $Q_{5/4} = \Omega = \sum_n k_n^2 |u_n|^2$ so that the model is the natural candidate for mimicking 2D fluids. For $\delta < 1$ the conserved quantity has a structure similar to helicity $H = \int \mathrm{d}\boldsymbol{x}(\boldsymbol{\nabla} \times \boldsymbol{v}(\boldsymbol{x})) \cdot \boldsymbol{v}(\boldsymbol{x}) = i \sum [\boldsymbol{k} \times \boldsymbol{u}(\boldsymbol{k})] \cdot \boldsymbol{u}(\boldsymbol{k})$ which is conserved by the three-dimensional Euler equation. In particular for $\delta = 1/2$, we have $Q_{1/2} = H = \sum_n (-1)^n k_n |u_n|^2$ and thus the GOY model is a candidate for 3D turbulence.

The great advantage of shell models is that the number of shells N necessary to mimic the cascade mechanism of fully developed turbulence is relatively small, because of the geometrical progression in k_n we roughly have $N \sim \log_2(\mathrm{Re})$. We have thus a chaotic dynamical system with a reasonably small number of degrees of freedom where methods of deterministic chaos can be used to link the statistical description to the dynamical properties.

The GOY (13.18) model with $\delta < 1$ qualitatively reproduces the direct energy cascade from large scales (where forcing is acting) to dissipative scales ($n \sim N$), with a statistically constant energy flux $\bar{\epsilon}$, balancing input and dissipation. A sort of K41 theory and "4/5" law can be written also for the GOY model and, moreover, the moments of velocity develop a power law scaling $\langle |u_n|^p \rangle \sim k_n^{-\zeta_p}$ with exponents ζ_p that deviate from K41 for an amount that depends on δ. Remarkably, for $\delta = 1/2$ the exponents are fairly close to the experimentally observed ones (Fig. B31.1). The behavior for $\delta > 1$, i.e. in the 2D regime, is more problematic and we refer to Bohr *et al.* (1998) for a detailed discussion.

We mention that the scaling properties of the GOY model are somehow spoiled by the presence of oscillations in n, due to an extra, unrelated to NSE, conservation law on the phases of the dynamical variables. Free from this extra conservation, another version of the shell model (called SABRA[17]) [Lv́ov *et al.* (1998)]

$$\left(\frac{\mathrm{d}}{\mathrm{d}t} + \nu k_n^2\right) u_n = i[k_{n+1} u_{n+1}^* u_{n+2} - \delta k_n u_{n-1}^* u_{n+1} - (1-\delta) k_{n-1} u_{n-2} u_{n-1}] + f_n \quad (13.19)$$

displays a much cleaner scaling behavior. Nevertheless, GOY and SABRA models differ only for negligibly small quantitative details [Biferale (2003)].

Over the years, shell models attracted the attention of many scientists with different aims: the possibility to perform detailed numerical computation on a model for the energy cascade to test ideas or conjectures, we shall see some examples of this approach in the context of predictability in the next Section; the investigation of analytic methods, e.g. to test some ideas for the closure problem [Benzi *et al.*

[17]The name originates from Sabra, a term used to describe a Jew born in Israel, as opposed to Goy, which translates as "nation" or "people" and it is a synonym for non-Jew.

(1993)]; developing rigorous results[18] [Constantin *et al.* (2006)]; understanding the link between dynamical properties in phase space and more standard quantities (in traditional turbulent literature) such as structure functions and velocity pdf. In this respect it is worth remarking that shell models well reproduce the main features of the multifractal model of turbulence (Box B.31), providing also a nice laboratory to test its predictions [Bohr *et al.* (1998); Biferale (2003)].

We conclude mentioning shell models, such as Eqs. (13.18)-(13.19), have been generalized to "simulate" other turbulent phenomena (see Bohr *et al.* (1998) and references therein) such as thermal convection, passive scalars, binary fluid mixtures, and also to model magnetohydrodynamics [Carbone (1993)] or polymers in turbulence [Benzi *et al.* (2003)].

13.4 Predictability in turbulent systems

The subject of predictability in turbulence largely overlaps with the problem of weather forecasting, which can be split into two main issues. On the practical side, we need to devise more and more refined models for weather evolution.[19] On the theoretical side, we have to understand the intrinsic limits of our ability to predict the future evolution of turbulent flows.

Despite the importance of the former aspect, we focus here on the latter. In principle, we should account for both our imperfect knowledge of the initial state and of the governing processes (the "true" equations of motion). However, as discussed in Chapter 10, these two sources of uncertainty have quite similar effects (see Boffetta *et al.* (2000b) for a discussion of such an aspect in the context of models for turbulence and other dynamical systems), so that we can restrict the discussion to the imperfect knowledge of the initial conditions.

Due to the presence of many characteristic temporal and spatial scales, when setting the predictability limits in turbulent flows, we should ask for predictability of what feature against which perturbation, or more specifically, against a perturbation (uncertainty) of what size. In particular, it is important to distinguish between the predictability of small (infinitesimal) and large scale features. Moreover, we can separate cases in which coherent structures are unimportant, as when we are interested in predicting statistical features, from those in which they are crucial, as if we ask where an hurricane will pass, with the possibility of destroying buildings.

In the following, we shall discuss these issues. Due to difficulty to test or illustrate basic ideas and processes for high Reynolds number turbulent flows, we

[18]For instance, the existence, in the case $N = \infty$, of a finite dimensional globally invariant manifold (called inertial manifold) which attracts all bounded sets in the phase space at an exponential rate and, consequently, contains the global attractor.

[19]NSE well accounts for fluid motion, but weather forecasting needs a detailed knowledge and modeling of important chemical and physical processes, such as energy exchanges between atmosphere and oceans, clouds formation, chemical reactions etc. Moreover, small scale modeling of the basic processes is mandatory.

shall often resort to phenomenological caricatures of turbulent flows such as the shell models or 2D flows, whose numerical investigations can be performed also for realistically high Re values.

13.4.1 *Small scales predictability*

Physically speaking, in turbulence, infinitesimal uncertainties correspond to perturbations below the Kolmogorov scale, meaning uncertainties on the velocity field smaller than $\delta v(\ell_D) \sim U\mathrm{Re}^{-1/4}$. Infinitesimally small perturbations are controlled by the tangent space evolution, and thus by the maximal Lyapunov exponent λ_1. Therefore, the first natural question is about the dependence of λ_1 on Re.

As seen in the previous sections, one of the most connotative aspect of 3D turbulence is the presence of many characteristic, hierarchically organized time scales (see Eq. (13.11)). The scales of interest here, i.e. the very small ones, are controlled by the smallest active eddies, whose time scale is given by the eddy turnover time at the Kolmogorov scale

$$\tau_D \sim \ell_D/\delta v(\ell_D) \sim \tau_L \mathrm{Re}^{-1/2}, \tag{13.20}$$

$\tau_L = L/U$ being the eddy turnover time of the energy injection scale. From the hierarchy (13.11), it is straightforward to realize that τ_D is the fastest time scale, hence the maximal LE should be roughly given by its inverse

$$\lambda_1 \sim {\tau_D}^{-1} \sim {\tau_L}^{-1}\mathrm{Re}^{1/2}\ . \tag{13.21}$$

This result, due to Ruelle (1979), provides a first answer to the question about the relationship between Lyapunov exponent and Reynolds number.

However, being essentially based on K41 theory which is only approximately valid, the scaling (13.21) might not be completely satisfactory. Therefore, in order to check to what extent intermittency changes the result (13.21), we can use the basic ideas of the multifractal model for turbulence (Box B.31) to elaborate Ruelle's argument [Crisanti *et al.* (1993a)].

As detailed in Box B.31, velocity fluctuations can be written as $\delta v(\ell) \sim U(\ell/L)^h$, with the exponent h that can differ from K41 value 1/3. In particular, different sets of the space, with fractal dimension $\mathcal{D}(h)$, express different h values. Accordingly to the variations of h, the Kolmogorov length fluctuates as $\ell_D(h) \sim L\mathrm{Re}^{-1/(1+h)}$ [Paladin and Vulpiani (1987)] and Eq. (13.20) becomes $\tau_D(h) \sim \tau_L \mathrm{Re}^{-(1-h)/(1+h)}$. As a consequence, to obtain the Lyapunov exponent, we must average over all possible h values which appear with probability $P_h(\ell_D) \sim (\ell_D/L)^{3-\mathcal{D}(h)}$ or, equivalently, $P_h(\ell_D) \sim \mathrm{Re}^{-(3-\mathcal{D}(h))/(1+h)}$. In this way the Lyapunov exponent is given by [Crisanti *et al.* (1993a)]

$$\lambda_1 \sim \int \mathrm{d}h\, \frac{P_h(\ell_D)}{\tau_D(h)} \sim \frac{1}{\tau_L}\int \mathrm{d}h\, \mathrm{Re}^{\frac{\mathcal{D}(h)-2-h}{1+h}}\ .$$

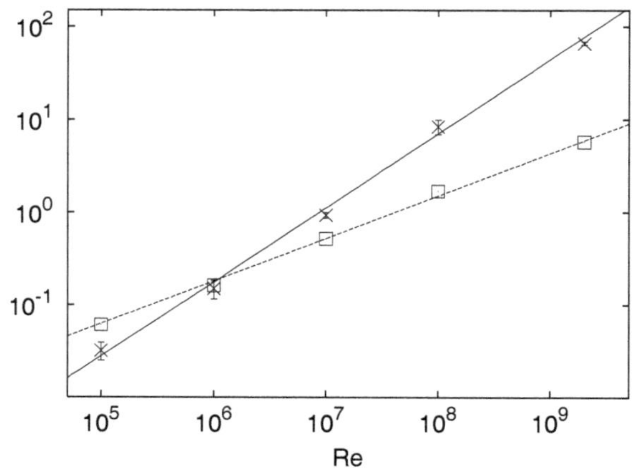

Fig. 13.7 λ_1 ($\square$) and variance σ^2 ($\times$) as a function of Re for the GOY model with $N = 27$ shells. The dashed line is the multifractal prediction $\lambda \sim \text{Re}^\alpha$ with $\alpha = 0.46$. The full line represents $\mu \sim \text{Re}^w$ with $w = 0.8$.

As we are interested in the Re $\to \infty$ limit, the above integral can be evaluated with the steepest descent method and we obtain

$$\lambda_1 \sim \text{Re}^\alpha \quad \text{with} \quad \alpha = \max_h \left\{ \frac{\mathcal{D}(h) - 2 - h}{1 + h} \right\} . \tag{13.22}$$

From experimental data on the structure function exponents, we can obtain a prediction for $\mathcal{D}(h)$ (Box B.31), which inserted in (13.22) gives $\alpha \simeq 0.46$ not far from, but slightly smaller than, Ruelle prediction $\alpha = 0.5$ [Crisanti *et al.* (1993a)].

Due to the intermittent dynamics characterizing small scales turbulent flows, it might be important to account for the fluctuations of the finite time Lyapunov exponent $\gamma(t)$ (FTLE) (Sec. 5.3.3). For instance, we can assume that $\gamma(t)$ follows a Gaussian statistics (see Eq. (5.35)) with mean λ_1 and variance $\sigma^2 = \lim_{t\to\infty} t \left[\langle \gamma(t)^2 \rangle - \langle \gamma(t) \rangle^2\right]$. Although physical arguments suggest a scaling behavior [Crisanti *et al.* (1993a)]

$$\sigma^2 \sim \frac{1}{\tau_L} \text{Re}^w , \tag{13.23}$$

the exponent w remains undetermined.

Now that we have a theoretical guess on the dependence of λ_1 and σ^2 on Re, it would be interesting to test it against direct numerical simulations of the three-dimensional NSE. This is an hard task due to the large values of Re required. However, the results (13.22)-(13.23) mostly rely on the phenomenology of the energy cascade and on intermittency, which are well reproduced, at least qualitatively, by the shell models for turbulence. Indeed, such models allow high Re behaviors to be tested by employing relatively few degrees of freedom (Sec. 13.3.5), and are thus best suited for verifying such ideas. With this aim, Crisanti *et al.* (1993a) numerically studied the GOY model (13.18) obtaining results (Fig. 13.7) in very good agreement with the multifractal prediction.

If we, roughly, estimate the predictability time as

$$T_p \sim 1/\lambda_1 \sim \mathrm{Re}^{-\alpha}\,,$$

as Re can be very high, we easily realize that T_p would results in a few seconds or minutes, which contrasts with the fact that weather forecasting is actually possible for a few days. This apparent paradox originates by the fact that we are not interested in infinitesimal errors but in forecasting the large scale features of the weather. In other terms, we handle errors on the velocity field of the atmosphere which are much larger than $\delta v(\ell_D)$. These scales are controlled by the physics of the inertial range for which standard Lyapunov exponent is irrelevant.

13.4.2 *Large scales predictability*

The classical theory of predictability in turbulence has been developed by Lorenz (1969) using physical arguments later confirmed with more refined treatments [Leith (1971); Leith and Kraichnan (1972)]. Lorenz approach stems from the assumption, natural in the energy cascade picture, that the time needed for a perturbation at scale $\ell/2$ to induce a complete uncertainty on the velocity field at scale ℓ is proportional to the eddy turn-over time of the scale ℓ, which from Eq. (13.11) reads

$$\tau(\ell) \sim \ell/\delta v(\ell) \sim \tau_L(\ell/L)^{2/3}\,.$$

Because of the algebraic progression, the time T_p to propagate an uncertainty from the Kolmogorov scale ℓ_D upwards to the energy containing eddies scale L is dominated by the slowest time scale, indeed

$$T_p \sim \tau(\ell_d) + \tau(2\ell_d) + \cdots + \tau(L) \sim \tau_L \sim \frac{L}{\delta v(L)}\,.$$

As an outcome of Lorenz approach, the predictability time results to be Reynolds independent, which may sound surprising because of Eq. (13.22), i.e., of the increase of the Lyapunov exponent with Re. However, as stated several times, there is no contradiction: in forecasting the large scales, i.e. non-infinitesimal perturbations, the Lyapunov exponent plays no role.

It is instructive to recast the above phenomenological considerations in the language of dynamical systems. The natural way is by considering the Finite Size Lyapunov Exponent (FSLE) $\lambda(\delta)$[20] introduced in Sec. 9.4 precisely with the aim of characterizing non-infinitesimal perturbations. In terms of the FSLE, the predictability time for an error δ and a given tolerance Δ reads [Boffetta *et al.* (1998)]

$$T_p(\delta,\Delta) = \int_\delta^\Delta \frac{\mathrm{d}\ln\delta'}{\lambda(\delta')}\,, \tag{13.24}$$

[20]Note that here the relevant variable is velocity, therefore $\delta = |\boldsymbol{v} - \boldsymbol{v}'|$ represents the velocity uncertainty.

i.e. we have to sum up the inverse of the error growth rate at all scales, from the initial perturbation δ to the chosen error tolerance Δ. Clearly, if both δ and Δ are infinitesimal $\lambda(\delta) \simeq \lambda_1$ and Eq. (13.24) reduces to the naive expectation

$$T_p = \frac{1}{\lambda_1} \ln\left(\frac{\Delta}{\delta}\right) . \tag{13.25}$$

However, in the (generic) case where $\lambda(\delta)$ is a decreasing function of δ (meaning that larger errors are characterized by smaller growth rates, as in Lorenz scenario), the predictability time given by Eq. (13.24) becomes much longer than the naive estimate (13.25).

Within the phenomenological framework of K41 theory and closely following Lorenz (1969) ideas, we can predict the scaling behavior of $\lambda(\delta)$ when the perturbation is in the inertial range $\delta v(\ell_D) \ll \delta \ll \delta v(L)$. According to Lorenz argument, the doubling time of an error of magnitude δ is proportional to the turn-over time $\tau(\ell)$ of an eddy with typical velocity difference $\delta v(\ell) \sim \delta$. Using $\delta v(\ell) \sim U(\ell/L)^{1/3}$ we have $\tau(\ell) \sim \tau_L(\ell/L)^{2/3} \sim \tau_L(\delta v(\ell)/U)^{-2}$. As a result, for velocity uncertainties in the inertial range, we obtain[21]

$$\lambda(\delta) \sim \delta^{-2} . \tag{13.26}$$

We may now wonder if refinements of K41 theory that take into account intermittency would change prediction (13.26). Remarkably, intermittency does not impinge the scaling behavior of the FSLE[22] (see Aurell *et al.* (1996, 1997) for details).

Testing the scaling (13.26) in DNS of 3D turbulence presents several difficulties and, above all, the problem that with current computers it is hard to simulate an inertial range wide enough to verify the δ^{-2} behavior. However, once again, we can resort to shell models where this scaling behavior can be tested using a reasonable number of degrees of freedom.

Figure 13.8 shows $\langle 1/\tau(\delta,\rho)\rangle_t$, which is proportional to $\lambda(\delta)$ (see Sec. 9.4), as a function of δ in the GOY model [Aurell *et al.* (1996, 1997)]. Precisely, $\tau(\delta,\rho)$ is the time necessary for a perturbation of size δ to increase by a factor ρ. When the perturbation is at the dissipative scales $\delta < \delta v(\ell_D) \sim U\mathrm{Re}^{-1/4}$, it can be considered infinitesimal, and $\tau(\delta,\rho)$ does not depend on δ, as it is essentially coincident with the inverse of the standard Lyapunov exponent. At larger errors, the scaling (13.26) is well reproduced. In the figure data obtained with different Re are compared. K41 theory suggests to perform such a comparison rescaling the time and velocity fluctuations at the Kolmogorov length as $\mathrm{Re}^{-1/2}$ and $\mathrm{Re}^{-1/4}$, respectively [Frisch (1995)].

[21] Note that Eq. (13.26) is formally the same as Eq. (9.21) for the FSLE behavior in the presence of diffusion. However, the origin of the δ^{-2} in the two cases is completely different.

[22] Indeed accounting for intermittency, in the framework of the multifractal approach, we can write

$$\lambda(\delta) \sim \tau_L^{-1} \int dh \, [\delta/U]^{[3-\mathcal{D}(h)]/h} (\delta/U)^{1-1/h} \sim \delta^{\min_h\left\{\frac{2+h-\mathcal{D}(h)}{h}\right\}},$$

where the second relation stems from a steepest descent estimation. As a direct consequence of 4/5 "law" we have $\mathcal{D}(h) \leq 3h+2$ (see Eq. (B.31.2)) and thus $(2+h-\mathcal{D}(h))/h \geq -2$ for all h implying $\lambda(\delta) \sim \delta^{-2}$ independently of intermittency corrections.

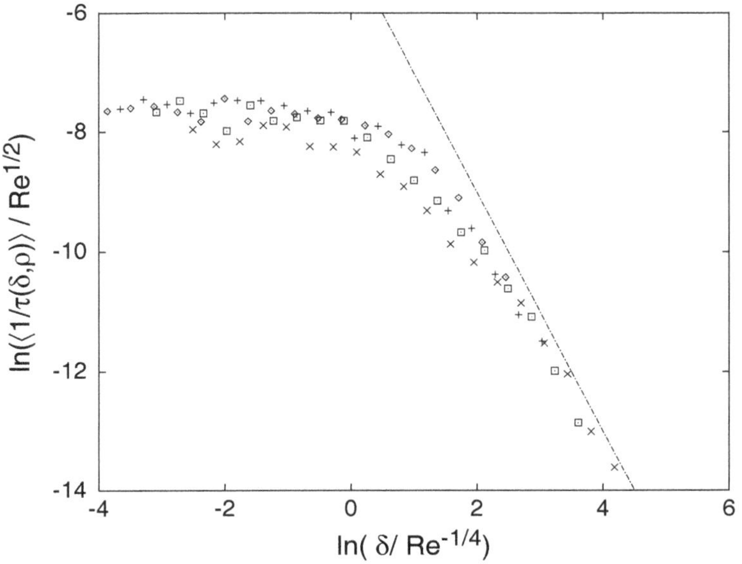

Fig. 13.8 $\ln\left(\langle 1/\tau(\delta,\rho)\rangle_t/\mathrm{Re}^{1/2}\right]$) versus $\ln\left(\delta/\mathrm{Re}^{-1/4}\right)$ at different Reynolds numbers $\mathrm{Re}=\nu^{-1}$. ($\Diamond$) $N=24$ and $\mathrm{Re}=10^8$; (+) $N=27$ and $\mathrm{Re}=10^9$; ($\Box$) $N=32$ and $\mathrm{Re}=10^{10}$; ($\times$) $N=35$ and $Re=10^{11}$. The straight line has slope -2.

In this way, data at different Re fairly collapse onto the same curve, as shown in the figure.[23] We can thus conclude that small-scale predictability, with small error amplitudes, behaves (apart from intermittency corrections) as predicted by Ruelle, whereas large-scale predictability, characterized by large error amplitudes, is well described by Lorenz argument.

It is interesting to observe that Lorenz argument is only based on the algebraic organization of the characteristic times $\tau(\ell) \sim \ell^{2/3}$. Not even the direction of the energy cascade (i.e. being it directed toward the large or the small scales) matter for the validity of the argument. The same is true also for the FSLE, whose scaling properties are not affected by intermittency. It is then clear that when considering two dimensional turbulent flows in the inverse cascade regime, where the behavior $\tau(\ell) \sim \ell^{2/3}$ is also verified, we should expect Lorenz theory to apply and, in particular, Eq. (13.26) to hold. DNS in 2D can be performed by current computers reaching very high Re numbers and thus constitutes a valid framework to further test the above results.

Figure 13.9 shows the FSLE as obtained in a high resolution DNS of the two-dimensional Navier-Stokes equation made stationary with the addition of a friction term (see Sec. 13.2.4) [Boffetta and Musacchio (2000)]. The constant value for

[23]A better data collapse can be obtained by accounting intermittency, for instance, with the multifractal model [Aurell *et al.* (1996, 1997)]

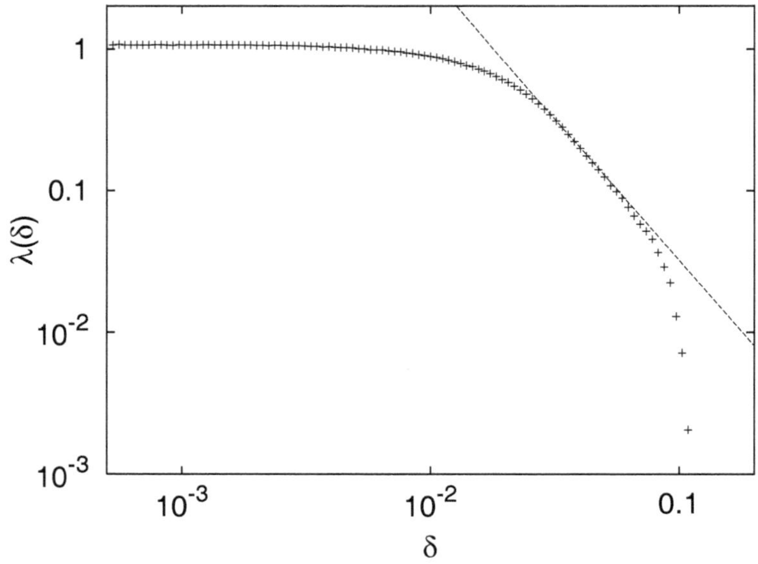

Fig. 13.9 FSLE $\lambda(\delta)$ as a function of velocity uncertainty δ in a DNS of 2D turbulence in the inverse cascade regime obtained with 1024^2 grid points. The asymptotic constant value for $\delta \to 0$ corresponds to the largest Lyapunov exponent. The dashed line has slope -2. [Courtesy of G. Boffetta]

$\delta \to 0$ corresponds to the largest Lyapunov exponent of the turbulent flow. While at inertial range scales the δ^{-2} scaling behavior is clearly detected. The large δ fall off is due to the saturation of the error at the largest available scale in the simulation. This result further confirms the validity of Lorenz approach. It is noteworthy the rather wide scaling range for the behavior $\lambda(\delta) \sim \delta^{-2}$ with respect to the shell model simulations (Fig. 13.8), which were obtained at much larger Re. The rationale for this behavior has to be searched in the absence of intermittency in 2D, which makes the transition from the infinitesimal regime $\lambda(\delta) \approx \lambda_1$ to the inertial range behavior $\lambda(\delta) \sim \delta^{-2}$ very sharp.

Although in realistic problems, like atmospheric forecasting, it is necessary to face a number of practical aspects, ranging from small scales modeling to the problem of determining the initial state from incomplete observations, the previously discussed dynamical systems methods had been rather useful (see, e.g., the recent review by Yoden (2007)).

It is worth concluding this section by mentioning a few examples of the successful use of the FSLE (or similar approaches) for the characterization of predictability as function of the resolution in geophysical data and realistic models. Basu *et al.* (2002) used the FSLE for the analysis of long records of high-resolution data of atmospheric boundary layer flows. Their results demonstrate the (expected) enhancement of the predictability at large scale, and quantify the dependence of the predictability on the atmospheric environment, a result which would be difficult to obtain from the analysis of the Lyapunov exponent. Kalnay and coworkers introduced the breeding method for the study of finite amplitude perturbations (see Kalnay (2002)). Such a technique, which has many similarities with the FSLE, consists in adding an initial perturbation of size δ to a reference trajectory, integrating forward both the perturbed and unperturbed trajectories, and periodically rescaling (every time interval ΔT) the amplitude of the perturbation to the initial value δ. With a proper choice of δ and ΔT (in the physically appropriate scales) it is possible to estimate the shape of the baroclinic instabilities [Toth and Kalnay (1993)]. In a similar way the method allows for the separation of fast and slow unstable modes in coupled systems of geophysical relevance with different time scales, e.g. in the El Niño-Southern Oscillation [Peña and Kalnay (2004)].

13.4.3 *Predictability in the presence of coherent structures*

Coherent structures are patches of activity remaining spatially correlated and organized for long times. A rather well known example is constituted by long-lived vortices in 2D fluids (Fig. 13.1), which dominates the dynamics when no forcing is acting. Forecasting in the presence of such structures requires to specify the features we would like to predict because, as illustrated by the following example, the resulting predictability time may strongly depend on used norm.

Consider a freely decaying 2D turbulent flow, i.e. we do not impose any external driving and leave the flow, which starts in a specific state, free to decay as a consequence of the viscous forces. Asymptotically the rest state will be reached with $\lambda_1 < 0$, i.e. the system will ultimately be attracted by the trivial fixed point $\omega(\boldsymbol{x}, t) = 0$. However, the transient before reaching this trivial asymptotic state is typically very long and dominated by the presence of vortices [McWilliams (1984)]. During this transient, from a practical point of view, we can consider the system to be in quasistationary regime, characterized by a positive Lyapunov exponent.

We then consider two realizations of the decaying flow starting from slightly different initial conditions, $\omega(\boldsymbol{x}, t)$ and $\omega'(\boldsymbol{x}, t)$, and, following classical theories of predictability in turbulence [Leith (1971)], we study the evolution of the error field[24]

$$\delta\omega(\boldsymbol{x}, t) = \frac{1}{\sqrt{2}} \left(\omega'(\boldsymbol{x}, t) - \omega(\boldsymbol{x}, t)\right) .$$

The "error" is computed from $\delta\omega$ and measured in terms of a given norm. Typically so-called enstrophy and energy norms are considered [Leith (1971)]

$$Z_\delta(t) = \frac{1}{2} \int \mathrm{d}\boldsymbol{x}\, |\delta\omega(\boldsymbol{x}, t)|^2 = \int_0^\infty \mathrm{d}k\, Z_\delta(k, t)$$

$$E_\delta(t) = \int_0^\infty \mathrm{d}k\, k^{-2}\, Z_\delta(k, t) = \int_0^\infty \mathrm{d}k\, E_\delta(k, t)$$

[24]The factor $1/\sqrt{2}$ is just for normalization convenience.

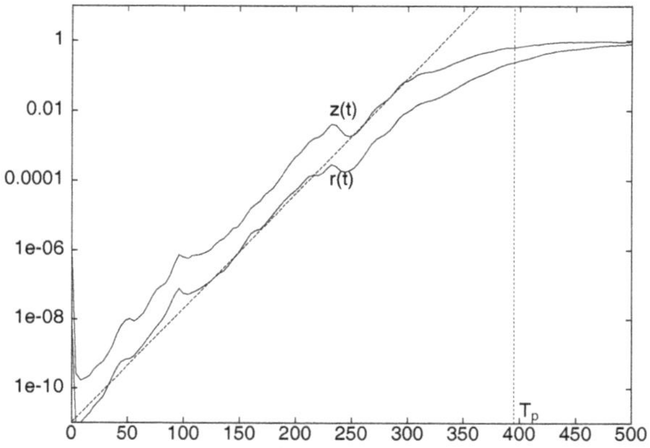

Fig. 13.10 Time evolution of the relative energy (r) and enstrophy (z) error in a DNS of the two dimensional Navier-Stokes equation, with 512^2 collocation points. T_p indicate the predictability time defined as $r(T_p) = 1/4$. The dashed line indicates the exponential regime $r(t) \sim \exp(0.08t)$.

where Z_δ and E_δ denote the enstrophy and energy error spectra, respectively. We can then introduce the relative errors,

$$r(t) = \frac{E_\delta(t)}{E(t)}, \qquad z(t) = \frac{Z_\delta(t)}{Z(t)} \tag{13.27}$$

with $E(t) = 1/2 \int \mathrm{d}\boldsymbol{x}\ |\boldsymbol{v}(\boldsymbol{x},t)|^2$ and $Z(t) = 1/2 \int \mathrm{d}\boldsymbol{x}\ \omega^2(\boldsymbol{x},t)$, and the relative error spectra

$$r(k,t) = \frac{E_\delta(k,t)}{E(k,t)} = \frac{Z_\delta(k,t)}{Z(k,t)} .$$

We now consider an initial error corresponding to complete uncertainty at scales smaller than a given k_0^{-1}, meaning that $r(k,0) = 0$ for $k < k_0$ and $r(k,0) = 1$ for $k > k_0$. Then we can define the predictability time by fixing the maximal tolerance Δ we can accept or, equivalently, a threshold on the relative errors (13.27): for instance the classical prescription is $r(T_p) = 1/4$ [Leith (1971)]. Figure 13.10 shows the relative errors (13.27) as a function of time obtained by DNS performed with the described protocol [Boffetta *et al.* (1997)]. For small times ($t < 250$) both $r(t)$ and $z(t)$ roughly grow exponentially with an effective Lyapunov exponent ≈ 0.08, while at larger times the error curves bend. Over all the predictability time given by the energy norm is $T_p \simeq 395$ while that obtained with the enstrophy norm is slightly larger.

Now it is interesting to see by direct inspection to what extent the fields $\omega(\boldsymbol{x}, T_p)$ and $\omega'(\boldsymbol{x}, T_p)$ differ. Although the two fields differ, by definition, by 25% in energy and about 65% in enstrophy, they look remarkably similar for what concerns the distribution of vortices. Most of coherent structures have maintained their shape and are in close locations (Fig. 13.11a,b). The difference field $\delta\omega(\boldsymbol{x}, T_p)$ (Figure 13.11c) indicates that the error is concentrated in correspondence of the vortices and that it is mainly due to small differences in the vortex positions in the two realizations.

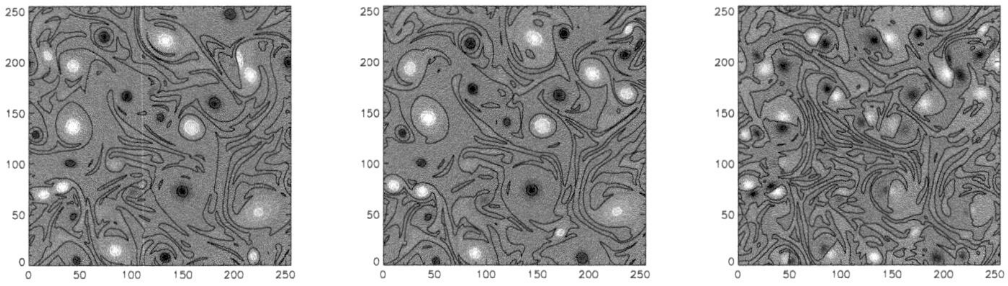

(a) (b) (c)

Fig. 13.11 Gray scale map of the vorticity fields (obtained by a 256^2 simulation) at time $T_p = 177$. White corresponds to positive vorticity regions, black to negative ones. (a) Reference field $\omega(\boldsymbol{x})$ (b) the perturbed one $\omega'(\boldsymbol{x})$ (c) the error filed $\delta\omega(\boldsymbol{x})$.

Clearly, this subtle difference cannot be revealed by $r(t)$ or $z(t)$, i.e. using the energy or enstrophy norms. The error is initially confined to small scales $k > k_0$, so that the vorticity-based norm is always larger than the energy-based norm, but the predictability time is essentially independent of the used norm. Actually we can reasonably expect that any Eulerian norm would give comparable result.[25] Therefore, if we are interested in emphasizing the differences among coherent structures, we have to search for another kind of norm. Figure 13.11 suggests that a Lagrangian measure of the error, based on the vortex positions, would be more suitable in this case.

The basic idea is then look at the distance among "vortex" centers, for this a vortex tracking algorithm should be used to recognize and follow vortices during the dynamics. This calls for a definition of vortex. For instance, we can call vortex any connected region D_α with vorticity maximum z_α larger (in absolute value) than a given threshold and average vorticity larger than a fraction (e.g. 0.2) of the vorticity peak [Boffetta *et al.* (1997)]. Given the vortex domains D_α, all the physical quantities are computed by integrating inside the domains. For example, vortex circulation is defined as $\Gamma_\alpha = \int_{D_\alpha} d\boldsymbol{x}\, \omega(\boldsymbol{x})$ and vortex center $\boldsymbol{x}_\alpha$ is the center of mass computed from the vorticity field. Finally, vortex trajectories are reconstructed by matching center positions at different times [Boffetta *et al.* (1997)].

We can then measure the error weighting the vortex distance with its intensity

$$d^2(t) = \frac{1}{\sum_\alpha |\Gamma_\alpha|} \sum_\alpha |\Gamma_\alpha| \, |\boldsymbol{x}'_\alpha - \boldsymbol{x}_\alpha|^2$$

where $\boldsymbol{x}_\alpha$ and $\boldsymbol{x}'_\alpha$ are the vortex positions respectively in the reference and perturbed field. Figure 13.12 shows d as obtained from the DNS by Boffetta *et al.* (1997). At the classical predictability time, the mean vortex separation is $d(T_{\rm p}) \simeq 0.5$, well

[25]Because the error propagates from small to large scales, a norm which emphasizes small scale features (as the enstrophy norm) saturates earlier than a large scale based norm (energy, in our example), but the results remain essentially the same.

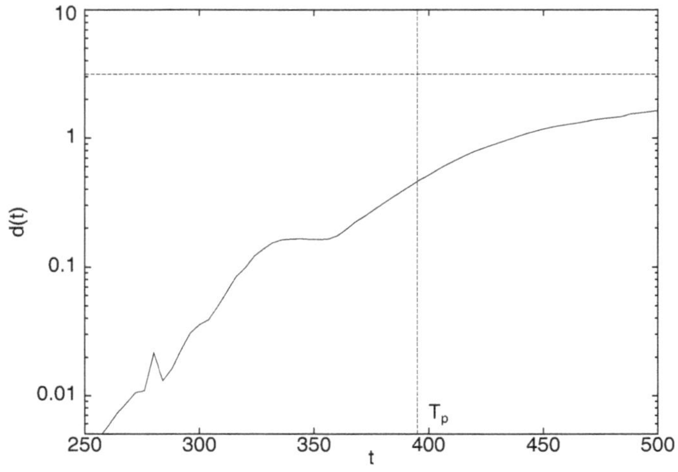

Fig. 13.12 Mean vortex separation $d(t)$ at resolution 512^2. At the classical predictability time T_p (vertical line), the mean vortex separation is about one-tenth of the saturation level (horizontal line), imposed by the domain boundaries.

below the saturation value ($d_{\max} \sim L/2 = \pi$ in the periodic computational domain). This result is a quantitative confirmation of the observations drawn from Fig. 13.12 i.e. the existence of an intermediate regime in which the (finite) error is ruled by the displacement of the strong coherent structures. In conclusion, if one is interested in predicting, with some tolerance, positions and intensities of coherent structures, it is possible to have a much larger predictability time than classically expected.

Chapter 14

Chaos and Statistical Mechanics: Fermi-Pasta-Ulam a Case Study

The biggest problem is not to let people accept new ideas, but to let them forget the old ones.

John Maynard Keynes (1883–1946)

This Chapter discusses, starting from the paper by Fermi, Pasta and Ulam (1955), usually indicated as FPU, the role of chaos on the foundations of statistical mechanics. Indeed FPU system epitomizes the necessity of a critical use of concepts such as chaos and ergodicity. Remarkably, this work was decisive also for the development of other important, long-lasting research fields, encompassing chaos and solitons. It was among the first computer experiments and, perhaps, the first showing how simulations can be used as a powerful instrument able to provide new physical insights and ideas, constituting an invaluable support for theoretical advancement. In this respect, beyond its conceptual interest, FPU work has a great pedagogical merit as it illustrates the typical way modern research proceeds in the understanding of a problem, namely a convoluted combination of physical conjectures, computer simulations, probabilistic arguments and theory.

14.1 An influential unpublished paper

As chaotic systems display irregular evolutions characterized by memory-loss of the initial state and resembling stochastic processes, it seems rather natural to expect that chaos may have a positive impact on the validity of statistical mechanics. In this perspective, Poincaré's result on the non existence of first integrals of motion (except for energy), in generic Hamiltonian systems, sounds rather positive. In addition, in 1923 Fermi, generalizing Poincaré's result, showed that, in a generic Hamiltonian system with $N > 2$ degrees of freedom, no smooth surface can divide the phase space into two regions containing open invariant sets. From this (correct) result, Fermi argued that non-integrable Hamiltonian systems are generically ergodic. At that time this conclusion was widely accepted, at least among physicists, even in the absence of a rigorous demonstration, so that the ergodic problem was considered essentially solved.[1]

After the Second World War, Fermi had the possibility to reconsider the problem of ergodicity thanks to the development of digital computers and their applications, as discussed in the introduction of the FPU paper by Ulam.[2] In 1952/3, Fermi, Pasta and Ulam, pioneering modern computer simulations, studied the time evolution of a system composed by N particles of mass m interacting with slightly nonlinear springs, described by the Hamiltonian

$$H = \sum_{i=0}^{N} \left[\frac{p_i^2}{2m} + \frac{K}{2}(q_{i+1} - q_i)^2 + \frac{\epsilon}{\alpha}(q_{i+1} - q_i)^\alpha \right], \tag{14.1}$$

where K controls the strength of the springs and ϵ the degree of anharmonicity with $\alpha = 3$ or $\alpha = 4$. They used fixed boundary conditions $q_0 = q_{N+1} = p_0 = p_{N+1} = 0$; later studies confirmed FPU results with different boundary conditions (e.g. periodic).

When $\epsilon = 0$ the Hamiltonian (14.1) is integrable. Indeed, in such a limit, using the normal modes:

$$a_k = \sqrt{\frac{2}{N+1}} \sum_n q_n \sin\left(\frac{n\, k\, \pi}{N+1}\right) \qquad (k = 1, \dots, N),$$

the system reduces to N non-interacting harmonic oscillators with frequencies

$$\omega_k = 2\sqrt{\frac{K}{m}} \sin\left(\frac{k\, \pi}{2(N+1)}\right)$$

and energies

$$E_k = \frac{1}{2}\left[\left(\frac{\mathrm{d}a_k}{\mathrm{d}t}\right)^2 + \omega_k^2 a_k^2\right] = const\,.$$

When $\epsilon \neq 0$, the Hamiltonian (14.1) is a typical example of perturbed integrable system.

Before examining FPU numerical experiment, it is worth recalling some basic issues of statistical mechanics and ergodic theory.

As previously discussed, the ergodic hypothesis is crucial for the statistical mechanics description of Hamiltonian systems. Ergodic hypothesis (Sec. 4.3) states that the time averages of an observable of an isolated system at equilibrium can be computed as phase averages over the constant-energy hyper-surface, i.e. over the microcanonical ensemble.[3] Clearly, whenever the ergodic hypothesis can be proved,

[1] One may wonder why Fermi was not very worried about the lacking rigor of his "proof". Likely, the reason is that, at that time, his main interest was the development of quantum physics.

[2] This paper, written as an internal report of the Los Alamos Laboratories, was completed in May 1955, after Fermi's death, but it was made public only in 1965 in the anthology of Fermi's writings *Note e Memorie* (Collected Papers).

[3] Notice that the microcanonical ensemble can be considered, at conceptual level, the basic one. For instance, the canonical ensemble, describing equilibrium statistics of a small but macroscopic subsystem, can be derived, under rather general conditions, from the microcanonical ensemble.

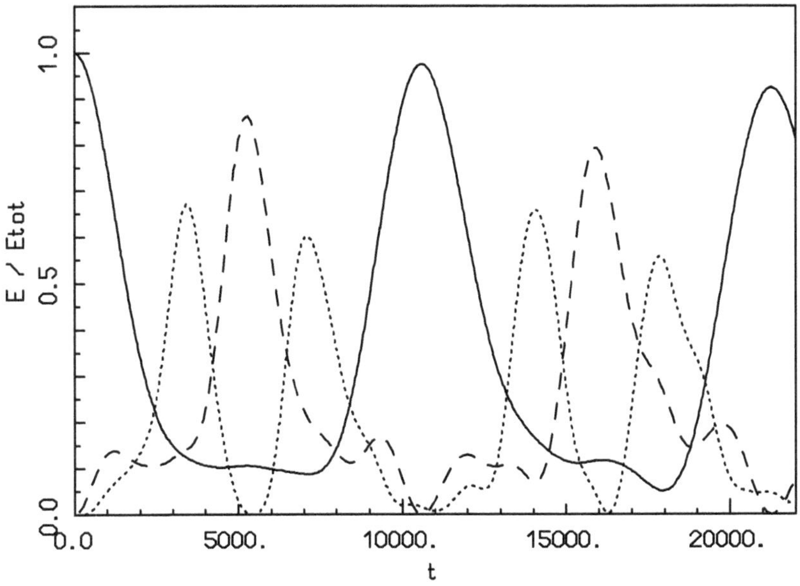

Fig. 14.1 Normalized modes energies $E_k(t)/E_{tot}$ for $k = 1$ (solid line), $k = 2$ (dashed line) and $k = 3$ (dotted line) obtained with $N = 32$, $\alpha = 3$ and $\epsilon = 0.1$. The initial condition is $E_1(0) = E_{tot} = 2.24$ and $E_k(0) = 0$ for $k = 2, \ldots, 32$. [Courtesy of G. Benettin]

it provides a dynamical justification to statistical ensembles. It should be stressed that the ergodic hypothesis is not a mere technical requirement but, unlike ensembles, has a strong physical motivation, as in experiments thermodynamic quantities are measured through a long time average. For instance, the pressure of a gas is measured by a manometer through a process that takes a time much longer than the microscopic one (e.g. the mean collision time). Therefore, from a physical point of view, time averages are the basic quantities, while the ensembles approach can be considered a useful algorithmic tool for computing averages without the knowledge of trajectories.

Back to the FPU system, for small values of ϵ, it is easy to compute the thermodynamic quantities using the microcanonical (or canonical) ensemble. In particular, in the integrable limit ($\epsilon = 0$), it is easy to obtain the equipartition law:

$$\langle E_k \rangle = \frac{E_{tot}}{N}, \tag{14.2}$$

where $\langle \cdot \rangle$ denotes the microcanonical (or canonical) ensemble average. In the presence of a small anharmonic perturbation ($\epsilon \neq 0$), we have small corrections to (14.2), $\langle E_k \rangle = E_{tot}/N + \mathcal{O}(\epsilon)$.

However in the integrable limit $\epsilon = 0$, normal modes are decoupled and thus cannot exchange energy, which remains constant for each mode, so that equipartition (14.2) is just a formal result of the ensemble approach, and the system is

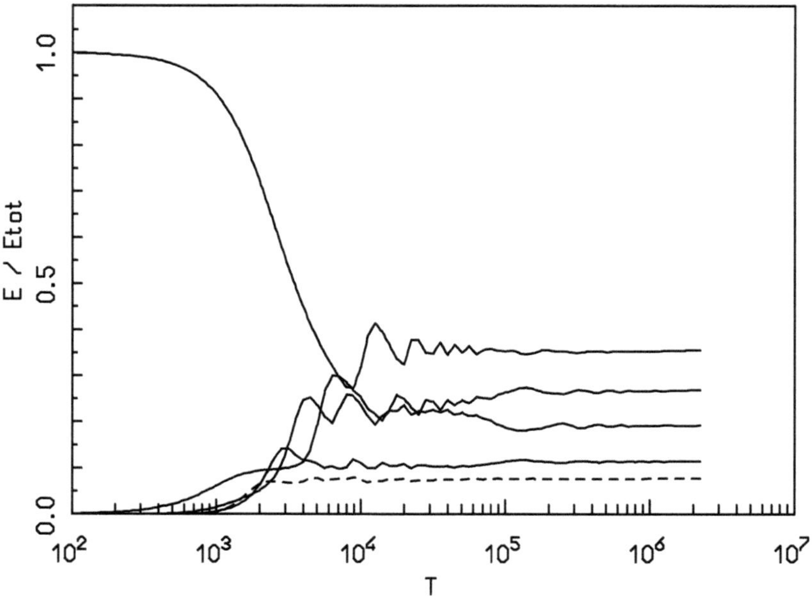

Fig. 14.2 Time averaged fraction of energy, in modes $k = 1, 2, 3, 4$ (bold lines, from top to below), the dashed line shows the time average of the sum from $k = 5$ to $N = 32$. The parameters of the system are the same as in Fig. 14.1. [Courtesy of G. Benettin]

not truly ergodic. This means that the statistical mechanics treatment can be well founded only in the non-integrable case, and that the time average computed along the trajectory over a long observation time $\mathcal{T}(\to \infty)$

$$\overline{E_k}^{\mathcal{T}} = \frac{1}{\mathcal{T}} \int_0^{\mathcal{T}} \mathrm{d}t\, E_k(t) \tag{14.3}$$

can coincide with the ensemble average $\langle E_k \rangle$ only when $\epsilon \neq 0$.

We can now appreciate the importance of FPU numerical experiment. What does it happen to the evolution of the system (14.1) with $\epsilon \neq 0$ if energy is initially concentrated only in a few normal modes, for instance $E_1(0) \neq 0$ and $E_k(0) = 0$ for $k > 1$? Before FPU, from Poincaré's result as well as Fermi's generalization, the general expectation would have been that the first normal mode should progressively transfer energy to the other modes and, after a thermalization time,[4] the energy of each mode $E_k(t)$ would fluctuate around the equilibrium value E_{tot}/N. Indeed, as Ulam wrote in the introduction to FPU, Fermi's *motivation* [...] *was to observe the rates of mixing and "thermalization"*.

FPU numerical experiment was performed, for small ϵ and system sizes $N = 16$, 32 and 64, with the energy initially concentrated in one or two normal modes.

[4]Basically the thermalization time is the characteristic relaxation (or mixing) time, see Sec. 4.4.

Unexpectedly, no tendency towards equipartition was observed, even for long times. In the words of Fermi, Pasta and Ulam

> *Instead of a gradual, continuous flow of energy from the first mode to the higher modes, all of the problems show an entirely different behavior* [...] *It is only the first few modes which exchange energy among themselves and they do this in a rather regular fashion.*

The phenomenon observed in FPU is clear from Fig. 14.1, where the time evolution of the energy E_k for a few modes is shown: energy is initially concentrated in the first mode and, after a long time, E_1 reverts back almost to its initial value. The resulting motion is thus rather regular, almost periodic with no memory loss of the initial condition and no equipartition among the modes as clear from Fig. 14.2. These observations imply a violation of ergodicity also for $\epsilon \neq 0$.

These results strongly contrasted with expectations. Fermi himself, writing to Ulam, said that he was surprised and stressed that they were dealing with an important discovery, which unambiguously shows how the prevalent opinion (at that time), on the generality of mixing and thermalization properties of nonlinear systems, might not always be justified.

14.1.1 *Toward an explanation: Solitons or KAM?*

A first[5] explanation for FPU findings come with Zabusky and Kruskal (1965) who, performing the continuous limit of the ODE associated to the Hamiltonian (14.1), obtained a partial differential equations which admits solitonic solutions, i.e. solitary waves that maintain their shape while traveling. From the finding of such regular solutions, they argued that the regular behavior observed in FPU could be attributed to the existence of these solitons. Let us briefly discuss the idea.

The equations which govern the evolution of the FPU system, for $\alpha = 3$, are

$$m\,\frac{\mathrm{d}^2 q_n}{\mathrm{d}t^2} = f(q_{n+1} - q_n) - f(q_n - q_{n-1})\,, \tag{14.4}$$

with $f(y) = Ky + \epsilon y^2$. The above second order ODEs can be interpreted as a lattice approximation of a PDE, where the variable $q_i(t)$ approximates a continuous field $\psi(x,t)$ at the point $i\,\Delta x$ of a lattice, Δx being the lattice spacing. Properly rescaling K, m and ϵ as function of Δx, it is indeed easy to derive a PDE for $\psi(x,t)$:

$$\frac{\partial^2 \psi}{\partial t^2} = c^2 \frac{\partial^2 \psi}{\partial x^2} + g\,\frac{\partial \psi}{\partial x}\frac{\partial^2 \psi}{\partial x^2}\,, \tag{14.5}$$

where $c^2 = \lim_{\Delta x \to 0} K\Delta x^2/m$ and $g = \lim_{\Delta x \to 0} \epsilon \Delta x^3/m$. Seeking for a solution of (14.4) which, in the continuous limit, slowly varies with t, when the quantity $x - ct$ is fixed, we obtain the equation

[5] Even if, to be precise, this explanation was not the first in the historical sense. Indeed KAM theorem, at least in the Kolmogorov formulation, was known to mathematicians since 1954.

$$\frac{\partial v}{\partial \tau} + \epsilon v \frac{\partial v}{\partial \xi} + \frac{1}{24}\frac{\partial^3 v}{\partial \xi^3} = 0\,, \tag{14.6}$$

where the variables ξ and τ are proportional to $x - ct$ and to t respectively and $v = \partial\psi/\partial\xi$ [Cercignani (1977)]. Remarkably, Eq. (14.6) is essentially a rewriting of the Korteweg-de Vries (KdV) equation (dating back to 1895) which was introduced to describe the propagation of surface waves in shallow water and admits a "solitonic wave" solution of the type $v = F(\xi - V\tau)$, where V is a constant and $F(z)$ is a localized function that decays to zero at large values of $|z| = |\xi - V\tau|$. Solitary waves have been considered for a long time as being a mere mathematical curiosity, but solitons now have turned to be fundamental in many physical phenomena [Cercignani (1977); Dauxois and Peyrard (2006)].

However, the original Zabusky and Kruskal (1965) explanation of the regularity of the FPU system in terms of solitary waves originating from the KdV equation is, for some aspects, not totally convincing. For instance, different ways to perform the continuous limit from the ODE $\{q_n(t)\}$ to PDE for the field $\psi(x,t)$ can lead to PDEs with very different features. For instance, while Eq. (14.6) possesses solitonic solutions Eq. (14.5) develops spatial discontinuities after a finite time $t_c \sim 1/|\psi_0|$, where ψ_0 is the maximum field amplitude at $t = 0$ [Cercignani (1977)]. Moreover, solitonic solutions are related to integrable, weakly nonlinear PDEs which are rare, in the sense that a generic perturbation destroys the integrability. Therefore, since a generic Hamiltonian systems is non integrable, it is not physically sound to associate behaviors such as those observed in the FPU to integrability.

The other explanation of the inefficient energy transfer among modes in FPU is based on KAM and Nekhoroshev theorems (Sec. 7.2), which establish that non-ergodic behaviors of non-integrable Hamiltonian systems are actually typical. In other terms the non-existence of global invariants of motion (i.e. Poincaré's result and Fermi's generalization) does not imply all trajectories obtained by a non-integrable perturbation of an integrable Hamiltonian system will be far from the unperturbed (integrable) ones. This fact was surely unknown to Fermi, Pasta and Ulam. Moreover, we should underline that the interpretation of FPU results in terms of KAM theorem is not straightforward [Gallavotti (2007)]. For instance, the periodic behavior of the energies $E_k(t)$ shown in Fig. 14.1 cannot be simply interpreted as a motion on a KAM torus corresponding to a deformation of the integrable one, for which $E_k = \text{const}$.

It is worth concluding this section by mentioning that several numerical investigations indicated that the two pictures in terms of solitons and KAM are not completely exclusive: it is matter of time scales. Up to a certain time, which depends on the initial condition and the parameters of the FPU, i.e. ϵ, m, K and N, for suitable initial states (only low frequency modes are excited), the KdV solitons are good approximations of the actual numerical solutions of the (discrete) FPU system. At larger time the non-integrable character of the system becomes dominant [Lichtenberg *et al.* (2007)].

14.2 A random walk on the role of ergodicity and chaos for equilibrium statistical mechanics

Even many years after FPU numerical experiment, the actual importance of ergodicity, chaos and the number of degrees of freedom as building bricks of equilibrium statistical mechanics is still under debate. Essentially, the contraposition between the role of ergodicity incidental to chaos on one side and the role of the large number of degrees of freedom on the other side can be seen as a "dynamical" versus "statistical" justification to statistical mechanics, respectively. The aliveness of this debate reflects in the different opinions reported in the scientific literature. On one side, Grad (1967) explicitly writes "*the single feature which distinguishes statistical mechanics is the large number of degrees of freedom*". A rather similar point of view is expressed in the Landau and Lifshitz (1980) book on statistical physics as well as in Khinchin (1949). On the opposite side, there is the opinion of those, as Prigogine (1994) and his school, who consider chaos to be a source of "randomness" crucial for the consistency of statistical mechanics. The demanding reader may find interesting the discussion on this debated issue made by Lebowitz (1993),[6] and Bricmont (1995).

Before illustrating some of these topics by referring to FPU and similar systems, it is worth discussing the importance of many degrees of freedom for ergodicity.

14.2.1 *Beyond metrical transitivity: a physical point of view*

In Sec. 4.3, we presented the ergodic problem from an abstract mathematical viewpoint. However, if we are interested to statistical mechanics of macroscopic systems, we can adopt a more flexible approach as best illustrated in the well known book *Mathematical Foundation of the Statistical Mechanics* by Khinchin (1949). The physical motivation roots in the following facts:

(a) statistical mechanics deals with macroscopic bodies composed by a huge number N of degrees of freedom;

(b) the important observables of statistical mechanics are not generic functions, so that it is enough to prove the equivalence of ensemble and time averages, $\overline{f} = \langle f \rangle$, just for the relevant observables;

(c) it is physically acceptable to admit a failure of such an equivalence, i.e. $\overline{f} \neq \langle f \rangle$, for initial conditions in a phase-space region of small measure, vanishing as $N \to \infty$.

Khinchin restricts his analysis to separable Hamiltonian systems:

$$H = \sum_{n=1}^{N} H_n(\boldsymbol{q}_n, \boldsymbol{p}_n) \tag{14.7}$$

[6]It is interesting to see also the comments, together with Lebowitz's replies: H. Barnum, C.M. Caves, C. Fuchs, and R. Schack, Physics Today **47**, 11 (1994); J. Driebe, Physics Today **47**, 13 (1994); W.G. Hoover, H. Posch, and B.L. Holian, Physics Today **47**, 15 (1994); R. Peierls, Physics Today **47**, 16 (1994).

and to a special class of observables, called *sum functions*, of the form

$$f(\boldsymbol{X}) = \sum_{n=1}^{N} f_n(\boldsymbol{q}_n, \boldsymbol{p}_n) \quad \text{with} \quad \boldsymbol{X} = (\boldsymbol{q}_1, \ldots, \boldsymbol{q}_N; \boldsymbol{p}_1, \ldots, \boldsymbol{p}_N),$$

where the functions f_n are assumed to be of the same order of magnitude, meaning that all degrees of freedom contribute the same. Interesting examples of sum functions are pressure and kinetic energy.

Under quite general hypothesis, and without invoking metrical transitivity (Sec. 4.3), if f is a sum function then

$$\text{Prob}\left(\frac{|\overline{f} - \langle f \rangle|}{|\langle f \rangle|} \geq K_1 N^{-1/4} \right) \leq K_2 N^{-1/4}, \tag{14.8}$$

where K_1 and K_2 are $\mathcal{O}(1)$ constants.[7] In words: provided the number of degrees of freedom is large ($N \to \infty$), time and ensemble averages differ, $\overline{f} \neq \langle f \rangle$, only in a phase-space set of negligible measure.

Technically speaking, Khinchin's approach concerns non-interacting systems (14.7), which poses some problems because we know that the exchange of energy among degrees of freedom is mandatory for the existence of statistical equilibrium. However, the physical wisdom would suggest that, remaining in the framework of the ensembles theory, a weak short range interaction among the degrees of freedom should contribute very little to statistical averages but should allow for the statistical equilibrium to be reached. This intuition has been rationalized by Mazur and van der Linden (1963), who extended Khinchin's result to systems of particles interacting through a short range potential.

14.2.2 *Physical questions and numerical results*

In spite of its conceptual importance, ergodic theory, even in the "weak version" considered by Khinchin, is not conclusive. From an experimental point of view, the main point, disregarded by the ergodic theory, is the understanding of the physically relevant times. When measuring a quantity through a time average as in Eq. (14.3), the observation time $\mathcal{T}$ cannot be infinite and thus it is natural to wonder how large it should be for $\overline{A}^{\mathcal{T}}$ to be close to $\langle A \rangle$. The answer to this question may be subtle and depend on both the observable A and the number of particles N. For instance, for some observable $\mathcal{T}$ grows exponentially with N.[8]

In Khinchin approach, as well as in the generalization of Mazur and van der Linden, the dynamics has a marginal role and the existence of good statistical properties basically follows from the fact that $N \gg 1$. Such a result (Eq. (14.8))

[7] Here probability is respect to the microcanonical measure.

[8] This had been already realized by Boltzmann in his answer to the criticisms on H-theorem [Cercignani (1998)]. If A is the characteristic function of a set in phase space, $\mathcal{T}$ scales as the return time to the set, which is exponentially large in N, see the discussion of Kac's lemma (B.7.1).

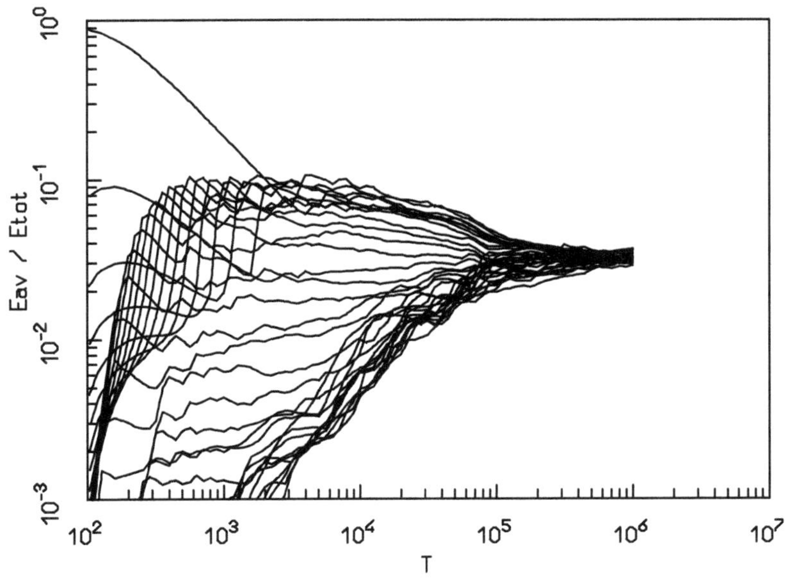

Fig. 14.3 $\overline{E}_k^{\mathcal{T}}/E_{tot}$ versus $\mathcal{T}$ for $k = 1, \ldots, 32$. The parameters of the system (14.1) are $N = 32$, $\alpha = 3$, $\epsilon = 0.1$ and energy density $\mathcal{E} = E_{tot}/N = 1.2$. [Courtesy of G. Benettin]

holds for generic, with respect to the microcanonical measure, initial conditions. On the contrary, both in experiments and in numerical simulations, the initial conditions can be far from equilibrium, as for instance those used in FPU. Therefore, it is important to understand how the thermalization (or mixing) time depends on the number of degrees of freedom N and the system parameters.

Let us now consider such issues on the light of some numerical results for the FPU system.

Izrailev and Chirikov (1966) first noticed that for high values of ϵ, where the effects of KAM tori are switched off, there is a good statistical behavior. This is shown in Fig. 14.3 where the energy, initially concentrated in the lowest frequency normal modes, equally spreads on all normal modes, and the time averages are in agreement with those of equilibrium statistical mechanics. This result is comforting since it shows that if the nonlinearity is strong enough dynamics and statistics (i.e. ergodicity and statistical ensembles) are in agreement.

More specifically, for a given number of particles N and energy density $\mathcal{E} = E/N$ there is a threshold ϵ_c for the strength of the perturbation such that [Livi *et al.* (1985); Kantz *et al.* (1994); Carati *et al.* (2005)]:

(a) if $\epsilon < \epsilon_c$ KAM tori play a major role and the system does not follow equipartition, even after a very long time;
(b) if $\epsilon > \epsilon_c$ KAM tori have a minor effect: the system follows equipartition and there is agreement with equilibrium statistical mechanics.

Similarly, if the value of the perturbation ϵ is given, as in actual physical situations, the energy density formally acts as a control parameter and a threshold $\mathcal{E}_c$ would exist separating regular from irregular behavior.

Given the above scenario, several physical questions arise

(1) whether the regular behavior for small nonlinearities, and irregular for large ones, is peculiar of FPU Hamiltonian;
(2) what is the dependence of ϵ_c on N (at fixed $\mathcal{E}$) or, equivalently, the dependence of $\mathcal{E}_c$ on N (at fixed ϵ)?
(3) what are the characteristic times of the equipartition process as function of N and ϵ?
(4) for a given N, how small is the part of the phase space with regular behavior.

Point 1 is clear: the mechanism of the transition to chaos at increasing the degree of non-integrability is typical for all Hamiltonian systems which (as FPU) are perturbations of harmonic systems. This behavior is present also in multidimensional lattices, for instance in two-dimensional Lennard-Jones systems, where the Hamiltonian can be written as an harmonic part plus an anharmonic perturbation [Benettin and Tenenbaum (1983)].

In spite of the great analytical and numerical efforts, there still is no general consensus on points 2 and 3 [Casartelli *et al.* (1976); Kantz *et al.* (1994); Carati *et al.* (2005)]. The dependence of ϵ_c (or, equivalently, of $\mathcal{E}_c$ at fixed ϵ) on N is obviously very important: if $\epsilon_c \to 0$ when $N \to \infty$, the ante-FPU point of view would be re-established. On the contrary, if ϵ_c is independent of N, there would be a serious discrepancy with the results expected by equilibrium statistical mechanics. Then, of course, even if a good statistical behavior is established it is important to understand on which time scales it is obtained.

We briefly mention some results. Casetti *et al.* (1997) show that in FPU with quadratic nonlinearity ($\alpha = 3$), for an energy density smaller than $\mathcal{E}_c = E_c/N \sim 1/N^2$ the motion is regular with solitonic behaviors as in Zabusky and Kruskal interpretation, even in the limit $N \to \infty$. Below $\mathcal{E}_c$ the relaxation to statistical equilibrium, if any, takes longer than observable on any available computer. Above the threshold the system possesses good statistical behaviors. However, the relaxation time $\tau_R(\mathcal{E})$ necessary to reach the equipartition, starting from a far-from-equilibrium initial condition may be very long: $\tau_R \sim \mathcal{E}^{-3}$. Similar results hold for cubic ($\alpha = 4$) nonlinearities [de Luca *et al.* (1999)]. The relaxation time, τ_R might also depend on the initial condition. Indeed, since Hamiltonian systems do not have attractors, the choice of initial conditions (particularly for large N) constitutes an important technical aspect. For instance, if the initially excited normal modes are always between k_1 and k_2 (with k_1 and k_2 fixed), at increasing N we have $\tau_R \sim N^{1/2}\mathcal{E}^{-1}$ [Ruffo (2001)]. For a very detailed, both numerical and analytical, recent study on the role of initiatl conditions see Benettin *et al.* (2008).

Point 4 will be analyzed in the framework of a numerical study of high dimensional symplectic maps (see next section). However, we anticipate that even if the portion of phase space where chaos resides is large, the problem of time scales — point 3 — can be still important.

As clear from the above non-exhaustive survey, even more than fifty years after the original paper, FPU system is still a subject of active investigations and far from being understood. In this respect we mention a recent special issue of Chaos [Campbell *et al.* (2005)] and a collective volume [Gallavotti (2007)] containing reviews on the different aspects of this topic.

We conclude mentioning that in two-dimensional FPU-like models, unlike the above described scenario in one dimension (where the dynamics is dominated for a long time scale by a weakly chaotic metastable regime), if the number of degrees of freedom is large enough the time scale energy redistribution among normal modes, although rather long, is drastically shorter than in one dimension. However, in the two-dimensional case, it has been found that boundary conditions play a role for both the dynamical and statistical properties [Benettin and Gradenigo (2008)].

14.2.3 *Is chaos necessary or sufficient for the validity of statistical mechanical laws?*

Naively, one could be tempted to say that the equivalence between time and ensemble averages requires chaos (in the sense of positive Lyapunov exponent) as a necessary ingredient. The actual scenario is more complex. Even if a system is chaotic and most KAM tori are destroyed, ergodicity cannot be taken for granted. In other terms the justification of equilibrium statistical mechanics on a dynamical basis is not automatically obtained [Livi *et al.* (1987)]. It should be stressed that the above concerns are not restricted to FPU-like systems. In the following, through the analysis of a few numerical examples, we will see that the identification of chaos as a necessary or sufficient ingredient for the validity of statistical mechanics is delicate, if not wrong.

We start considering a high dimensional system of symplectic coupled maps

$$\begin{aligned} \phi_n(t+1) &= \phi_n(t) + \mathrm{I}_n(t) && \mod 2\pi \\ \mathrm{I}_n(t+1) &= \mathrm{I}_n(t) + \epsilon \tfrac{\partial F(\phi(t+1))}{\partial \phi_n} && \mod 2\pi\,, \end{aligned}$$

with $n = 1, ..., N$. The above map is a non-integrable perturbation, controlled by ϵ, of a system of independent rotators. If the number N of maps is large enough, even for small ϵ, chaotic trajectories become dominant [Falcioni *et al.* (1991); Hurd *et al.* (1994)]. Specifically, the measure of the phase-space region occupied by KAM tori is exponentially small in N. Although this sounds like good news for having robust statistical behaviors, it turns out that very long time-scales are actually involved, namely individual trajectories diffuse in phase space very slowly and forget their initial conditions after a typical time $\mathcal{O}(10^5) - \mathcal{O}(10^6)$ larger than the Lyapunov time $(1/\lambda_1)$ [Falcioni *et al.* (1991)]. The main lesson of the above result is that the time scales given by chaos (i.e. $1/\lambda_1$) can be completely unrelated to those relevant to statistical mechanics (relaxation or diffusion times).

Livi *et al.* (1987) studied the connection between chaos and equilibrium statistical mechanics in high dimensional nonlinear Hamiltonian systems. Their concern was to compare the prediction of equilibrium statistical mechanics and time averages for quantities like internal energy U and specific heat C_V. In particular, they performed such a comparison using different values of energy density $\mathcal{E}$, that is exploring regimes where chaos is weak or dominant. Since the averages involved in the computation of U and C_V are easier in the canonical ensemble, before discussing the results, it is useful to describe the numerical strategy employed to mimic, in a deterministic way, the canonical ensemble.

A dynamical simulation of the canonical ensemble, which avoids the use of stochastic heat baths, can be realized studying the fluctuations of suitable observables in a small part of a large system. For instance, in FPU this can be done subdividing a chain of N particles into N_1 subsystems composed of $N_2 = N/N_1$ particles each, with $N_1 \gg 1$ and $N_2 \gg 1$. In such a way, the time average of a certain observable, defined in the subsystems, can be computed and compared with the corresponding predictions of the canonical ensemble. For example, we can define the internal energy U as the time average of the energy of the j-th subsystem E_j normalized by the number of particles N_2 composing the subsystem, i.e.:

$$U = \frac{\overline{E_j}}{N_2} .$$

Analogously, for the specific heat C_V we have:

$$C_V = \frac{1}{T^2}\frac{\overline{E_j^2} - \overline{E_j}^2}{N_2} ,$$

where $T = \overline{p^2}/m$ is the temperature.

Detailed numerical computations show that both the internal energy U and the specific heat C_V are in good agreement with the prediction of the canonical ensemble at large energy densities $\mathcal{E}$ (high temperatures), where chaos is overwhelming. Surprisingly, this agreement holds also for small $\mathcal{E}$ (low temperatures), where the KAM tori are dominant and the system is mostly regular [Livi *et al.* (1987)]. This result seems to indicate that chaos is not a necessary ingredient for good statistical behaviors. This may be seen as a confirmation of Khinchin ideas on the marginal role of dynamics, although, strictly speaking, the considered observables are not in the class of the sum functions.

However, this is not the end of the story. Livi *et al.* (1987) repeated the same analysis for a system of coupled rotators described by the Hamiltonian

$$H = \sum_{i=0}^{N}\left[\frac{p_i^2}{2m} + \gamma\Big(1 - \cos(q_{i+1} - q_i)\Big)\right] ,$$

where the variables $\{q_i\}$ are angles, and γ sets the coupling strength. Fixing γ, we can identify two different integrable limits:

(a) for $\mathcal{E} \to 0$ we recover an harmonic chain of oscillators;
(b) for $\mathcal{E} \to \infty$, the system reduces to perturbed rotators.

Numerical simulations show that U agrees with the canonical predictions at any temperature. While at high temperatures, in spite of the rather large Lyapunov exponent, C_V strongly deviates from canonical ensemble expectations, suggesting that chaos is not sufficient for the validity of the ensemble approach.[9]

14.3 Final remarks

Let us summarize the main aspects discussed in this Chapter. At first we stress that the ergodic approach, even with all the examined caveats, appears the natural way to introduce probability in a deterministic context. In particular, as ergodicity is equivalent to the condition that different trajectories have identical asymptotic properties, ergodic theory poses the ground for the frequentistic interpretation of probability, where the probability of an event is defined through the relative frequency with which it happens. Therefore, assuming ergodicity, it is possible to obtain an empirical notion of probability which is an objective property of the trajectory [von Plato (1994)].

Physically one deals with a unique system (although with many degrees of freedom) and not with an ensemble of (identical) systems. It thus seems natural, according to Boltzmann, to assume that the only physically consistent (at conceptual level) statistical method is in terms of time averages obtained following the time evolution of the system.[10] In this interpretation, the ensembles are just a practical tool to compute explicitly the averages, avoiding the determination of the trajectory. We should, however, mention that there is not a complete consensus on this viewpoint. For example, Landau and Jaynes share the opposite opinion that ergodicity is simply not relevant for the ensembles' method [Jaynes (1967); Landau and Lifshitz (1980)]. Moreover, ergodicity is, at the same time, an extremely demanding property — time and phase averages must be equal for almost all initial conditions — and physically not very accessible — because of the infinite time request for the average. The latter point is particularly severe when considering far-from-equilibrium initial conditions, due to the, possibly, very long mixing times involved.

[9]The different behavior of C_V in the two systems (FPU and coupled rotators) can be understood as follows. For the FPU system and for the low temperature rotators the "natural" variables are the normal modes. Even if the normal modes are almost decoupled, the energy in a subsystem is defined in real space and depends on all the normal modes, so there are non-negligible fluctuations of the "local" energy. On the contrary, in the high T limit, in the chain of rotators the normal modes are now the "local" variables $\{q_j\}$ themselves, therefore the exchange of energy among the subsystems is weak and fluctuations of their energies are strongly depressed.

[10]This aspect, with some obvious changes, basically holds even for non Hamiltonian systems

As far as chaos is concerned, the analyzed examples of high dimensional Hamiltonian systems provided indications that chaos is neither a necessary nor a sufficient ingredient for justifying, on a dynamical basis, equilibrium statistical mechanics. On the one hand, even when chaos is very weak (or absent), we may have a good agreement between time and ensemble averages. On the other hand, chaotic behaviors do not necessarily imply such an agreement.

The scenario is even more controversial when passing to non-equilibrium statistical mechanics. Several simulations and theoretical works have shown that, in hyperbolic systems, there exists a close relationship between transport coefficients, such as viscosity, thermal and electrical conductivity, and chaos indicators, such as Lyapunov exponents, KS entropy and escape rates [Gaspard and Nicolis (1990); Ueshima *et al.* (1997); Gaspard (1998); Dorfman (1999)]. However, others have shown that chaos, in the technical sense of positive Lyapunov exponent, is not necessary for non-equilibrium phenomena such as diffusion and conduction, which can be obtained also in systems having zero (maximal) Lyapunov exponent [Dettmann and Cohen (2000, 2001); Cecconi *et al.* (2003); Jepps and Rondoni (2006); Cecconi *et al.* (2007)]. In Box B.32, we illustrate an example of non-chaotic system with irregular motions — often the term *pseudochaos* is used for systems of this kind [Vega *et al.* (1993); Mantica (2000)] — displaying diffusive behavior.

Box B.32: Pseudochaos and diffusion

Irregular dynamics in systems with non-positive Lyapunov exponents are rather intriguing from a conceptual and practical point of view. We have seen an example of such systems in Box B.29 where we discussed the phenomenon of stable chaos. In this Box we consider another class of non-chaotic systems, having zero Lyapunov exponent, but displaying irregular behaviors, for which often, in the literature, the term *pseudochaos* is used [Vega *et al.* (1993); Mantica (2000)]. In particular, we aim to show an example of non-chaotic system able to generate large scale diffusive behaviors, so to illustrate that, in deterministic systems, well defined non-equilibrium phenomena are not necessarily related to chaos.

Consider the $1D$ map already investigated in Sec. 10.3.3

$$x(t+1) = [x(t)] + F(x(t) - [x(t)])\,,$$

where $[\,.\,]$ indicates the integer part and

$$F(u) = \begin{cases} 2(1+a)u & \text{if } u \in [0:1/2[\\ 2(1+a)(u-1)+1 & \text{if } u \in [1/2:1] \end{cases} \tag{B.32.1}$$

with $a > 0$. Such a system is chaotic ($\lambda_1 = \ln(2+a) > 0$). As $a > 0$ implies the possibility for a trajectory that is, at time t, in the cell $C_\alpha \equiv [\alpha : \alpha+1]$ (with α integer) to reach, at time $t+1$, one of the adjacent cells $C_{\alpha\pm1}$. So that at scales larger than the size of the elementary cell, i.e. $\gg 1$, and times long enough a Brownian-like motion is observed with diffusion coefficient: $D = \lim_{t\to\infty}\langle(x(t) - x(0))^2\rangle/(2t)$.

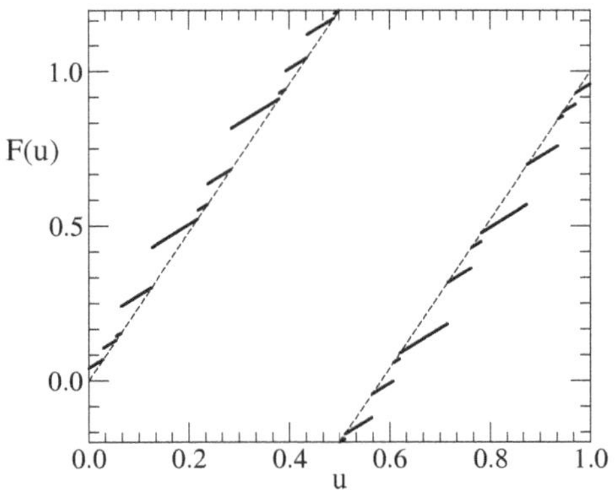

Fig. B32.1 Sketch of the random staircase map in the unitary cell, with $a = 0.23$. Half domain $[0:1/2]$ is divided into $N = 12$ micro-intervals of random size.

We now consider a non-chaotic variant of the above map. As exemplified in Fig. B32.1, the function F is replaced by a step-wise version generated as follows. The first-half of each elementary cell C_α is partitioned in N intervals $[\alpha + \xi_{n-1} : \alpha + \xi_n[\ , n = 1, \ldots, N$, with $\xi_0 = 0 < \xi_1 < \xi_2 < \cdots < \xi_{N-1} < \xi_N = 1/2$. In each interval the map is defined as

$$F_N(u) = u - \xi_n + f(\xi_n) \qquad \text{if } u \in [\xi_{n-1}, \xi_n[\,.$$

The map in the second half of the elementary cell C_α is then obtained by imposing the anti-symmetry condition with respect to the center of the cell. The quenched random variables $\{\xi_k\}_{k=1}^{N-1}$ are uniformly distributed in the interval $[0:1/2]$, i.e. the intervals have a *random* extension. Furthermore, they are chosen independently in each cell C_α (so that, more properly, we should write $\xi_n^{(\alpha)}$). All cells are partitioned into the same number N of randomly chosen micro-intervals (of mean size $\Delta = 1/(2N)$).

As the step-wise approximate F_N has slope 1 almost everywhere, the map is no longer chaotic, and the continuous chaotic map (B.32.1) is recovered in the limit $N \to \infty$. Such a limit, however, is singular as for any finite N chaos is absent.

In spite of the absence of chaos, when a quasi-periodic external perturbation is added, i.e.

$$x(t+1) = [x(t)] + F_N(x(t) - [x(t)]) + \gamma \cos(\omega t)\,, \tag{B.32.2}$$

diffusion can still be observed [Cecconi *et al.* (2003)]. The strength of the external forcing is controlled by γ and ω defines its frequency, while N indicates a specific quenched disorder realization. Numerical computations show that for $\gamma > \gamma_c \sim 1/N$, the diffusion coefficient is close to that obtained using the chaotic map (B.32.1). These findings are robust and do not depend much on the details of forcing. Therefore, we have an example of a non-chaotic model in the Lyapunov sense by construction, performing diffusion. See also Jepps and Rondoni (2006); Cecconi *et al.* (2007) for other examples of diffusion without chaos.

We conclude mentioning that the system (B.32.2) is somehow the analogue in $1D$ of the wind-tree Ehrenfest's model [Dettmann and Cohen (2000)]: a particle (wind) scatters elastically against square obstacles (trees) randomly distributed in the plane with a fixed orientation. Such a non chaotic system shows the same diffusive feature of the Lorentz (1905) gas (where the obstacles are circular) which is chaotic [Bunimovich and Sinai (1981); Dorfman (1999)].

Epilogue

> I am of the opinion that the task of the theory consists in constructing a picture of the external world that exists purely internally and must be our guiding star in all thought and experiment.
>
> *Ludwig Eduard Boltzmann (1844–1906)*

By way of conclusion, we would like to draw some remarks on the use of dynamical systems and the role of chaos in the broader context of model building and computer experiments; rather than an exhaustive account of the topic, the following pages should be read as a summary of our perspective mainly based on our experience as physicists.

Chaos, modeling, computer experiments and complex systems

The classical "initial value" problem — given the initial state of a given differential equation find the solution, or an approximate solution, after a certain time — is often not viable in several interesting phenomena characterized by complex (or complicated) behaviors. As a matter of facts, even when we have (or presume to have) detailed knowledge of the evolution laws, the initial value approach is spoiled by the presence of chaos and/or by the huge number of degrees of freedom involved, which lead to the need of having an arbitrarily large amount of information about the initial state. If we do not know the equations governing the phenomenon, the problem shifts to that of model building that will be covered below.

In all such circumstances, we are typically forced to change our approach from understanding the specific trajectory originated from a given initial state to the statistical character of an ensemble of trajectories (Chap. 4). This task is usually accomplished by means of numerical computations. In this respect, the computer has played a key role in developing the theory of dynamical systems and chaos. In fact, it is only thanks to the fast computations and visualization tools made possible by computers that we have been able to discover/understand the existence of the wealth of behaviors in nonlinear systems, and to advance in their systematic quantitative characterization.

In this book we discussed several instances of dynamical systems which have their counterpart in the way computer simulations can be used. Roughly speaking, we can identify two main categories of numerical simulations, although their boundaries are sometimes blurred:

(i) Accurate numerical simulations able to approximate the solution of equations representing, or assumed to represent, a given system.
(ii) Numerical implementations of models which, retaining the basic principles of a real system, are a crude simplification (or phenomenological caricature) of the primary model. Similarly, we can also consider models that are completely free from the necessity to represent, even as a caricature, a real system and only aim at highlighting some universal mechanisms.

Class *i)* includes, for example, direct numerical simulations of the Navier-Stokes equation, or the full N-body gravitational problem in celestial mechanics. Perhaps, this computational approach is the most obvious one, and reflects the etymological origin of the term "computer": from Latin *computare* "to count, sum up".[11] The idea underlying this way of using computers is that systems can be completely "known" and "reproduced" *in silico* once the equations representing it are numerically solved. Pushing forward this line of reasoning, we arrive to von Neumann's prophecy that automatic computing machines, being able to solve for instance the Navier-Stokes equation, would make laboratory experiments in fluid mechanics obsolete, at least when computers become large and powerful enough to integrate a sufficient number of degrees of freedom (Sec. 13.3). Although von Neumann's prophecy is not yet reality [Celani (2007)], we report that many experimentalists tune their instruments using the results of careful simulations [Moin and Mahesh (1998)]. This approach to computation is clearly constrained by fact that it requires (true or presumed) knowledge of the complete mathematical representation of a given system.

Class *ii)* implies some kind of model-building activity. As an explicit connection between models and reality is not there or, more generally, is not required, the results of the numerical computation only concern the abstract mathematical structures of the model and, as a consequence, can be considered just a metaphor of the original phenomenon. Typical examples are: shell models (Sec. 13.3.5) that are a phenomenological caricature of the Navier-Stokes equation; coupled map lattices (CML, see Chap. 12) that constitute a prototype for spatially extended systems, but far from representing any of them; the Lotka-Volterra equations (Sec. 11.3), that incorporate some basic mechanisms of competition between a prey and a predator species, whose "true" dynamics is unknown.

[11] It is noteworthy to follow the evolution of the term: in 1646 computer stands for the person who computes; in 1897 for mechanical calculating machine, and in 1946 or 1941 for an electronic machines. The modern meaning, "programmable digital electronic computer" is from 1945 (while its theoretical sense is from 1937, as Turing machine).

The above classification prompts us to reflect upon the role of models. Scientific models can be grouped into different categories depending on the scientific discipline of interest, or on the methodological approach. In our opinion, within the framework of dynamical systems, two broad classes of scientific models can be identified: *predictive models* and *interpretative* or *explanatory models.*

Predictive models, as the name suggests, allow us to make definite predictions on a particular phenomenon. If we assume that the ultimate mathematical representation of a fluid in motion is the Navier-Stokes equation, that equation does not explain a particular flow, but it can be used to predict it by solving the equation (with a computer experiment) when initial and boundary conditions are specified.

Predictive models pertain to class *i).* Their predictive power may be limited by the presence of chaos, but they can still be valuable in statistical forecasting, and we will come back to this aspect at the end of this discussion.

Explanatory models are typically developed with different aims.

In Chapter 13, we introduced proper orthogonal decomposition and shell models, which make it possible to describe some specific aspects of fluid motion,[12] namely the evolution of coherent structures and the phenomenology of the energy cascade in turbulence, respectively. This is an interesting example, because although both coherent structures and turbulence are, in principle, deducible from the Navier-Stokes equation, we felt the need to introduce simplified and abstract (explanatory) models that are able to capture the underlying physical mechanisms, otherwise obscured by the difficulty of the NSE.

Another very instructive example is coupled map lattices, better illustrating the role of interpretative models which *carve nature at its joints*, namely de- and reconstructing parts or mechanisms. The strategy behind CMLs modeling is mainly based on the decomposition of the processes underlying the phenomenon into independent "components", see Kaneko (1992). For instance, Jensen (1989), considering the convective and the diffusive components, was able to build a CLM which captures the basic phenomenology of boundary-layer instabilities (including the emission of plumes from the bottom boundary). In a similar way, Yanagita and Kaneko (1993) introduced and studied a CML system for thermal convection, reproducing many complex spatiotemporal features observed in experiments [Castaing *et al.* (1989)]. Therefore, a CML does not approximate a specific system, which can eventually have its predictive model in terms of a PDE, but represents an idealized model whose behavior is significant for the spatiotemporal features of the system. In particular, CMLs offer a theoretical lab to extract features expected to be universal, e.g. the asymptotic behavior of the Lyapunov Exponents or of the KS-entropy in the thermodynamic limit (Sec. 12.2) or the propagation and growth of perturbations in generic spatially distributed systems (Sec. 12.3). Similarly, the study of globally coupled maps gave some insights on the emergence of coherent collective behavior as observed in system with a huge number of components (Sec. 12.5.2).

[12]Even if we assumed that it suffices to account for some aspects of fluid motion, i.e. that it is a predictive model for the motion of fluids.

The added value of computer experiments, unlike real (laboratory) ones, is thus in the freedom the scientist has to change some aspects of the investigated system (changing the model). Numerical simulations make it possible to extend the realm of experimental science, performing experiments in a hypothetical world, where algorithmic rules constitute the "governing laws". This allows scientists to isolate the aspects of a phenomenon believed to be primary, and thus to implement Galileo's procedure of *difalcare gli impedimenti*[13] by performing *in silico* the *Gedanken* experiments, that, in the history of science, often made it possible to progress in the understanding of many aspects of nature [Irvine (1991)]. However, computer simulations should not become an activity completely disjointed from reality. Empirical inputs from natural phenomena are essential if one wants to transform abstract computer models into powerful tools to identify specific or universal mechanisms underlying a class of phenomena. Valid numerical experiments should thus always live at boundaries between imagination and reality in order to avoid becoming a sterile game of "you get what you put".

Within the above identified boundaries, science and, in particular, physics should always be seen as based on three pillars: experiments, theory (and thus model building) and computer simulations. Sometimes, we tend to forget or overlook, and we should not, the intrinsic limitations of computer simulations and thought experiments which are based on models, and models (especially the good ones) only involve a few selected empirical inputs. The selection encompasses both the pros and cons of model building as it requires to neglect, rightly or wrongly, some of the facts present in the real system [Irvine (1991)]. The rightness or wrongness of such procedure can only be assessed *a posteriori* via experiments. In this respect, even if a model reproduces empirical observations, we cannot be sure that the mechanism incorporated in the model is enough to explain the phenomenon. There are many roads to Rome: we can imagine many mechanisms that can produce a given phenomenon, but only experiments can determine which, among the various candidates, is the right one. This is even more true when models are developed in disciplines such as biology, ecology or social sciences (see, for instance, the eye-opening essay on the role of models in ecology by Levin (1992), which has been a source of inspiration for this discussion).

However partial and incomplete, we think that the discussion summarized above on the role of computer simulations and modeling reflects the way in which scientific investigations proceed today, as a balanced mixture of observation, intuition, model building, computer experiments and "guided" experiments. In contrast with the past, however, nowadays the first step does not necessarily come from laboratory experiments or the observation of naturally occurring phenomena, but sometimes is inherent to computer experiments. Indeed, the recognition of the ubiquity and importance of chaos owes much to the pioneers of computer experiments, as the history of the discovery of chaos in the Lorenz model taught us (Sec. 3.2).

[13] Eliminate any hindrances.

As a tribute to Edward Norton Lorenz (1917–2008), who passed away while we were writing this book, we conclude with two general lessons we inherited from his pioneering studies:

- complex (unpredictable) behaviors are not necessarily the outcome of the complicated structure of a system (e.g. the presence of many components), but can be present even in "innocent" low dimensional systems;
- the methodological approach (say "micro-reductionism" [Smith (1998)]) which seeks to understand (and control) system behaviors by determining the equations ruling the interactions of its parts can fail, summarizing this in a motto we can say that *knowing the Navier-Stokes equation does not solve the problem of understanding turbulence.*

The awareness of the practical and conceptual limits associated to chaotic behaviors has induced a drastic change in the way in which scientific questions are explored, with a shift towards probabilistic or, sometimes, "qualitative" aspects (even in low dimensional, simple according to the old nomenclature, systems).

To better appreciate the difference from a not too remote past, as a paradigmatic example of the pre-chaos approach to complex systems, we mention von Neumann, who believed that powerful computers and a clever use of numerical analysis should enable us to accurately forecast (and even also a control) of weather and climate. Despite the fast growth of computer power, the forecasting skills[14] of the largest centers for weather forecast are advancing rather slowly [Yoden (2007)]. Modern weather forecasting balances its activity between two poles: the first attempting detailed predictions and the second aiming to cognition and qualitative understanding.

However, even the operational activity is carried out with a different perspective compared to von Neumann's expectations. The intrinsic limits on predictability, imposed by the chaotic nature of the atmospheric flow, require meteorologists to run a series of forecasts —*ensemble forecasting*—, each starting from a slightly different initial condition, and to collect the outputs to deduce probabilistic predictions. This can sound pessimistic, as a surrender to the possibility of predictions and detailed descriptions. In fact, we simply believe that these changes are dictated by the strong evidence that, in the field of complex systems, we can only wonder about problems which are "physically" well posed.

[14]The forecasting skill is judged measuring the forecast errors defined by the difference between the forecast and analysis for a given verification time (3, 4 and 7 days).

Bibliography

Abarbanel, H. D. I. (1996). *Analysis of Observed Chaotic Data* (Springer-Verlag, New York).

Abbott, L. F. (1999). Lapique's introduction of the integrate-and-fire model neuron (1907), *Brain Res. Bull.* **50**, p. 303.

Abel, M., Biferale, L., Cencini, M., Falcioni, M., Vergni, D. and Vulpiani, A. (2000a). Exit-Time Approach to ε-entropy, *Phys. Rev. Lett.* **84**, p. 6002.

Abel, M., Biferale, L., Cencini, M., Falcioni, M., Vergni, D. and Vulpiani, A. (2000b). Exit-times and ε-entropy for dynamical systems, stochastic processes, and turbulence, *Physica D* **147**, p. 12.

Adler, R. (1946). A study of locking phenomena in oscillators, *Proc. IRE* **34**, p. 351.

Ahlers, G. (1998). Experiments on spatio-temporal chaos, *Physica A* **249**, p. 18.

Ahlers, V. and Pikovsky, A. (2002). Critical Properties of the Synchronization Transition in Space-Time Chaos, *Phys. Rev. Lett.* **88**, p. 254101.

Aizawa, Y. and Murakami, C. (1983). Generalization of Baker's transformation — chaos and stochastic process on a Smale's horseshoe, *Prog. Theor. Phys.* **69**, p. 1416.

Albrecht, H.-E., Damaschke, N., Borys, M. and Tropea, C. (2002). *Laser Doppler and Phase Doppler Measurement Techniques (Experimental Fluid Mechanics)*, 1st edn. (Springer-Verlag, Berlin).

Alekseev, V. M. and Yakobson, M. V. (1981). Symbolic dynamics and hyperbolic dynamic-systems, *Phys. Rep.* **75**, p. 287.

Allen, M. P. and Tildesly, T. J. (1993). *Computer simulation of Liquids* (Clarendon Press, Oxford, UK).

Anderson, P. W. (1958). Absence of Diffusion in Certain Random Lattices, *Phys. Rev.* **109**, p. 1492.

Anselmet, F., Gagne, Y., Hopfinger, E. J. and Antonia, R. A. (1984). High order velocity structure functions in turbulent shear flow, *J. Fluid. Mech.* **140**, p. 63.

Appleton, E. V. (1922). The automatic synchronization of triode oscillator, *Proc. Cambridge Phil. Soc. (Math. and Phys. Sci.)* **21**, p. 231.

Aranson, I. S., Gaponov-Grekhov, A. V. and Rabinovich, M. I. (1988). The onset and spatial development of turbulence in flow systems, *Physica D* **33**, p. 1.

Aranson, I. S. and Kramer, L. (2002). The world of the complex Ginzburg-Landau equation, *Rev. Mod. Phys.* **74**, p. 99.

Arecchi, F. T. (1988). Instabilities and chaos in optics, *Physica Scripta* **T23**, p. 1604.

Arecchi, F. T., Giacomelli, G., Lapucci, A. and Meucci, R. (1991). Dynamics of a CO_2 laser with delayed feedback: The short-delay regime, *Phys. Rev. A* **43**, p. 4997.

Arecchi, F. T., Giacomelli, G., Lapucci, A. and Meucci, R. (1992). Two-dimensional representation of a delayed dynamical system, *Phys. Rev. A* **45**, p. R4225.

Arecchi, F. T., Meucci, R., Puccioni, G. and Tredicce, J. (1982). Experimental Evidence of Subharmonic Bifurcations, Multistability, and Turbulence in a Q-Switched Gas Laser, *Phys. Rev. Lett.* **49**, p. 1217.

Aref, H. (1983). Integrable, and turbulent vortex motion in two-dimensional flows, *Ann. Rev. Fluid Mech.* **15**, p. 345.

Aref, H. (1984). Stirring by chaotic advection, *J. Fluid Mech.* **143**, p. 1.

Aref, H. and Siggia, E. D. (1980). Vortex dynamics of the two-dimensional turbulent shear layer, *J. Fluid Mech.* **100**, p. 705.

Arnèodo, A., Baudet, C., Belin, F., Benzi, R., Castaing, B., Chabaud, B., Chavarria, R., Ciliberto, S., Camussi, R., Chilla, F., Dubrulle, B., Gagne, Y., Hebral, B., Herweijer, J., Marchand, M., Maurer, J., Muzy, J. F., Naert, A., Noullez, A., Peinke, J., Roux, F., Tabeling, P., van de Water, W. and Willaime, H. (1996). Structure functions in turbulence, in various flow configurations, at Reynolds number between 30 and 5000, using extended self-similarity, *Europhys. Lett.* **34**, p. 411.

Arnèodo, A., Benzi, R., Berg, J., Biferale, L., Bodenschatz, E., Busse, A., Calzavarini, E., Castaing, B., Cencini, M., Chevillard, L., Fisher, R. T., Grauer, R., Homann, H., Lamb, D., Lanotte, A. S., Lévèque, E., Lüthi, B., Mann, J., Mordant, N., Müller, W.-C., Ott, S., Ouellette, N. T., Pinton, J.-F., Pope, S. B., Roux, S. G., Toschi, F., Xu, H. and Yeung, P. K. (2008). Universal Intermittent Properties of Particle Trajectories in Highly Turbulent Flows, *Phys. Rev. Lett.* **100**, p. 254504.

Arnèodo, A., Coullet, P., Peyraud, J. and Tresser, C. (1982). Strange attractors in Volterra equations for species in competition, *J. Math. Biol.* **14**, p. 153.

Arnold, L. (1998). *Random Dynamical Systems* (Springer-Verlag, Berlin).

Arnold, V. I. (1963a). Proof of a theorem by A. N. Kolmogorov on the invariance of quasi-periodic motions under small perturbation of the Hamiltonian, *Russ. Math. Surveys* **18**, p. 9.

Arnold, V. I. (1963b). Small denominators and problems of stability of motion in classical and celestial mechanics, *Russ. Math. Surv.* **18**, p. 85.

Arnold, V. I. (1964). Instability of dynamical systems with many degrees of freedom, *Dokl. Akad. Nauk SSSR* **156**, p. 9.

Arnold, V. I. (1965). Sur la topologie des ecoulements stationnaires des fluides parfaits, *C. R. Acad. Sci. Paris A* **261**, p. 17.

Arnold, V. I. (1978). *Ordinary Differential Equations* (The MIT Press, Cambdridge, USA).

Arnold, V. I. (1989). *Mathematical Methods of Classical Mechanics*, 2nd edn. (Springer-Verlag, Berlin).

Arnold, V. I. and Avez, A. (1968). *Ergodic Problems in Classical Mechanics* (Benjamin, New York, USA).

Artale, V., Boffetta, G., Celani, A., Cencini, M. and Vulpiani, A. (1997). Dispersion of passive tracers in closed basins: Beyond the diffusion coefficient, *Phys. Fluids* **9**, p. 3162.

Artuso, R., Casati, G. and Lombardi, R. (1993). Periodic orbit theory of anomalous diffusion, *Phys. Rev. Lett.* **71**, p. 62.

Arya, S. P. (1998). *Air Pollution Meteorology and Dispersion* (Oxford University Press, New York, USA).

Atkins, P. and Jones, L. (2004). *Chemical Principles: The Quest for Insight*, 3rd edn. (W. H. Freeman and Co., New York, USA).

Aubin, D. and Dalmedico, A. D. (2002). Writing the History of Dynamical Systems and Chaos: Longue Durée and Revolution, Disciplines and Cultures, *Historia Mathematica* **29**, p. 273.

Aubry, N., Holmes, P., Lumley, J. L. and Stone, E. (1988). The dynamics of coherent structures in the wall region of a turbulent boundary layer, *J. Fluid Mech.* **192**, p. 115.

Aurell, E., Boffetta, G., Crisanti, A., Paladin, G. and Vulpiani, A. (1996). Growth of Non-infinitesimal Perturbations in Turbulence, *Phys. Rev. Lett.* **77**, p. 1262.

Aurell, E., Boffetta, G., Crisanti, A., Paladin, G. and Vulpiani, A. (1997). Predictability in the large: an extension of the concept of Lyapunov exponent, *J. Phys. A: Math. Gen.* **30**, p. 1.

Avellaneda, M. and Majda, A. J. (1991). An integral representation and bounds on the effective diffusivity in passive advection by laminar and turbulent flows, *Comm. Math. Phys.* **138**, p. 339.

Avellaneda, M. and Vergassola, M. (1995). Stieltjes integral representation of effective diffusivities in time-dependent flows, *Phys. Rev. E* **52**, p. 3249.

Babiano, A., Boffetta, G., Provenzale, A. and Vulpiani, A. (1994). Chaotic advection in point vortex models and two-dimensional turbulence, *Phys. Fluids A* **6**, p. 2465.

Badii, R. and Politi, A. (1997). *Complexity: hierarchical structures and scaling physics* (Cambridge University Press, Cambridge, UK).

Bagnoli, F. and Cecconi, F. (2000). Synchronization of non-chaotic dynamical systems, *Phys. Lett. A* **282**, p. 9.

Balkovsky, E., Falkovich, G. and Fouxon, A. (2001). Intermittent Distribution of Inertial Particles in Turbulent Flows, *Phys. Rev. Lett.* **86**, p. 2790.

Bär, M., Hegger, R. and Kantz, H. (1999). Fitting partial differential equations to space-time dynamics, *Phys. Rev. E* **59**, p. 337.

Baroni, L., Livi, R. and Torcini, A. (2001). Transition to stochastic synchronization in spatially extended systems, *Phys. Rev. E* **63**, p. 036226.

Bartuccelli, M. V., Gentile, G. and Georgiou, K. V. (2001). On the dynamics of a vertically driven damped planar pendulum, *Proc. Royal Soc. London A* **457**, p. 3007.

Basu, S., Foufoula-Georgiou, E. and Porté-Agel, F. (2002). Predictability of atmospheric boundary-layer flows as a function of scale, *Geophys. Res. Lett.* **29**, p. 2038.

Batchelor, G. K. (1959). Small-scale variation of convected quantities like temperature in turbulent fluid. Part 1. General discussion and the case of small conductivity, *J. Fluid Mech.* **5**, p. 113.

Batygin, K. and Laughlin, G. (2008). On the dynamical stability of the solar system, *Astr. J.* **683**, p. 1207.

Bec, J. (2003). Fractal clustering of inertial particles in random flows, *Phys. Fluids* **15**, p. L81.

Bec, J., Biferale, L., Boffetta, G., Cencini, M., Musacchio, S. and Toschi, F. (2006). Lyapunov exponents of heavy particles in turbulence, *Phys. Fluids* **18**, p. 091702.

Bec, J., Biferale, L., Cencini, M., Lanotte, A., Musacchio, S. and Toschi, F. (2007). Heavy Particle Concentration in Turbulence at Dissipative and Inertial Scales, *Phys. Rev. Lett.* **98**, p. 084502.

Beck, C. and Schlögl, F. (1997). *Thermodynamics of Chaotic Systems* (Cambridge University Press, Cambridge, UK).

Belousov, B. P. (1959). A periodic reaction and its mechanism, in *Collection of short papers on Radiation Medicine* (Med. Publ. Moscow), p. 145.

Benczik, I. J., Toroczkai, Z. and Tél, T. (2002). Selective Sensitivity of Open Chaotic Flows on Inertial Tracer Advection: Catching Particles with a Stick, *Phys. Rev. Lett.* **89**, p. 164501.

Bendixon, I. (1901). Sur les courbes définié par des équations différentielles, *Acta Math.* **24**, p. 1.

Benedicks, M. and Young, L. S. (1993). Sinai-Bowen-Ruelle measures for certain Hénon maps, *Inventiones Mathematicae* **112**, p. 541.

Benettin, G., Casartelli, M., L. Galgani, A. G. and Strelcyn, J. M. (1978a). On the reliability of numerical studies on stochasticity. I: Existence of time averages, *Il Nuovo Cimento B* **44**, p. 183.

Benettin, G., Galgani, L. and Giorgilli, A. (1985). A proof of Nekhoroshev's theorem for the stability times in nearly integrable Hamiltonian systems, *Celestial Mechanics* **37**, p. 1.

Benettin, G., Galgani, L., Giorgilli, A. and Strelcyn, J. M. (1978b). Lyapunov characteristic exponents for smooth dynamical systems and for Hamiltonian systems: a method for computing all of them, *C. R. Acad. Sci. Paris A* **206**, p. 431.

Benettin, G., Galgani, L., Giorgilli, A. and Strelcyn, J.-M. (1980). Lyapunov characteristic exponents for smooth dynamical systems and for Hamiltonian systems: A method for computing all of them. Part I: Theory, and Part II: Numerical application, *Meccanica* **15**, p. 9.

Benettin, G. and Gradenigo, G. (2008). A study of the Fermi–Pasta–Ulam problem in dimension two, *Chaos* **18**, p. 013112.

Benettin, G., Henrard, J. and Kuksin, S. (1999). *Hamiltonian Dynamics Theory and Applications* (Springer-Verlag, Berlin).

Benettin, G., Livi, R. and Ponno, A. (2008). The Fermi-Pasta-Ulam Problem: Scaling Laws vs. Initial Conditions, *J. Stat. Phys* , p. 174.

Benettin, G. and Strelcyn, J.-M. (1978). Numerical experiments on the free motion of a point mass moving in a plane convex region: Stochastic transition and entropy, *Phys. Rev. A* **17**, p. 773.

Benettin, G. and Tenenbaum, A. (1983). Ordered and stochastic behavior in a two-dimensional Lennard-Jones system, *Phys. Rev. A* **28**, p. 3020.

Benincà, E., Huisman, J., Heerkloss, R., Jöhnk, K. D., Branco, P., Van-Nes, E. H., Scheffer, M. and Ellner, S. P. (2008). Chaos in a long-term experiment with a plankton community, *Nature* **415**, p. 822.

Bennet, C. H. (1990). How to define complexity in physics and why, in W. H. Zurek (ed.), *Complexity, entropy and the physics of information* (Addison-Wesley, Redwood City), p. 137.

Benzi, R., Biferale, L., Paladin, G., Vulpiani, A. and Vergassola, M. (1991). Multifractality in the statistics of the velocity gradients in turbulence, *Phys. Rev. Lett.* **67**, p. 2299.

Benzi, R., Biferale, L. and Parisi, G. (1993). On intermittency in a cascade model for turbulence, *Physica D* **65**, p. 163.

Benzi, R., De Angelis, E., Govindarajan, R. and Procaccia, I. (2003). Shell model for drag reduction with polymer additives in homogeneous turbulence, *Phys. Rev. E* **68**, p. 016308.

Benzi, R., Paladin, G., Parisi, G. and Vulpiani, A. (1984). On the multifractal nature of fully developed turbulence and chaotic systems, *J. Phys. A: Math. Gen.* **17**, p. 3521.

Benzi, R., Paladin, G., Parisi, G. and Vulpiani, A. (1985). Characterisation of intermittency in chaotic systems, *J. Phys. A: Math. Gen.* **18**, p. 2157.

Bergé, P., Pomeau, Y. and Vidal, G. (1987). *Order within Chaos: Towards a deterministic approach to turbulence*, 2nd edn. (John Wiley & Sons, Canada).

Berger, T. and Gibson, J. D. (1998). Lossy Source Coding, *IEEE Transact. Inf. Theor.* **44**, p. 2693.

Berkooz, G. (1994). An observation probability density equations, or, when simulations reproduce statistics? *Nonlinearity* **7**, p. 313.

Bernard, D., Boffetta, G., Celani, A. and Falkovich, G. (2006). Conformal invariance in two-dimensional turbulence, *Nature Phys.* **2**, p. 124.
Berry, M. V. (1978). Regular and Irregular Motion, in S. Jorna (ed.), *Topics in Nonlinear Mechanics, AIP Conf. Proc No.46*, p. 16.
Biferale, L. (2003). Shell models of energy cascade in turbulence, *Ann. Rev. Fluid Mech.* **35**, p. 441.
Biferale, L., Cencini, M., Vergni, D. and Vulpiani, A. (1999). Exit time of turbulent signals: A way to detect the intermediate dissipative range, *Phys. Rev. E* **60**, p. 6295.
Billingsley, P. (1965). *Ergodic Theory and Information* (Wiley, New York, USA).
Birkhoff, G. D. (1927). On the periodic motions of dynamical systems, *Acta Math.* **50**, p. 359.
Birkhoff, G. D. (1931). Proof of the Ergodic Theorem, *Proc. Nat. Acad. Sci.* **17**, p. 656.
Birkhoff, G. D. (1966). *Dynamical Systems*, AMS Colloquium Publications, Vol. IX (American Mathematical Society, Providence, USA).
Biskamp, D. (1993). *Nonlinear Magnetohydrodynamics* (Cambridge University Press, Cambridge).
Bodenschatz, E., Pesch, W. and Ahlers, G. (2000). Recent Developments in Rayleigh-Bénard Convection, *Ann. Rev. Fluid Mech.* **32**, p. 709.
Boffetta, G. (2007). Energy and enstrophy fluxes in the double cascade of two-dimensional turbulence, *J. Fluid Mech.* **589**, p. 253.
Boffetta, G., Celani, A., Cencini, M., Lacorata, G. and Vulpiani, A. (2000a). Nonasymptotic properties of transport and mixing, *Chaos* **10**, p. 50.
Boffetta, G., Celani, A., Cencini, M., Lacorata, G. and Vulpiani, A. (2000b). The predictability problem in systems with an uncertainty in the evolution law, *J. Phys. A: Math. Gen.* **33**, p. 1313.
Boffetta, G., Celani, A., Crisanti, A. and Vulpiani, A. (1997). Predictability in Two Dimensional Decaying Turbulence, *Phys. Fluids A* **9**, p. 724.
Boffetta, G., Celani, A. and Vergassola, M. (2000c). Inverse energy cascade in two-dimensional turbulence: Deviations from Gaussian behavior, *Phys. Rev. E* **61**, p. R29.
Boffetta, G., Cencini, M., Espa, S. and Querzoli, G. (2000d). Chaotic advection and relative dispersion in a convective flow, *Phys. Fluids* **12**, p. 3160.
Boffetta, G., Cencini, M., Falcioni, M. and Vulpiani, A. (2002). Predictability: a way to characterize complexity, *Phys. Rep.* **356**, p. 367.
Boffetta, G., Giuliani, P., Paladin, G. and Vulpiani, A. (1998). An Extension of the Lyapunov Analysis for the Predictability Problem, *J. Atmos. Sci.* **55**, p. 3409.
Boffetta, G., Mazzino, A. and Vulpiani, A. (2008). Twenty-five years of multifractals in fully developed turbulence: a tribute to Giovanni Paladin, *J. Phys. A: Math. Gen.* **41**, p. 363001.
Boffetta, G. and Musacchio, S. (2000). Predictability of the energy cascade in 2D turbulence, *Phys. Fluids A* **13**, p. 1060.
Boffetta, G., Paladin, G. and Vulpiani, A. (1996). Strong Chaos without Butterfly Effect in Dynamical Systems with Feedback, *J. Phys. A: Math. Gen.* **29**, p. 2291.
Boffetta, G. and Sokolov, I. M. (2002). Relative Dispersion in Fully Developed Turbulence: The Richardson's Law and Intermittency Corrections, *Phys. Rev. Lett.* **88**, p. 094501.
Bohr, T., Jensen, M. H., Paladin, G. and Vulpiani, A. (1998). *Dynamical systems approach to turbulence* (Cambridge University Press, Cambridge, UK).
Bohr, T. and Rand, D. A. (1991). A mechanism for localised turbulence, *Physica D* **52**, p. 532.

Bohr, T., van Hecke, M., Mikkelsen, R. and Ipsen, M. (2001). Breakdown of Universality in Transitions to Spatiotemporal Chaos, *Phys. Rev. Lett.* **86**, p. 5482.

Boldrighini, C. and Franceschini, V. (1979). A Five-Dimensional Truncation of the Plane Navier-Stokes equations, *Comm. Math. Phys.* **64**, p. 159.

Bollt, E. M., Stanford, T., Lai, Y.-C. and Zyczkowski, K. (2001). What symbolic dynamics do we get with a misplaced partition? On the validity of threshold crossings analysis of chaotic time-series, *Physica D* **154**, p. 259.

Bonaccini, R. and Politi, A. (1997). Chaotic-like behaviour in chains of stable nonlinear oscillators, *Physica D* **103**.

Bouchaud, J. P. and Georges, A. (1990). Anomalous diffusion in disordered media: Statistical mechanisms, models and physical applications, *Phys. Rep.* **195**, p. 127.

Bower, A. S. (1991). A simple kinematic mechanism for mixing fluid parcels across a meandering jet, *J. Phys. Oceanogr.* **21**, p. 173.

Bower, A. S., Rossby, H. T. and Lillibridge, J. L. (1985). The Gulf Stream: barrier or blender? *J. Phys. Oceanogr.* **15**, p. 24.

Bower, J. M. and Beeman, D. (1995). *The Book of GENESIS: exploring realistic neural models with the GEneral NEural SImulation System* (Springer, New York).

Bracco, A., Chavanis, P. H., Provenzale, A. and Spiegel, E. A. (1999). Particle aggregation in a turbulent Keplerian flow, *Phys. Fluids* **11**, p. 2280.

Brandenburg, A., Klapper, I. and Kurths, J. (1995). Generalized entropies in a turbulent dynamo simulation, *Phys. Rev. E* **52**, p. R4602.

Bricmont, J. (1995). Science of Chaos or Chaos in Science? *Physicalia Mag.* **17**, p. 159.

Briggs, K. M. (1997). Feigenbaum Scaling in Discrete Dynamical Systems, Ph.D. thesis. Melbourne, Australia: University of Melbourne.

Broadbent, S. . R. and Hammersley, J. . M. (1957). Percolation processes i and ii, *Proc. Cambridge Philos. Soc.* **53**, p. 629.

Brudno, A. A. (1983). Entropy and the complexity of the trajectories of a dynamical system, *Trans. Moscow Math. Soc.* **44**, p. 127.

Brusch, L., Zimmermann, M. G., van Hecke, M., Bär, M. and Torcini, A. (2000). Modulated Amplitude Waves and the Transition from Phase to Defect Chaos, *Phys. Rev. Lett.* **85**, p. 86.

Bunimovich, L. A., Livi, R., Martínez-Mekler, G. and Ruffo, S. (1992). Coupled trivial maps, *Chaos* **2**, pp. 283–291.

Bunimovich, L. A. and Sinai, Y. G. (1981). Statistical properties of Lorentz gas with periodic configuration of scatterers, *Comm. Math. Phys.* **78**, p. 479.

Bunimovich, L. A. and Sinai, Y. G. (1993). Statistical mechanics of coupled map lattices, in K. Kaneko (ed.), *Theory and applications of Coupled Map Lattices* (John Wiley & Sons, Chichester, UK), p. 169.

Buric, N., Rampioni, A., Turchetti, G. and Vaienti, S. (2003). Weak chaos and Poincaré recurrences for area preserving maps, *J. Phys. A: Math. Gen.* **36**, p. L209.

Cakmur, R. V., Egolf, D. A., Plapp., B. B. and Bodenschatz, E. (1997). Bistability and Competition of Spatiotemporal Chaotic and Fixed Point Attractors in Rayleigh-Bénard Convection, *Phys. Rev. Lett.* **79**, p. 1853.

Calzavarini, E., Cencini, M., Lohse, D. and Toschi, F. (2008). Quantifying turbulence induced segregation of inertial particles, *Phys. Rev. Lett.* **101**, p. 084504.

Campbell, D. K., Rosenau, P. and Zaslavsky, G. M. (2005). Introduction: The Fermi-Pasta-Ulam problem– The first fifty years, *Chaos* **15**, p. 015101.

Carati, A., Galgani, L. and Giorgilli, A. (2005). The Fermi-Pasta-Ulam problem as a challenge for the foundations of physics, *Chaos* **15**, p. 015105.

Carbone, V. (1993). Cascade model for intermittency in fully developed magnetohydrodynamic turbulence, *Phys. Rev. Lett.* **71**, p. 1546.
Casartelli, M., Casati, G., Galgani, L. and Scotti, A. (1976). Stochastic transition in a classical nonlinear dynamical system: A Lennard-Jones chain, *Theor. Math. Phys.* **29**, p. 205.
Casati, G., Guarneri, I. and Shepelyansky, D. L. (1988). Hydrogen atom in monochromatic field: chaos and dynamical photonic localization, *IEEE J. Quantum Electron.* **24**, p. 1420.
Casdagli, M. and Roy, J. (1991). Chaos and Deterministic versus Stochastic Non-linear Modelling, *Statist. Soc. Ser. B* **54**, p. 303.
Casetti, L., Cerruti-Sola, M., Pettini, M. and Cohen, E. G. D. (1997). The Fermi-Pasta-Ulam problem revisited: Stochasticity thresholds in nonlinear Hamiltonian systems, *Phys. Rev. E* **55**, p. 6566.
Castaing, B., Gunaratne, G., Heslot, F., Libchaber, L. K. A., Thomae, S., Wu, X.-Z., Zaleski, S. and Zanetti, G. (1989). Scaling of hard thermal turbulence in Rayleigh-Bénard convection, *J. Fluid Mech.* **204**, p. 1.
Castiglione, P., Crisanti, A., Mazzino, A., Vergassola, M. and Vulpiani, A. (1998). Resonant enhanced diffusion in time-dependent flow, *J. Phys. A: Math. Gen.* **31**, p. 7197.
Castiglione, P., Mazzino, A., Muratore-Ginanneschi, P. and Vulpiani, A. (1999). On strong anomalous diffusion, *Physica D* **134**, p. 75.
Cecconi, F., Cencini, M. and Vulpiani, A. (2007). Transport properties in chaotic and non-chaotic many particle systems, *J. Stat. Mech.: Th. and Exp.* , p. P12001.
Cecconi, F., del Castillo-Negrete, D., Falcioni, M. and Vulpiani, A. (2003). The origin of diffusion: the case of non-chaotic systems, *Physica D* **180**, p. 129.
Cecconi, F., Livi, R. and Politi, A. (1998). Fuzzy phase transition in a 1D coupled stable-map lattice, *Phys. Rev. E* **57**, p. 2703.
Cecconi, F. and Politi, A. (1997). n-tree approximation for the largest Lyapunov exponent of a coupled-map lattice, *Phys. Rev. E* **56**, p. 4998.
Cecconi, F. and Vulpiani, A. (1995). Approximation of chaotic systems in terms of Markovian processes, *Phys. Lett. A* **201**, p. 326.
Celani, A. (2007). The frontiers of computing in turbulence: challenges and perspectives, *J. Turb.* **8**, p. 1.
Celani, A., Cencini, M., Mazzino, A. and Vergassola, M. (2004). Active and passive fields face to face, *New J. Phys.* **6**, p. 72.
Cencini, M., Falcioni, M., Kantz, H., Olbrich, E. and Vulpiani, A. (2000). Chaos or Noise –Difficulties of a Distinction, *Phys Rev. E.* **62**, p. 427.
Cencini, M., Falcioni, M., Vergni, D. and Vulpiani, A. (1999a). Macroscopic chaos in globally coupled maps, *Physica D* **130**, p. 58.
Cencini, M., Lacorata, G., Vulpiani, A. and Zambianchi, E. (1999b). Mixing in a Meandering Jet: a Markovian Approximation, *J. Phys. Oceanogr.* **29**, p. 2578.
Cencini, M., López, C. and Vergni, D. (2003). Reaction-Diffusion Systems: Front Propagation and Spatial Structures, in R. Livi and A. Vulpiani (eds.), *The Kolmogorov Legacy in Physics* (Springer, Berlin, Heidelberg), p. 187.
Cencini, M., Tessone, C. J. and Torcini, A. (2008). Chaotic synchronizations of spatially extended systems as nonequilibrium phase transitions, *Chaos* **18**, p. 037125.
Cencini, M. and Torcini, A. (2001). Linear and Nonlinear information flow in spatially extended systems, *Phys. Rev. E* **63**, p. 056201.
Cencini, M. and Torcini, A. (2005). Nonlinearly driven transverse synchronization in coupled chaotic systems, *Physica D* **208**, p. 191.
Cercignani, C. (1977). Solitons-Theory and application, *Riv. Nuovo Cimento* **7**, p. 429.

Cercignani, C. (1998). *Ludwig Boltzmann: the man who trusted atoms* (Oxford University Press, Oxford UK).
Chaitin, G. J. (1966). On the Length of Programs for Computing Finite Binary Sequences, *J. Assoc. Comp. Mach.* **13**, p. 547.
Chaitin, G. J. (1974). Information-theoretic limitations of formal systems, *J. Assoc. Comp. Mach.* **21**, p. 403.
Chaitin, G. J. (1982). Gödel theorem and information, *Int. J. Theor. Phys.* **22**, p. 941.
Chaitin, G. J. (1990). *Information, randomness and incompleteness*, 2nd edn. (World Scientific, Singapore).
Chandrasekhar, S. (1943). Stochastic problems in Physics and Astronomy, *Rev. Mod. Phys.* **15**, p. 1.
Chaté, H. (1994). Spatiotemporal intermittency regimes of the one-dimensional complex Ginzburg-Landau equation, *Nonlinearity* **7**, p. 185.
Chaté, H. and Manneville, P. (1988). Spatio-temporal intermittency in coupled map lattices, *Physica D* **32**, p. 409.
Chaté, H. and Manneville, P. (1992). Collective Behaviors in Spatially Extended Systems with Local Interactions and Synchronous Updating, *Prog. Theor. Phys.* **87**, p. 1.
Chaté, H. and Manneville, P. (1996). Phase diagram of the two-dimensional complex Ginzburg-Landau equation, *Physica A* **224**, p. 348.
Chazottes, J. R. and Fernandez, B. (eds.) (2004). *Dynamics of Coupled Map Lattices and of Related Spatially Extended Systems* (Springer, Berlin).
Chirikov, B. V. (1979). A universal instability of many dimensional oscillator systems, *Phys. Rep.* **52**, p. 263.
Chirikov, B. V. (1988). Particle Confinement and Adiabatic Invariance, *Proc. Royal Soc. London A* **413**, p. 145.
Chirikov, B. V. and Vecheslavov, V. V. (1989). Chaotic dynamics of Comet Halley, *Astron. Astrophys.* **221**, p. 146.
Chong, M. S., Perry, A. E. and Cantwell, B. J. (1990). A general classification of three dimensional flow field, *Phys. Fluids A* **2**, p. 765.
Chorin, A. J. (1994). *Vorticity and Turbulence* (Springer-Verlag, New York).
Ciccotti, G. and Hoover, W. G. (eds.) (1986). *Molecular dynamics simulation of statistical-mechanical systems* (North-Holland, Amsterdam).
Ciliberto, S. and Bigazzi, P. (1988). Spatiotemporal Intermittency in Rayleigh-Bénard Convection, *Phys. Rev. Lett.* **60**, p. 286.
Ciliberto, S., Pampaloni, E. and Pérez-García, C. (1991). The role of defects in the transition between different symmetries in convective patterns, *J. Stat. Phys.* **64**, p. 1045.
Ciliberto, S. and Rubio, M. A. (1987). Chaos and order in the temperature field of Rayleigh-Bénard convection, *Physica Scripta* **36**, p. 920.
Cohen, A. and Procaccia, I. (1985). Computing the Kolmogorov entropy from time signals of dissipative and conservative dynamical systems, *Phys. Rev. A* **31**, p. 1872.
Cohen, J. and Stewart, I. (1994). *The Collapse of Chaos: Discovering Simplicity in a Complex World*, 1st edn. (Viking Penguin, New York, USA).
Collet, P. and Eckmann, J.-P. (1980). *Iterated Maps on the Interval as Dynamical System* (Birkhäuser, Basel).
Collet, P. and Eckmann, J.-P. (1999). Extensive Properties of the Complex Ginzburg-Landau Equation, *Comm. Math. Phys.* **200**, p. 699.
Collet, P. and Eckmann, J.-P. (2006). *Concepts and Results in Chaotic Dynamics: A Short Course* (Springer-Verlag, Berlin).
Constantin, P., Levant, B. and Titi, E. S. (2006). Analytic study of shell models of turbulence, *Physica D* **219**, p. 120.

Contopoulos, G., Voglis, N., Efthymiopoulos, C., Froeschlé, C., Gonczi, R., Lega, E., Dvorak, R. and Lohinger, E. (1997). Transition spectra of dynamical systems, *J. Cel. Mech. Dyn. Astr.* **67**, p. 293.

Cornfeld, I., Fomin, S. and Sinai, Y. G. (1982). *Ergodic Theory* (Springer-Verlag, Berlin).

Coste, J. and Hénon, M. (1986). Invariant cycles in the random mapping of N integers onto themselves. Comparison with Kauffman binary network, in M. Y. Bienenstock, F. F. Soulié and G. Weisbuch (eds.), *Disordered Systems and Biological Organization, NATO ASI Series*, Vol. F20 (Springer-Verlag, Heidelberg), p. 361.

Courant, R. and Hilbert, D. (1989). *Methods of Mathematical Physics, Wiley Classics Library*, Vol. I and II (Wiley-Interscience, New York, USA).

Cover, T. M., Gacs, P. and Gray, R. M. (1989). Kolmogorov's Contributions to Information Theory and Algorithmic Complexity, *Ann. Probab.* **17**, p. 840.

Cover, T. M. and Thomas, J. A. (1991). *Elements of Information Theory* (John Wiley & Sons, Inc., USA).

Cressman, J. R., Davoudi, J., Goldburg, W. I. and Schumacher, J. (2004). Eulerian and Lagrangian studies in surface flow turbulence, *New J. Phys.* **6**, p. 53.

Crisanti, A., Falcioni, M., Paladin, G. and Vulpiani, A. (1991). Lagrangian chaos: transport, mixing and diffusion in fluids, *Riv. Nuovo Cimento* **14**, p. 1.

Crisanti, A., Falcioni, M. and Vulpiani, A. (1989). On the effects of an uncertainty on the evolution law in dynamical systems, *Physica A* **160**, p. 482.

Crisanti, A., Jensen, M. H., Paladin, G. and Vulpiani, A. (1993a). Intermittency and predictability in turbulence, *Phys. Rev. Lett.* **70**, p. 166.

Crisanti, A., Paladin, G. and Vulpiani, A. (1993b). *Products of Random Matrices in Statistical Physics* (Spinger-Verlag, Berlin).

Cross, M. C. and Hohenberg, P. C. (1993). Pattern formation outside of equilibrium, *Rev. Mod. Phys.* **65**, p. 851.

Crowe, C. T., Sommerfeld, M. and Tsuji, Y. (1998). *Multiphase Flows with Particles and Droplets* (CRC Press, New York, USA).

Crutchfield, J. P. and Kaneko, K. (1988). Are Attractors Relevant to Turbulence? *Phys. Rev. Lett.* **60**, p. 2715.

Dauxois, T. and Peyrard, M. (2006). *Physics of Solitons* (Cambridge University Press, Cambridge, UK).

de Luca, J., Lichtenberg, A. J. and Ruffo, S. (1999). Finite times to equipartition in the thermodynamic limit, *Phys. Rev. E* **60**, p. 3781.

de Pater, I. and Lissauer, J. (2001). *Planetary Science* (Cambridge University Press, Cambridge).

Decroly, O. and Goldbeter, A. (1982). Birhythmicity, chaos, and other patterns of temporal self-organization in a multiply regulated biochemical system, *Proc. Natl. Acad. Sci. USA* **79**, p. 6917.

Deissler, R. (1987). Spatially growing waves, intermittency, and convective chaos in a flow system, *Physica D* **25**, p. 233.

Deissler, R. J. and Kaneko, K. (1987). Velocity dependent Lyapunov exponents as a measure of chaos for open flow systems, *Phys. Lett. A* **119**, p. 397.

Derrida, B., Grevois, A. and Pomeau, Y. (1979). Universal metric properties of bifurcations of endomorphisms, *J. Phys. A: Math. Gen.* **12**, p. 269.

Dettmann, C. P. and Cohen, E. G. D. (2000). Microscopic chaos and diffusion, *J. Stat. Phys.* **101**, p. 775.

Dettmann, C. P. and Cohen, E. G. D. (2001). Note on chaos and diffusion, *J. Stat. Phys.* **103**, p. 589.

Ding, J. and Li, T.-Y. (1991). Markov finite approximation of Frobenius-Perron operator, *Nonlin. Anal.: Theor., Meth. Appl.* **8**, p. 759.
Doering, C. R. (2009). The 3D Navier-Stokes Problem, *Annu. Rev. Fluid Mech.* **41**, p. 109.
Doering, C. R. and Gibbon, J. D. (1995). *Applied Analysis of the Navier-Stokes Equations* (Cambridge University Press, Cambridge, UK).
Domany, E. and Kinzel, W. (1984). Equivalence of Cellular Automata to Ising Models and Directed Percolation, *Phys. Rev. Lett.* **53**, pp. 311–314.
Dombre, T., Frisch, U., Greene, J. M., Hénon, M., Mehr, A. and Soward, A. M. (1986). Chaotic streamlines in the ABC flows, *J. Fluid Mech.* **167**, p. 353.
Dorfman, J. R. (1999). *An introduction to chaos in nonequilibrium statistical mechanics* (Cambridge University Press, Cambridge).
d'Ovidio, F., Fernández, V., Hernández-García, E. and López, C. (2004). Mixing structures in the Mediterranean Sea from finite-size Lyapunov exponents, *Geophys. Res. Lett.* **31**, p. 17203.
d'Ovidio, F., Isern-Fontanet, J., López, C., Hernández-García, E. and García-Ladona, E. (2009). Comparison between Eulerian diagnostics and finite-size Lyapunov exponents computed from altimetry in the Algerian basin, *Deep Sea Res. Part I: Oceanogr. Res. Papers* **56**, p. 15.
Dressler, U. and Farmer, J. D. (1992). Generalized Lyapunov exponents corresponding to higher derivatives, *Physica D* **59**, p. 365.
Dritschell, D. G. and Legras, B. (1993). Modelling oceanic and atmospheric vortices, *Phys. Today* **46**, p. 44.
Droz, M. and Lipowski, A. (2003). Dynamical properties of the synchronization transition, *Phys. Rev. E* **67**, p. 056204.
Durrett, R. and Levin, S. (1994). The importance of being discrete (and spatial), *Theor Popul. Biol.* **46**, p. 363.
Dutkiewicz, S., Griffa, A. and Olson, D. B. (1993). Particle Diffusion in a Meandering Jet, *J. Geophys. Res.* **98**, p. 16478.
Eaton, J. K. and Fessler, J. R. (1994). Preferential concentration of particles by turbulence, *Int. J. Multiphase Flow* **20**, p. 169.
Eckmann, J.-P. (1981). Roads to turbulence in dissipative dynamical systems, *Rev. Mod. Phys.* **53**, p. 643.
Eckmann, J.-P., Oliffson Kamphorst, S., Ruelle, D. and Ciliberto, S. (1986). Lyapunov exponents from time series, *Phys. Rev. A* **34**, p. 4971.
Eckmann, J.-P. and Ruelle, D. (1985). Ergodic Theory of Chaos and Strange Attractors, *Rev. Mod. Phys.* **57**, p. 617.
Egolf, D. A. and Greenside, H. S. (1994). Relation between fractal dimension and spatial correlation length for extensive chaos, *Nature* **369**, p. 129.
Einstein, A. (1956). *Investigation on the Theory of the Brownian Motion* (Dover Publications INC., New York, USA).
Ellis, R. S. (1999). The theory of large deviations: from Boltzmann's 1877 calculation to equilibrium macrostates in 2D turbulence, *Physica D* **133**, p. 106.
Ershov, S. V. and Potapov, A. B. (1992). On the nature of nonchaotic turbulence, *Phys. Lett. A* **167**.
Erzan, A., Pietronero, L. and Vespignani, A. (1995). The fixed-scale transformation approach to fractal growth, *Rev. Mod. Phys.* **67**, p. 545.
Espa, S., Boffetta, G., Cencini, M. and Querzoli, G. (1999). Experimental evidence of chaotic advection in a convective flow, *Europhys. Lett.* **48**, p. 629.
Essex, C. and Nerenberg, M. A. H. (1991). Comment on Deterministic chaos: the science and the fiction, *Proc. Royal Soc. London A* **435**, p. 287.

Eyink, G. and Goldenfeld, N. (1994). Analogies between scaling in turbulence, field theory, and critical phenomena, *Phys. Rev. E* **50**, p. 4679.

Eyink, G. L. and Spohn, H. (1993). Negative-temperature states and large-scale, long-lived vortices in two-dimensional turbulence, *J. Stat. Phys.* **70**, p. 833.

Falcioni, M., Marini Bettolo Marconi, U. and Vulpiani, A. (1991). Ergodic properties of high-dimensional symplectic maps, *Phys. Rev. A* **44**, p. 2263.

Falcioni, M., Paladin, G. and Vulpiani, A. (1988). Regular and chaotic motion of fluid-particles in two-dimensional fluids, *J. Phys. A: Math. Gen.* **21**, p. 3451.

Falcioni, M., Palatella, L., Pigolotti, S. and Vulpiani, A. (2005). Properties making a chaotic system a good pseudo random number generator, *Phys. Rev. E* **72**, p. 016220.

Falcioni, M., Vergni, D. and Vulpiani, A. (1999). Characterization of the spatial complex behavior and transition to chaos in flow systems, *Physica D* **125**, p. 652.

Falconer, K. (2003). *Fractal Geometry: Mathematical Foundations and Applications*, 2nd edn. (John Wiley & Sons Ltd, West Sussex, UK).

Falkovich, G., Fouxon, A. and Stepanov, M. G. (2002). Acceleration of rain initiation by cloud turbulence, *Nature* **419**, p. 151.

Falkovich, G., Gawędzki, K. and Vergassola, M. (2001). Particles and fields in fluid turbulence, *Rev. Mod. Phys.* **73**, p. 913.

Farge, M. (1992). Wavelet transforms and their applications to turbulence, *Ann. Rev. Fluid Mech.* **24**, p. 395.

Farmer, J. D. (1982). Chaotic attractors of an infinite-dimensional dynamical system, *Physica D* **4**, p. 366.

Feigenbaum, M. J. (1978). Quantitative universality for a class of nonlinear transformations, *J. Stat. Phys.* **19**, p. 25.

Feigenbaum, M. J. (1979). The universal metric properties of nonlinear transformations, *J. Stat. Phys.* **21**, p. 669.

Feingold, M., Kadanoff, L. P. and Piro, O. (1988). Passive scalars, three-dimensional volume-preserving maps, and chaos, *J. Stat. Phys.* **50**, p. 529.

Feller, W. (1968). *An Introduction to Probability Theory and Its Applications*, Vol. I (John Wiley & Son, New York, USA).

Fermi, E. (1923). Dimostrazione che in generale un sistema meccanico normale è quasi ergodico, *Nuovo Cimento* **25**, p. 267.

Fermi, E., Pasta, J. and Ulam, S. (1955). Studies of non linear problems, Tech. Rep. LA-1940, Los Alamos Sci. Lab.

Feudel, U., Kuznetsov, S. and Pikovsky, A. (2006). *Strange nonchaotic attractors: Dynamics between Order and Chaos in Quasiperiodically Forced Systems* (World Scientific, Singapore).

Field, R. J., Köros, E. and Noyes, R. M. (1972). Oscillations in chemical systems 2. Thorough analysis of temporal oscillation in bromate-cerium-malonic acid system, *J. Amer. Chem. Soc.* **94**, p. 8649.

Fisher, R. A. (1937). The wave of advance of advantageous genes, *Ann. Eugenics* **7**, p. 353.

Ford, J. (1983). How Random is a Coin Tossing? *Phys. Today* **36**, p. 40.

Ford, J. (1986). Chaos: Solving the Unsolvable, Predicting the Unpredictable, in M. F. Barnaley and S. Demko (eds.), *Chaotic Dynamics and Fractals* (Academic Press, New York, USA), p. 1.

Franceschini, V. and Tebaldi, C. (1979). Sequences of infinite bifurcations and turbulence in a five-mode truncation of the Navier-Stokes equations, *J. Stat. Phys.* **21**, p. 707.

Franceschini, V. and Tebaldi, C. (1981). A seven-mode truncation of the plane incompressible Navier-Stokes equations, *J. Stat. Phys.* **25**, p. 397.

Fraser, A. M. and Swinney, H. L. (1986). Independent coordinates for strange attractors from mutual information, *Phys. Rev. A* **33**, p. 1134.
Frisch, U. (1995). *Turbulence; the Legacy of A. N. Kolmogorov* (Cambridge University Press, Cambridge).
Frisch, U., Matsumoto, T. and Bec, J. (2004). Singularities of Euler Flow? Not Out of the Blue! *J. Stat. Mech.* **113**, p. 761.
Frisch, U. and Vergassola, M. (1991). A Prediction of the Multifractal Model: the Intermediate Dissipation Ra nge, *Europhys. Lett.* **14**, p. 439.
Fujisaka, H. (1983). Statistical dynamics generated by fluctuations of local Lyapunov exponents, *Prog. Theor. Phys.* **70**, p. 1264.
Fujisaka, H. and Inoue, M. (1987). Statistical-Thermodynamics Formalism of Self-Similarity, *Prog. Theor. Phys.* **77**, p. 1334.
Fujisaka, H. and Yamada, T. (1983). Stability theory of synchronized motion in coupled-oscillator systems, *Prog. Theor. Phys.* **69**, p. 32.
Fujisaka, H. and Yamada, T. (1985). A new intermittency in coupled dynamical systems, *Prog. Theor. Phys.* **74**, p. 918.
Fujisaka, H. and Yamada, T. (1986). Stability theory of synchronized motion in coupled-oscillator systems. IV. Instability of synchronized chaos and new intermittency, *Prog. Theor. Phys.* **75**, p. 1045.
Gallavotti, G. (1983). *The Elements of Mechanics* (Springer-Verlag, New York, USA).
Gallavotti, G. (ed.) (2007). *The Fermi-Pasta-Ulam Problem: a status report*, *Lect. Note Phys.*, Vol. 728 (Springer, Berlin).
Gardiner, C. W. (1982). *Handbook of Stochastic Methods for Physics, Chemistry and the Natural Sciences* (Springer-Verlag, Berlin).
Gaspard, P. (1994). Comment on dynamical randomness in quantum systems, *Prog. Theor. Phys. Suppl.* **116**, p. 369.
Gaspard, P. (1998). *Chaos, Scattering and Statistical Mechanics* (Cambridge University Press, Cambridge UK).
Gaspard, P. and Nicolis, G. (1990). Transport properties, Lyapunov exponents, and entropy per unit time, *Phys. Rev. Lett.* **65**, p. 1693.
Gaspard, P. and Wang, X.-J. (1988). Sporadicity: Between Periodic and Chaotic Dynamical Behaviors, *Proc. Natl. Acad. Soc. USA* **85**, p. 4591.
Gaspard, P. and Wang, X.-J. (1993). Noise, chaos, and (ε, τ)-entropy per unit time, *Phys. Rep.* **235**, p. 291.
Geisel, T. and Thomae, S. (1984). Anomalous Diffusion in Intermittent Chaotic Systems, *Phys. Rev. Lett.* **52**, p. 1936.
Gelfand, I. M., Kolmogorov, A. N. and Yaglom, A. M. (1958). Amount of information and entropy for continuous distributions, in A. N. Shiryeyev (ed.), *Selected works of A. N. Kolmogorov*, Vol. III (Kluwer Academic Publishing (1993)), p. 33.
Gershenfeld, N. A. and Weigend, A. S. (1994). The future of time series: Learning and understanding, in A. S. Weigend and N. A. Gershenfeld (eds.), *Time Series Prediction: Forecasting the Future and Understanding the Past* (Addison-Wesley Publ. Comp., Reading), p. 1.
Giacomelli, G. and Politi, A. (1991). Spatio-Temporal Chaos and Localization, *Europhys. Lett.* **15**, p. 387.
Giacomelli, G. and Politi, A. (1996). Relationship between Delayed and Spatially Extended Dynamical Systems, *Phys. Rev. Lett.* **76**, p. 2686.
Giglio, M., Musazzi, S. and Perini, U. (1981). Transition to Chaotic Behavior via a Reproducible Sequence of Period-Doubling Bifurcations, *Phys. Rev. Lett.* **47**, p. 243.
Ginelli, F., Livi, R., Politi, A. and Torcini, A. (2003). Relationship between directed

percolation and the synchronization transition in spatially extended systems, *Phys. Rev. E* **67**, p. 046217.
Giovannini, F. and Politi, A. (1992). Generating partitions in Hénon-type maps, *Phys. Lett. A* **161**, p. 332.
Gledzer, E. B. (1973). Systems of hydrodynamic type admitting two quadratic integrals of motion. *Sov. Phys. Dokl.* **18**, p. 216.
Glendinning, P. (2001). Milnor attractors and topological attractors of a piecewise linear map, *Nonlinearity* **14**, p. 239.
Gnedenko, B. V. and Ushakov, I. A. (1997). *Theory of Probability* (Gordon and Breach Science Publishers, Amsterdam).
Goel, N. S., Maitra, S. C. and Montroll, E. W. (1971). On the Volterra and Other Nonlinear Models of Interacting Populations, *Rev. Mod. Phys.* **43**, p. 231.
Goldbeter, A. (1996). *Biochemical Oscillations and Cellular Rhythms: The molecular bases of periodic and chaotic behaviour* (Cambridge University Press, Cambridge UK).
Goldbeter, A. and Lefever, R. (1972). Dissipative structures for an allosteric model. Application to glycolytic oscillations, *Biophys. J.* **12**, p. 1302.
Goldhirsch, I., Sulem, P.-L. and Orszag, S. A. (1987). Stability and Lyapunov stability of dynamical systems: A differential approach and a numerical method, *Physica D* **27**, p. 311.
Goldstein, H., Poole, C. and Safko, J. (2002). *Classical Mechanics*, 3rd edn. (Addison-Wesley, Cambridge, USA).
Gollub, J. P. and Benson, S. V. (1980). Many routes to turbulent convection, *Phys. Rev. Lett.* **100**, p. 449.
Gollub, J. P. and Swinney, H. L. (1975). Onset of turbulence in a rotating fluid, *Phys. Rev. Lett.* **35**, p. 927.
Grad, H. (1967). Levels of Description in Statistical Mechanics and Thermodynamics, in M. Bunge (ed.), *Delaware Seminar in the Foundations of Physics* (Springer-Verlag, Berlin), p. 49.
Grassberger, P. (1982). On phase transitions in Schlögl's second model, *Z. Phys. B* **47**, p. 365.
Grassberger, P. (1986). Toward a quantitative theory of self-generated complexity, *Int. J. Theor. Phys.* **25**, p. 907.
Grassberger, P. (1989). Information Content and Predictability of Lumped and Distributed Dynamical Systems, *Physica Scripta* **40**, p. 107.
Grassberger, P. (1991). Randomness, Information and Complexity, in F. Ramos-Gómez (ed.), *Proceedings of the Fifth Mexican School on Statistical Physics* (World Scientific, Singapore), p. 59.
Grassberger, P. (1997). Directed Percolation: results and open problems, in S. Puri and S. Dattagupta (eds.), *Nonlinearities in Complex Systems* (Narosa Publishing House, New Delhi), p. 61.
Grassberger, P. (1999). Synchronization of coupled systems with spatiotemporal chaos, *Phys. Rev. E* **59**, p. R2520.
Grassberger, P., Badii, R. and Politi, A. (1988). Scaling laws for invariant measures on hyperbolic and non-hyperbolic attractors, *J. Stat. Phys.* **51**, p. 135.
Grassberger, P. and Kantz, H. (1985). Generating partitions for the dissipative Hnon map, *Phys. Lett. A* **113**, p. 235.
Grassberger, P. and Procaccia, I. (1983a). Estimation of the Kolmogorov entropy from a chaotic signal, *Phys. Rev. A* **28**, p. 2591.
Grassberger, P. and Procaccia, I. (1983b). Measuring the strangeness of strange attractors, *Physica D* **9**, p. 189.

Grassberger, P. and Procaccia, I. (1983c). On the characterization of strange attractors, *Phys. Rev. Lett.* **50**, p. 346.
Grassberger, P. and Schreiber, T. (1991). Phase transitions in coupled map lattices, *Physica D* **50**, p. 177.
Grebogi, C., Ott, E. and Yorke, J. A. (1988). Roundoff-induced periodicity and the correlation dimension of chaotic attractors, *Phys. Rev. A* **38**, p. 3688.
Green Jr., B. F., Smith, J. E. K. and Klem, L. (1959). Empirical Tests of an Additive Random Number Generator, *J. ACM* **6**, p. 527.
Grimmett, G. R. and Stirzaker, D. R. (2001). *Probability and random processes*, 3rd edn. (Oxford University Press, Oxford).
Grinstein, G., Muñoz, M. A. and Tu, Y. (1996). Phase Structure of Systems with Multiplicative Noise, *Phys. Rev. Lett.* **76**, p. 4376.
Gutzwiller, M. C. (1990). *Chaos in classical and quantum mechanics* (Springer-Verlag, New York).
Gutzwiller, M. C. (1998). Moon-Earth-Sun: The oldest three-body problem, *Rev. Mod. Phys.* **70**, p. 589.
Guzzo, M., Lega, E. and Froeschlé, C. (2002). On the numerical detection of the effective stability of chaotic motions in quasi-integrable systems, *Physica D* **163**, p. 1.
Guzzo, M., Lega, E. and Froeschlé, C. (2005). First numerical evidence of global Arnold diffusion in quasi–integrable systems, *Disc. Cont. Dyn. Syst. - Series B* **5**, p. 687.
Hald, O. H. (1976). Constants of motion in models of two-dimensional turbulence, *Phys. Fluids* **19**, p. 914.
Halliwell, G. R. and Mooers, C. N. K. (1983). Meanders of the Gulf Strean downstream Cape Hatteras, *J. Phys. Oceanogr.* **14**, p. 1275.
Halpin-Healy, T. and Zhang, Y.-C. (1995). Kinetic roughening phenomena, stochastic growth, directed polymers and all that. aspects of multidisciplinary statistical mechanics, *Phys. Rep.* **254**, p. 215.
Hammel, S. M., Yorke, J. A. and Grebogi, C. (1987). Do numerical orbits of chaotic dynamical processes represent true orbits? *J. Complexity* **3**, p. 136.
Heagy, J. F., Platt, N. and Hammel, S. M. (1994). Characterization of on-off intermittency, *Phys. Rev. E* **49**, p. 1140.
Hegger, R., Kantz, H., Schmuser, F., Diestelhorst, M., Kapsch, R. P. and Beige, H. (1998). Dynamical properties of a ferroelectric capacitor observed through nonlinear time series analysis, *Chaos* **8**, p. 727.
Hegger, R., Kantz, H. and Schreiber, T. (1999). Practical implementation of nonlinear time series methods: The TISEAN package, *Chaos* **9**, p. 413.
Hénon, M. (1966). Sur la topologie des lignes de courant dans un cas particulier, *C. R. Acad. Sci. Paris A* **262**, p. 312.
Hénon, M. (1976). A two-dimensional mapping with a strange attractor, *Comm. Math. Phys.* **50**, p. 69.
Hénon, M. (1988). Chaotic scattering modelled by an inclined billiard, *Physica D* **33**, p. 132.
Hénon, M. and Heiles, C. (1964). The applicability of the third integral of motion: Some Numerical Experiments, *Astr. J.* **69**, p. 73.
Hinrichsen, H. (2000). Non-equilibrium critical phenomena and phase transitions into absorbing states, *Adv. Phys.* **49**, p. 815.
Hirsch, J. E., Huberman, B. A. and Scalapino, D. J. (1982). Theory of intermittency, *Phys. Rev. A* **25**, p. 519.
Hirsch, M. W., Smale, S. and Devaney, R. L. (2003). *Differential equations*, 2nd edn. (Academic Press, New York, USA).

Hodgkin, A. and Huxley, A. (1952). A quantitative description of membrane current and its application to conduction and excitation in nerve, *J. Physiol.* **117**, p. 500.

Hohenberg, P. C. and Shraiman, B. I. (1989). Chaotic behavior of an extended system, *Physica D* **37**, p. 109.

Holling, C. S. (1965). The functional response of predator to prey density and its role in mimicry and population regulation, *Mem. Ent. Canada* **45**, p. 1.

Holmes, P. (1990). Poincaré, celestial mechanics, dynamical-systems theory, and "chaos", *Phys. Rep.* **193**, p. 137.

Holmes, P. J., Lumley, J. L., Berkooz, G., Mattingly, J. C. and Wittenberg, R. W. (1997). Low-dimensional models of coherent structures in turbulence, *Phys. Rep.* **287**, p. 337.

Hopf, E. (1943). Abzweigung einer periodischen Lösung von einer stationären eines Lösung Differentialsystems, *Ber. Verh. Sächs. Akad. Wiss. Leipzig. Math.-Nat. Kl.* **95**, p. 3.

Hu, Y., Ecke, R. E. and Ahlers, G. (1995). Transition to Spiral-Defect Chaos in Low Prandtl Number Convection, *Phys. Rev. Lett.* **74**, p. 391.

Huang, K. (1987). *Statistical Mechanics* (Wiley, New York, USA).

Hudson, J. L. and Mankin, J. C. (1981). Chaos in the Belousov-Zhabotinsky reaction, *J. Chem. Phys* **74**, p. 6171.

Hudson, J. L. and Rössler, O. E. (1986). *Chaos and Complex Oscillations in Stirred Chemical Reactors* (Gordon & Breach Science Pub, New York).

Hurd, L., Grebogi, C. and Ott, E. (1994). On the tendency toward ergodicity with increasing number of degrees of fredom in Hamiltonian systems, in J. Siemenis (ed.), *Hamiltonian Mechanics* (Plenum, New York, USA), p. 123.

Hut, P., Alvarez, W., Elder, W. P., Hansen, T., Kauffman, E. G., Keller, G., Shoemaker, E. M. and Weissman, P. (1987). Comet showers as a cause of mass extinctions, *Nature* **329**, p. 118.

Huygens, C. (1673). *Horologium Oscillatorium* (Apud F. Muguet, Parisiis, France), English translation: (1986) *The Pendulum Clock* (Ames: Iowa State University Press).

Ikeda, K. and Matsumoto, K. (1987). High-dimensional chaotic behavior in systems with time-delayed feedback, *Physica D* **29**, p. 223.

Irvine, A. D. (1991). On the Nature of Thought Experiments in Scientific Reasoning, in T. Horowitz and G. J. Massey (eds.), *Thought experiments in science and philosophy* (Rowman & Littlefield Publisher, Inc., Savage, Maryland USA), p. 16.

Isola, S., Politi, A., Ruffo, S. and Torcini, A. (1990). Lyapunov spectra of coupled map lattices, *Phys. Lett. A* **143**, 8, p. 365.

Izrailev, F. M. and Chirikov, B. V. (1966). Statistical properties of a nonlinear string, *Sov. Phys. Dokl.* **166**, p. 57.

Janssen, H. K. (1981). On the nonequilibrium phase transition in reaction-diffusion systems with an absorbing stationary state, *Z. Phys. B* **42**, p. 151.

Jaynes, E. T. (1957a). Information Theory and Statistical Mechanics, *Phys. Rev* **106**, p. 620.

Jaynes, E. T. (1957b). Information Theory and Statistical Mechanics, *Phys. Rev* **107**, p. 171.

Jaynes, E. T. (1967). Foundations of Probability Theory and Statistical Mechanics, in M. Bunge (ed.), *Delaware Seminar in the Foundations of Physics* (Springer-Verlag, Berlin), p. 77.

Jeffries, C. and Perez, J. (1982). Observation of a Pomeau-Manneville intermittent route to chaos in a nonlinear oscillator, *Phys. Rev. A* **26**, p. 2117.

Jensen, M. H. (1989). Fluctuations and scaling in a model for boundary-layer-induced turbulence, *Phys. Rev. Lett.* **62**, p. 1361.

Jepps, O. G. and Rondoni, L. (2006). Thermodynamics and complexity of simple transport phenomena, *J. Phys. A: Math. Gen.* **39**, p. 1311.

Jost, J. (2005). *Dynamical Systems: Examples of Complex Behaviour* (Springer Verlag, Berlin).

Judd, K. and Mees, A. (1995). On selecting models for nonlinear time-series, *Physica D* **82**, p. 426.

Jullien, M.-C., Castiglione, P. and Tabeling, P. (2000). Experimental Observation of Batchelor Dispersion of Passive Tracers, *Phys. Rev. Lett.* **85**, p. 3636.

Kac, M. (1959). *Probability and Related Topics in Physical Sciences* (Interscience, New York).

Kadanoff, L. P. (1999). *Statistical Physics: Statics, Dynamics and Renormalization* (World Scientific, Singapore).

Kalnay, E. (2002). *Atmospheric Modeling, Data Assimilation and Predictability* (Cambridge University Press, Cambridge UK).

Kaneko, K. (1984). Period-doubling of kink-antikink patterns quasiperiodicity in antiferro-like structures, and spatial intermittency in coupled logistic lattice, *Prog. Theor. Phys.* **72**, p. 480.

Kaneko, K. (1986). Lyapunov analysis and information flow in coupled map lattices, *Physica D* **23**, p. 436.

Kaneko, K. (1992). Overview of coupled map lattices, *Chaos* **2**, p. 279.

Kaneko, K. (ed.) (1993). *Theory and applications of Coupled Map Lattices* (John Wiley & Sons, Chichester, UK).

Kaneko, K. (1995). Remarks on the mean field dynamics of networks of chaotic elements, *Physica D* **86**, p. 158.

Kaneko, K. and Konishi, T. (1989). Diffusion in Hamiltonian dynamical systems with many degrees of freedom, *Phys. Rev. A* **40**, p. 6130.

Kantz, H. and Letz, T. (2000). Quasi-chaos and quasi-regularity- the breakdown of linear stability analysis, *Phys. Rev E* **61**, p. 2533.

Kantz, H., Livi, R. and Ruffo, S. (1994). Equipartition Thresholds in Chains of Anharmonic Oscillators, *J. Stat. Phys.* **76**, p. 627.

Kantz, H. and Schreiber, T. (1997). *Nonlinear time series analysis* (Cambridge University Press, Cambridge, UK).

Kaplan, D. T. and Glass, L. (1992). Direct Test for Determinism in a Time Series, *Phys. Rev. Lett.* **68**, p. 427.

Kaplan, D. T. and Glass, L. (1993). Coarse-grained embeddings of time series: Random walks, Gaussian random processes, and deterministic chaos, *Physica D* **64**, p. 431.

Kaplan, J. L. and Yorke, J. A. (1979). Chaotic Behavior of Multidimensional Difference Equations, in H.-O. Peitgen and H.-O. Walther (eds.), *Functional Differential Equations and Approximations of Fixed Points*, Lecture Notes in Mathematics (Springer-Verlag, Berlin), p. 204.

Kapral, R. (1985). Pattern formation in two-dimensional arrays of coupled, discrete-time oscillators, *Phys. Rev. A* **31**, p. 3868.

Kapral, R. and Showalter, K. (eds.) (1995). *Chemical Waves and Patterns* (Kluwer Academic Publisher, Dordrecht, The Neatherlands).

Kardar, M., Parisi, G. and Zhang, Y.-C. (1986). Dynamic Scaling of Growing Interfaces, *Phys. Rev. Lett.* **56**, p. 889.

Katok, A. and Hasselblatt, B. (1995). *Introduction to the Modern Theory of Dynamical Systems* (Cambridge University Press, Cambridge UK).

Kells, L. C. and Orszag, S. A. (1978). Randomness of low-order models of two-dimensional inviscid dynamics, *Phys. Fluids* **21**, p. 162.

Kennel, M. B., Brown, R. and Abarbanel, H. D. I. (1992). Determining embedding dimension for phase-space reconstruction using a geometrical construction, *Phys. Rev. A* **45**, p. 3403.

Khinchin, A. I. (1949). *Mathematical Foundations of Statistical Mechanics* (Dover Publications Inc., New York).

Khinchin, A. I. (1957). *Mathematical foundations of Information Theory* (Dover, London, UK).

Khinchin, A. I. (1997). *Continued Fractions*, new edn. (Dover Publications, New York, USA).

Kirkwood, D. (1888). *The Asteroids, or Minor Planets Between Mars and Jupiter* (J. B. Lippencott, Philadelphia).

Klages, R. and Dorfman, J. R. (1995). Simple Maps with Fractal Diffusion Coefficients, *Phys. Rev. Lett.* **74**, p. 387.

Kolmogorov, A. N. (1936). Sulla teoria di Volterra della lotta per l'esistenza, (in Italian), *Giorn. Istituto Ital. d. Attuari* **7**, p. 74.

Kolmogorov, A. N. (1941). The local structure of turbulence in incompressible viscous fluid for very large Reynold number, *Dokl. Akad. Nauk. SSSR* **30**, p. 299.

Kolmogorov, A. N. (1954). On the conservation of conditionally periodic motions for a small change in Hamilton's function, *Dokl. Akad. Nauk SSSR* **98**, p. 527.

Kolmogorov, A. N. (1956). On the Shannon theory of information transmission in the case of continuous signals, *IRE Trans. Inform. Theor.* **IT-2**, p. 102.

Kolmogorov, A. N. (1958). New metric invariant of transitive dynamical systems and automorphism of Lebesgue spaces, *Dokl. Akad. Nauk SSSR* **119**, p. 861.

Kolmogorov, A. N. (1965). Three approaches to the quantitative definition of information, *Prob. Info. Trans.* **1**, p. 1.

Kolmogorov, A. N. and Fomin, S. V. (1999). *Elements of the Theory of Functions and Functional Analysis*, i edn. (Dover Publications, Mineola, New York).

Kolmogorov, A. N., Petrovskii, I. and Piskunov, N. (1937). A study of the diffusion equation with increase in the amount of substance and its application to a biology problem, *Bull. Univ. Moscow, Ser. Int. A* **1**, p. 1, reprinted in Selected works of A. N. Kolmogorov, ed. V. M. Tikhomirov, Vol. I pg. 242.

Kolmogorov, A. N. and Tikhomirov, V. M. (1959). ε-entropy and ε-capacity of sets in functional spaces, *Uspekhi. Math. Nauk.* **14**, p. 3.

Koon, W. S., Lo, M. W., Marsden, J. E. and Ross, S. D. (2000). Heteroclinic connections between periodic orbits and resonance transitions in celestial mechanics, *Chaos* **10**, p. 427.

Koppel, M. and Atlan, H. (1991). An almost machine-independent theory of program length complexity, sophistication and induction, *Inf. Sci.* **56**, p. 23.

Kostelich, E. J. and Lathrop, D. P. (1994). Time series prediction by using the method of analogues, in A. S. Weigend and N. A. Gershenfeld (eds.), *Time Series Prediction: Forecasting the Future and Understanding the Past* (Addison-Wesley Publ. Comp., Reading), p. 283.

Kraichnan, R. H. (1958). Irreversible Statistical Mechanics of Incompressible Hydromagnetic Turbulence, *Phys. Rev.* **111**, p. 1747.

Kraichnan, R. H. (1967). Inertial ranges in two-dimensional turbulence, *Phys. Fluids* **10**, p. 1417.

Kraichnan, R. H. and Montgomery, D. (1980). Two-dimensional turbulence, *Rep. Prog. Phys.* **43**, p. 547.

Krieger, W. (1970). On entropy and generators of measure preserving transformations, *Transact. Am. Math. Soc.* **149**, p. 453.

Krug, J. and Meakin, P. (1990). Universal finite-size effects in the rate of growth processes, *J. Phys. A: Math. Gen.* **23**, p. L987.
Krylov, N. S. (1979). *Works on the foundations of statistical physics* (Princeton University Press, Princeton, USA).
Kubin, G. (1995). What is a chotic signal? in *IEEE Workshop on Nonlinear Signal and Image Processing, vol. 1: Halkidiki, Greece, 1995* (IEEE), p. 141.
Kuramoto, Y. (1984). *Chemical Oscillations, Waves and Turbulence* (Springer Verlag, Berlin).
Kuramoto, Y. and Tsuzuki, T. (1976). Persistent Propagation of Concentration Waves in Dissipative Media Far from Thermal Equilibrium, *Progr. Theor. Phys.* **55**, p. 356.
Kurths, J. (2000). A focus issue on phase synchronization in chaotic systems, *Int. J. Bifurc. Chaos* **10**, p. 2289.
Kuznetsov, S. P. (1983). On critical behavior of one-dimensional lattices, *Pis'ma J. Technich. Fiz.* **9**, p. 94, in Russian.
Lacorata, G., Aurell, E., Legras, B. and Vulpiani, A. (2004). Evidence for a $k^{-5/3}$ spectrum from the EOLE Lagrangian balloons in the low stratosphere, *J. Atmos. Sci.* **61**, p. 2936.
Landau, L. D. (1944). On the problem of turbulence, *Dokl. Akad. Nauk SSSR* **44**, p. 339.
Landau, L. D. and Lifshitz, E. M. (1980). *Statistical Physics*, 3rd edn. (Butterworth-Heinemann, Oxford, UK).
Lanford, O. E. (1998). Some informal remarks on the orbit structure of discrete approximations to chaotic maps, *Experimental Mathematics* **7**, p. 317.
Langevin, P. (1908). Sur la theorie du mouvement Brownien, *C. R. Acad. Sci. (Paris)* **146**, p. 530, translated in Am. J. Phys. 65, 1079 (1997).
Laskar, J., Joutel, F. and Boudin, F. (1993). Orbital, precessional, and insolation quantities for the Earth from -20 MYR to +10 MYR, *Astron. Astrophys.* **270**, p. 522.
Lasota, A. and Mackey, M. C. (1985). *Probabilistic properties of deterministic systems* (Cambridge University Press, Cambridge, UK).
Lasota, A. and Yorke, J. A. (1982). Exact Dynamical Systems and the Frobenius-Perron Operator, *Trans. Am. Math. Soc.* **273**, p. 375.
Lebowitz, J. (1993). Boltzmann's entropy and time's arrow, *Phys. Today* **46**, p. 32.
Ledrappier, F. (1981). Some relations between dimension and Lyapounov exponents, *Comm. Math. Phys.* **81**, p. 229.
Lee, J. (1987). Triad-angle locking in low-order models of the 2D Navier-Stokes equations, *Physica D* **24**, p. 54.
Leith, C. E. (1971). Atmospheric predictability and two-dimensional turbulence, *J. Atmos. Sci.* **28**, p. 144.
Leith, C. E. and Kraichnan, R. H. (1972). Predictability of turbulent flows, *J. Atmos. Sci.* **29**, p. 1041.
Lepri, S., Livi, R. and Politi, A. (2003). Thermal conduction in classical low-dimensional lattices, *Phys. Rep.* **377**, p. 1.
Lepri, S., Politi, A. and Torcini, A. (1996). Chronotopic Lyapunov analysis:(I) a Detailed Characterization of 1D Systems, *J. Stat. Phys.* **82**, p. 1429.
Lepri, S., Politi, A. and Torcini, A. (1997). Chronotopic Lyapunov analysis: II. Toward a unified approach, *J. Stat. Phys.* **88**, p. 31.
Leskovac, V. (2003). *Comprehensive Enzyme Kinetics* (Springer, Berlin).
Levin, S. A. (1992). The Problem of Pattern and Scale in Ecology: The Robert H. MacArthur Award Lecture, *Ecology* **73**, p. 1943.
Li, M. and Vitányi, P. (1997). *An introduction to Kolmogorov complexity and its applications* (Springer-Verlag, Berlin).

Li, T. Y. (1976). Finite approximation for Frobenius-Perron operator. A solution to Ulam's conjecture, *J. Approx. Theor.* **17**, p. 177.

Libchaber, A., Fauve, S. and La-Roche, C. (1983). Two-parameter study of the routes to chaos, *Physica D* **7**, p. 73.

Lichtenberg, A. J. and Aswani, A. M. (1998). Arnold diffusion in many weakly coupled mappings, *Phys. Rev. E* **57**, p. 5325.

Lichtenberg, A. J. and Lieberman, M. A. (1992). *Regular and Chaotic Dynamics*, 2nd edn. (Springer-Verlag, Berlin).

Lichtenberg, A. J., Livi, R., Pettini, M. and Ruffo, S. (2007). Dynamics of Oscillator Chains, in G. Gallavotti (ed.), *The Fermi-Pasta-Ulam Problem: A Status Report* (Springer-Verlag, New York), p. 21.

Liverani, C. and Wojtkowski, M. P. (1995). Ergodicity in Hamiltonian systems, *Dynamics Reported* **4**, p. 131.

Livi, R., Pettini, M., Ruffo, S., Sparpaglione, M. and Vulpiani, A. (1985). Equipartition threshold in nonlinear large Hamiltonian systems: The Fermi-Pasta-Ulam model, *Phys. Rev. A* **31**, p. 1039.

Livi, R., Pettini, M., Ruffo, S. and Vulpiani, A. (1987). Chaotic behaviour in nonlinear Hamiltonian systems and equilibrium statistical mechanics, *J. Stat. Phys.* **48**, p. 539.

Livi, R., Politi, A. and Ruffo, S. (1986). Distribution of characteristic exponents in the thermodynamic limit, *J. Phys. A: Math. Gen.* **19**, p. 2033.

Livi, R., Ruffo, S. and Shepelyansky, D. (2003). Kolmogorov pathways from integrability to chaos and beyond, in R. Livi and A. Vulpiani (eds.), *The Kolmogorov Legacy in Physics*, Lecture Note in Physics (Springer-Velag, Berlin), p. 3.

Lochak, P. and Neishtadt, A. I. (1992). Estimates of stability time for nearly integrable systems with a quasiconvex Hamiltonian, *Chaos* **2**, p. 495.

Lorentz, H. A. (1905). The motion of electrons in metallic bodies, *Proc. R. Acad. Amsterdam* **7**, pp. 438, 585, 684.

Lorenz, E. N. (1963). Deterministic nonperiodic flows, *J. Atmos. Sci.* **20**, p. 130.

Lorenz, E. N. (1969). The predictability of a flow which possesses many scales of motion, *Tellus* **21**, p. 3.

Lorenz, E. N. (1996). Predictability – a problem partly solved, in *Predictability*, ECMWF Seminar Proceedings 4-5 Sept. 1995 (ECMWF, Reading, UK), p. 1.

Losson, J., Vannitsem, S. and Nicolis, G. (1998). Aperiodic mean-field evolutions in coupled map lattices, *Phys. Rev. E* **57**, p. 4921.

Lotka, A. (1910). Zur Theorie der periodischen Raktionen, *Z. Phys. Chemie* **72**, p. 508.

Lozi, R. (1978). Un attracteur étrange du type attracteur de Hénon, *J. Phys. Coll. 5* **39**, p. 9.

Lumley, P. H. J. and Berkooz, G. (1996). *Turbulence, Coherent Structures, Dynamical Systems and Symmetry*, 2nd edn. (Cambridge University Press, Cambridge, UK).

Lv́ov, V. S., Podivilov, E., Pomyalov, A., Procaccia, I. and Vandembroucq, D. (1998). Improved shell model of turbulence, *Phys. Rev. E* **58**, p. 1811.

Mackey, D. S. and Mackey, N. (2003). On the Determinant of Symplectic Matrices, URL `citeseer.ist.psu.edu/mackey03determinant.html`, numerical Analysis Report No. 422, Manchester Centre for Computational Mathematics, Manchester, England. February 2003.

Mackey, M. C. and Glass, L. (1977). Oscillation and chaos in physiological control systems, *Science* **197**, p. 287.

Majda, A. J. and Kramer, P. R. (1999). Simplified models for turbulent diffusion: Theory, numerical modelling, and physical phenomena, *Phys. Rep.* **314**, p. 237.

Malraison, B., Atten, P., Berge, P. and Dubois, M. (1983). Dimension of strange attractors: An experimental determination for the chaotic regime of two convective systems, *J. Physique Lettres* **44**, p. 897.

Mandelbrot, B. (1977). *Fractals: Form, Chance, and Dimension* (Freeman, New York, USA).

Mandelbrot, B. (1985). Self-affine fractals and fractal dimension, *Physica Scripta* **32**, p. 257.

Manneville, P. (1990). *Dissipative structures and weak turbulence* (Academic Press, Boston, USA).

Mantica, G. (2000). Quantum algorithmic integrability: The metaphor of classical polygonal billiards, *Phys. Rev. E* **61**, p. 6434.

Marchioro, C. and Pulvirenti, M. (1994). *Mathematical theory of incompressible nonviscous fluids* (Springer-Verlag, New York).

Marsden, J. E. and McCracken, M. (1976). *The Hopf bifurcation and its applications* (Springer-Verlag, New York).

Martin-Löf, P. (1966). The definition of random sequences, *Inform. Contr.* **9**, p. 602.

Maurer, J. and Libchaber, A. (1979). Rayleigh-Bénard experiment in liquid-helium frequency locking and the onset of turbulence, *J. Phys.(Paris) Lett.* **40**, p. L419.

Maurer, J. and Libchaber, A. (1980). Effect of the Prandtl number on the onset of turbulence in liquid-He-4, *J. Phys.(Paris) Lett.* **41**, p. L515.

Maxey, M. R. and Riley, J. J. (1983). Equation of motion for a small rigid sphere in a nonuniform flow, *Phys. Fluids* **26**, p. 883.

May, R. M. (1974). Biological Populations with Nonoverlapping Generations: Stable Points, Stable Cycles, and Chaos, *Science* **186**, p. 645.

May, R. M. (1976). Simple mathematical models with very complicated dynamics, *Nature* **261**, p. 459.

Mazur, P. and Montroll, E. (1960). Poincaré Cycles, Ergodicity, and Irreversibility in Assemblies of Coupled Harmonic Oscillators, *J. Math. Phys.* **1**, p. 70.

Mazur, P. and van der Linden, J. (1963). Asymptotic form of the structure function for real systems, *J. Math. Phys.* **4**, p. 271.

McDonald, S. W., Grebogi, C., Ott, E. and Yorke, J. A. (1985). Fractal Basin Boundaries, *Physica D* **17**, p. 125.

McMillan, B. (1953). The Basic Theorems of Information Theory, *Ann. Math. Stat.* **24**, p. 196.

McQuain, M. K., Seale, K., Peek, J., Fisher, T. S., Levy, S., Stremler, M. A. and Haselton, F. R. (2004). Chaotic mixer improves microarray hybridization, *Analyt. Biochem.* **325**, p. 215.

McWilliams, J. C. (1984). The emergence of isolated coherent vortices in turbulent flows, *J. Fluid. Mech.* **146**, p. 21.

Meiss, J. D. (1992). Symplectic maps, variational principles, and transport, *Rev. Mod. Phys.* **64**, p. 795.

Melnikov, V. K. (1963). On the stability of the center for periodic perturbation of time, *Tran. Moskov Math. Soc.* **12**, p. 1.

Metzler, R. and Klafter, J. (2000). The random walk's guide to anomalous diffusion: a fractional dynamics approach, *Phys. Rep.* **339**, p. 1.

Meyer, C. W., Ahlers, G. and Cannell, D. S. (1987). Initial stages of pattern formation in Rayleigh-Bénard convection, *Phys. Rev. Lett.* **59**, p. 1577.

Milnor, J. (1985). On the concept of attractor, *Comm. Math. Phys.* **99**, p. 177.

Moeng, C. H. (1984). A Large-Eddy Simulation Model for the Study of Planetary Boundary-Layer turbulence, *J. Atmos. Sci.* **41**, p. 2052.

Moin, P. and Mahesh, K. (1998). Direct Numerical Simulation: A Tool in Turbulence Research, *Annu. Rev. Fluid Mech.* **30**, p. 539.

Monin, A. and Yaglom, A. (1975). *Statistical Fluid Dynamics*, Vol. I and II (MIT Press, Cambridge MA).

Monod, J., Wyman, J. and Chateaux, J. P. (1965). On the nature of allosteric transitions: a plausible model, *J. Mol. Biol.* **12**, p. 88.

Morbidelli, A. (2002). *Modern Celestial Mechanics; Aspects of Solar System Dynamics* (Taylor & Francis, London: UK).

Moser, J. (1962). On invariant curves of area preserving mappings of an annulus, *Nachr. Akad. Wiss. Gött, Math. Phys.* **K1**, p. 1.

Muñoz, M. A. (2004). Multiplicative Noise in Non-equilibrium Phase Transitions: A Tutorial, in E. Korutcheva and R. Cuerno (eds.), *Advances in Condensed Matter and Statistical Mechanics* (Nova Science Publishers, New York), p. 37.

Muñoz, M. A. and Pastor-Satorras, R. (2003). Stochastic Theory of Synchronization Transitions in Extended Systems, *Phys. Rev. Lett.* **90**, p. 204101.

Murray, J. D. (2002). *Mathematical Biology: I. An Introduction*, Interdisciplinary Applied Mathematics (Springer-Verlag, Berlin Heidelberg).

Murray, J. D. (2003). *Mathematical Biology: II. Spatial Models and Biomedical Applications*, 3rd edn., Interdisciplinary Applied Mathematics (Springer-Verlag, Berlin Heidelberg).

Nayfeh, A. H. and Balachandran, B. (1995). *Applied Nonlinear Dynamics: Analytical, Computational, and Experimental Methods* (John Wiley & Sons, New York, USA).

Nekhoroshev, N. N. (1977). An exponential estimate of the time of stability of nearly integrable Hamiltonian systems, *Russ. Math. Surv.* **32**, p. 1.

Nepomnyashchy, A. A. (1974). Stability of wave regimes in a film flowing down an inclined plane, *Izv. Akad. Nauk SSSR, Mekh. Zhidk. Gaza* **3**, p. 28.

Newhouse, R. T., Ruelle, D. and Takens, F. (1978). Occurrence of strange Axiom A attractors near quasi periodic flows on Tm, $m \geq 3$, *Comm. Math. Phys.* **64**, p. 35.

Newton, P. K. (2001). *The N-Vortex Problem: Analytical Techniques* (Springer-Verlag, New York).

Nicolis, C. and Nicolis, G. (1984). Is there a climatic attractor, *Nature* **311**, p. 529.

Nicolis, G. and Nicolis, C. (1988). Master-equation approach to deterministic chaos, *Phys. Rev. A* **38**, p. 427.

Ódor, G. (2004). Universality classes in nonequilibrium lattice systems, *Rev. Mod. Phys.* **76**, p. 663.

Ohkitani, K. and Yamada, M. (1989). Temporal intermittency in the energy cascade process and local lyapunov analysis in fully-developed model turbulence, *Prog. Theor. Phys.* **89**, p. 329.

Olbrich, E., Hegger, R. and Kantz, H. (1998). Analyzing local observations of weakly coupled maps, *Phys. Lett. A* **244**, p. 538.

Olbrich, E. and Kantz, H. (1997). Inferring chaotic dynamics from time-series: On which length scale determinism becomes visible, *Phys. Lett. A* **232**, p. 63.

Onsager, L. (1949). Statistical Hydrodynamics, *Nuovo Cimento (Supp.)* **6**, p. 279.

Orszag, S. A. (1969). Numerical Methods for the Simulation of Turbulence, *Phys. Fluids* **12**, pp. II–250.

Orszag, S. A. and Patterson Jr, G. S. (1972). Numerical Simulation of Three-Dimensional Homogeneous Isotropic Turbulence, *Phys. Rev. Lett.* **28**, p. 76.

Osborne, A. R. and Provenzale, A. (1989). Finite correlation dimension for stochastic systems with power-law spectra, *Physica D* **35**, p. 357.

Oseledec, V. I. (1968). A multiplicative ergodic theorem. Lyapunov characteristic numbers for dynamical systems, *Trans. Mosc. Math. Soc.* **19**, p. 197.
Ott, E. (1993). *Chaos in Dynamical Systems* (Cambridge University Press, Cambridge, UK).
Ott, E., Sauer, T. and Yorke, J. A. (1994). *Coping With Chaos* (Wiley, Series in Nonlinear Science, New York).
Ottino, J. M. (1990). Mixing, chaotic advection, and turbulence, *Ann. Rev. Fluid Mech.* **22**, p. 207.
Ouyang, Q. and Flesselles, J. M. (1996). Transition from spirals to defect turbulence driven by a convective instability, *Nature* **379**, p. 143.
Ouyang, Q. and Swinney, H. L. (1991). Transition to chemical turbulence, *Chaos* **1**, p. 411.
Ouyang, Q., Swinney, H. L. and Li, G. (2000). Transition from Spirals to Defect-Mediated Turbulence Driven by a Doppler Instability, *Phys. Rev. Lett.* **84**, p. 1047.
Paladin, G. and Vaienti, S. (1988). Looking at the equilibrium measures in dynamical systems, *J. Phys. A: Math. Gen.* **21**, p. 4609.
Paladin, G. and Vulpiani, A. (1986). Intermittency in chaotic systems and Rényi entropies, *J Phys. A: Math. Gen.* **19**, p. L997.
Paladin, G. and Vulpiani, A. (1987). Anomalous scaling laws in multifractal objects, *Phys. Rep.* **156**, p. 147.
Paladin, G. and Vulpiani, A. (1994). Predictability in spatially extended systems, *J. Phys. A: Math. Gen.* **27**, p. 4911.
Parisi, G. and Frisch, U. (1985). On the singularity structure of fully developed turbulence, in M. Ghil, G. Parisi and R. Benzi (eds.), *Turbulence and predictability of geophysical fluid dynamics* (North-Holland, Amsterdam), p. 84.
Pasmanter, R. A. (1994). On long-lived vortices in 2-D viscous flows, most probable states of inviscid 2-D flows and a soliton equation, *Phys. Fluids* **6**, p. 1236.
Paul, M. R., Einarsson, M. I., Fischer, P. F. and Cross, M. C. (2007). Extensive chaos in Rayleigh-Bénard convection, *Phys. Rev. E* **75**, p. 045203(R).
Pecora, L. M. and Carroll, T. L. (1990). Synchronization in chaotic systems, *Phys. Rev. Lett.* **64**, p. 821.
Peña, M. and Kalnay, E. (2004). Separating fast and slow modes in coupled chaotic systems, *Nonl. Proc. in Geophys* **11**, p. 319.
Perez, G. and Cerdeira, H. A. (1992). Instabilities and nonstatistical behavior in globally coupled systems, *Phys. Rev. A* **46**, p. 7492.
Pesin, Y. B. (1976). Lyapunov characteristic exponent and ergodic properties of smooth dynamical systems with an invariant measure, *Sov. Math. Dokl.* **17**, p. 196.
Petersen, K. (1990). *Ergodic Theory* (Cambridge University Press, Cambridge, UK).
Petrosky, T. Y. (1986). Chaos and cometary clouds in the solar system, *Phys. Lett. A* **117**, p. 328.
Pikovsky, A. (1993). Local Lyapunov exponents for spatiotemporal chaos, *Chaos* **3**, p. 225.
Pikovsky, A. and Kurths, J. (1994a). Roughening interfaces in the dynamics of perturbations of spatiotemporal chaos, *Phys. Rev. E* **49**, p. 898.
Pikovsky, A. and Politi, A. (1998). Dynamic localization of Lyapunov vectors in spacetime chaos, *Nonlinearity* **11**, p. 1049.
Pikovsky, A. and Politi, A. (2001). Dynamic localization of Lyapunov vectors in Hamiltonian lattices, *Phys. Rev. E* **63**, p. 036207.
Pikovsky, A., Rosenblum, M. and Kurths, J. (2000). Phase synchronization in regular and chaotic systems, *Int. J. Bifurc. Chaos* **10**, p. 2291.
Pikovsky, A., Rosenblum, M. and Kurths, J. (2001). *Synchronization: a universal concept in nonlinear dynamics* (Cambridge University Press, Cambridge, UK).

Pikovsky, A. S. (1989). Spatial development of chaos in nonlinear media, *Phys. Lett. A* **137**, p. 121.
Pikovsky, A. S. (1992). Spatial development of chaos, in S. Vohra, M. Spano, M. Schlesinger, L. Pecora and W. Ditto (eds.), *Proc. of the 1st Experiment. Chaos Conf., Arlington V., Oct. 1-3, 1991* (World Scientific, Singapore), p. 382.
Pikovsky, A. S. and Grassberger, P. (1991). Symmetry breaking bifurcation for coupled chaotic attractors, *J. Phys. A: Math. Gen.* **24**, p. 4587.
Pikovsky, A. S. and Kurths, J. (1994b). Do globally coupled maps really violate the law of large numbers? *Phys. Rev. Lett.* **72**, p. 1644.
Poincaré, H. (1881). Mémoire sur les courbes définié par une équation différentielle, *J. Math. Pura Appl.* **7**, p. 375.
Poincaré, H. (1890). Sur le probléme des trois corps et les équations de la dynamique, *Acta Math.* **13**, p. 1.
Poincaré, H. (1892, 1893, 1899). *Les méthods nouvelles de la mécanique céleste*, Vol. I, II, III (Gauthier-Villars, Paris), (English Translation, New methods of celestial mechanics, American Institute of Physics, 1993).
Politi, A., Livi, R., Oppo, G.-L. and Kapral, R. (1993). Unpredictable behavior of stable systems, *Europhys. Lett.* **22**, p. 571.
Politi, A. and Torcini, A. (1992). Periodic orbits in coupled Hénon maps: Lyapunov and multifractal analysis, *Chaos* **2**, p. 293.
Politi, A. and Torcini, A. (1994). Linear and nonlinear mechanisms of information propagation, *Europhys. Lett.* **28**.
Politi, A. and Torcini, A. (2009). Stable Chaos, Preprint arXiv:0902.2545v1 [nlin.CD].
Pomeau, Y. (1986). Front motion, metastability and subcritical bifurcations in hydrodynamics, *Physica D* **23**, p. 3.
Pomeau, Y. and Manneville, P. (1980). Intermittent transition to turbulence in dissipative dynamical systems, *Comm. Math. Phys.* **74**, p. 189.
Porter, M. A. and Cvitanovic, P. (2005). Ground Control to Niels Bohr: Exploring Outer Space with Atomic Physics, *Notices Am. Math. Soc* **52**, p. 1020.
Pöschel, J. (2001). A lecture on classical KAM theorem, *Proc. Symp. Pure Math.* **69**, p. 707.
Prigogine, I. (1994). *Les lois du chaos* (Flammarion, Paris).
Primas, H. (2002). Hidden determinism, probability and time's arrow, in H. Atmanspacher and R. Bishop (eds.), *Between chance and choice* (Imprint Academic), p. 89.
Primo, C., Szendro, I. G., Rodríguez, M. A. and Gutiérrez, J. M. (2007). Error Growth Patterns in Systems with Spatial Chaos: From Coupled Map Lattices to Global Weather Models, *Phys. Rev. Lett.* **98**, p. 108501.
Primo, C., Szendro, I. G., Rodríguez, M. A. and López, J. M. (2005). Dynamic scaling of bred vectors in spatially extended chaotic systems, *Europhys. Lett.* **76**, p. 767.
Pruppacher, H. R. and Klett, J. D. (1996). *Microphysics of Clouds and Precipitation* (Kluwer Academic Publishers, Dordrecht).
Puglisi, A., Benedetto, D., Caglioti, E., Loreto, V. and Vulpiani, A. (2003). Data compression and learning in time sequences analysis, *Physica D* **180**, p. 92.
Raghunathan, M. S. (1979). A proof of Oseledec's Multiplicative Ergodic Theorem, *Israel J. Math.* **32**, p. 356.
Ramsey, N. F. (1956). Thermodynamics and Statistical Mechanics at Negative Absolute Temperatures, *Phys. Rev.* **103**, p. 20.
Rényi, A. (1960). On measures of Entropy and Information, in J. Neyman (ed.), *Proceedings of the Fourth Berkeley Symposium on Mathematical Statistics and Probability*, Vol. 1 (University of California Press, Berkeley, USA), p. 547.

Rényi, A. (1970). *Probability Theory* (North-Holland, Amsterdam).
Richardson, L. F. (1926). Atmospheric Diffusion Shown on a Distance-Neighbour Graph, *Proc. Royal Soc. London. Ser A* **110**, p. 709.
Ricker, W. E. (1954). Stock and Recruitment, *J. Fish. Res. Canada* **II**, p. 559.
Rio, M. H., Poulain, P. M., Pascual, A., Mauri, E., Larnicol, G. and Santoleri, R. (2007). A Mean Dynamic Topography of the Mediterranean Sea computed from altimetric data, in-situ measurements and a general circulation model, *J. Mar. Syst.* **65**, p. 484.
Rissanen, J. (1989). *Stochastic Complexity in Statistical Inquiry Theory* (World Scientific Publishing Co., Singapore).
Robert, R. and Sommeria, J. (1991). Statistical equilibrium states for 2-dimensional flows, *J. Fluid Mech.* **229**, p. 291.
Robinson, J. C. (2007). Parametrization of global attractors, experimental observations, and turbulence, *J. Fluid Mech.* **578**, p. 495.
Rose, H. A. and Sulem, P. L. (1978). Fully developed turbulence and statistical mechanics, *J. Phys. (Paris)* **39**, p. 441.
Rosenblum, M. G., Pikovsky, A. S. and Kurths, J. (1996). Phase synchronization of chaotic oscillators, *Phys. Rev. Lett.* **76**, p. 1804.
Rössler, O. E. (1976). An Equation for Continuous Chaos, *Phys. Lett. A* **57**, p. 397.
Roux, J. C. (1983). Experimental studies of bifurcations leading to chaos in the Belousov-Zhabotinsky reaction, *Physica D* **7**, p. 57.
Roy, R. and Thornburg, K. S. (1994). Experimental synchronization of chaotic lasers, *Phys. Rev. Lett.* **72**, p. 2009.
Ruelle, D. (1978a). An inequality for the entropy of differentiable maps, *Bol. Soc. Bras. Mat.* **9**, p. 83.
Ruelle, D. (1978b). *Thermodynamic formalism: the mathematical structures of classical equilibrium statistical mechanics* (Addison-Wesley, Reading, USA).
Ruelle, D. (1979). Microscopic fluctuations and turbulence, *Phys. Lett. A* **72**, p. 81.
Ruelle, D. (1982). Large volume limit of the distribution of characteristic Lyapunov exponents in turbulence, *Comm. Math. Phys.* **87**, p. 287.
Ruelle, D. (1989). *Chaotic evolution and strange attractors: The statistical analysis of time series for deterministic nonlinear systems* (Cambridge University Press, Cambridge, UK).
Ruelle, D. (1990). The Claude Bernard Lecture, 1989. Deterministic chaos: The Science and the Fiction, *Proc. R. Soc. London Ser. A, Math. Phys. Sci.* **427**, p. 241.
Ruelle, D. and Takens, F. (1971). On the nature of turbulence, *Comm. Math. Phys.* **20**, p. 167.
Ruffo, S. (2001). Time-Scales for the Approach to Thermal Equilibrium, in J. Bricmont, D. Dürr, G. Galavotti, G. Ghirardi, F. Petruccione and N. Zanghi (eds.), *Chance in Physics, Lecture Notes in Physics*, Vol. 574 (Springer-Verlag, Berlin), p. 243.
Saltzman, B. (1962). Finite amplitude free convection as an initial value problem, *J. Atmos. Sci.* **19**, p. 329.
Samelson, R. M. (1992). Fluid exchange across a meandering Jet, *J. Phys. Oceanogr.* **22**, p. 431.
Sapoval, B., Baldassarri, A. and Gabrielli, A. (2004). Self-Stabilized Fractality of Seacoasts through Damped Erosion, *Phys. Rev. Lett.* **93**, p. 098501.
Sauer, T., Yorke, J. A. and Casdagli, M. (1991). Embedology, *J. Stat. Phys.* **65**, p. 579.
Saw, E. W., Shaw, R. A., Ayyalasomayajula, S., Chuang, P. Y. and Gylfason, A. (2008). Inertial Clustering of Particles in High-Reynolds-Number Turbulence, *Phys. Rev. Lett.* **100**, p. 214501.

Schell, M., Fraser, S. and Kapral, R. (1982). Diffusive dynamics in systems with translational symmetry: A one-dimensional-map model, *Phys. Rev. A* **26**, p. 504.

Schena, M., Shalon, D., Davis, R. W. and Brown, P. O. (1995). Quantitative monitoring of gene expression patterns with a complementary DNA microarray, *Science* **270**, p. 467.

Schmitz, R. A., Graziani, K. R. and Hudson, J. L. (1977). Experimental Evidence of Chaotic States in the Belousov–Zhabotinsky Reaction, *J. Chem. Phys.* **67**, p. 3040.

Schürmann, T. and Grassberger, P. (1996). Entropy estimation of symbol sequences, *Chaos* **6**, p. 414.

Schuster, H. G. and Just, W. (2005). *Deterministic Chaos* (Wiley-VCH, Weinheim).

Schuster, H. G., Martin, S. and Martienssen, W. (1986). A new method for determining the largest Lyapunov exponent in simple nonlinear systems, *Phys. Rev. A* **33**, p. 3547.

Shannon, C. E. (1948). A Mathematical Theory of Communication, *Bell System Tech. J.* **27**, pp. 379, 623.

Shannon, C. E. (1959). Coding Theorems for a Discrete Source with a Fidelity Criterion, *Inst. Radio Eng. Int. Conv. Rec.* **7**, p. 142, reprinted with changes in Information and Decision Processes, edited by R. E. Machol, McGraw-Hill, NY, 1960, p. 93. Reprinted in D. Slepian, editor, Key Papers in the Development of Information Theory, IEEE Press, NY, 1974. Included in Part A.

Shannon, C. E. and Weaver, W. (1949). *The Mathematical Theory of Communication* (The University of Illinois Press, Urbana, Illinois).

She, Z. S. and Lévêque, E. (1994). Universal scaling laws in fully developed turbulence, *Phys. Rev. Lett.* **72**, p. 336.

Shibata, T. and Kaneko, K. (1998). Collective Chaos, *Phys. Rev. Lett.* **81**, p. 4116.

Shraiman, B. I., Pumir, A., van Saarloos, W., Hohenberg, P. C., Chaté, H. and Holen, M. (1992). Spatiotemporal chaos in the one-dimensional complex Ginzburg-Landau equation, *Physica D* **57**, p. 241.

Shtern, V. N. (1983). Attractor dimension for the generalized Baker's transformation, *Phys. Lett. A* **99**, p. 268.

Siggia, E. D. and Aref, H. (1981). Point-vortex simulation of the inverse energy cascade in two-dimensional turbulence, *Phys. Fluids* **24**, p. 171.

Simonet, J., Warden, M. and Brun, E. (1994). Locking and Arnold tongues in an infinite-dimensional system: The nuclear magnetic resonance laser with delayed feedback, *Phys. Rev. E* **50**, p. 3383.

Sinai, Y. G. (1959). On the concept of entropy for a dynamic system, *Dokl. Akad. Nauk. SSSR* **124**, p. 768.

Sinai, Y. G. (1996). A remark concerning the thermodynamical limit of the Lyapunov spectrum, *Int. J. Bifur. Chaos* **6**, p. 1137.

Sivashinsky, G. I. (1977). Nonlinear analysis of hydrodynamic instability in laminar flames. I- Derivation of basic equations, *Acta Astron* **4**, p. 1177.

Smale, S. (1965). Diffeomorphisms with many periodic points, in *Differential and combinatorial topology* (Univ. Press, Princeton, USA), p. 63.

Smale, S. (1976). On the differential equations of species in competition, *J. Math. Biol.* **3**, p. 5.

Smith, L. A. (1988). Intrinsic limits on dimension calculations, *Phys. Lett. A* **133**, p. 283.

Smith, L. M. and Yakhot, V. (1993). Bose condensation and small-scale structure generation in a random force driven 2D turbulence, *Phys. Rev. Lett.* **71**, p. 352.

Smith, P. (1998). *Explaining Chaos* (Cambridge University Press, Cambridge UK).

Smith, T. R., Moehlis, J. and Holmes, P. (2005). Low-dimensional modelling of turbulence using the proper orthogonal decomposition: A tutorial, *Nonl. Dyn.* **41**, p. 275.

Solomon, T. H. and Gollub, J. P. (1988). Chaotic particle transport in time-dependent Rayleigh-Bénard convection, *Phys. Rev. A* **38**, p. 6280.
Solomonoff, R. J. (1964). A formal theory of inductive inference, *Inform. Contr.* **7**, p. 1; 224.
Sommerer, J. C. and Ott, E. (1993). Particles Floating on a Moving Fluid: A Dynamically Comprehensible Physical Fractal, *Science* **259**, p. 335.
Stein, P. R. and Ulam, S. M. (1964). Non-linear transformation studies on electronic computers, *Rozprawy Matematyczne* **39**, p. 1.
Strait, E. J., Lao, L. L., Mauel, M. E., Rice, B. W., Taylor, T. S., Burrell, K. H., Chu., M. S., Lazarus, E. A., Osborne, T. H., Thompson, S. J. and Turnbull, A. D. (1995). Enhanced Confinement and Stability in DIII-D Discharges with Reversed Magnetic Shear, *Phys. Rev. Lett.* **75**, p. 4421.
Stroock, A. D., Dertinger, S. K. W., Ajdari, A., Mezic, I., Stone, H. A. and Whitesides, G. M. (2002). Chaotic Mixer for Microchannels, *Science* **295**, p. 647.
Sugihara, G. and May, R. (1990). Nonlinear forecasting as a way of distinguishing chaos from measurement errors in time series, *Nature* **344**, p. 734.
Sussman, G. J. and Wisdom, J. (1992). Chaotic Evolution of the Solar System, *Science* **257**, p. 56.
Szebehely, V. (1967). *Theory of Orbits: The Restricted Problem of Three Bodies* (Academic Press).
Szendro, I. G. and López, J. M. (2005). Universal critical behavior of the synchronization transition in delayed chaotic systems, *Phys. Rev. E* **71**, p. 055203.
Tabeling, P. (2002). Two-dimensional turbulence: a physicist approach, *Phys. Rep.* **362**, p. 1.
Tabeling, P. and Cheng, S. (2005). *Introduction to Microfluidics* (Oxford University Press, Oxford).
Tabor, M. (1989). *Chaos and integrability in nonlinear dynamics* (John Wiley & Sons, New York, USA).
Takens, F. (1981). Detecting strange attractors in turbulence, in D. A. Rand and L. S. Young (eds.), *Dynamical Systems and Turbulence, Lecture Notes in Mathematics*, Vol. 898 (Springer-Verlag, Berlin), p. 366.
Takeuchi, K. A., Kuroda, M., Chaté, H. and Sano, M. (2007). Directed Percolation Criticality in Turbulent Liquid Crystals, *Phys. Rev. Lett.* **99**, p. 234503.
Tam, W. Y. and Swinney, H. L. (1990). Spatiotemporal patterns in a one-dimensional open reaction-diffusion system, *Physica D* **46**, p. 10.
Tél, T. and Lai, Y.-C. (2008). Chaotic transients in spatially extended systems, *Phys. Rep.* **460**, p. 245.
Theiler, J. (1991). Some comments on the correlation dimension of 1/f-alpha noise, *Phys. Lett. A* **155**, p. 480.
Timberlake, T. (2004). A computational approach to teaching conservative chaos, *Am. J. Phys.* **72**, p. 1002.
Torcini, A., Frauenkron, H. and Grassberger, P. (1997). A Novel Integration Scheme for Partial Differential Equations: an Application to the Complex Ginzburg-Landau Equation, *Physica D* **103**, p. 605.
Torcini, A., Grassberger, P. and Politi, A. (1995). Error propagation in extended chaotic systems, *J. Phys. A: Math. Gen.* **28**, p. 4533.
Torcini, A. and Lepri, S. (1997). Disturbance propagation in chaotic extended systems with long-range coupling, *Phys. Rev. E* **55**, p. R3805.
Toth, A. and Kalnay, E. (1993). Ensemble forecasting at nmc- the generation of perturbations, *Bull. Am. Meteo. Soc* **74**, p. 2317.

Tritton, D. J. (1988). *Physical fluid dynamics* (Oxford Science Publ., Oxford UK).
Tsonis, A. A., Elsner, J. B. and Georgakakos, K. P. (1993). Estimating the dimension of weather and climate attractors, *J. Atmos. Sci.* **50**, p. 2549.
Tucker, W. (2002). A Rigorous ODE Solver and Smale's 14th Problem, *Found. Comput. Math.* **2**, p. 53.
Turing, A. M. (1936). On Computable Numbers, with an application to the Entscheidungsproblem, *Proc. London Math. Soc.* **2**, p. 230.
Turing, A. M. (1953). The chemical basis of morphogenesis, *Phil. Trans. Royal Soc. B* **237**, p. 37.
Tyson, J. J. (1983). Periodic enzyme synthesis and oscillatory repression: why is the period of oscillation close to the cell cycle time? *J. Theor. Biol.* **103**, p. 313.
Ueshima, Y., Nishihara, K., Barnett, D. M., Tajima, T. and Furukawa, H. (1997). Relation between Lyapunov Exponent and Dielectric Response Function in Dilute One Component Plasmas, *Phys. Rev. Lett.* **79**, p. 2249.
Ulam, S. M. and von Neumann, J. (1947). On combination of stochastic and deterministic processes - preliminary report, *Bull. Am. Math. Soc.* **53**, p. 1120.
van de Water, W. and Bohr, T. (1993). Critical properties of lattices of diffusively coupled quadratic maps, *Chaos* **3**, p. 747.
van der Pol, B. (1927). Forced oscillations in a circuit with nonlinear resistance (Reception with relative triode), *Phil. Mag.* **3**, p. 65.
van Saarloos, W. (1988). Front propagation into unstable states: Marginal stability as a dynamical mechanism for velocity selection, *Phys. Rev. A* **37**, p. 211.
van Saarloos, W. (1989). Front propagation into unstable states. II. Linear versus nonlinear marginal stability and rate of convergence, *Phys. Rev. A* **39**, p. 6367.
Varadhan, S. R. S. (1987). *Large Deviations and Applications*, CBMS-NSF Regional Conference Series in Applied Mathematics 46 (SIAM, Philadelphia, USA).
Vastano, J. A., Russo, T. and Swinney, H. L. (1990). Bifurcation to spatially induced chaos in a reaction-diffusion system, *Physica D* **46**, p. 23.
Vega, J. L., Uzer, T. and Ford, J. (1993). Chaotic billiards with neutral boundaries, *Phys. Rev. E* **48**, p. 3414.
Vergni, D., Falcioni, M. and Vulpiani, A. (1997). Spatial complex behavior in nonchaotic flow systems, *Phys. Rev. E* **56**, p. 6170.
Volterra, V. (1926a). Fluctuations in the abundance of a species considered mathematically, *Nature* **118**, p. 558.
Volterra, V. (1926b). Variazioni e fluttuazioni del numero d'individui in specie animali conviventi, *Mem. R. Accad. Naz. dei Lincei* **2**, p. 31.
von Neumann, J. (1932). Proof of the quasi-ergodic hypothesis, *Proc. Nat. Acad. Sci.* **18**, p. 70.
von Plato, J. (1994). *Creating Modern Probability* (Cambridge University Press, Cambridge, UK).
Voss, H. U., Bünner, M. J. and Abel, M. (1998). Identification of continuous, spatiotemporal systems, *Phys. Rev. E* **57**, p. 2820.
Voss, H. U., Kolodner, P., Abel, M. and Kurths, J. (1999). Amplitude Equations from Spatiotemporal Binary-Fluid Convection Data, *Phys. Rev. Lett.* **83**, p. 3422.
Wagon, S. (1985). Is π Normal? *The Math. Intelligencer* **7**, p. 65.
Wang, X.-J. (1989). Statistical physics of temporal intermittency, *Phys. Rev. A* **40**, p. 6647.
Welsh, D. (1989). *Codes and Cryptography* (Clarendon Press, Oxford, UK).
White, H. (1993). Algorithmic complexity of Points in Dynamical Systems, *Erg. Theor. Dyn. Syst.* **13**, p. 807.

Wiggins, S. and Holmes, P. (1987). Homoclinic Orbits in Slowly Varying Oscillators, *SIAM J. Math. An.* **18**, p. 612.
Wilson, K. G. (1975). The renormalization group: critical phenomena and the Kondo problem, *Rev. Mod. Phys.* **47**, p. 773.
Winfree, A. T. (1980). *The geometry of biological time* (Springer-Verlag, New York, Heidelberg and Berlin, 19).
Wisdom, J. (1982). The origin of the Kirkwood gaps - A mapping for asteroidal motion near the 3/1 commensurability, *Astr. J.* **87**, p. 132.
Wisdom, J., Peale, S. J. and Mignard, F. (1984). The Chaotic Rotation of Hyperion, *Icarus* **58**, p. 137.
Wisdom, J., Peale, S. J. and Mignard, F. (1987). The Chaotic Rotation of Hyperion, in R. S. MacKay and J. D. Meiss (eds.), *Hamiltonian Dynamical Systems* (Adam Hilger), p. 660.
Wolf, A., Swift, J. B., Swinney, H. L. and Vastano, J. A. (1985). Determining Lyapunov exponents from a time series, *Physica D* **16**, p. 285.
Wolfram, S. (ed.) (1986). *Theory and Application of Cellular Automata* (Addison-Wesley, Reading, USA).
Yakhot, V. (1999). Two-dimensional turbulence in the inverse cascade range, *Phys. Rev. E* **60**, p. 5544.
Yanagita, T. and Kaneko, K. (1993). Coupled map lattice model for convection, *Phys. Lett. A* **175**, p. 415.
Yeomans, K. and Kiang, T. (1981). The long-term motion of comet Halley, *Royal Astr. Soc., Monthly Notices* **197**.
Yoden, S. (2007). Atmospheric predictability, *J. Met. Soc. Jap.* **85 B**, p. 77.
Young, L.-S. (1982). Dimension, entropy, and Lyapunov exponents, *Ergod. Theor. Dyn. Syst.* **2**, p. 109.
Yuan, G.-C., Nam, K., Antonsen, T. M., Ott, E. and Guzdar, P. N. (2000). Power spectrum of passive scalars in two dimensional chaotic flows, *Chaos* **10**, p. 39.
Zabusky, N. J. and Kruskal, M. D. (1965). Interaction of solitons in a collisionless plasma and the recurrence of initial states, *Phys. Rev. Lett.* **15**, p. 240.
Zaslavsky, G. M. (2005). *Hamiltonian Chaos and Fractional Dynamics* (Oxford University Press, Oxford).
Zhabotinsky, A. M. (1991). A history of chemical oscillations and waves, *Chaos* **1**, p. 379.
Zhan, M. and Kapral, R. (2006). Destruction of spiral waves in chaotic media, *Phys. Rev. E* **73**, p. 026224.
Zhang, D., Györgyi, L. and Peltier, W. R. (1993). Deterministic chaos in the Belousov-Zhabotinsky reaction: experiments and simulations, *Chaos* **3**, p. 723.
Ziv, J. and Lempel, A. (1977). A Universal Algorithm for Sequential Data Compression, *IEEE Transact. Inf. Theor.* **23**, p. 337.
Ziv, J. and Lempel, A. (1978). Compression of individual sequences via variable-rate coding, *IEEE Transact. Inf. Theor.* **24**, p. 530.

Index

www.ingramcontent.com/pod-product-compliance
Ingram Content Group UK Ltd.
Pitfield, Milton Keynes, MK11 3LW, UK
UKHW052114170926
8624IPUK00026B/378

9 789814 277655